T0245101

ARTIFICIAL INTELLIGENCE IN CLINICAL PRACTICE

ARTIFICIAL INTELLIGENCE IN CLINICAL PRACTICE

How AI Technologies Impact Medical Research and Clinics

Edited by

CHAYAKRIT KRITTANAWONG
Cardiology Division, NYU Langone Health and NYU School of Medicine, New York, NY, United States

Academic Press is an imprint of Elsevier
125 London Wall, London EC2Y 5AS, United Kingdom
525 B Street, Suite 1650, San Diego, CA 92101, United States
50 Hampshire Street, 5th Floor, Cambridge, MA 02139, United States
The Boulevard, Langford Lane, Kidlington, Oxford OX5 1GB, United Kingdom

ISBN: 978-0-443-15688-5

For Information on all Academic Press publications
visit our website at https://www.elsevier.com/books-and-journals

Publisher: Stacy Masucci
Acquisitions Editor: Linda Versteeg-Buschman
Editorial Project Manager: Billie Jean Fernandez
Production Project Manager: Fahmida Sultana
Cover Designer: Christian J. Bilbow

Typeset by MPS Limited, Chennai, India

Contents

17. Artificial intelligence in neglected tropical diseases 155

Girish Thunga, Sohil Khan, Pooja Gopal Poojari, Asha K. Rajan, Muhammed Rashid, Harsimran Kaur and Viji Pulikkel Chandran

18. Artificial intelligence in psychiatry: current practice and major challenges 163

Ali Amer Hazime and Marc Fakhoury

19. Application of artificial intelligence frameworks in the clinical practice of neurology: recent advances and future directions 169

Nick Corriveau-Lecavalier, Filip Mivalt and David T. Jones

20. Artificial intelligence in rheumatology 177

Junjie Peng, George Robinson, Elizabeth C. Jury, Pierre Dönnes and Coziana Ciurtin

21. Artificial intelligence in endocrinology 187

Ethan D.L. Brown, Fady Hannah-Shmouni and Skand Shekhar

22. Artificial intelligence in sleep medicine 195

Anuja Bandyopadhyay and Cathy Goldstein

23. Artificial intelligence in nephrology 201

24. Artificial intelligence in surgery 211

25. Artificial intelligence in cardiothoracic surgery: current applications and future perspectives 217

26. Artificial intelligence in orthopedics 235

27. Artificial intelligence in plastic surgery 245

48. Artificial intelligence in dermatology 437

Samuel Yeroushalmi, Alexander Ildardashty, Mimi Chung, Erin Bartholomew, Marwa Hakimi, Tina Bhutani and Wilson Liao

49. Artificial intelligence in gastroenterology and hepatology 443

Joseph C. Ahn and Vijay H. Shah

50. Artificial intelligence in nutrition research 465

Mélina Côté and Benoît Lamarche

51. Artificial intelligence in cardiac electrophysiology 475

Sulaiman S. Somani, Sanjiv M. Narayan and Albert J. Rogers

List of contributors

Joseph C. Ahn Division of Gastroenterology and Hepatology, Mayo Clinic, Rochester, MN, United States

Shankara Anand Boston University Chobanian & Avedisian School of Medicine, Boston, MA, United States

O. Arnaout Department of Neurosurgery, Brigham and Women's Hospital, Harvard Medical School, Boston, MA, United States

Anuja Bandyopadhyay Indiana University School of Medicine, Indianapolis, IN, United States

Erin Bartholomew Department of Dermatology, University of California, San Francisco, San Francisco, CA, United States

David W. Bates Division of General Internal Medicine, Brigham and Women's Hospital, Boston, MA, United States; Harvard Medical School, Boston, MA, United States; Department of Health Policy and Management, Harvard T.H. Chan School of Public Health, Boston, MA, United States

Sarah M.L. Bender University of Michigan Law School, Ann Arbor, MI, United States

Puneet K. Bhullar Mayo Clinic Alix School of Medicine, Scottsdale, AZ, United States

Tina Bhutani Department of Dermatology, University of California, San Francisco, San Francisco, CA, United States

A. Boaro Section of Neurosurgery, Department of Neurosciences, Biomedicine and Movement Sciences, University of Verona, Verona, Italy

Edward W. Boyer Department of Emergency Medicine, Harvard Medical School, Boston, MA, United States; Emergency Medicine, Ohio State University, Columbus, OH, United States

Ethan D.L. Brown National Institute of Environmental Health Sciences, National Institutes of Health (NIH), Research Triangle Park, NC, United States

Stephanie Carreiro Department of Emergency Medicine, University of Massachusetts Chan Medical School, Worcester, MA, United States

Jorge Chahla Rush University Medical Center, Chicago, IL, United States

Viji Pulikkel Chandran Department of Pharmacy Practice, Manipal College of Pharmaceutical Sciences, Manipal Academy of Higher Education, Manipal, Karnataka, India

Ching-Yu Cheng Singapore Eye Research Institute, Singapore National Eye Centre, Singapore; Department of Ophthalmology, Yong Loo Lin school of Medicine, National University of Singapore, Singapore

Carol Y. Cheung Department of Ophthalmology and Visual Sciences, The Chinese University of Hong Kong, Hong Kong, P.R. China

Anirudh Choudhary Department of Computer Science, University of Illinois, Urbana-Champain, IL, United States

Avishek Choudhury Industrial and Management Systems Engineering, West Virginia University, Morgantown, WV, United States

Mimi Chung Department of Dermatology, University of California, San Francisco, San Francisco, CA, United States

Coziana Ciurtin Department of Medicine, Centre for Adolescent Rheumatology Versus Arthritis, University College London, London, United Kingdom

Nneka I. Comfere Department of Dermatology, Mayo Clinic, Rochester, MN, United States; Department of Laboratory Medicine and Pathology, Mayo Clinic, Rochester, MN, United States; Mayo Clinic Dermatology Digital Health, Artificial Intelligence and Innovations Program, Rochester, MN, United States

Nick Corriveau-Lecavalier Department of Neurology, Mayo Clinic, Rochester, MN, United States

Mélina Côté Centre Nutrition, santé et société (NUTRISS), Institut sur la nutrition et les aliments fonctionnels (INAF), Laval University, Québec, QC, Canada; School of Nutrition, Laval University, Québec, QC, Canada

Diego del Alamo GSK R&D, Upper Providence, PA, United States

Roger D. Dias Department of Emergency Medicine, Harvard Medical School, Boston, MA, United States; STRATUS Center for Medical Simulation, Mass General Brigham, Boston, MA, United States

Pierre Dönnes Scicross AB, Skövde, Sweden

Dat Duong National Human Genome Research Institute, Bethesda, MD, United States

Mahdi Ebnali Department of Emergency Medicine, Harvard Medical School, Boston, MA, United States; STRATUS Center for Medical Simulation, Mass General Brigham, Boston, MA, United States

Adham El Sherbini Faculty of Health Sciences, Queen's University, Kingston, ON, Canada

M. Khair ElZarrad Office of Medical Policy, Center for Drug Evaluation and Research, US Food and Drug Administration, Silver Spring, MD, United States

Tala H. Fakhouri Office of Medical Policy, Center for Drug Evaluation and Research, US Food and Drug Administration, Silver Spring, MD, United States

Marc Fakhoury Department of Natural Sciences, School of Arts and Sciences, Lebanese American University (LAU), Beirut, Lebanon

Joshua A. Fein Department of Medicine, Weill Cornell Medicine, New York, NY, United States

Uwe M. Fischer Yale University School of Medicine, New Haven, CT, United States

Benjamin S. Glicksberg Hasso Plattner Institute for Digital Health, Icahn School of Medicine at Mount Sinai, New York, NY, United States

Cathy Goldstein University of Michigan Sleep Disorders Center, Ann Arbor, MI, United States

Sherif Gonem Department of Respiratory Medicine, Nottingham University Hospitals NHS Trust, Nottingham, United Kingdom; Division of Respiratory Medicine, University of Nottingham, Nottingham, United Kingdom

Darren V.S. Green GSK R&D, Stevenage, United Kingdom

Raghav Gupta Department of Urology, Icahn School of Medicine at Mount Sinai, New York, NY, United States

Marwa Hakimi Department of Dermatology, University of California, San Francisco, San Francisco, CA, United States

John D. Halamka Mayo Clinic Platform, Mayo Clinic, Rochester, MN, United States

Christina S. Han Division of Maternal-Fetal Medicine, Department of Obstetrics and Gynecology, David Geffen School of Medicine, University of California at Los Angeles (UCLA), Los Angeles, CA, United States

Fady Hannah-Shmouni National Institute of Child Health and Human Development, National Institutes of Health (NIH), Bethesda, MD, United States

Stefan Harrer Digital Health Cooperative Research Centre Ltd., Strategic Business Insights, Melbourne, VIC, Australia

Ali Amer Hazime Department of Natural Sciences, School of Arts and Sciences, Lebanese American University (LAU), Beirut, Lebanon

Michael A. Howard Division of Plastic Surgery, Northwestern Medicine, Chicago, IL, United States

Herbert Y.H. Hui Department of Ophthalmology and Visual Sciences, The Chinese University of Hong Kong, Hong Kong, P.R. China

Alexander Ildardashty Department of Dermatology, University of California, San Francisco, San Francisco, CA, United States

Hashim J.F. Shaikh Department of Orthopaedics, University of Rochester Medical Center, Rochester, NY, United States

Jeliazko R. Jeliazkov GSK R&D, Upper Providence, PA, United States

David T. Jones Department of Neurology, Mayo Clinic, Rochester, MN, United States; Department of Radiology, Mayo Clinic, Rochester, MN, United States

Young J. Juhn Department of Pediatric and Adolescent Medicine, Mayo Clinic, Rochester, MN, United States

Elizabeth C. Jury Department of Medicine, Centre for Adolescent Rheumatology Versus Arthritis, University College London, London, United Kingdom; Department of Medicine, Centre for Rheumatology Research, University College London, London, United Kingdom

Swaminathan Kandaswamy Department of Pediatrics, Emory University School of Medicine, Atlanta, GA, United States

Yanna Kang U.S. Food and Drug Administration, Center for Devices and Radiological Health, Office of Product Evaluation and Quality, Silver Spring, MD, United States

Benjamin H. Kann Artificial Intelligence in Medicine (AIM) Program, Mass General Brigham, Harvard Medical School, Boston, MA, United States; Department of Radiation Oncology, Brigham and Women's Hospital, Dana-Farber Cancer Institute, Harvard Medical School, Boston, MA, United States

Scott Kaplin Cardiology Division, NYU Langone Health and NYU School of Medicine, New York, NY, United States

Joel Karpiak GSK R&D, Upper Providence, PA, United States

Elias Kassir Division of Maternal-Fetal Medicine, Department of Obstetrics and Gynecology, David Geffen School of Medicine, University of California at Los Angeles (UCLA), Los Angeles, CA, United States

Harsimran Kaur Department of Pharmacy Practice, Manipal College of Pharmaceutical Sciences, Manipal Academy of Higher Education, Manipal, Karnataka, India

Sohil Khan School of Pharmacy and Medical Sciences, Griffith University, Gold Coast Campus, Queensland, Australia

Reem Khatib Department of General Anesthesiology, Anesthesiology Institute, Cleveland Clinic, Cleveland, OH, United States

Yousra Kherabi Infectious Diseases Department, Bichat-Claude Bernard Hospital, Assistance-Publique Hôpitaux de Paris, Paris, France

Lindsey A. Knake Department of Pediatrics, Division of Neonatology, University of Iowa, IA, United States

Chayakrit Krittanawong Cardiology Division, NYU Langone Health and NYU School of Medicine, New York, NY, United States

Veronica C. Kuhn University of Virginia, Charlottesville, VA, United States

Kyle Kunze Hospital for Special Surgery, New York, NY, United States

Benoît Lamarche Centre Nutrition, santé et société (NUTRISS), Institut sur la nutrition et les aliments fonctionnels (INAF), Laval University, Québec, QC, Canada; School of Nutrition, Laval University, Québec, QC, Canada

Simon Laplante Surgical Artificial Intelligence Research Academy, University Health Network, Toronto, ON, Canada; Department of Surgery, University of Toronto, Toronto, ON, Canada

Lisa Soleymani Lehmann Google, LLC, Mountain View, CA, United States; Harvard Medical School and Brigham and Women's Hospital, Boston, MA, United States

Dawei Li College of Future Technology, Peking University, Beijing, P.R. China

Wilson Liao Department of Dermatology, University of California, San Francisco, San Francisco, CA, United States

Jirapat Likitlersuang Artificial Intelligence in Medicine (AIM) Program, Mass General Brigham, Harvard Medical School, Boston, MA, United States; Department of Radiation Oncology, Brigham and Women's Hospital, Dana-Farber Cancer Institute, Harvard Medical School, Boston, MA, United States

Qi Liu Office of Clinical Pharmacology, Office of Translational Sciences, Center for Drug Evaluation and Research, US Food and Drug Administration, Silver Spring, MD, United States

Amin Madani Surgical Artificial Intelligence Research Academy, University Health Network, Toronto, ON, Canada; Department of Surgery, University of Toronto, Toronto, ON, Canada; Division of General Surgery, University Health Network, Toronto, ON, Canada

Momin M. Malik The Center for Digital Health, Mayo Clinic, Rochester, MN, United States

Matthias Mann Department of Proteomics and Signal Transduction, Max Planck Institute of Biochemistry, Martinsried, Germany; Proteomics Program, NNF Center for Protein Research, Faculty of Health Sciences, University of Copenhagen, Copenhagen, Denmark

Sameer Masood Division of Emergency Medicine, University Health Network, Department of Medicine, University of Toronto, Toronto, ON, Canada

Piyush Mathur Department of General Anesthesiology, Anesthesiology Institute, Cleveland Clinic, Cleveland, OH, United States; Department of Intensive Care and Resuscitation, Anesthesiology Institute, Cleveland Clinic, Cleveland, OH, United States

Sean McManus Department of General Anesthesiology, Anesthesiology Institute, Cleveland Clinic, Cleveland, OH, United States

Jeffrey Menard Roche Diagnostics, Rotkreuz, Switzerland

Filip Mivalt Department of Neurology, Mayo Clinic, Rochester, MN, United States

Dennis Murphree Department of Dermatology, Mayo Clinic, Rochester, MN, United States; Mayo Clinic Dermatology Digital Health, Artificial Intelligence and Innovations Program, Rochester, MN, United States

Thomas G. Myers Department of Orthopaedics, University of Rochester Medical Center, Rochester, NY, United States

Sanjiv M. Narayan Department of Medicine, Stanford University School of Medicine, Stanford, CA, United States; Cardiovascular Institute, Stanford University School of Medicine, Stanford, CA, United States

Vivek Natarajan Google, LLC, Mountain View, CA, United States

Krunal Pandav Department of Urology, Icahn School of Medicine at Mount Sinai, New York, NY, United States

Anil V. Parwani Department of Pathology, The Ohio State University Wexner Medical Center, Columbus, OH, United States

Adriana Marcela Pedraza Bermeo Department of Urology, Icahn School of Medicine at Mount Sinai, New York, NY, United States

Nathan Peiffer-Smadja Infectious Diseases Department, Bichat-Claude Bernard Hospital, Assistance-Publique Hôpitaux de Paris, Paris, France; Université Paris Cité, INSERM, IAME, Paris, France; National Institute for Health Research Health Protection Research Unit in Healthcare Associated Infections and Antimicrobial Resistance, Imperial College London, London, United Kingdom

Junjie Peng Department of Medicine, Centre for Adolescent Rheumatology Versus Arthritis, University College London, London, United Kingdom

Lily Peng Google, LLC, Mountain View, CA, United States; Verily Life Sciences, South San Francisco, CA, United States

Margot S. Peters Department of Dermatology, Mayo Clinic, Rochester, MN, United States; Department of Laboratory Medicine and Pathology, Mayo Clinic, Rochester, MN, United States

Nicholas Petrick U.S. Food and Drug Administration, Center for Devices and Radiological Health, Office of Product Evaluation and Quality, Silver Spring, MD, United States

Evan Polce Rush University Medical Center, Chicago, IL, United States

Pooja Gopal Poojari Department of Pharmacy Practice, Manipal College of Pharmaceutical Sciences, Manipal Academy of Higher Education, Manipal, Karnataka, India

W. Nicholson Price, II University of Michigan Law School, Ann Arbor, MI, United States

Lavanya Raghavan Singapore Eye Research Institute, Singapore National Eye Centre, Singapore

Asha K. Rajan Department of Pharmacy Practice, Manipal College of Pharmaceutical Sciences, Manipal Academy of Higher Education, Manipal, Karnataka, India

An Ran Ran Department of Ophthalmology and Visual Sciences, The Chinese University of Hong Kong, Hong Kong, P.R. China

Sowmith Rangu Division of Pediatric Cardiology, Department of Pediatrics, Stanford University School of Medicine, Lucile Packard Children's Hospital at Stanford, Stanford University, Palo Alto, CA, United States

Muhammed Rashid Department of Pharmacy Practice, Manipal College of Pharmaceutical Sciences, Manipal Academy of Higher Education, Manipal, Karnataka, India

Charitha D. Reddy Division of Pediatric Cardiology, Department of Pediatrics, Stanford University School of Medicine, Lucile Packard Children's Hospital at Stanford, Stanford University, Palo Alto, CA, United States

Nicholas L. Rider Division of Clinical Informatics, Liberty University College of Osteopathic Medicine, Lynchburg, VA, United States

Marc Rigatti Department of Emergency Medicine, University of Massachusetts Chan Medical School, Worcester, MA, United States

Michael Rivers Roche Diagnostics, Rotkreuz, Switzerland; Roche Diagnostics, Santa Clara, CA, United States

George Robinson Department of Medicine, Centre for Adolescent Rheumatology Versus Arthritis, University College London, London, United Kingdom; Department of Medicine, Centre for Rheumatology Research, University College London, London, United Kingdom

Albert J. Rogers Department of Medicine, Stanford University School of Medicine, Stanford, CA, United States; Cardiovascular Institute, Stanford University School of Medicine, Stanford, CA, United States

Colin M. Rogerson Department of Pediatrics, Division of Critical Care, Indiana University, IN, United States

Euijung Ryu Department of Quantitative Health Sciences, Mayo Clinic, Rochester, MN, United States

Berkman Sahiner U.S. Food and Drug Administration, Center for Devices and Radiological Health, Office of Product Evaluation and Quality, Silver Spring, MD, United States

Kunaal Sarnaik Case Western Reserve University School of Medicine, Cleveland, OH, United States

Saba Shafi Department of Pathology, The Ohio State University Wexner Medical Center, Columbus, OH, United States

Vijay H. Shah Division of Gastroenterology and Hepatology, Mayo Clinic, Rochester, MN, United States

Maxim V. Shapovalov GSK R&D, Upper Providence, PA, United States

Samin K Sharma Cardiac Catheterization Laboratory of the Cardiovascular Institute, Mount Sinai Hospital, New York, NY, United States

Skand Shekhar National Institute of Environmental Health Sciences, National Institutes of Health (NIH), Research Triangle Park, NC, United States

Harold Shin Division of Clinical Informatics, Liberty University College of Osteopathic Medicine, Lynchburg, VA, United States

Roni Shouval Department of Medicine, Weill Cornell Medicine, New York, NY, United States; Department of Medicine, Adult Bone Marrow Transplantation Service, Memorial Sloan Kettering Cancer Center, New York, NY, United States

Kenneth Smith Infectious Disease Diagnostics Laboratory, The Children's Hospital of Philadelphia, Philadelphia, PA, United States; Department of Pathology and Laboratory Medicine, Perelman School of Medicine at the University of Pennsylvania, Philadelphia, PA, United States

Olayemi Sokumbi Department of Dermatology, Mayo Clinic, Jacksonville, FL, United States; Department of Laboratory Medicine and Pathology, Mayo Clinic, Jacksonville, FL, United States

Benjamin D. Solomon National Human Genome Research Institute, Bethesda, MD, United States

Sulaiman S. Somani Department of Medicine, Stanford University School of Medicine, Stanford, CA, United States

Matt C. Sternke GSK R&D, Upper Providence, PA, United States

Maximillian T. Strauss Novo Nordisk Foundation Center for Protein Research, University of Copenhagen, Copenhagen, Denmark

Ania Syrowatka Division of General Internal Medicine, Brigham and Women's Hospital, Boston, MA, United States; Harvard Medical School, Boston, MA, United States

W. H. Wilson Tang Kaufman Center for Heart Failure Treatment and Recovery, Heart, Vascular and Thoracic Institute, Cleveland Clinic, Cleveland, OH, United States

Chad M. Teven Division of Plastic Surgery, Northwestern Medicine, Chicago, IL, United States

Ashutosh Kumar Tewari Department of Urology, Icahn School of Medicine at Mount Sinai, New York, NY, United States

Girish Thunga Department of Pharmacy Practice, Manipal College of Pharmaceutical Sciences, Manipal Academy of Higher Education, Manipal, Karnataka, India

Estefania Urena Registered Nurse, Intensive Critical Unit, Lincoln Medical and Mental Health Centre, Bronx, NY, United States

Ashish Verma Section of Nephrology, Department of Medicine, Boston University Avedisian & Chobanian School of Medicine, Boston, MD, United States

Jay Vietas Emerging Technologies Branch, Division of Science Integration, the National Institute for Occupational Safety and Health, Centers for Disease Control and Prevention, Washington DC, United States

Rebekah L. Waikel National Human Genome Research Institute, Bethesda, MD, United States

Chung-Il Wi Department of Pediatric and Adolescent Medicine, Mayo Clinic, Rochester, MN, United States

Meredith C. Winter Department of Anesthesiology, Critical Care Medicine, Children's Hospital Los Angeles, CA, United States; Department of Pediatrics, Keck School of Medicine, University of Southern California, CA, United States

Melissa S. Wong Division of Maternal-Fetal Medicine, Department of Obstetrics and Gynecology, Cedars-Sinai Medical Center Department of Obstetrics and Gynecology, Los Angeles, CA, United States; Division of Informatics, Department of Biomedical Sciences, Cedars-Sinai Medical Center, Los Angeles, CA, United States

Tien Yin Wong School of Clinical Medicine, Tsinghua Medicine, Tsinghua University, Beijing, P.R. China; Singapore Eye Research Institute, Singapore National Eye Centre, Singapore

Chao-Ping Wu Department of Critical Care, Respiratory Institute, Cleveland Clinic, Cleveland, OH, United States

Samuel Yeroushalmi Department of Dermatology, University of California, San Francisco, San Francisco, CA, United States

Marco A. Zenati Department of Surgery, Harvard Medical School, Boston, MA, United States; Division of Cardiac Surgery, VA Boston Healthcare System, West Roxbury, MA, United States

Wen-Feng Zeng Department of Proteomics and Signal Transduction, Max Planck Institute of Biochemistry, Martinsried, Germany

Yingfeng Zheng State Key Laboratory of Ophthalmology, Zhongshan Ophthalmic Center, Sun Yat-sen University, Guangzhou, P.R. China

Foreword

The public uptake of ChatGPT beginning in late 2022 was historic, with over 1 billion unique users in 90 days, remarkably faster than any previous new technology, and an indicator of the keen current keen interest in artificial intelligence (AI). These large language models (LLMs), also known as generative AI, are pluripotent. The list of capabilities seems unlimited: from reading, writing, editing, summarizing, and discussing text, to writing poems, music, essays, stories, and code apps. These capabilities were all present in ChatGPT but were advanced with the release of GPT-4 in early 2023. That was the first LLM that was multimodal, able to take text, speech, images, and video as inputs. This sets the stage for multimodal AI, which will become an important driver of the future of medicine.

Until recently, deep learning in healthcare, via the use of deep neural networks, achieved milestones in AI that were largely segmented to medical image interpretation—the full gamut of scans from X-rays, CT, MRI, ultrasound, and nuclear, but also including pathology slides, electrocardiograms (ECGs), retinal photos, and skin lesions. In the AI world, these are considered "narrow" unimodal tasks that capitalize on supervised learning, with tens of thousands to millions of image inputs that train the AI to have extraordinary "machine eye" vision, superseding that of physician specialists. So long as there is a human-in-the-loop to provide oversight, the promise of more accurate interpretation of scans has begun to get actualized in the clinic.

Surprisingly, the narrow AI interpretation of medical images got far broader than anticipated. Well beyond providing information about the eye, retinal photos have been shown to be a window to many organs and disease processes throughout the body, including chronic kidney disease, early Alzheimer's disease, hepatobiliary diseases, blood glucose and blood pressure control, hyperlipidemia, and the risk of heart disease. Similarly, the 12-lead ECG can give an accurate estimate of a person's hemoglobin, age and gender, ejection fraction, the filling pressure of the left ventricle, the presence of diabetes or prediabetes, prediction of subsequent atrial fibrillation, valve disease, and kidney disease. Pathology slides can indicate the driver mutation of a tumor, the tissue of origin, structural genomic variants, and prognosis. All these examples emphasize the power of machine eyes and the ability to train algorithms to see things that humans cannot.

Now we are at the beginning of a new era based on transformer model architecture that goes well beyond images, integrating all forms of data inputs. This has enabled processing so many of the layers of data that make each of us unique, including our genome and related omics such as our microbiome, physiology (via sensors), anatomy, immunome, and environment—no less all our medical records. Add to that the corpus of medical knowledge which is occurring with LLMs that have specific medical training. With this progress, we can move from the medical image interpretation "sweet spot" to wider AI, with deep individual phenotyping and multimodal AI to help prevent, diagnose, and treat patients. When the evidence is compelling from prospective studies and randomized clinical trials, the era of "wide" AI will be eventually ready for clinical practice. A first look of GPT-4 led Dr. Isaac Kohane to write: "How well does the AI perform clinically? And my answer is, I'm stunned to say: Better than many doctors I've observed."

In this textbook, edited by the brilliant Chayakrit Krittanawong, a comprehensive assessment of AI in clinical practice is provided: it spans the gamut from primary care to all medical specialties. It even goes beyond that ambitious goal with chapters on AI for global health, regulatory considerations, adverse drug events, toxicology, and nutrition. It is the first such book of its kind, drilling down on our present fund of medical knowledge in each area, knowing full well that we are still in the early stages of AI in medical practice.

One of the specialties that first initiated randomized clinical trials was gastroenterology. Multiple endoscopy and colonoscopy trials that compared standard-of-care physician detection of polyps versus physician plus machine vision demonstrated the superiority of the combined approach. It was even shown that fatigue among gastroenterologist late in the day added to the benefit of support with machine vision. Not only were more colonic polyps detected, but also the likelihood of being cancerous, requiring a biopsy, was provided by real-time algorithmic output. This early experience in gastroenterology exemplifies the synergy of strengths between physicians and AI. Still pending are clinical trials that show improved outcomes, beyond detection of polyps, for the combined approach of human and algorithmic detection.

That brings us to a critical lack in the field of AI in medicine: the lack of compelling evidence, which is a prerequisite for major change in clinical practice. There are currently about 70 completed randomized clinical trials of AI and nearly half are in gastroenterology, chiefly involving endoscopy procedures. A Mayo Clinic randomized using deep learning of the ECG for determination of ejection fraction helped primary care physicians make accurate diagnoses. This led to routine AI readouts for ECG detection of low ejection fraction throughout the Mayo health system. But there are limited examples of clinical implementation due to lack of randomized trials or prospective studies. Instead, of the more than 500 Food and Drug Administration–cleared or –approved AI commercial algorithms, very few companies have ever published their results. Most are retrospective studies, particularly in radiology scan interpretation, in relatively small number of patients. This lack of transparency has held back uptake of AI in clinical practice, reflected by penetration of AI in the United States imaging is only 2%.

Added to the lack of transparency are concerns about bias of algorithms, which are typically related to input data rather than the neural network per se. Other important issues that are obstacles to implementation include cost (and lack of reimbursement), the lack of diversity of patients and venues for training and validating models, the potential for exacerbation of health inequities, privacy and security of the data, and the need for post-deployment monitoring. It will take time to squarely address or mitigate these issues, but the cornerstone will be establishing incontrovertible evidence that health outcomes are significantly improved.

The patient empowered with AI support is a vital perspective that is all too frequently discounted or missed. There are already several virtual health coaching apps for chronic conditions like hypertension, diabetes, and depression in wide use. With LLMs, the capabilities of the chatbots will improve and gradually go beyond a single condition and progress to a more holistic virtual health assistant. We have already seen preliminary evidence that LLMs can answer patient questions with higher quality and empathy than doctors, setting up the potential of using AI as a "front door" to handle routine questions from patients, decompressing clinical workloads. With increasing ability for patients to aggregate their data from their health system portal and wearable sensors, generative AI will come more into play for providing rapid, doctorless interpretation. Of course, all these functions are dependent on the accuracy of the algorithms, which must be firmly established through the same rigorous medical research outlined for clinicians.

In the next couple of years, there is likely to be a movement towards keyboard liberation, such that the data clerk work that is a heavy burden for physicians will be supported by LLMs. There is mounting evidence for high-quality synthetic notes from the patient–doctor conversation, along with the downstream functionality that includes scheduling follow-up appointments and test, prescriptions, billing codes, and preauthorization for insurance carriers. Should this be proven to be accurate, cost-saving, and popular, it would greatly reduce the time clinicians are currently dedicated to such tasks. Furthermore, it will help bring back face-to-face conversation during clinic visits, which have been adversely affected by the need for keyboard clicking. No less, giving the gift of time to clinicians to have more time actually seeing, listening to, and examining patients.

This brings us to the culmination of where AI can take us in the years ahead—restoring the patient-doctor relationship, with presence, trust, empathy, all part of the human–human bond which is the essential component of caring for patients. Recently, Aaron Neinstein, an endocrinologist, wrote a piece entitled "Our AI future is better than you think." He wrote, "The use of AI in diabetes care has already irreversibly shifted my role as an endocrinologist, but I no longer fear that I will become obsolete. AI will not come between me and my patient but will instead relieve me of the burden of analyzing the data. AI can restore human-to-human connections, allowing me to be a coach, supporter, and healer, giving me space to gain a broader perspective into a patient and their life."

This is the vital, overarching objective that we must keep in mind. This book *Artificial Intelligence in Clinical Practice* will hopefully have many new editions in the years ahead, each giving updates of our fund of AI knowledge. Hopefully, that knowledge will ultimately lead us to better *care* of patients--not just making more accurate diagnoses or better prevention and treatment plans but restoring the precious patient–doctor relationship.

Eric Topol

Scripps Research, La Jolla, California

CHAPTER

1

Artificial intelligence in primary care

Adham El Sherbini[1], *Benjamin S. Glicksberg*[2] *and Chayakrit Krittanawong*[3]

[1]Faculty of Health Sciences, Queen's University, Kingston, ON, Canada [2]Hasso Plattner Institute for Digital Health, Icahn School of Medicine at Mount Sinai, New York, NY, United States [3]Cardiology Division, NYU Langone Health and NYU School of Medicine, New York, NY, United States

Introduction

The collaboration of artificial intelligence (AI) and primary care to blossom a captivating new field of study is worth understanding. How the subsets of AI, machine learning (ML), and even deeper, deep learning (DL) interact with the countless facets of healthcare is significant for a progressive future. Over the course of a half-century, AI has made tremendous strides in many areas of high-value primary care clinical practice (e.g., annual physical screening, preoperative risk stratification, vaccination), as will be discussed in this chapter. Nonetheless, there are limitations and challenges that have prolonged its implementation. This chapter briefly covers AI's potential, limitations, and challenges in primary care transformation.

Potentials

In nearly all, research has been conducted to evaluate AI's potential within its implementation. Although certain sectors of healthcare are more promising than others, ML typically provides a unique outlook on healthcare practices and opens doors for new strategies to be tested. To date, and speaking to its ever-lasting effects, AI's applications in primary care can be broken down into three sections: screening, preoperative management, and detection. In all three, AI has made progressive strides, as seen through research studies and innovative healthcare start-ups (Fig. 1.1).

Artificial Intelligence in Clinical Practice
DOI: https://doi.org/10.1016/B978-0-443-15688-5.00039-5

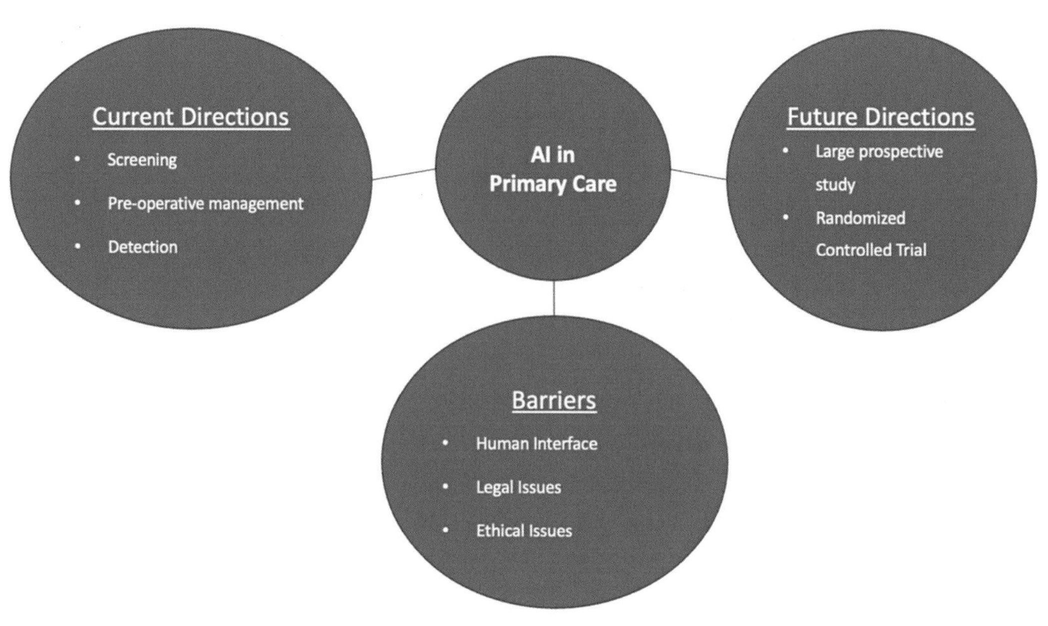

FIGURE 1.1 Barriers, current, and future direction of artificial intelligence in primary care.

Screening

Regarding screening, ML models have been applied to predict, diagnose, and assess the risk of cancer, cardiovascular diseases (CVD), sexually transmitted diseases (STDs), osteoporosis/osteomalacia, diabetes mellitus (DM), hypertension, hyperlipidemia, ocular diseases, and obstructive sleep apnea (OSA).

Cancer screening

Cancer screening measures have been relatively positive with the introduction of ML into breast, lung, prostate, colorectal, and cervical cancer. Certain ML algorithms, such as CervDetect [composed of Pearson correlation, random forest (RF), and shallow neural networks (NNs)] and DeepCervix (DL-based framework), are independently developed and assessed on cervical cancer screening and have achieved exceptionally high accuracies ($>90\%$) [1–3]. Similarly, numerous studies have developed or compared a number of analytical models (ML and DL) in the screening and prediction of lung cancer [4,5]. The general conclusion is that most of these models achieve high accuracy and outperform traditional or previously established models, such as the 2012 Prostate, Lung, Colorectal, and Ovarian Cancer Screening Trial risk mode (mPLCOm2012) [6]. Physicians utilize several types of diagnostic modalities, such as electronic health records (EHRs), computed tomography (CT) scans, and low-dose CT scans to train these models to predict lung cancer. Comparatively, mammograms are the focal images utilized for training models to screen for breast cancer [7,8]. Breast cancer-focused studies presented similar results, in which AI models had great accuracy in the screening for breast cancer and even outperformed the established BOAIDICEA model [8]. As for colorectal cancer, fecal microbiotas and additional modalities were utilized to compare numerous models predicting colorectal cancer, and models typically had higher accuracy (>90%) [9,10]. ML has also been utilized in chemotherapy for pretreatment prediction. In one study, multiple ML models [penalized linear regression, support vector machine (SVM), and RF] were utilized to predict left ventricular dysfunction in lymphoma or breast cancer patients. The RF was the best performing, with an accuracy of 0.94, a sensitivity of 0.81, and a specificity of 0.98. Similarly, prostate biopsies, EHRs, and PSA levels were applied to countless ML models to predict prostate cancer, and all studies concluded that the AI was effective in its screening [11,12].

Cardiovascular disease predictions or surveillance

Studies have evaluated AI applications on cardiovascular risk predictions (e.g., hypertension, CVD, coronary calcium scoring) through a number of ML models [RF, logistic regression (LR), gradient boosting (GB), NN, convolutional neural network (CNN), long short-term memory, etc.] [13–15]. Generally, ML models are high performing and have outperformed traditional scores, such as the Framingham score and a Cox PH model [16]. Similar findings are concluded in hypertension, resistant hypertension (defined as uncontrolled blood pressure (BP) despite the use of ≥5 antihypertensive medication, including diuretics), refractory hypertension (defined as uncontrolled BP despite the use of ≥5 antihypertensive medication, including diuretics and mineralocorticoid receptor antagonist), and estimate BPs, where models have reported area under the curve (AUC) scores as high as 0.92 [17,18]. Wearable technology or mobile health device with ML could be used in ambulatory BP monitoring and lifestyle recommendations. Given the significance of ambulatory monitoring for cardiovascular management, the introduction of novel monitoring devices can be effective given ML's assistance to increase accuracy and actionability. To date, the combination of portable sensors alongside ML models allows for near-real-time diagnosis [19]. Specifically, wearable sensors detecting relevant biosignals can be interpreted through ML algorithms for heightened accuracy. Similarly, the combination of wearable technology and ML can be utilized for personalized care, such as improvements in BP. In Chiang et al., RF with Shapley-Value-based Feature Selection was used to apply a highly accurate and personalized BP model [20]. Comparatively, a systematic review was conducted on ML techniques combined with smartphones for the monitoring of BP [21]. ANN and SVMs were the most commonly used input features ranging from five to 233 [21].

AI research also allows for the identification of significant clinical factors, such as hypertension research, where they found educational status, tobacco use, mental health, socioeconomic status, lifestyle (e.g., consumption of fruits and vegetables, physical activity), occupation, age, gender, history of diabetes, abdominal obesity, and family history (e.g., mothers history of hypertension, and history of high cholesterol) [22]. Comparatively, when investigating DM, ML models have outputted very similar results [23–25]. A number of approaches have been utilized for training the models, such as cardiorespiratory fitness records and administrative data, which assisted in outperforming established models and detecting adverse complications [22–25]. Coronary calcium scoring, a noninvasive tool for risk assessment of coronary artery disease, can utilize AI to enhance its clinical applications. Eng et al. developed two DL models capable of automating coronary artery calcium (CAC) scoring [26]. One model, a gated coronary CT model intended for CAC scoring, presented with a significantly high agreement (mean difference in scores = −2.86; Cohen's kappa = 0.89, $P < 0.0001$) [26]. The other model, a DL algorithm, was trained on the MESA study to undergo CAC scoring [26]. Across all levels of CAC, sensitivity ranged from 80%–100%, and PPV ranged from 87%–100% [26]. A similar study by van Velzen et al. trained and validated. DL model on CT examinations for automatic calcium scoring [27]. The DL model's intraclass correlation coefficients for CAC are 0.79–0.97 [27]. Similarly, Mu et al. trained and validated a DL model to quantify CAC scores from a singular coronary CT angiography (CTA) [28]. The risk categorization agreement between DL CTA and noncontrast CT CAC scores was exceptional (weight $k = 0.94$) [95% confidence interval (CI): 0.91, 0.97] [28]. To add, the positive correlation between the semiautomatic noncontrast CT CAC score and the automatic DL CTA was considered excellent (Pearson correlation = 0.96; $r^2 = 0.92$) [28]. AI may also have the potential to conduct risk prediction of CAC scores from retinal photographs, as observed in Rim et al. [29]. Five datasets from Singapore, the United Kingdom, and South Korea were used to train and validate DL models' ability to predict CAC (RetiCAC) [29]. RetiCAC was performed between all single clinical parameter models, and an AUROC of 0.742 was achieved [29]. In Siva Kumar et al., the prediction of ASCVD through an ML-centric ECG risk score in combination with or without CAC scores was measured [30]. The M_{ecg} had a strong association and discrimination of MACE (*C*-index: 0.7), whereas the $M_{ecg+cac}$ was associated with MACE (*C*-index: 0.71) [30].

In general, the pooled cohort equations (PCE) could predict patients' atherosclerotic cardiovascular disease risk [31]. However, PCE performance may over- or underestimate in certain ethnicity, and PCE performance among Asians or Hispanics remain inconclusive [31]. ML could potentially improve risk prediction in certain ethnic population [31]. In Ward et al., ML models (LR, GBM, and extreme GB) were trained on EHR for the ACSVD risk assessment of multiethnic patients [32]. GBM was well performing in the cohort and achieved an AUC score of 0.835, which was better than the PCE (AUC 0.775) [32]. Integration of additional EHR data had minimal effect on GBM's performance (before: AUC 0.784, after AUC 0.790) [32].

Another CVD tool, the American College of Cardiology/American Heart Association (ACC/AHA) Pooled Cohort Equations Risk Calculator, may possibly underestimate the risk of atherosclerotic CVD in specific individuals [33]. With that, Kakadiaris et al., set to compare the ACC/AHA CVD Risk Calculator to an SVM-based ML Risk Calculator using the Multi-Ethnic Study of Atherosclerosis (MESA) [33]. The ML Risk Calculator had AUC, sensitivity, and specificity of 0.92, 0.86, and 0.95 [33]. This was better performing than the ACC/AHA CVD Risk Calculator, which resulted in an AUC of 0.71, a sensitivity of 0.76, and a specificity of 0.56 [33]. Although more studies are required to validate these findings, the ML Risk Calculator recommended fewer drug therapies but did miss fewer events [33]. Notably, the combination of an ML model with the established CAC score evoked a high AUC score in the prediction of coronary artery disease (CAD) [30–32,34–36]

In Petrazzini et al., an ML algorithm was developed based on clinical features extracted from HER for the prediction of CAD [37]. The algorithm was applied in the UK Biobank (population-based cohort) and BioMe (multiethnic clinical care cohort) [37]. Relative to PCE, the ML model improved CAD prediction in UK Biobank and BioMe by 9 and 12%, respectively [37]. Similarly, Cho et al. evaluated preexisting cardiovascular risk prediction models while developing ML-centric algorithms on a healthy Korean population [38].

Overall, the preexisting models were deemed moderate to good, with PCE having a *C*-statistic of 0.738 [38]. Regarding the ML models, an NN model produced the greatest *C*-statistic of 0.751 [38].

Sexually transmitted disease screening

As for STDs, studies have been conducted in rural areas of South Africa, Uganda, and Kenya with reported outcomes on human immunodeficiency virus (HIV), gonorrhea, chlamydia, and syphilis [39,40]. Overall, studies have been generally positive, with high AUC scores across a number of models and diagnostic training tools. For instance, in Turbe et al., DL was trained and validated on rapid HIV tests to predict them as positive or negative [39]. The model achieved high specificity (100%) and sensitivity (97.8%) [39]. Similarly, in Balzer et al., ML was applied to detect individuals at high risk of HIV across Rural Kenya and Uganda [40]. Regarding efficiency, for a fixed sensitivity of 50%, the model-based strategy (based on LR) targeted 27%, whereas the ML targeted only 18%, and it improved sensitivity [40].

Ocular diseases

Similarly, the assessment of ocular diseases, such as glaucoma, age-related macular degeneration, and retinopathy, have been promising [41–43]. Models have shown exceptionally high accuracy scores, especially when trained on retinal images. Additionally, using retinal fundus images has allowed models to surmount eye diseases and precisely predict cardiovascular risk, age, sex, systolic BP, adverse cardiac events, and smoking status. The unmatched ML models have also shown progress in CAD, OSA, and osteoporosis through CTs, questionaries, and EHRs [41–43].

Osteoporosis screening

AI has also been assessed in osteoporosis screening through several diagnostic modalities. For example, in Sebro et al., SVMs were trained on the wrist and forearm CT scans of 196 patients [44]. Radial-basis-function kernel SVM was the best-performing model for the prediction of osteoporosis (AUC = 0.818) [44]. Similarly, Hsieh et al., assessed the risk of fracture and bone mineral density through training in DL on plain radiographs [36]. The model was high performing for spine osteoporosis, hip osteoporosis, high hip fracture risk, and high 10-year major fracture risk, with accuracies of 86.2%, 91.7%, 90.0%, and 95.0%, respectively [36]. In Zhang et al., another diagnostic, lumbar spine X-ray images, were utilized for training a deep convolutional NN model in screening for osteopenia and osteoporosis [45]. The model was trained on 1616 X-rays and tested on two test datasets (396 and 348 images) and a validation dataset (204 images) [45]. This model may have the potential to screen for osteoporosis (AUC 0.767) and osteopenia (AUC 0.787) [45]. Comparatively, another study Lim et al., utilized ML for the screening of femoral osteoporosis based on abdomen-pelvic CT and extracted radiomic features [46]. The ML model achieved accuracies of 92.9% and 92.7% in training and validation datasets, respectively. AUC scores of 95.9% and 96.0% in training and validation datasets were also reported [46]. Similar findings were reported by Pickhardt et al. where a DL tool was evaluated for its osteoporosis prediction based on CT images [47]. In regard to success rate, the DL model outperformed the traditional image processing BMD algorithm (99.3% vs 89.4%)

[47]. AI has the potential to screen for osteoporosis through several modalities and outperform established models [47].

Obstructive sleep apnea screening

Given the increasing prevalence of OSA, preliminary studies evaluating the potential for AI to screen OSA have been noted. For one, in Alvarez et al., the assessment of the diagnostic utility of airflow recordings in combination with at-home oximetry through ML was evaluated [48]. Measurements are captured through at-home polysomnography, and regression SVMs were utilized to predict the apnea-hypopnea index [48]. When combining the two metrics, the model was better performing (kapaa:0.71; 4-class accuracy: 81.3%) than airflow (kappa: 0.42, 4-class accuracy: 61.5%) and oximetry (kappa: 0.61; 4-class accuracy: 75.0%) [48]. Similar findings were observed in Stretch et al., where ML models were trained on home sleep apnea testing for OSA [49]. All ML algorithms prompted higher partial AUPRC scores than LR (0.574), and the RF model was the best-perform (pAUPRC 0.862) [49]. Comparatively, in Kelly et al., in-home polysomnography was compared to the mandibular movement in combination with ML regarding the diagnosis of OSA [50]. Forty patients underwent both diagnostics, and there was good agreement between the two tools (median bias 0.00; 95% CI −23.25 to +9.73 events/hour) [50]. On a broader scale, Ferreira-Santos et al. conducted a systematic review on early OSA screening through ML [51]. Of 63 included studies, 23 studies focused on diagnostic development, and 26 studies included internal validation [51]. Studies included a wide range of ML algorithms, including LR, SVM, linear regression, NN, and DT, and RF, to name a few [51]. More notably, the highest reported AUROC was 0.98 (0.96–0.99). That said, there remains an absence in external validation within larger cohorts [51].

Preoperative risk assessment

Another sector of primary care where AI has asserted grounds is preoperative management, where pre- and intraoperative data are utilized for training ML models to predict postoperative mortality, postoperative complications, and prolonged length of stay in the intensive care unit (ICU) [52–54]. Functional capacity is the most important part of preoperative risk assessment for cardiovascular complications [55]. ML has evaluated functional capacity through intelligence robotics and wearable monitoring. Intelligent robotic systems may improve functional capacity evaluations in combination with ML methods [56]. ML allows for the fusion of expert human intelligence and robotic systems, which can allow for increased quantification [56]. As for wearable technology, several studies have used this ML-centric approach for the monitoring of rehabilitation. In Canniere et al., cardiac rehabilitation patients were equipped with a multiparameter sensor throughout activity for the interpretation of functional capacity [57]. In combination with chronotropic response and effort, SVM presented a mean absolute error of 42.8 m (±36.8 m) [57]. In Rens et al., CVD patients underwent a home-based 6-minute walk test (6MWT) which was evaluated by LR with forward feature selection for prediction of functional capacity [58]. The model could evaluate "frailty" with high specificity (85%) and sensitivity (90%) [58]. Another study used wearable monitoring devices for the prediction of metabolic equivalents (METs) [59]. Five multiple-regression models were used, and all presented significant improvement in the mean absolute percentage error of METs within the high-intensity group [59]. Similarly, Fuller et al. evaluated how wearable devices (Apple Watch Series 2, Fitbit Charge HR2, and iPhone 6S) could elevate ML to detect lying, sitting, walking, and running [60]. Rotation Forest models produced the highest accuracies for Fitbit and Apple Watch (90.8 and 82.6%, respectively) [60]. Overall, ML and wearable technology integration to evaluate functional capacity could be used in preoperative risk assessment.

The Revised Cardiac Risk Index (RCRI) is also recommended to use in preoperative risk assessment to predict perioperative cardiac complications. However, RCRI did not perform well at predicting cardiac events after vascular noncardiac surgery or at predicting death [61]. Another study showed that the RCRI score failed to accurately predict the risk of cardiac complications in patients undergoing elective resection of lung cancer [62]. Hofer et al. evaluated the ability of a rules-based algorithm to identify patients with diseases from the RCRI [63]. The algorithm produced a higher incidence of definite or likely disease than that of the anesthesiologist [63]. The changed RCRI from the algorithms achieved an AUROC score of 0.70 (0.67–0.73) in predicting in-hospital mortality and took 12.64 ± 1.20 to go through 1.4 million patients [63].

Another ML-based preoperative risk assessment metric is the Cardia Comorbidity Risk Score (CCoR), which screens for major adverse cardiac events (MACE). This is typically conducted following elective knee and hip arthroplasty. The tool was well-performing for those with no history of RCRI conditions and even out-performed RCRI in detecting patients undergoing knee and hip arthroplasty [64] at a high risk of MACE [65]. The risk score utilizes ML to predict MACE through EHRs [65]. In Onishchenko et al. cohort, the CCoR achieved an AUROC score of 80.1% and 80.0% for males and women, respectively [65].

Notable studies have been conducted on countless surgeries in several populations, including those infected with SARS-CoV-2. The COVIDSurg Cohort Study was an international multicentered cohort of patients diagnosed with SARS-CoV-2 either 30 days postsurgical operation or seven days presurgical operation [64]. Through linear and nonlinear modeling, five relevant features were combined in 26 predictor sets that were fitted with either LR, DT, or RF [64]. The COVIDSurg Mortality Score achieved AUROC scores of 0.73 and 0.80 in the discrimination and validation sets, respectively [64]. This was comparable to that of previously published nonsurgical COVID risk assessments [64]. In another study, Chiew et al. compared ML models to the Combined Assessment of Risk Encountered in Surgery model and the American Society of Anaesthesiologists-Physical Status in the prediction of 30-day postoperative mortality and ICU stay longer than 24 hours [66]. The ML models (RF, adaptive boosting (AB), GB, and SVM), which were trained on EHR data, prompted high AUROCs but low sensitivities [66]. Specifically, GB was the highest-performing model with AUPRC scores of 0.23 and 0.38 for postoperative mortality and ICU admissions, respectively [66]. It was noted that the ML models predicted all negatives within the dataset, which was mostly negative [66]. In another study, Sahara et al. concluded that an ML algorithm outperformed the established American College of Surgeons National Surgery Quality Improvement Program surgical risk calculator in predicting 30-day mortality in hepatopancreatic surgical candidates [67]. Regardless of the training tools, nearly all studies have performed exceptionally accurately [67]. Such research is instrumental as it allows for forecasting clinical decisions and primary care resources, such as ICU beds and additional costs. It also can allow for cost-effective analysis of the patients and their providers.

Vaccination

ML may also have the potential to aid in prioritizing, screening and increasing adherence to vaccination. For instance, in Couto et al., ML was utilized to determine priority vaccination groups to reduce mortality rates. A GB model was trained on health variables (sex, age, and chronic health conditions) on a cohort of individuals hospitalized in one of 336 Brazilian hospitals [68]. Based on in-hospital death, the GB models achieved an AUROC score of 0.80 [68]. Regarding other vaccines, Hada et al. set out to develop a hybrid ML model to recommend vaccines for patients based on host-based factors (age, sex, medical history, postvaccination recovery rate, mortality rate, and symptoms). Sixteen experiments were conducted on COVID-19 and FLU3 vaccines, and there were clear differences ins cores based on the fore-mentioned factors. In regards to adherence, Kim et al., evaluated the potential of ML models to detect Korean adult CVD patients with low adherence to the influenza vaccination [69]. A total of 815 adults from the Fifth Korea National Health and Nutrition Examination Survey (KNHANES V) were utilized to assess several ML algorithms (LR, RF, SVM, XGB) [69]. The dataset was divided into <65 and ≥ 65 years old, as the elderly were provided with free immunization. For the ≥ 65 years old cohort, SVM had the highest accuracy (68.4%), and in the <65-year-old group, XGB and RF had the highest accuracies (84.7%) [69].

Detection

The final central part of primary care that ML has exceptionally investigated is the detection of COVID-19, flu, influenza, chronic coughs, pharyngitis, and atrial fibrillation (AF) [70,71]. Research has garnered rather promising results for all the listed conditions, as AUC scores are at an all-time high. For example, in Zhou et al., an XGBoost algorithm was utilized to differentiate cases of influenzas and COVID-19. The model achieved an AUC score of 0.85 in the external validation and 0.93 in the testing dataset. Similar findings follow through for all previously listed conditions. Another noteworthy mention is in the detection of AF, where ML models installed onto wearable devices can be trained on the health records, waveform data, RR intervals, W-PPG, and pulse oximetry data to accurately detect AF [72,73]. In Perez et al., a clinical study was conducted utilizing Apple Watch optical sensors were utilized to detect AF [74]. In the study, if participants had an irregular pulse that indicated possible AF, they were mailed an electrocardiography (ECG) patch [74]. Of those that returned an ECG ($n = 450$), 34% of participants had AF [74].

Limitations and challenges

While favorable outcomes have been discussed regarding its applications, no randomized clinical trials are available. Indeed, AI suffers from several limitations and drawbacks that are difficult to address and resolve [75–77]. First, the introduction of AI has concurrently introduced a number of issues that were not previously an obstacle. For example, an escalation in the lack of trust by patients toward their healthcare services can arise. As bedside manners and trust are key elements in medical education and patient interaction, reliance on technology could break that trust. Frontiers in AI research and its applications should focus on developing a human interface that is capable of providing qualities that typical healthcare practitioners would otherwise deliver. Second, education on technology and AI would have to be implemented into medical education for upcoming and ongoing healthcare workers. This would incur additional costs and extensive time, as it is not a primary focus of these employees. Third, the implementation of AI into the workplace would be exceptionally costly, and a cost-effective analysis of all aspects of its application would be required. Fifth, a lack of accountability persists in primary care once AI is utilized. Typically, a wrong diagnosis or risk assessment would fall in the hands of the physician or healthcare provider in charge. If this specific task is to be replaced by AI, there serves a lack of responsibility. Even if the physician and AI are utilized for the same assessment, healthcare workers will rely on the model and question their individual opinions. This raises numerous legal and ethical concerns as the absence of accountability is a predicament. To be addressed, legal identification of ownership over the algorithms and healthcare practices should be strictly identified. Sixth, there are several limitations to AI in clinical research. For one, many studies suffer from a lack of external validation, making the findings questionable. Lastly, additional limitations that are quite common include a single center, retrospective data, missing data, and a lack of comparison to traditional scores. Although these drawbacks are more difficult to address, it's necessary to acknowledge them when interpreting the results and conclusions of studies.

Conclusion

AI applications in primary care, though early, are up-and-coming. Study after study, ML models typically outperform established scores or formulas in cancer screening, preoperative risk prediction, vaccination guidance, detection, and prediction of nearly all aspects of primary care. However, its implementation in day-to-day primary care is still in its infancy. Large prospective studies and randomized clinical trials focusing on remote home monitoring and lifestyle monitoring are needed.

References

[1] Mehmood M, Rizwan M, Gregus ml M, Abbas S. Machine learning assisted cervical cancer detection. Front Public Health 2021;9:788376. Available from: https://doi.org/10.3389/fpubh.2021.788376.

[2] Rahaman MM, Li C, Yao Y, Kulwa F, Wu X, Li X, et al. DeepCervix: a deep learning-based framework for the classification of cervical cells using hybrid deep feature fusion techniques. Comput Biol Med 2021;136:104649. Available from: https://doi.org/10.1016/j.compbiomed.2021.104649.

[3] Hou X, Shen G, Zhou L, Li Y, Wang T, Ma X. Artificial intelligence in cervical cancer screening and diagnosis. Front Oncol 2022;12:851367. Available from: https://doi.org/10.3389/fonc.2022.851367.

[4] Huang B, Sollee J, Luo YH, Reddy A, Zhong Z, Wu J, et al. Prediction of lung malignancy progression and survival with machine learning based on pre-treatment FDG-PET/CT. eBioMedicine. 2022;82:104127. Available from: https://doi.org/10.1016/j.ebiom.2022.104127.

[5] Wang S, Zhang H, Liu Z, Liu Y. A novel deep learning method to predict lung cancer long-term survival with biological knowledge incorporated gene expression images and clinical data. Front Genet 2022;13:800853. Available from: https://doi.org/10.3389/fgene.2022.800853.
[6] Gould MK, Huang BZ, Tammemagi MC, Kinar Y, Shiff R. Machine learning for early lung cancer identification using routine clinical and laboratory data. Am J Respir Crit Care Med 2021;204(4):445–53. Available from: https://doi.org/10.1164/rccm.202007-2791OC.
[7] Cardoso MJ, Houssami N, Pozzi G, Séroussi B. Artificial intelligence (AI) in breast cancer care—leveraging multidisciplinary skills to improve care. Breast 2021;56:110–13. Available from: https://doi.org/10.1016/j.breast.2020.11.012.
[8] Ming C, Viassolo V, Probst-Hensch N, Dinov ID, Chappuis PO, Katapodi MC. Machine learning-based lifetime breast cancer risk reclassification compared with the BOADICEA model: impact on screening recommendations. Br J Cancer 2020;123(5):860–7. Available from: https://doi.org/10.1038/s41416-020-0937-0.
[9] Ho C, Zhao Z, Chen XF, Sauer J, Saraf SA, Jialdasani R, et al. A promising deep learning-assistive algorithm for histopathological screening of colorectal cancer. Sci Rep 2022;12(1):2222. Available from: https://doi.org/10.1038/s41598-022-06264-x.
[10] Wang KS, Yu G, Xu C, Meng XH, Zhou J, Zheng C, et al. Accurate diagnosis of colorectal cancer based on histopathology images using artificial intelligence. BMC Med 2021;19(1):76. Available from: https://doi.org/10.1186/s12916-021-01942-5.
[11] Bashashati A, Goldenberg SL. AI for prostate cancer diagnosis—hype or today's reality? Nat Rev Urol 2022;19(5):261–2. Available from: https://doi.org/10.1038/s41585-022-00583-4.
[12] Janowczyk A, Leo P, Rubin MA. Clinical deployment of AI for prostate cancer diagnosis. Lancet Digit Health 2020;2(8):e383–4. Available from: https://doi.org/10.1016/S2589-7500(20)30163-1.
[13] Adedinsewo DA, Pollak AW, Phillips SD, Smith TL, Svatikova A, Hayes SN, et al. Cardiovascular disease screening in women: leveraging artificial intelligence and digital tools. Circ Res 2022;130(4):673–90. Available from: https://doi.org/10.1161/CIRCRESAHA.121.319876.
[14] Aryal S, Alimadadi A, Manandhar I, Joe B, Cheng X. Machine learning strategy for gut microbiome-based diagnostic screening of cardiovascular disease. Hypertension. 2020;76(5):1555–62. Available from: https://doi.org/10.1161/HYPERTENSIONAHA.120.15885.
[15] Lou YS, Lin CS, Fang WH, Lee CC, Ho CL, Wang CH, et al. Artificial intelligence-enabled electrocardiogram estimates left atrium enlargement as a predictor of future cardiovascular disease. J Pers Med 2022;12(2):315. Available from: https://doi.org/10.3390/jpm12020315.
[16] Alaa AM, Bolton T, Di Angelantonio E, Rudd JHF, van der Schaar M. Cardiovascular disease risk prediction using automated machine learning: a prospective study of 423,604 UK Biobank participants. PLoS One 2019;14(5):e0213653. Available from: https://doi.org/10.1371/journal.pone.0213653.
[17] Chaikijurajai T, Laffin LJ, Tang WHW. Artificial intelligence and hypertension: recent advances and future outlook. Am J Hypertens 2020;33(11):967–74. Available from: https://doi.org/10.1093/ajh/hpaa102.
[18] Hung MH, Shih LC, Wang YC, Leu HB, Huang PS, Wu TC, et al. Prediction of masked hypertension and masked uncontrolled hypertension using machine learning. Front Cardiovasc Med 2021;8:778306. Available from: https://doi.org/10.3389/fcvm.2021.778306.
[19] Krittanawong C, Rogers AJ, Johnson KW, Wang Z, Turakhia MP, Halperin JL, et al. Integration of novel monitoring devices with machine learning technology for scalable cardiovascular management. Nat Rev Cardiol 2021;18(2):75–91. Available from: https://doi.org/10.1038/s41569-020-00445-9.
[20] Chiang PH, Wong M, Dey S. Using wearables and machine learning to enable personalized lifestyle recommendations to improve blood pressure. IEEE J Transl Eng Health Med 2021;9:1–13. Available from: https://doi.org/10.1109/JTEHM.2021.3098173.
[21] Haugg F, Elgendi M, Menon C. Assessment of blood pressure using only a smartphone and machine learning techniques: a systematic review. Front Cardiovasc Med 2022;9:894224. Available from: https://doi.org/10.3389/fcvm.2022.894224.
[22] Zhao H, Zhang X, Xu Y, Gao L, Ma Z, Sun Y, et al. Predicting the risk of hypertension based on several easy-to-collect risk factors: a machine learning method. Front Public Health 2021;9:619429. Available from: https://doi.org/10.3389/fpubh.2021.619429.
[23] Ravaut M, Sadeghi H, Leung KK, Volkvos M, Kornas K, Harish V, et al. Predicting adverse outcomes due to diabetes complications with machine learning using administrative health data. npj Digit Med 2021;4(1):24. Available from: https://doi.org/10.1038/s41746-021-00394-8.

[24] Yuan A, Lee AY. Artificial intelligence deployment in diabetic retinopathy: the last step of the translation continuum. Lancet Digit Health 2022;4(4):e208–9. Available from: https://doi.org/10.1016/S2589-7500(22)00027-9.

[25] Xue J, Min F, Ma F. Research on diabetes prediction method based on machine learning. J Phys Conf Ser 2020;1684:012062. Available from: https://doi.org/10.1088/1742-6596/1684/1/012062.

[26] Eng D, Chute C, Khandwala N, Rajpurkar P, Long J, Shleifer S, et al. Automated coronary calcium scoring using deep learning with multicenter external validation. npj Digit Med 2021;4(1):88. Available from: https://doi.org/10.1038/s41746-021-00460-1.

[27] van Velzen SGM, Lessmann N, Velthuis BK, Bank IEM, van den Bongard DHJG, Leiner T, et al. Deep learning for automatic calcium scoring in CT: validation using multiple cardiac CT and chest CT protocols. Radiology. 2020;295(1):66–79. Available from: https://doi.org/10.1148/radiol.2020191621.

[28] Mu D, Bai J, Chen W, et al. Calcium scoring at coronary CT angiography using deep learning. Radiology. 2022;302(2):309–16. Available from: https://doi.org/10.1148/radiol.2021211483.

[29] Rim TH, Lee CJ, Tham YC, et al. Deep-learning-based cardiovascular risk stratification using coronary artery calcium scores predicted from retinal photographs. Lancet Digit Health 2021;3(5):e306–16. Available from: https://doi.org/10.1016/S2589-7500(21)00043-1.

[30] Siva Kumar S, Al-Kindi S, Tashtish N, et al. Machine learning derived ECG risk score improves cardiovascular risk assessment in conjunction with coronary artery calcium scoring. Front Cardiovasc Med 2022;9:976769. Available from: https://doi.org/10.3389/fcvm.2022.976769.

[31] DeFilippis AP, Young R, Carrubba CJ, et al. An analysis of calibration and discrimination among multiple cardiovascular risk scores in a modern multiethnic cohort. Ann Intern Med 2015;162(4):266–75. Available from: https://doi.org/10.7326/M14-1281.

[32] Ward A, Sarraju A, Chung S, et al. Machine learning and atherosclerotic cardiovascular disease risk prediction in a multi-ethnic population. Npj Digit Med 2020;3(1):125. Available from: https://doi.org/10.1038/s41746-020-00331-1.

[33] Kakadiaris IA, Vrigkas M, Yen AA, Kuznetsova T, Budoff M, Naghavi M. Machine learning outperforms ACC/AHA CVD risk calculator in MESA. J Am Heart Assoc 2018;7(22):e009476. Available from: https://doi.org/10.1161/JAHA.118.009476.

[34] Brennan HL, Kirby SD. Barriers of artificial intelligence implementation in the diagnosis of obstructive sleep apnea. J Otolaryngol Head Neck Surg 2022;51(1):16. Available from: https://doi.org/10.1186/s40463-022-00566-w.

[35] Tanphiriyakun T, Rojanasthien S, Khumrin P. Bone mineral density response prediction following osteoporosis treatment using machine learning to aid personalized therapy. Sci Rep 2021;11(1):13811. Available from: https://doi.org/10.1038/s41598-021-93152-5.

[36] Hsieh CI, Zheng K, Lin C, et al. Automated bone mineral density prediction and fracture risk assessment using plain radiographs via deep learning. Nat Commun 2021;12(1):5472. Available from: https://doi.org/10.1038/s41467-021-25779-x.

[37] Petrazzini BO, Chaudhary K, Márquez-Luna C, et al. Coronary risk estimation based on clinical data in electronic health records. J Am Coll Cardiol 2022;79(12):1155–66. Available from: https://doi.org/10.1016/j.jacc.2022.01.021.

[38] Cho SY, Kim SH, Kang SH, et al. Pre-existing and machine learning-based models for cardiovascular risk prediction. Sci Rep 2021;11(1):8886. Available from: https://doi.org/10.1038/s41598-021-88257-w.

[39] Turbé V, Herbst C, Mngomezulu T, et al. Deep learning of HIV field-based rapid tests. Nat Med 2021;27(7):1165–70. Available from: https://doi.org/10.1038/s41591-021-01384-9.

[40] Balzer LB, Havlir DV, Kamya MR, et al. Machine learning to identify persons at high-risk of human immunodeficiency virus acquisition in rural Kenya and Uganda. Clin Infect Dis 2020;71(9):2326–33. Available from: https://doi.org/10.1093/cid/ciz1096.

[41] Dong X, Du S, Zheng W, Cai C, Liu H, Zou J. Evaluation of an artificial intelligence system for the detection of diabetic retinopathy in Chinese Community Healthcare Centers. Front Med 2022;9:883462. Available from: https://doi.org/10.3389/fmed.2022.883462.

[42] Mursch-Edlmayr AS, Ng WS, Diniz-Filho A, et al. Artificial intelligence algorithms to diagnose glaucoma and detect glaucoma progression: translation to clinical practice. Transl Vis Sci Technol 2020;9(2):55. Available from: https://doi.org/10.1167/tvst.9.2.55.

[43] Ting DSW, Cheung CYL, Lim G, et al. Development and validation of a deep learning system for diabetic retinopathy and related eye diseases using retinal images from multiethnic populations with diabetes. JAMA. 2017;318(22):2211. Available from: https://doi.org/10.1001/jama.2017.18152.
[44] Sebro R, De la Garza-Ramos C. Machine learning for opportunistic screening for osteoporosis from CT scans of the wrist and forearm. Diagnostics. 2022;12(3):691. Available from: https://doi.org/10.3390/diagnostics12030691.
[45] Zhang B, Yu K, Ning Z, et al. Deep learning of lumbar spine X-ray for osteopenia and osteoporosis screening: a multicenter retrospective cohort study. Bone. 2020;140:115561. Available from: https://doi.org/10.1016/j.bone.2020.115561.
[46] Lim HK, Ha HI, Park SY, Han J. Prediction of femoral osteoporosis using machine-learning analysis with radiomics features and abdomen-pelvic CT: a retrospective single center preliminary study. PLoS One 2021;16(3):e0247330. Available from: https://doi.org/10.1371/journal.pone.0247330.
[47] Pickhardt PJ, Nguyen T, Perez AA, et al. Improved CT-based osteoporosis assessment with a fully automated deep learning tool. Radiol Artif Intell 2022;4(5):e220042. Available from: https://doi.org/10.1148/ryai.220042.
[48] Álvarez D, Cerezo-Hernández A, Crespo A, et al. A machine learning-based test for adult sleep apnoea screening at home using oximetry and airflow. Sci Rep 2020;10(1):5332. Available from: https://doi.org/10.1038/s41598-020-62223-4.
[49] Stretch R, Ryden A, Fung CH, et al. Predicting nondiagnostic home sleep apnea tests using machine learning. J Clin Sleep Med 2019;15(11):1599–608. Available from: https://doi.org/10.5664/jcsm.8020.
[50] Kelly JL, Ben Messaoud R, Joyeux-Faure M, et al. Diagnosis of sleep apnoea using a mandibular monitor and machine learning analysis: one-night agreement compared to in-home polysomnography. Front Neurosci 2022;16:726880. Available from: https://doi.org/10.3389/fnins.2022.726880.
[51] Ferreira-Santos D, Amorim P, Silva Martins T, Monteiro-Soares M, Pereira Rodrigues P. Enabling early obstructive sleep apnea diagnosis with machine learning: systematic review. J Med Internet Res 2022;24(9):e39452. Available from: https://doi.org/10.2196/39452.
[52] Ouyang D., Theurer J., Stein N.R., et al. Electrocardiographic deep learning for predicting post-procedural mortality. 2022. Available from: https://doi.org/10.48550/ARXIV.2205.03242.
[53] Fernandes MPB, Armengol de la Hoz M, Rangasamy V, Subramaniam B. Machine learning models with preoperative risk factors and intraoperative hypotension parameters predict mortality after cardiac surgery. J Cardiothorac Vasc Anesth 2021;35(3):857–65. Available from: https://doi.org/10.1053/j.jvca.2020.07.029.
[54] Alghatani K, Ammar N, Rezgui A, Shaban-Nejad A. Predicting intensive care unit length of stay and mortality using patient vital signs: machine learning model development and validation. JMIR Med Inf 2021;9(5):e21347. Available from: https://doi.org/10.2196/21347.
[55] Fleisher LA, Fleischmann KE, Auerbach AD, et al. 2014 ACC/AHA guideline on perioperative cardiovascular evaluation and management of patients undergoing noncardiac surgery: a report of the American College of Cardiology/American Heart Association Task Force on Practice Guidelines. Circulation 2014;130(24). Available from: https://doi.org/10.1161/CIR.0000000000000106.
[56] Fong J, Ocampo R, Gross DP, Tavakoli M. Intelligent robotics incorporating machine learning algorithms for improving functional capacity evaluation and occupational rehabilitation. J Occup Rehabil 2020;30(3):362–70. Available from: https://doi.org/10.1007/s10926-020-09888-w.
[57] De Cannière H, Corradi F, Smeets CJP, et al. Wearable monitoring and interpretable machine learning can objectively track progression in patients during cardiac rehabilitation. Sensors. 2020;20(12):3601. Available from: https://doi.org/10.3390/s20123601.
[58] Rens N, Gandhi N, Mak J, et al. Activity data from wearables as an indicator of functional capacity in patients with cardiovascular disease. PLoS One 2021;16(3):e0247834. Available from: https://doi.org/10.1371/journal.pone.0247834.
[59] Nakanishi M, Izumi S, Nagayoshi S, et al. Estimating metabolic equivalents for activities in daily life using acceleration and heart rate in wearable devices. Biomed Eng OnLine 2018;17(1):100. Available from: https://doi.org/10.1186/s12938-018-0532-2.
[60] Fuller D, Anaraki JR, Simango B, et al. Predicting lying, sitting, walking and running using Apple Watch and Fitbit data. BMJ Open Sport Exerc Med 2021;7(1):e001004. Available from: https://doi.org/10.1136/bmjsem-2020-001004.
[61] Ford MK. Systematic review: prediction of perioperative cardiac complications and mortality by the revised cardiac risk index. Ann Intern Med 2010;152(1):26. Available from: https://doi.org/10.7326/0003-4819-152-1-201001050-00007.

[62] Wotton R, Marshall A, Kerr A, et al. Does the revised cardiac risk index predict cardiac complications following elective lung resection? J Cardiothorac Surg 2013;8(1):220. Available from: https://doi.org/10.1186/1749-8090-8-220.

[63] Hofer IS, Cheng D, Grogan T, et al. Automated assessment of existing patient's revised cardiac risk index using algorithmic software. Anesth Analg 2019;128(5):909–16. Available from: https://doi.org/10.1213/ANE.0000000000003440.

[64] COVIDSurg Collaborative, Dajti I, Valenzuela JI, et al. Machine learning risk prediction of mortality for patients undergoing surgery with perioperative SARS-CoV-2: the COVIDSurg mortality score. Br J Surg 2021;108(11):1274–92. Available from: https://doi.org/10.1093/bjs/znab183.

[65] Onishchenko D, Rubin DS, van Horne JR, Ward RP, Chattopadhyay I. Cardiac comorbidity risk score: zero-burden machine learning to improve prediction of postoperative major adverse cardiac events in hip and knee arthroplasty. J Am Heart Assoc 2022;11(15):e023745. Available from: https://doi.org/10.1161/JAHA.121.023745.

[66] Chiew CJ, Liu N, Wong TH, Sim YE, Abdullah HR. Utilizing machine learning methods for preoperative prediction of postsurgical mortality and intensive care unit admission. Ann Surg 2020;272(6):1133–9. Available from: https://doi.org/10.1097/SLA.0000000000003297.

[67] Sahara K, Paredes AZ, Tsilimigras DI, et al. Machine learning predicts unpredicted deaths with high accuracy following hepatopancreatic surgery. Hepatobiliary Surg Nutr 2021;10(1):20–30. Available from: https://doi.org/10.21037/hbsn.2019.11.30.

[68] Couto RC, Pedrosa TMG, Seara LM, et al. Covid-19 vaccination priorities defined on machine learning. Rev Saúde Pública 2022;56:11. Available from: https://doi.org/10.11606/s1518-8787.2022056004045.

[69] Kim M, Kim YJ, Park SJ, et al. Machine learning models to identify low adherence to influenza vaccination among Korean adults with cardiovascular disease. BMC Cardiovasc Disord 2021;21(1):129. Available from: https://doi.org/10.1186/s12872-021-01925-7.

[70] Su K, Xu L, Li G, et al. Forecasting influenza activity using self-adaptive AI model and multi-source data in Chongqing, China. EBioMedicine. 2019;47:284–92. Available from: https://doi.org/10.1016/j.ebiom.2019.08.024.

[71] Alqudaihi KS, Aslam N, Khan IU, et al. Cough sound detection and diagnosis using artificial intelligence techniques: challenges and opportunities. IEEE Access 2021;9:102327–44. Available from: https://doi.org/10.1109/ACCESS.2021.3097559.

[72] Christopoulos G, Attia ZI, Van Houten HK, et al. Artificial intelligence—electrocardiography to detect atrial fibrillation: trend of probability before and after the first episode. Eur Heart J Digit Health 2022;3(2):228–35. Available from: https://doi.org/10.1093/ehjdh/ztac023.

[73] Lown M, Brown M, Brown C, et al. Machine learning detection of atrial fibrillation using wearable technology. PLoS One 2020;15(1):e0227401. Available from: https://doi.org/10.1371/journal.pone.0227401.

[74] Perez MV, Mahaffey KW, Hedlin H, et al. Large-scale assessment of a smartwatch to identify atrial fibrillation. N Engl J Med 2019;381(20):1909–17. Available from: https://doi.org/10.1056/NEJMoa1901183.

[75] Kelly CJ, Karthikesalingam A, Suleyman M, Corrado G, King D. Key challenges for delivering clinical impact with artificial intelligence. BMC Med 2019;17(1):195. Available from: https://doi.org/10.1186/s12916-019-1426-2.

[76] Jiang L, Wu Z, Xu X, et al. Opportunities and challenges of artificial intelligence in the medical field: current application, emerging problems, and problem-solving strategies. J Int Med Res 2021;49(3). Available from: https://doi.org/10.1177/03000605211000157 030006052110001.

[77] London AJ. Artificial intelligence in medicine: overcoming or recapitulating structural challenges to improving patient care? Cell Rep Med 2022;3(5):100622. Available from: https://doi.org/10.1016/j.xcrm.2022.100622.

CHAPTER

2

Artificial intelligence in general internal medicine

Adham El Sherbini[1], *Benjamin S. Glicksberg*[2] *and Chayakrit Krittanawong*[3]

[1]Faculty of Health Sciences, Queen's University, Kingston, ON, Canada [2]Hasso Plattner Institute for Digital Health, Icahn School of Medicine at Mount Sinai, New York, NY, United States [3]Cardiology Division, NYU Langone Health and NYU School of Medicine, New York, NY, United States

Introduction of artificial intelligence in general medicine

The concept of artificial intelligence (AI) is based on the sole human capability of making decisions while concurrently problem-solving can be digitalized [1]. The development of a intelligence similar to the human mind that is free of the human flaws observed daily can have limitless applications. AI applications of medicine show promise in delivering a standard of care that is free of human judgment or variability [2]. More importantly, the possibility of implementing an AI that can provide more accuracy serves as a benefit to the patient and healthcare workers, specifically in timely tasks. In the past 50 years, the applications of AI in hospital medicine have extended far beyond what was expected and believed by scientists. Its advancements in risk assessment, care managements, and screening have been heavily assessed by researchers with riveting conclusions to be made. This chapter briefly explores the use of AI in all facets of hospital medicine. In a global stage that suffers from a shortage of physicians, the applications of AI in medicine should be taken seriously.

High-value care

With all attempts to apply change in a hospital setting, two factors take high priority: (1) how does it help? and (2) is it cost efficient? [3] To answer this, it is best to observe the

DOI: https://doi.org/10.1016/B978-0-443-15688-5.00025-5

common contentious argument of AI: its replacement of radiologists. The sole belief that AI will unreservedly replace radiologists is not only a misconception, but rather a significant stride in providing high-value care [4]. Radiologists are necessary for confirmation, judgment ruling, and conducting human components of the job. To add, the leap in efficiency from radiologists when using AI can allow for minimized imaging costs and waiting times. It is only those who refuse to work alongside AI that may feel excluded as their degree of work is unparallel to their AI-driven counterparts. However, such high-value care comes with several drawbacks. For one, the implementation of AI allows for the absence of empathy, a quality that is essential in a hospital setting [3]. With this come countless ethical-centric concerns and a lack of patient trust, raising several problems that may have not existed before the AI. Luckily, corporations within this field, such as Google and IBM have experience in the development of human interaction, allowing a great potential to be explored in the near future.

Perioperative management

The applications of AI are found to be most effective when administered in the early stages of care. Such an upstream intervention has found its ways around risk assessment, venous thromboembolism (VTE) prophylaxis, and the management of anticoagulant therapy, to name a few. One of the most leading specialties of risk assessment is cardiovascular, as its significance and impact in the hospital setting is well understood. One study was capable of assessing the survival of right ventricular failure in 256 patients with pulmonary hypertension [5]. Utilizing supervised machine learning (ML), a category of AI, the 3D patterns of the systolic cardiac motion were utilized to determine the risk, independent of individual risk factors. Such potential extends into pulmonary risk assessment as well, where previous and current chest computed tomography (CT) scans were used by a dense neural network (DNN), another type of AI, to predict the likelihood of lung cancer [6]. The AI achieved an area under the curve (AUC) of 0.944. To put that into context, when only the current CT scans of the patient were available, this analytic model was more accurate than six board-certified radiologists. Even in the presence of both images, the DNN model was just as accurate as the radiologists. However, research works on AI applications in risk assessment, as is other applications of AI, contain several limitations that are hard to overcome. For one, since the ML is on the basis of training the model, it requires a large database to do so [2]. To add, additional data is encouraged to assess the model in phases, a term known as cross-validation [7]. Challenges to AI implementation can also be observed by specific fields, such as pulmonary risk assessment, where there remain many obstacles in assessing chronic obstructive pulmonary diseases or asthma [8,9]. Since these respiratory diseases require a number of tests, including a patient history, it makes it exceptionally difficult for AI to be useful. However, there remains trouble in the mistreatment of these conditions, making for a great potential avenue for AI. Other applications of AI in preoperative management are VTE prophylaxis treatment. One exemplary study of this was a systematic review conducted on 12 studies that utilized AI to predict VTE, which found the test study to have an AUC of 0.98 and a specificity of 0.96 [10]. The 2021 review concluded that the AI was effective in diagnosing VTE. Such risk stratification procedures can allow for an early diagnosis of deep vein thrombosis, while concurrently improving the use of prophylactic anticoagulants [11,12]. With respect to anticoagulant therapy, AI has played a large role in adherence. One study found a mean adherence of 90.5% when utilizing an AI to monitor ischemic stroke patients on anticoagulant therapy on a daily basis [13]. The AI is practical in that it can identify the medication, ingestion, and patient all simultaneously. Although there are numerous studies of AI practices on all three DOACs (dabigatran, rivaroxaban, and apixaban) and warfarin, there are some limitations of this focused research [14]. Similar to other research processes, there are limited databases, a higher rate of selection bias, and the retrospective databases make the analysis and conclusion rely on accuracy and completeness of original data entry [14].

Patient safety and quality issues

A number of studies have assessed the applications of AI on crucial safety measures faced in the hospital, such as diagnostic errors, medication errors, transitions of care, and AI quality improvement models. Additional findings of patient safety and quality issues utilizing AI include infection control, hand-hygiene alerts, adverse drug reactions, postoperative bleeding risk, prolonged ventilation prediction, steps of laparoscopic sleeve gastrectomy procedure, risk of pressure ulcers, risk of falls, and sepsis detection to name a few [15–23]. Although AI applications have been proven to provide standardized care with high accuracy in countless medical specialties, there are diagnostic issues surrounding such implementations. For one, in the case that AI conducts a misdiagnosis, there is a lack of accountability that persists, as no one can be held at fault [24]. Another issue is the rising of new diagnostic errors, such as inconsistency between the training dataset for the AI and real-life clinical scene, as there remains a limit of effective training data [25]. However, AI can also be advantageous in diagnostic errors, as it provides harmonization between cases, an exceptionally difficult skill for physicians as they may have biases [3]. In terms of medication self-administration errors, the literature has shown its associations with poor adherence, increased healthcare costs, and increased hospitalizations [26]. One study presented a self-effacing AI that was capable of accurately detecting the use of inhalers and insulin pens and their respective appropriate steps by analysis of wireless signals within the home [27]. Another implementation of AI is in the transitions of care, where patients are moved across institutions or from an inpatient to an outpatient setting. Transitions of care is very risky, and a qualitative study in Denmark found that it was challenging, mainly because hospital discharge papers and nursing care plans were time-consuming and difficult to read [28]. This same study introduced a number of ways that AI could potentially be introduced into the process of transition. Some ideas include using a confidence score to predict the most likely section for a certain bit of information, targeting information to different groups (nursing home nurses, social care assistants, and so on) to alleviate overload, or use an AI to summarize the information [28]. Advantages of the AI would be that it is capable of analyzing the complex segments of the patient's dataset, but a drawback would be that the complex design of an AI would be nontransparent if not specifically designed to provide nuanced communication [29]. Finally, patient safety and quality issues can be combatted through the practice of AI quality improvement models. To truly integrate AI into hospital medicine, appropriate assessments and implementation science practices must answer the demanding questions about AI [30]. Some of these topics are, but not limited to, cost, human effort, expertise, model bias, ongoing model maintenance, and the support needed to detect and allay for a decline in model performance. This is a severe disadvantage of AI, as other introductions, such as new medical devices or drugs have exceptionally developed research methods in place to ensure success. On the other hand, the effectiveness of an AI is completely dependent on the training data and its integration into its surroundings. Using the quality improvement methodology, we can understand the variability that persists in the current environment, and how the AI can withstand normal variation.

Artificial intelligence screening

The most prominent application of AI in hospital medicine is in the screening for a number of diseases. These include abdominal aortic aneurysm, cardiovascular and cerebrovascular disease, depression and anxiety, diabetes mellitus, dyslipidemia, hypertension, obesity, obstructive sleep apnea (OSA), thyroid disease, and cancer. In terms of abdominal aortic aneurysm (AAA), a study attempted to utilize a CT angiography and ML to predict the expansion of AAA [31]. This study found that the ML was effective in predicting the AAA expansion through the area and major axis data [31]. Another significant application of AI in AAA has been the assessment of the biomechanical wall stresses [32]. Regarding cardiovascular and cerebrovascular disease, the screening potential through AI is endless. One trial tested if implementing AI to screen electrocardiogram (ECG) could predict asymptotic left ventricular dysfunction (ALVD), a disease highly associated with a decline in quality of living [33]. The study found that utilization of AI made ECG a far more useful tool for the screening of ALVD [33]. AI has limitless applications in cardiology, especially cardiac imaging, though there remain limitations that hinder such advancements. One main concern is that since AI is quite complex with strict parameters, making it liable to overfitting, and thus far more unsuitable for nontraining datasets [34]. With mental health, AI has shown great potential by utilizing mood scales, brain scans, social media, and electronic health records (EHRs) to predict mental health conditions such as anxiety or depression. A study conducted in India concluded that depression could be accurately predicted with an artificial neural network analytic model within a geriatric population [35]. Therefore the use of AI on mental health conditions can allow for an earlier diagnosis, allowing for interventions to be more effective. Regarding diabetes, a review on 141 papers for AI on diabetes management found that the use of AI showed progressive success in self-management and daily clinical practice of diabetes [36]. However, a number of limitations of AI on diabetes have been addressed, including technical, data, and research design issues. The cost, datasets, and the use of retrospective datasets on diabetes-focused studies have made the conclusions difficult to validate [37]. For dyslipidemia, a study utilized recurrent neural network (RNN) and long short-term memory (LSTM) to predict dyslipidemia in steel workers and found that LSTM performed better than RNN with an accuracy of 95% [38]. Another interesting study utilized AI to predict dyslipidemia, hyperglycemia, and hypertension through retinal fundus images

[39]. The study concluded that the deep learning is feasible in predicting all three conditions and further chronic diseases [39]. Speaking of hypertension, a retrospective study was conducted using XGBoost on EHR data to predict hypertension and was later validated and deployed in the state of Maine [40]. Although the potential of AI applications on hypertension is quite high, the research is still early and their remain many challenges that need to be addressed. In terms of obesity, unique methods have been taken to implement AI such as the utilization of fuzzy logic interference or the use of mobile devices for the self-management of obesity [41,42]. Both studies were effective as the fuzzy logic interference, a subtype of intelligence, assisted in the understanding of obesity, and the mobile device intervention was determined to be a great resource for managing obesity [41,42]. Another condition, OSA, which is difficult to diagnose, was assessed by genetic algorithm models through completed questionnaires by moderate-to-severe OSA patients and presented satisfactory results [43]. Similarly, thyroid disease can also be difficult to diagnose as symptoms can be muddled for other conditions. Although the research output on AI applications on thyroid disease is relatively small, a comprehensive review reported that several studies have attempted to utilize AI to classify between benign and malignant thyroid nodules successfully [44]. Finally and arguably, the most important disease for AI applications is cancer, as studies have been conducted on breast cancer, prostate cancer, colorectal cancer, cervical cancer, and additional cancer screening tests. One review claimed that although the potential for AI to predict breast cancer through mammograms or tomosynthesis images is quite high, it remains in its early stages, as no large case screening trials have been conducted [45]. This note follows for the other types of cancer, as the clinical potential is substantial, yet far more assessments are still required [46]. Although cancer screening through AI has been proven to be as equal to human performance, such as in the case of colorectal cancer, cancer screening suffers from many of the AI limitations that other diseases are subject to [47].

Artificial intelligence vaccine recommendations

In the case of vaccines, AI can be utilized to predict the appropriate candidates to combat a particular viral infection [48]. In the case of SARS-CoV-2 coronavirus, a study utilized ML to predict the types of Covid-19 vaccine candidates. In this study, the S, nsp3, and nsp8 proteins were predicted by the ML to cause high protective antigenicity [49]. These predications can assist in the development of a vaccine. Other applications of AI and vaccine development include the use of AI in drug repurposing [50]. One study was capable of identifying older drugs in the market that contained anticoronavirus activities using differing databases [50]. They found 80 drugs with the potential, and 8 of them presented in vitro activities against proliferation of feline infectious peritonitis (FIP) virus [50].

Telehealth

With the Covid-19 virus causing a pandemic, the role of telehealth has exceptionally increased, allowing for the applications of AI to be assessed. Since telehealth is conducted remotely, it allows for AI to be utilized for diagnosis far more easily. One study that utilized

AI in telehealth to screen for diabetic retinopathy had decreased patient visits by 14,000 visits [51]. This allows for a reduction in physician burnout and cost and makes the job far more efficient [51]. Other benefits of AI applications in telehealth include better elderly care, as the development of assistive robots can provide healthcare straight to the home, benefiting those who are at risk for constant travel [52]. To add, constant patient monitoring from home can be conducted when combining AI and telehealth, allowing for reduced physician work, burnout, and even an earlier detection of epidemics [53]. The common limitation across telehealth is that accessing technology can be difficult, especially for seniors, and the usage of telemedicine will increase if AI makes it more beneficial [54].

Current challenges preventing artificial intelligence application

Although AI applications are quite endless, especially in the realm of hospital medicine, there are numerous challenges, limitations, and drawbacks that have caused for their implementation to be delayed. For one, the applications of AI require hefty modifications within the hospital setting to work [55]. This includes integration to EHRs, education to clinicians, updates over time, and payment from public or private individuals [55]. Although all the listed challenges are solvable, such problems require time to resolve and will delay the introduction of AI into healthcare. Besides the deployment concerns, there remain more severe ethical, legal, community, and security issues that are yet to be appropriately addressed [56]. In terms of the ethical issues, a focal opposing argument is that since AI is developed on the data sources that likely contain a level of bias (gender, race, socioeconomic status (SES), and so on.), it makes the AI biased when applied to the real clinical setting [57]. To add, in a complex clinical decision-making setting, an individual with ethics is more trustworthy, as AI does not vary based on specific interactions [57]. Finally and arguably, the most ethical issue with AI is that the development and operation of an AI is on the basis of personal health data, which raises an ethical dilemma [55]. Regarding security, there are numerous hardware and software security issues when utilizing AI, especially in a healthcare setting that contains highly sensitive data. In terms of legality, no laws are yet in place concerning AI and its utilization in hospital medicine [56]. Similar to healthcare workers, who are held to a standard by the law and their requirements, AI requires a set of laws to ensure its safety and responsibility in the hospital. Although not as prominent, AI development is also liable to human error, that, if not immediately updated and fixed, could cause substantial health problems to patients [57]. On the topic of humans, clinicians can begin to rely on the AI that has a high accuracy and not question when mistakes may arise. Finally, patients are far more likely to trust, rely on, and believe healthcare workers rather than an AI diagnosis, especially if their diagnosis differs from one another.

Future directions

Although a number of concerns have been raised regarding AI, there are several paths that can be taken to resolve and even advance the use of AI in hospital medicine.

Whether AI can be implemented into the hospital setting or not boils down to whether the legal and ethics concerns will be addressed or not. To add, the drawback of security should be taken as a high priority, as the possibility of hacking into personal health data will shut down the possibility of AI applications immediately [9]. To add, training on ML and AI should be available for all medical students and physician residents. One survey found that medical students and physician residents who wished to be educated on ML were incapable of doing so [58]. For the reality of AI to be truth, a true legal system must be conducted that extends all the way from the engineering laboratory to the clinical setting. To add, the "three fairs" (equal allocation, equal outcomes, and equal performance) must all be met to ensure that marginalized populations are advocated for within the AI [59]. Finally, a human interface of AI should be explored to combat the patient-centric issues that may arise from patients [56].

Major takeaway points

The applications of AI in all fields of hospital medicine are substantial and there remains potential to expand into other fields. Also, there are several limitations on AI implementation regarding legality, ethics, community, cost, education, research methodology, and clinical decisions. Future AI research should focus on addressing the main addressed limitations to increase the potential of AI applications into hospital medicine.

References

[1] Kersting K. Machine learning and artificial intelligence: two fellow travelers on the quest for intelligent behavior in machines. Front Big Data 2018;1. Available from: https://doi.org/10.3389/fdata.2018.00006.

[2] Ahuja AS. The impact of artificial intelligence in medicine on the future role of the physician. PeerJ. 2019;7: e7702. Available from: https://doi.org/10.7717/peerj.7702.

[3] Bohr A, Memarzadeh K. The rise of artificial intelligence in healthcare applications. Artif Intell Healthc 2020;25–60. Available from: https://doi.org/10.1016/b978-0-12-818438-7.00002-2. Published online.

[4] Gampala S, Vankeshwaram V, Gadula SSP. Is artificial intelligence the new friend for radiologists? A review article. Cureus. 2020;. Available from: https://doi.org/10.7759/cureus.11137. Published online October 24.

[5] Dawes TJW, de Marvao A, Shi W, et al. Machine learning of three-dimensional right ventricular motion enables outcome prediction in pulmonary hypertension: a cardiac MR imaging study. Radiology. 2017;283 (2):381–90. Available from: https://doi.org/10.1148/radiol.2016161315.

[6] Ardila D, Kiraly AP, Bharadwaj S, et al. End-to-end lung cancer screening with three-dimensional deep learning on low-dose chest computed tomography. Nat Med 2019;25(6):954–61. Available from: https://doi.org/10.1038/s41591-019-0447-x.

[7] Little MA, Varoquaux G, Saeb S, et al. Using and understanding cross-validation strategies. Perspectives on Saeb et al. GigaScience. 2017;6(5). Available from: https://doi.org/10.1093/gigascience/gix020.

[8] Kaplan A, Cao H, FitzGerald JM, et al. Artificial intelligence/machine learning in respiratory medicine and potential role in asthma and COPD diagnosis. J Allergy ClImmunology: Pract 2021;9(6):2255–61. Available from: https://doi.org/10.1016/j.jaip.2021.02.014.

[9] Khemasuwan D, Sorensen JS, Colt HG. Artificial intelligence in pulmonary medicine: computer vision, predictive model and COVID-19. Eur Respiratory Rev 2020;29(157):200181. Available from: https://doi.org/10.1183/16000617.0181-2020.

[10] Wang Q, Yuan L, Ding X, Zhou Z. Prediction and diagnosis of venous thromboembolism using artificial intelligence approaches: a systematic review and meta-analysis. Clin Appl Thrombosis/Hemostasis 2021;27. Available from: https://doi.org/10.1177/10760296211021162 107602962110211.

[11] Ryan L, Mataraso S, Siefkas A, et al. A machine learning approach to predict deep venous thrombosis among hospitalized patients. Clin Appl Thrombosis/Hemostasis 2021;27. Available from: https://doi.org/10.1177/1076029621991185 107602962199118.

[12] Nafee T, Gibson CM, Travis R, et al. Machine learning to predict venous thrombosis in acutely ill medical patients. Res Pract Thrombosis Haemost 2020;4(2):230–7. Available from: https://doi.org/10.1002/rth2.12292.

[13] Labovitz DL, Shafner L, Reyes Gil M, Virmani D, Hanina A. Using artificial intelligence to reduce the risk of nonadherence in patients on anticoagulation therapy. Stroke. 2017;48(5):1416–19. Available from: https://doi.org/10.1161/strokeaha.116.016281.

[14] Gordon J, Norman M, Hurst M, et al. Using machine learning to predict anticoagulation control in atrial fibrillation: a UK Clinical Practice Research Datalink study. Inform Med Unlocked 2021;25:100688. Available from: https://doi.org/10.1016/j.imu.2021.100688.

[15] Beeler C, Dbeibo L, Kelley K, et al. Assessing patient risk of central line-associated bacteremia via machine learning. Am J Infect Control 2018;46(9):986–91. Available from: https://doi.org/10.1016/j.ajic.2018.02.0211.

[16] Geilleit R, Hen ZQ, Chong CY, et al. Feasibility of a real-time hand hygiene notification machine learning system in outpatient clinics. J Hospital Infect 2018;100(2):183–9. Available from: https://doi.org/10.1016/j.jhin.2018.04.004.

[17] Dey S, Luo H, Fokoue A, Hu J, Zhang P. Predicting adverse drug reactions through interpretable deep learning framework. BMC Bioinforma 2018;19(S21). Available from: https://doi.org/10.1186/s12859-018-2544-0.

[18] Richard Nedelcu E, Bai Y, et al. Post-operative bleeding risk stratification in cardiac pulmonary bypass patients using artificial neural network. Ann Clin & Laboratory Sci 2015;45(2):181–6. Available from: http://www.annclinlabsci.org/content/45/2/181.long. Accessed June 21, 2022.

[19] Wise ES, Stonko DP, Glaser ZA, et al. Prediction of prolonged ventilation after coronary artery bypass grafting: data from an artificial neural network. Heart Surg Forum 2017;20(1):007. Available from: https://doi.org/10.1532/hsf.1566.

[20] Hashimoto DA, Rosman G, Witkowski ER, et al. Computer vision analysis of intraoperative video. Ann Surg 2019;270(3):414–21. Available from: https://doi.org/10.1097/sla.0000000000003460.

[21] Alderden J, Pepper GA, Wilson A, et al. Predicting pressure injury in critical care patients: a machine-learning model. Am J Crit Care 2018;27(6):461–8. Available from: https://doi.org/10.4037/ajcc2018525.

[22] Howcroft J, Lemaire ED, Kofman J. Wearable-sensor-based classification models of faller status in older adults. In: Gao ZK, editor. PLOS ONE, 11. 2016. p. e0153240 (4).

[23] Ward L, Paul M, Andreassen S. Automatic learning of mortality in a CPN model of the systemic inflammatory response syndrome. Math Biosci 2017;284:12–20. Available from: https://doi.org/10.1016/j.mbs.2016.11.004.

[24] Habli I, Lawton T, Porter Z. Artificial intelligence in health care: accountability and safety. Bull World Health Organ 2020;98(4):251–6. Available from: https://doi.org/10.2471/blt.19.237487.

[25] Kelly CJ, Karthikesalingam A, Suleyman M, Corrado G, King D. Key challenges for delivering clinical impact with artificial intelligence. BMC Med 2019;17(1). Available from: https://doi.org/10.1186/s12916-019-1426-2.

[26] Syrowatka A, Song W, Amato MG, et al. Key use cases for artificial intelligence to reduce the frequency of adverse drug events: a scoping review. Lancet Digital Health 2022;4(2):e137–48. Available from: https://doi.org/10.1016/s2589-7500(21)00229-6.

[27] Zhao M, Hoti K, Wang H, Raghu A, Katabi D. Assessment of medication self-administration using artificial intelligence. Nat Med 2021;27(4):727–35. Available from: https://doi.org/10.1038/s41591-021-01273-1.

[28] Schneider-Kamp A. The potential of AI in care optimization: insights from the user-driven co-development of a care integration system. INQUIRY: J Health Care Organization, Provision, Financing 2021;58. Available from: https://doi.org/10.1177/00469580211017992 004695802110179.

[29] Lynn LA. Artificial intelligence systems for complex decision-making in acute care medicine: a review. Patient Saf Surg 2019;13(1). Available from: https://doi.org/10.1186/s13037-019-0188-2.

[30] Smith M, Sattler A, Hong G, Lin S. From code to bedside: implementing artificial intelligence using quality improvement methods. J Gen Intern Med 2021;36(4):1061–6. Available from: https://doi.org/10.1007/s11606-020-06394-w.

[31] Hirata K, Nakaura T, Nakagawa M, et al. Machine learning to predict the rapid growth of small abdominal aortic aneurysm. J Computer Assist Tomography 2020;44(1):37–42. Available from: https://doi.org/10.1097/rct.0000000000000958.

[32] Chung TK, Liang NL, Vorp DA. Artificial intelligence framework to predict wall stress in abdominal aortic aneurysm. Appl Eng Sci 2022;10:100104. Available from: https://doi.org/10.1016/j.apples.2022.100104.

[33] Attia ZI, Kapa S, Lopez-Jimenez F, et al. Screening for cardiac contractile dysfunction using an artificial intelligence–enabled electrocardiogram. Nat Med 2019;25(1):70–4. Available from: https://doi.org/10.1038/s41591-018-0240-2.

[34] Eche T, Schwartz LH, Mokrane FZ, Dercle L. Toward generalizability in the deployment of artificial intelligence in radiology: role of computation stress testing to overcome underspecification. Radiology: Artif Intell 2021;3(6). Available from: https://doi.org/10.1148/ryai.2021210097.

[35] Sau A. Artificial neural network (ANN) model to predict depression among geriatric population at a slum in Kolkata, India. J Clin Diagnostic Res 2017;. Available from: https://doi.org/10.7860/jcdr/2017/23656.9762. Published online.

[36] Contreras I, Vehi J. Artificial intelligence for diabetes management and decision support: literature review. J Med Internet Res 2018;20(5):e10775. Available from: https://doi.org/10.2196/10775.

[37] Abhari S, Niakan Kalhori SR, Ebrahimi M, Hasannejadasl H, Garavand A. Artificial intelligence applications in type 2 diabetes mellitus care: focus on machine learning methods. Healthc Inform Res 2019;25(4):248. Available from: https://doi.org/10.4258/hir.2019.25.4.248.

[38] Cui S, Li C, Chen Z, Wang J, Yuan J. Research on risk prediction of dyslipidemia in steel workers based on recurrent neural network and LSTM neural network. IEEE Access 2020;8:34153–61. Available from: https://doi.org/10.1109/access.2020.2974887.

[39] Zhang L, Yuan M, An Z, et al. Prediction of hypertension, hyperglycemia and dyslipidemia from retinal fundus photographs via deep learning: a cross-sectional study of chronic diseases in central China. In: Chen L, editor. PLOS ONE, 15. 2020. p. e0233166 (5).

[40] Ye C, Fu T, Hao S, et al. Prediction of incident hypertension within the next year: prospective study using statewide electronic health records and machine learning. J Med Internet Res 2018;20(1):e22. Available from: https://doi.org/10.2196/jmir.9268.

[41] Bouharati S. Prevention of obesity using artificial intelligence techniques. ResearchGate. Available from: https://www.researchgate.net/publication/295920133_Prevention_of_Obesity_using_Artificial_Intelligence_Techniques; November 2012 [accessed 22.05.22].

[42] Sefa-Yeboah SM, Osei Annor K, Koomson VJ, Saalia FK, Steiner-Asiedu M, Mills GA. Development of a mobile application platform for self-management of obesity using artificial intelligence techniques Hu F, edInt J Telemed Appl 2021;2021:1–16. Available from: https://doi.org/10.1155/2021/6624057.

[43] Sun LM, Chiu HW, Chuang CY, Liu L. A prediction model based on an artificial intelligence system for moderate to severe obstructive sleep apnea. Sleep Breath 2010;15(3):317–23. Available from: https://doi.org/10.1007/s11325-010-0384-x.

[44] Bini F, Pica A, Azzimonti L, et al. Artificial intelligence in thyroid field—a comprehensive review. Cancers. 2021;13(19):4740. Available from: https://doi.org/10.3390/cancers13194740.

[45] Sechopoulos I, Teuwen J, Mann R. Artificial intelligence for breast cancer detection in mammography and digital breast tomosynthesis: state of the art. SemCancer Biol 2021;72:214–25. Available from: https://doi.org/10.1016/j.semcancer.2020.06.002.

[46] Mitsala A, Tsalikidis C, Pitiakoudis M, Simopoulos C, Tsaroucha AK. Artificial intelligence in colorectal cancer screening, diagnosis and treatment. A new era. Curr Oncol 2021;28(3):1581–607. Available from: https://doi.org/10.3390/curroncol28030149.

[47] East J, Vleugels J, Roelandt P, et al. Advanced endoscopic imaging: European Society of Gastrointestinal Endoscopy (ESGE) Technology Review. Endoscopy. 2016;48(11):1029–45. Available from: https://doi.org/10.1055/s-0042-118087.

[48] Ong E, Wong MU, Huffman A, He Y. COVID-19 coronavirus vaccine design using reverse vaccinology and machine learning. Front Immunology 2020;11. Available from: https://doi.org/10.3389/fimmu.2020.01581.

[49] Ong E, Wong MU, Huffman A, He Y. COVID-19 coronavirus vaccine Design using reverse vaccinology machine learning. bioRxiv 2020;. Available from: https://doi.org/10.1101/2020.03.20.000141. Published online March 21.

[50] Ke YY, Peng TT, Yeh TK, et al. Artificial intelligence approach fighting COVID-19 with repurposing drugs. Biomed J 2020;43(4):355–62. Available from: https://doi.org/10.1016/j.bj.2020.05.001.

[51] Daskivich LP, Vasquez C, Martinez C, Tseng CH, Mangione CM. Implementation and evaluation of a large-scale teleretinal diabetic retinopathy screening program in the Los Angeles County Department of Health Services. JAMA Intern Med 2017;177(5):642. Available from: https://doi.org/10.1001/jamainternmed.2017.0204.

[52] Kuziemsky C, Maeder AJ, John O, et al. Role of artificial intelligence within the telehealth domain. Yearb Med Inform 2019;28(01):035–40. Available from: https://doi.org/10.1055/s-0039-1677897.

[53] Arslan J, Benke KK. Artificial intelligence and telehealth may provide early warning of epidemics. Front Artif Intell 2021;4. Available from: https://doi.org/10.3389/frai.2021.556848.
[54] Hoffman DA. Increasing access to care: telehealth during COVID-19. J Law Biosci 2020;7(1). Available from: https://doi.org/10.1093/jlb/lsaa043.
[55] Davenport T, Kalakota R. The potential for artificial intelligence in healthcare. Future Healthc J 2019;6(2):94–8. Available from: https://doi.org/10.7861/futurehosp.6-2-94.
[56] Jiang L, Wu Z, Xu X, et al. Opportunities and challenges of artificial intelligence in the medical field: current application, emerging problems, and problem-solving strategies. J Int Med Res 2021;49(3). Available from: https://doi.org/10.1177/03000605211000157.
[57] Naik N, Hameed BMZ, Shetty DK, et al. Legal and ethical consideration in artificial intelligence in healthcare: who takes responsibility? Front Surg 2022;9. Available from: https://doi.org/10.3389/fsurg.2022.862322.
[58] Blease C, Kharko A, Bernstein M, et al. Machine learning in medical education: a survey of the experiences and opinions of medical students in Ireland. BMJ Health & Care Inform 2022;29(1):e100480. Available from: https://doi.org/10.1136/bmjhci-2021-100480.
[59] Rajkomar A, Hardt M, Howell MD, Corrado G, Chin MH. Ensuring fairness in machine learning to advance health equity. Ann Intern Med 2018;169(12):866. Available from: https://doi.org/10.7326/m18-1990.

CHAPTER

3

Artificial intelligence devices and assessment in medical imaging

Yanna Kang, Berkman Sahiner and Nicholas Petrick

U.S. Food and Drug Administration, Center for Devices and Radiological Health, Office of Product Evaluation and Quality, Silver Spring, MD, United States

Introduction

Devices enabled by artificial intelligence (AI) and/or machine learning (ML) models are seeing ever-increasing clinical use in all fields of medicine, including medical imaging. Incorporation of AI/ML-enabled models into medical imaging devices can be traced back to the early days of computer-aided detection (CADe) algorithms that were designed to help clinicians detect lesions or conditions on medical images [1]. Today, AI is applied in medical imaging in a much broader context, from aiding clinicians in the acquisition of higher-quality images to helping prioritize image interpretation in busy imaging clinics. These applications are expected to have a profound impact on clinical practice, with AI/ML algorithms integrated into the workflow to enhance clinicians' focus on the patient and to augment human intelligence.

Over the past few years, the US Food and Drug Administration (FDA) has reviewed and authorized a number of AI/ML-enabled devices to be marketed in the United States and provides a list of such devices based on publicly available information.[1] This list of AI/ML-enabled devices spans a range of devices intended for use across many different fields of medicine. Interestingly, about two thirds of these devices use medical images as their primary input, emphasizing the importance of medical imaging applications in the quickly advancing field of medical AI.

In this Chapter, we first provide the regulatory categorization of different types of AI/ML-enabled medical imaging devices, and briefly discuss the framework through which different types of AI/ML-enabled devices are authorized into the US market by the FDA.

[1] See FDA Website on Artificial Intelligence and Machine Learning (AI/ML)-Enabled Medical Devices | FDA. Available at https://www.fda.gov/medical-devices/software-medical-device-samd/artificial-intelligence-and-machine-learning-aiml-enabled-medical-devices?msclkid = 4043b736cd7e11ecbbe04411c5228ef1.

Artificial Intelligence in Clinical Practice
DOI: https://doi.org/10.1016/B978-0-443-15688-5.00002-4

We then summarize the major components in the evaluation of medical imaging-based AI/ML devices and discuss some of the key elements that are required to perform an adequate evaluation. We also review some of the ways in which key information can be conveyed to stakeholders to ensure an acceptable level of transparency.

Examples of artificial intelligence implementation in medical imaging

There are different types of AI/ML-enabled devices in medical imaging that have been authorized for marketing in the United States. The most established category, and discussed frequently, is CADe devices that mark, or in some manner direct the intended user's attention to portions of an image that may reveal abnormalities during interpretation of patient radiology images. Computer algorithms designed to provide an assessment in terms of the likelihood of the presence or absence of disease, or are intended to specify disease type, severity, or stage, are categorized as computer-aided diagnosis (CADx) devices. Computerized systems designed to provide both CADe and CADx functionality are classified as CADe and diagnosis (CADe/x) devices. An example of CADe/x device is an AI model designed to identify regions suspicious for breast cancer on mammography exams and also to provide a likelihood of malignancy score to the clinician for each detected finding as additional information. Most CAD devices are designed as software as a medical device (SaMD) such that the AI/ML is independent from the imaging hardware device.[2]

Another category of AI/ML device is computer-assisted triage and notification (CADt) devices. CADt devices are intended to aid in prioritization and triage of radiological medical images by notifying designated clinicians of the availability of time sensitive images for review based on computer image analysis. These devices do not provide detection or diagnostic information but instead are limited to only providing notification information. An example of a CADt device is an AI model that analyzes CT angiogram images of the brain and sends notifications to a neurovascular specialist that a suspected large vessel occlusion has been identified and recommends priority review of those images.

Additionally, there are image analysis and processing algorithms implemented through AI/ML with functionalities including quantitative imaging, image reconstruction, denoising and various other types of image processing. These algorithms can be designed as SaMD or can be directly incorporated into the image acquisition system or image review platform (such as a PACS—picture archiving and communications system).

In addition to AI/ML-enabled devices that analyze and process images after acquisition by the imaging platform, there are radiological acquisition and optimization guidance devices. These AI/ML devices are specifically intended to aid in the acquisition and/or optimization of radiological images. These devices interface with the acquisition system, analyze its output, and provide guidance and feedback to the operator to help the operator improve the image/signal quality of the radiological image acquisition.

[2] The term Software as a Medical Device is defined by the International Medical Device Regulators Forum (IMDRF) as "software intended to be used for one or more medical purposes that perform these purposes without being part of a hardware medical device."

If we consider the work performed by the AL/ML-enabled device on behalf of the physician, the examples mentioned above can be categorized as either assistive or augmentative AI [2]. A final category of AI/ML-enabled devices we will discuss are autonomous AI. There are different levels of autonomous AI with varying degrees of physician involvement [2]. An example is an AI model where the output indicates that a subset of patients (i.e., one or more patients in the target population) are normal and therefore their radiological data does not require interpretation by a clinician.

Regulatory framework

The FDA has classified CADe, CADx, and CADe/x devices as Class II, for which general controls and special controls together can provide a reasonable assurance of the safety and effectiveness of a device. Those devices are regulated under 21 CFR 892.2070 (some CADe devices are regulated under 21 CFR 892.2050), 21 CFR 892.2060, and 21 CFR 892.2090, respectively. Computer-aided triage devices are also Class II, regulated under 21 CFR 892.2080. Additionally, the FDA has classified radiological acquisition and optimization guidance devices as Class II and regulates them under 21 CFR 892.2100 with the restriction that a radiological acquisition and optimization guidance device regulated under 21 CFR 892.2100 may not include acquisition guidance devices intended to aid a lay user in acquiring radiological images.

Other types of AI-based image analysis and processing algorithms, depending on their intended use (and indications for use), the risk profile, and whether they are used as an accessory to a Class III device or are part of a Class III image acquisition hardware, could be regulated as Class II or Class III. FDA classifies a device based on the risk the device poses to the patient or the user. The International Medical Device Regulators Forum (IMDRF) Framework may be used by the FDA to inform the risk categories of software especially for novel AI models and tasks [3].

The FDA provides the device development community with Guidance documents that represent FDA's current thinking on a topic [4]. Guidance documents describe FDA's interpretation of our policy on a regulatory issue [21 CFR 10.115(b)] usually related to the design, production, labeling, promotion, manufacturing, and testing of regulated products. Guidance documents may also relate to the processing, content, and evaluation or approval of submissions as well as to inspection and enforcement policies. The agency has two relevant guidance documents related to medical imaging-based AI/ML devices. These relate the recommendations for Premarket Notification [510(k)] submissions [5] and clinical performance assessment [6] of computer-assisted detection devices applied to radiology images and device data. New guidance documents are being developed and released so refer to the FDA Guidance webpage [4] to see if a relevant guidance many be available for other medical image analysis tasks.

Evaluation framework

Evaluation components

AI/ML-enabled devices typically include a strong software component, and many of them are considered SaMDs. As such, software verification and validation are indispensable components in the evaluation of these devices. In addition, (i) standalone evaluation of device performance and (ii) clinical evaluation in the hands of physicians are necessary components for many device evaluations. In this subsection, we briefly discuss these three important evaluation components.

Software verification refers to the confirmation that the output of a particular phase of development meets all the input requirements for that phase. Software verification is typically performed as the software is being developed, and includes testing, such as static and dynamic analyses, code and document inspections, module level testing and integration testing. Software validation refers to establishing that the software specifications conform to user needs and intended uses, and that the particular requirements implemented through software can be consistently fulfilled. Software validation is a part of the overall design validation for a finished device. A device developer is expected to document their activities during software verification and validation, including software description requirements, design specification, testing, architecture design chart, risk management plan, revision history, and unresolved anomalies.

Standalone performance evaluation refers to the measurement of the intrinsic performance of the AI model independent of human factors or human interactions with the model. In standalone performance evaluation, the model is applied to a set of test cases that are representative of the intended task for the model. The metrics used for performance evaluation will depend on the task and the details of the AI model and implementation. For example, two primary standalone performance metrics for a device intended to identify diseased patients may be the sensitivity and specificity of case classification, whereas for a device intended to segment/outline some part of anatomy, a DICE similarity coefficient may be more appropriate. For every task it is important to consider whether the ground truth can be derived from the images alone or additional clinical information is needed.

Most types of radiology devices with disease specific outputs that use the AI technologies discussed above are intended to improve the performance of the radiologist or other clinicians in a given task (e.g., detection, classification, image acquisition). For these devices, it is important to characterize whether the use of the device by the clinician results in an improved performance for the clinical task. Clinical evaluation studies are thus typically designed as two-arm studies where the device is used by the clinician in one arm (AI-aided mode) and not used in the other control arm (AI-unaided mode). Since the potential performance improvement in the AI-aided mode will vary among clinicians, these studies are typically designed as multiple reader multiple case (MRMC) studies that account for both reader and case variability [7].

As described above, the FDA's regulatory framework used in device evaluation is risk-based. As a result, not all three components described above may be necessary for every device type or regulatory evaluation. Typically, software verification/validation and standalone performance estimation are expected for most AI/ML-enabled imaging devices. Devices that claim to aid clinicians in performing accurate diagnosis, such as CADe and CADx devices, are also often expected to go through a two-arm MRMC evaluation. Other device types, such as CADt devices, may not need an evaluation in the hands of clinicians since the intended use of the device is for workflow improvement rather than improvement in diagnostic accuracy. Likewise, devices with tool claims, such as AI/ML-enabled quantitative imaging devices, typically do not include MRMC evaluation studies as part of a submission but again this will depend on the clinical task and claims being made. AI/ML-enabled devices, depending on their functionality, use and claims, may also require other types of performance testing as part of a full assessment.

Challenges in artificial intelligence/machine learning device evaluation studies

Representativeness

In many situations, the purpose of the evaluation study is to characterize the performance of the AI model in clinical deployment. This is often the case for standalone evaluation studies. To achieve this, cases in the performance assessment study data should be representative of the population that the AI system is targeted for. This can be achieved by random sampling from the target population, and since performance metrics (such as sensitivity and specificity) in classification problems do not depend on prevalence, it is possible to perform random sampling for each class (diseased and nondiseased for a binary classification problem). When the data set size needs to be limited and important factors that impact performance are known in advance, stratified sampling may offer advantages. In either case, it is desirable to collect data from multiple independent institutions or sites, because differences in patient populations and clinical practices lead to differences in case difficulty and variability among sites.

Independence

When the data set used for evaluating an AI/ML model is not independent of that used for training the model, the test results are typically optimistically biased [8]. To avoid this bias, developers are expected to collect an independent and dedicated test data set, to be accessed only for evaluation purposes. It is important to assure that the collection method leads to independence between training and test data sets. An easily recognizable example of collecting a test data set that violates this principle is having image sets from one patient included in both the training and test data sets. Even when the images in training and test data sets from the same patient have been acquired on different occasions or dates, their structural and potentially pathological similarity will introduce a dependence or correlation in the data. Another, slightly more subtle example is using data sets acquired at one institution or site in both the training and test data sets. In this example, the similarities of the patient population, clinical practice and image acquisition conditions within the site may result in a model that provides good test performance for the cases from that site but that does not generalize to cases acquired at other sites. It is therefore strongly recommended that the test data set be collected from multiple sites that are different from the training set collection sites.

Reference standard

When developing an AI model, the training process most often employs supervised learning that requires categorical and/or localization labels for all the data. The process of determining the reference standard labels is called the truthing process. Even when other AI learning techniques are employed for training (e.g., semi-supervised, weakly-supervised, or unsupervised training), labeled data is still required for algorithm evaluation. The reference standard and truthing process needed for AI training and evaluation data will depend on the AI task and while the truthing process is often the same between training and test data, this is not a requirement. It can often be more practical and efficient to use a less rigorous truthing process (e.g., radiologist interpretation) to label the medical images used in training compared to what is required from an independent evaluation of AI performance (e.g., pathology confirmation).

While the literature often refers to these labels as the truth, gold standard or reference standard, it is appropriate to think of these labels as including some level of uncertainty. Ideally, the reference standard would be obtained from a modality independent from that in which the AI model is being applied. For example, AI designed to identify colon polyps during colonoscopy may use histopathology characterization as the reference standard in characterizing suspected polyps. When the bias and variability from the reference standard is much smaller than that associated with the input imaging modality, which is often the case when pathology is used as the reference standard for radiological AI models, it is common to ignore any small biases or variability coming from the reference standard in an AI model evaluation.

However, the reference standard may not have negligible uncertainty in many medical imaging AI applications. An example of this is lung nodule identification for lung CT. Many radiologically identified lung nodules do not undergo biopsy and pathologic assessment because of the risk associated with collecting tissue in the lungs. Therefore the reference standard for lung nodule identification is often fully or partly based on radiologist interpretation, which is subject to reader variability, especially for smaller nodules [9]. When the reference standard is based on clinician interpretation, it is critical to include multiple truthers in determining the reference standard so that the uncertainty in the reference can be estimated. Likewise, the uncertainty in the reference standard, if not found to be negligible, needs to be accounted for in the AI model evaluation to avoid underestimating the confidence interval for a selected summary performance metric.

Limitations in artificial intelligence/machine learning assessment

A few additional limitations in medical Imaging-based AI model evaluation are worthwhile mentioning. We discuss below three important limitations but there are obviously many others. The first is a lack of prospective or real-world studies of medical AI/ML device performance. Kim et al. [10] conducted a PubMed MEDLINE and Embase databases search to identify articles investigating the performance of medical AI algorithms. Of 516 identified studies published between January 1, 2018, and August 17, 2018, only 6% (31 studies) performed external validation (i.e., use of validation datasets collected either from newly recruited patients or at institutions other than those that provided training data) and only 4 used prospective data collection. Prospective assessment studies are in the literature for diabetic retinopathy, congenital cataracts, lymph node assessment, wrist fracture analysis and colonic polyp identification [11,12]. While Kim et al. [10] looked at the literature from 2018 and as such it does not define the current extent of prospective medical AI studies, prospective assessment still remains a rarity in medical AI validation. A second limitation is a lack of theoretical and often practical knowledge of deep-learning AI failure modes which can have an impact on AI fairness. The concept of convolutional neural networks (CNNs) for medical-imaging based AI have been around for decades [13] but modern deep learning CNN implementations are not well understood since these models may be multilayered and highly complex. This provides the potential to fit highly nonlinear functions because of the huge numbers of trainable parameters (weights). However, this can also lead to overfitting and the potential for unpredictable failures even for small changes to the patient population, disease presentation, input image acquisition, etc. Work is underway to improve generalizability of deep learning CNN methods through data argumentation, improved training methods and ensemble approaches, to name a few, and to improve AI fairness through improved training and validation dataset collection methods. Rajkomar et al., advocates for proactively designing and using medical AI to advance health equity by having a participatory process that involves key stakeholders, including frequently marginalized populations, and considering distributive justice to ensure AI fairness [14]. Much work is still needed to ensure AI model fairness and to develop both the fundamental understanding of AI failure modes and best practices to ensure generalizability. A final limitation is a lack of understanding of the clinical impact of AI models when implemented. While developers often report technical and maybe even diagnostic efficacy, there is very little information on higher levels of efficiency, such as therapeutic, patient outcome, and societal impact [15], making it difficult to determine where medical AI models may prove most efficacious.

Transparency

There are calls for improved transparency of medical AI models within and outside the medical and AI communities [16]. The meaning and context of AI transparency needs to be clearly defined, as it may mean different things to different users and groups. One approach to transparency has related to explainable AI where the "blackbox" of the deep learning algorithm is opened such that the mechanism or at least the basis for how the medical AI model came to a decision is transparent to clinicians or patients. Others discuss transparency in terms of AI validation such that study design, data and patient population, statiscal methods, etc. used to benchmark the AI performance is transparent to users and outside auditors. Other groups may see transparency as providing open-source access to the AI model and code.

The FDA held a virtual public workshop on the Transparency of Artificial Intelligence/ Machine Learning-enabled Medical Devices on October 14, 2021 [17]. The purpose of the workshop was to (1) identify unique considerations in achieving transparency for users of AI/ML-enabled medical devices and ways in which transparency might enhance the safety and effectiveness of these devices; and (2) gather input from various stakeholders on the types of information that would be helpful for a manufacturer to include in the labeling of and public facing information of AI/ML-enabled medical devices, as well as other potential mechanisms for information sharing. At the start of this workshop, FDA provided a working definition of AI transparency as the degree to which appropriate information about a device — including its intended use, development, performance, and, when available, logic — is clearly communicated to stakeholders. The Agency also indicated that transparency is fundamental to a patient-centered approach because it:

- Allows patients, providers, and caregivers to make informed decisions;
- Supports proper use of a device;
- Promotes health equity;
- Facilitates evaluation and monitoring of device performance; and
- Fosters trust and promotes adoption.

This workshop not only discussed transparency in the medical device regulation processes, but brought together medical AI developers and industry, academics, medical professionals, regulatory agencies, professional and support societies and patients to discuss their views on AI transparency including what is most important in terms of transparency from their perspective and how it might be effectively implemented in AI development, AI regulatory processes, and clinical and patient settings. This workshop not only discussed the meaning and role of transparency but also approaches to promoting transparency. This includes way to promote transparency through device labeling and FDA public facing documents as well as how industry, clinicians, patients and outside groups also have important roles in promoting medical AI transparency.

Medical AI transparency is an area of growing interest and research with much more work needed to provide a flexible and functional transparency framework that meets the need of industry, regulatory agencies, medical professionals, patients, and other stakeholders.

Conclusions

There is growing interest in AI/ML-enabled medical devices, and especially medical imaging devices. The FDA has classified AI/ML-enabled medical imaging devices into different device categories and has established classifications and regulatory controls for a number of different medical imaging AI/ML devices. Evaluation of AI/ML-enabled medical devices is important to ensure they are safe, effective, and reliable when the devices are used on patients. Although the general evaluation framework and evaluation components are known for AI/ML devices, there are challenges and limitations in AI/ML device evaluation that need to be addressed by all stakeholders. Finally, there is a need to establish a framework to meet stakeholder expectations for communicating device information and outcome of an AI model properly.

References

[1] Chan H-P, Doi K, Galhotra S, Vyborny CJ, MacMahon H, Jokich PM. Image feature analysis and computer-aided diagnosis in digital radiography. 1. Automated detection of microcalcifications in mammography. Med Phys 1987;14:538–48.

[2] Association AM. CPT Appendix S: artificial intelligence taxonomy for medical services and procedures. <https://www.ama-assn.org/system/files/cpt-appendix-s.pdf>; 2021.

[3] U.S. Food and Drug Administration. Software as a Medical Device (SAMD): clinical evaluation. Guidance for Industry and Food and Drug Administration Staff; 2020.

[4] U.S. Food and Drug Administration. Guidances Silver Spring, MD: U.S. Food and Drug Administration. <https://www.fda.gov/industry/fda-basics-industry/guidances>; 2022.

[5] U.S. Food and Drug Administration. Computer-assisted detection devices applied to radiology images and radiology device data—premarket notification [510(k)] submissions. Silver Spring, MD: U.S. Food and Drug Administration; 2012.

[6] U.S. Food and Drug Administration. Clinical performance assessment: considerations for computer-assisted detection devices applied to radiology images and radiology device data in premarket notification (510(k)) submissions—Guidance for Industry and Food and Drug Administration Staff; 2020.

[7] Gallas BD, Chan HP, D'Orsi CJ, Dodd LE, Giger ML, Gur D, et al. Evaluating imaging and computer-aided detection and diagnosis devices at the FDA. Acad Radiol 2012;19(4):463–77.

[8] Chan H-P, Sahiner B, Wagner RF, Petrick N. Classifier design for computer-aided diagnosis: effects of finite sample size on the mean performance of classical and neural network classifiers. Med Phys 1999;26(12):2654–68.

[9] Meyer CR, Johnson TD, McLennan G, Aberle DR, Kazerooni EA, Macmahon H, et al. Evaluation of lung MDCT nodule annotation across radiologists and methods. Acad Radiol 2006;13(10):1254–65.

[10] Kim DW, Jang HY, Kim KW, Shin Y, Park SH. Design characteristics of studies reporting the performance of artificial intelligence algorithms for diagnostic analysis of medical images: results from recently published papers. Korean J Radiol 2019;20(3):405–10.

[11] Kelly CJ, Karthikesalingam A, Suleyman M, Corrado G, King D. Key challenges for delivering clinical impact with artificial intelligence. BMC Med 2019;17(1):195.

[12] Repici A, Badalamenti M, Maselli R, Correale L, Radaelli F, Rondonotti E, et al. Efficacy of real-time computer-aided detection of colorectal neoplasia in a randomized trial. Gastroenterology. 2020;159(2):512–20.

[13] Lo SC, Lin JS, Freedman MT, Mun SK. Computer-assisted diagnosis of lung nodule detection using artificial convolution neural network. Proc SPIE Med Imaging 1993;1898:859–69.

[14] Rajkomar A, Hardt M, Howell MD, Corrado G, Chin MH. Ensuring fairness in machine learning to advance health equity. Ann Internal Med 2018;169(12):866–72.

[15] Fryback DG, Thornbury JR. The efficacy of diagnostic imaging. Med Decis Mak 1991;11(2):88–94.

[16] Walmsley J. Artificial intelligence and the value of transparency. AI Soc 2021;36(2):585–95.

[17] Virtual public workshop—transparency of artificial intelligence/machine learning-enabled medical devices: U.S. Food and Drug Adminstration. <https://www.fda.gov/medical-devices/workshops-conferences-medical-devices/virtual-public-workshop-transparency-artificial-intelligencemachine-learning-enabled-medical-devices>; 2021.

CHAPTER

4

Artificial intelligence in anatomical pathology

Saba Shafi and Anil V. Parwani

Department of Pathology, The Ohio State University Wexner Medical Center, Columbus, OH, United States

Introduction

Digital pathology (DP) is the process of digitizing histopathology slides to generate whole slide images (WSIs) and the subsequent analysis of these WSIs using various computational methods. Interestingly, the first forays into the use of digital image processing and computerized image analysis were for the analysis of cell and microscopy images. In the 1950s McCarthy et al. coined the term, "artificial intelligence (AI)" for the branch of computer science which uses machine-based approaches to arrive at certain predictions, thereby simulating human intelligence while solving complex problems [1,2]. In the 1960s Prewitt and Mendelsohn scanned images from a blood smear and devised a method of converting the optical data into a matrix of optical density values while preserving spatial and gray-scale relationships. Prewitt and her colleagues wrote some of the very first papers on the use of computerized image analysis of cell image [3–5]. The last decade has witnessed a dramatic and promising increase in the application of AI tools in pathology [6]. DP and AI have the potential of revolutionizing the practice of pathology as not only does it transfer an image from the glass slide to the monitor, but so does it augment the pathologist's eye with information that is impossible to be gleaned by human examination [7]. Machine learning (ML)-based approaches are based on the machine "learning" to make predictions based on the input data and algorithms and falls within the broad ambit of AI [8,9]. In the 1980s the advancement of artificial neural networks resulted in the development of an ML approach called deep learning (DL). A DL network consists of an input layer, an output layer, and multiple hidden layers, thereby recapitulating an artificial representation of the human neural architecture. The hidden layers can regenerate new representations of the image and with sufficient training iterations can identify representations that best differentiate between categories of interest [10,11].

DOI: https://doi.org/10.1016/B978-0-443-15688-5.00028-0

Transition and management of "images" in a digital workflow can be accomplished using various approaches: an LIS-based approach, a scanner vendor approach, or an intermediate software approach (e.g., Linköping University) [12–15]. Regardless of the type of strategy chosen to switch toward a digital visualization of images, the new system should be able to integrate every possible instrument (e.g., one or more scanners from same or different vendors with the possibility to manage different images from a variety of sources), preferably associated with a tracking system because of automation and innovation. Implementation models highlight the cost-effectiveness of DP, including the scope of investment, the potential return on investment, and cost-savings, as well as any proposed income deriving from the adoption of WSIs. Moreover, the adequate adaptation of a routine clinical workflow can finally lead to an optimization of resources (e.g., space, time, personnel, and equipment) [16].

Overview of digital anatomical pathology workflow

A fully digital laboratory uses electronic record system and is totally paperless and glass slides are substituted by WSIs. Transitioning from glass to digital workflows in anatomic pathology (AP) requires new DP equipment, image management systems, improved data management and storage capacities, and additional trained technicians [17]. For increasing efficiency and ensuring cost effectiveness, optimization of resources, such as time, space, people, and instruments, is paramount. The LEAN approach represents a valuable strategy to optimize the workflow, leading to a more logical distribution of the spaces to minimize staff and sample traffic inside the laboratory. Though not a prerequisite for adopting DP, it facilitates better allocation of resources [18]. Unlike the old, "analog" workflow, a fully digitized AP laboratory requires stringent efficient arrangement of the physical space for smooth crosstalk among the different components of the process. For instance, the scanning workstation should be located near the staining and mounting instruments, accelerating the production line but far from the microtome area to avoid the interference of paraffin with the scanning mechanisms. Before implementing DP in the workflow, it is critical to analyze the preexisting workflow. A careful analysis of the preexisting analogic workflow before the transition should consider the flow of the samples (workstation location) in the laboratory and time intervals (hands-on and waiting times) for each workstation, verifying the information technology support and establishment of adequate quality control (QC) checkpoints. After this retrospective analysis and reorganization of the structure, the optimal choices for the automation of each workstation must be made, namely by the introduction of a reliable tracking system, and different instruments would preferably work in a coordinated fashion, connected (mono-or bi-directionally) to the laboratory information system (LIS or LIMS). DP implementation is complex and interdisciplinary and requires a close collaboration of various key players like IT, laboratory staff, MDs, and institute management. It is crucial to involve all units at an early stage of development. Most DP teams include project managers, pathologists, specialized technicians, IT specialists responsible for LIS and LEAN specialists that meet regularly with the head of the pathology lab and representatives of LIS and image management system (IMS) providers [15,19].

The Ohio State University Wexner Medical Center is home to one of the world's leading DP programs. Daily over 2,400 slides are digitized and made instantly available to our pathologists who use them for primary diagnosis, consultation, education, and research. In 2018 Ohio State became the first site in the United States to utilize DP for primary diagnosis. Today, most of our sign-outs are digital. Because of these endeavors, we manage one of the world's largest collections WSIs with almost 2.7 million WSIs, representing over 280,000 patients, and every imaginable tissue and disease type. The DP group also works closely with computational pathology researchers. However, the main goal is the continuous development of DP for diagnostic routine in the department.

Slide scanners and whole slide images

The introduction of whole slide scanners in 1990 was an important milestone in DP and computational pathology [20,21]. In 2017 Philips received De Novo pathway clearance to market the IntelliSite Pathology Solution as a comprehensive DP system [22]. While the use of advanced high-resolution hardware with multiple graphical processing units can speed up training, it can become prohibitively expensive. The DP system scans traditional glass slides using a slide scanner to produce a WSI, which is then stored and transmitted to servers [23]. WSI which has an average of 1.6 billion pixels and occupies 4600 megabytes (MB) per unit, thus taking up much more space than a DICOM (digital imaging and communications in medicine) format used in radiography [24]. With the recent advances in technology, scanners, servers, WSIs can be rapidly processed, allowing pathologists to inspect images on their computers [25].

Certain technical aspects need to be considered while incorporating DP in the AP workflow. File sizes can vary considerably (up to over fourfold) among different scanners and will therefore have a major impact on the storage space needed. Scanning speed can be a limiting factor and a potential bottleneck in the diagnostic workflow. Scan times may vary up to a minute per slide on different scanners at the same magnification. Among the various determinants include the type of specimen (biopsies vs. resections), quality of slides (tissue folds, staining intensity, missing tissue), scan failures, software, hardware, or slide-related interruptions due to overnight scanning. Quality control checks must be performed to establish the proportion of slides that need to be rescanned with attempts to minimize rescan time. Another key consideration is the continuous autofocusing which may cause different areas of the slide being out of focus. Despite this limitation, scanners using continuous autofocus have lower rates of completely out of focus WSIs compared to scanners using focus points. Different scanners use different algorithms for detecting tissue on a slide. Faintly stained tissue (e.g., myxoid substance or fat) can be missed and only partially scanned by some scanners. The scanned file format should be readily convertible into other formats using various image analysis tools [19].

Scanner user friendliness is necessary to ensure smooth DP workflow, with the laboratory staff playing an important role in evaluating the usability of scanner software and hardware. The slides can either manually loaded one-by-one or racks can be directly loaded from the stainers. Care must be taken to ensure slides are dried as those with wet mounting medium tend to stick to the rack and will not be scanned. Some scanners stop to open, others continue with the process even if opened to reload, thereby impacting the speed of the workflow. Seamless integration with LIS is crucial to ensure robust maintenance of work lists and case management especially the remote setting. Additionally, pathologists should feel comfortable using the IMS, which should provide certain tools (measurement, area calculation, regions of interest, snapshot etc.). Interobserver comparisons among pathologists in ensures robust quality of scans and diagnostic confidence [19].

Applications of artificial intelligence in anatomic pathology

Integrating AI into the workflow can help in incorporating diagnosis with clinical information, ordering necessary pathology studies including IHC and molecular studies, automating repetitive tasks, on-demand consultations, QC of the preanalytic, analytic, and postanalytic phases of the work process, allowing QC of scan images and cloud server management [7,26–32].

Role of artificial intelligence in objective diagnosis

One of the advantages of DP is that it allows the simultaneous inspection of histopathology images along with patient metadata, such as demographic, gene sequencing or expression data, and progression and treatment outcomes. Several attempts are being made to integrate patient pathological tissue images and one or more metadata to obtain novel information that may be used for diagnosis and prediction, as it was discovered that predicting survival using merely pathologic tissue images was challenging and inaccurate [33]. An Israel-based company, Ibex received clinical approval for two AI-software's, Galen™ Breast and Galen™ Prostate [34,35]. In 2022 Galen™ Gastric by Ibex became the first AI software in the world to receive the CE mark for the detection of gastric cancer [36]. Paige, a US-based company introduced Paige Prostate Detect which is thus far the only AI pathology diagnostic software to have received FDA approval [37–39]. Nearly all these AI-based pathology diagnostic solutions have the ability to analyze patient WSIs, as well as enable tumor detection and classification including tumor grading with high accuracy [40]. Several validation studies have indicated better diagnostic performance of pathologists using AI- powered diagnostic solutions, with reduced interobserver variability, diagnostic uncertainty, and turn-around-time. AI applications hold great promise in replacing time-consuming, laborious, and repetitive components that take up a significant amount of pathologists' time in daily practice. Screening lymph nodes for possible metastatic cancer is one such tedious task. Pathologists spend a lot of time daily to manually screen for nodal metastasis and the process is prone to misdiagnosis which in turn can have huge implications in tumor staging and therapeutic strategies. AI assistance for nodal assessment in over 2800 breast cancers showed a 24% overall change with upstaging in 18% and downstaging in 6% [41]. In a related study, Steiner et al. developed a lymph node-assisted screening model, LYNA trained to identify regions with a high likelihood of tumor. While the specificity for negative lymph nodes was similar with and without LYNA, AI-assistance increased sensitivity for detecting micrometastasis. Additionally, AI assistance considerably reduced the average review time per image [42]. The CAMYLEON16 challenge used H&E WSI from sentinel lymph nodes of breast cancer patients for identifying metastasis. With no time constraint, DL algorithms showed comparable performance to a pathologist in detection of micrometastasis. To simulate a clinical practice setting, time constraint was imposed, which demonstrated an outperformance by algorithm over manual evaluation by 11 pathologists [28].

Supervised DL models based on cellular features have also been used to differentiate between benign and malignant breast tumors using WSIs of FNAC specimens [43]. More recently, Wu et al. developed a lymph node metastases diagnostic model (LNMDM) on WSIs to assess its efficiency. The area under the curve for accurate diagnosis of the LNMDM ranged from 0.978 to 0.998 in the five validation sets, demonstrating improved accuracy in detecting lymph node metastases, particularly micrometastases [44]. AI algorithms can also be used to predict the origin of a tumor in cancers of unknown primary source using only a histopathology image of the metastatic site [45]. Digital image analysis (DIA) tools can also be used for exact quantification of predicative IHC biomarkers with improved accuracy, such as estrogen receptors, progesterone receptors, human epidermal growth factor receptor 2, programmed death ligand 1, and the Ki-67 proliferative index. Numerous algorithms can quantify mitotic indexes in various cancers like breast or neuroendocrine tumors with high accuracy [46–50]. The MICCAI MIDOG 2021 challenge was the creation of scanner-agnostic mitotic figure detection algorithms. The challenge used a training set of 200 cases, split across four scanning systems. The winning algorithm yielded an *F*1 score of 0.748 [95% confidence interval (CI), 0.704–0.781], exceeding the performance of six experts on the same task [51].

Artificial intelligence for quality control

AI tools can be embedded within a pathology laboratory workflow for QC at the preanalytic, analytic, and postanalytic stage. AI tools can aid the process of triaging cases before they are sent to the pathologist for review to ensure cancer cases are prioritized. AI can provide QC by reviewing pathologists' grading for some cancers and avoid possible misclassifications. A study demonstrated the utility of the AI software, Galen™ Prostate as a second read QC system within a clinical workflow. AI was run in parallel to the usual pathologist review, and alerts were generated if slides that were determined to be benign by pathologists received high algorithmic cancer scores. An alert was also generated if a slide received a pathologist-determined grade group 1 but an algorithm-determined grade group ≥ 2. These alerts directed the pathologists to perform a second review, paying specific attention to the regions that triggered the alert, thereby serving as "second pair of eyes" [35,52]. Furthermore, preanalytic AI implementation can also affect the process of molecular pathology. QC of samples is required to confirm the adequacy of tumor tissue to proceed with further molecular analysis [53]. The digitization of slides provides a valuable way to archive, preserve, and retrieve important information when needed [54].

Artificial intelligence for cancer prediction and prognostication

AI can be used to predict prognosis and therapeutic responses based on histological features [55,56]. Directly linking images with several features of tumor, surrounding microenvironment and genetic profiles with survival outcomes and treatment response for adjuvant/neoadjuvant therapy could provide important information in a concise manner. Integrating multiple morphological features, such as tumor histological patterns and tumor microenvironment patterns into a single prognostic index, can be difficult for humans [6,57]. However, AI tools can provide a novel classification system depicting clinical outcome, probability of recurrence or metastases and therapeutic response by correlating important histological features, such as tumor morphology, stromal architecture, nuclear texture, and lympho-vascular invasion. Convolutional neural networks (CNNs) have been integrated with a Cox proportional hazards model to predict the overall survival of patients with gliomas using WSIs and genetic biomarkers, such as chromosome deletion and gene mutation [58].

Graphical approaches can prediction clinical outcome by evaluation of architectural organization and spatial configuration of different types of tissues [26]. AI approaches can also be used for quantifications, such as assessment of tumor-infiltrating lymphocytes (TILs), which are subject to interobserver variability [59–61]. Spatial organization analysis of tumor microenvironment containing multiple cell types, rather than only TILs, has been explored, and it is expected to yield information on tumor progression, metastasis, and treatment outcomes [62]. Image analysis of the relative amount of area of tumor and intratumoral stroma, such as the tumor–stroma ratio, is a widely studied prognostic factor in several cancers, including breast, colorectal, and lung [63–67]. Recently AI has been used to predict lymph node metastasis or risk of lymph node metastasis in T1 colorectal cancer. The risk of predicting nodal metastasis had an accuracy 81.8%–86.3% using a DL model on biopsy samples. The study demonstrated that AI models suggest an association between biopsy specimens and lymph node metastases in T1 CRC and may contribute to increased accuracy of preoperative diagnosis [68]. More recently, a study used an AI pipeline to predict the effect of preoperative chemotherapy employing H&E images of breast cancer tissue obtained from needle biopsies prior to chemotherapy. Using CNN model, support vector machines and random forest models, response to neoadjuvant chemotherapy was predicted with 95% accuracy on a test set of 103 unseen cases [69].

Artificial intelligence for grading tumors

The PANDA challenge, the largest histopathology competition to date, joined by 1,290 developers was organized to catalyze development of reproducible AI algorithms for Gleason grading using 10,616 digitized prostate biopsies. A diverse set of algorithms were validated to reach performance and accuracy levels similar to pathologists. The algorithms achieved agreements of 0.862 (quadratically weighted κ, 95% CI, 0.840–0.884) and 0.868 (95% CI, 0.835–0.900) with expert uropathologists [40]. Precise postop model developed by Donovan et al. could enhance Gleason grading for the prediction of clinical failure after radical prostatectomy [70]. Following this, several other AI-tools are being developed to facilitate autonomous grading of prostate cancers [34,40,71–74] (Figs. 4.1 and 4.2).

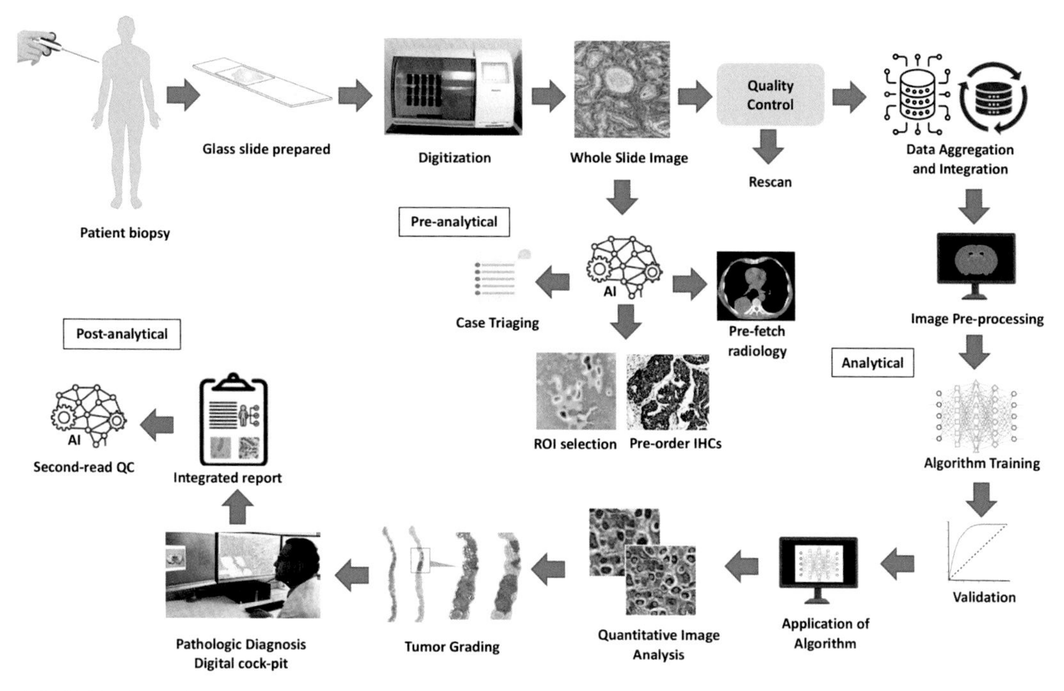

FIGURE 4.1 Schematic representation depicting integration of artificial intelligence in anatomic pathology workflow.

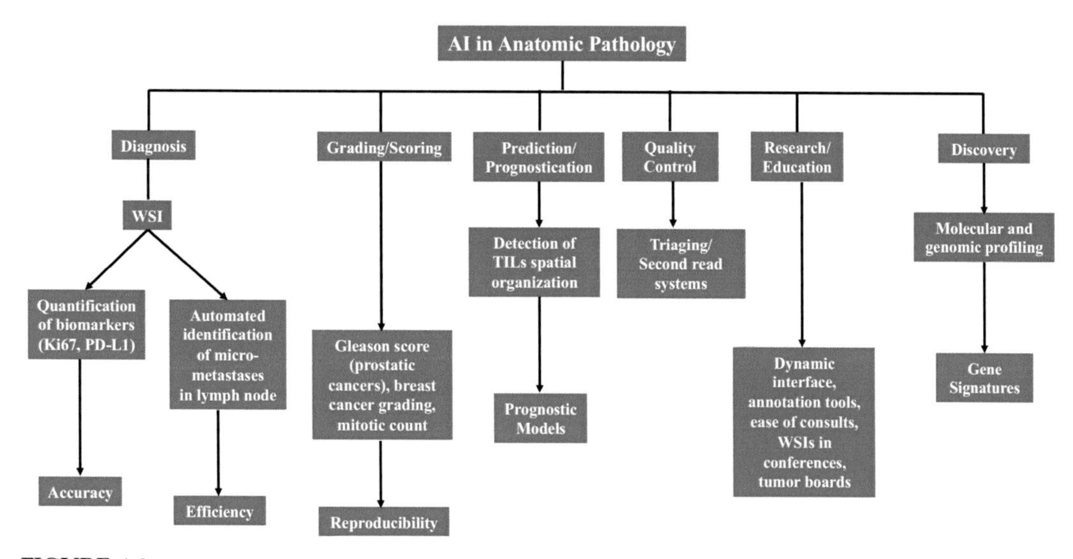

FIGURE 4.2 Overview of the utility of artificial intelligence in anatomic pathology (abbreviations: *WSIs*, whole slide imaging; *TILs*, tumor-infiltrating lymphocytes).

Promising discoveries with artificial intelligence

WSIs can be used in conjunction with other data modalities, such as radiology, transcriptomics, and microbiomics to train algorithms for detecting associations between genotypes, phenotypes, prognosis, and therapeutics in various cancers. Algorithms are being developed and validated to predict prognosis and therapeutic response in gastric cancers using a 32-gene signature [75]. Potential multiomics approach to model the relationship between gastric cancer and the gastric microbiome are other examples of AI being used for new discoveries [76]. AI algorithms can predict the activation of signature genes associated with increased sensitivity to immunotherapy in patients with advanced hepatocellular carcinoma, thereby highlighting the promise of AI for patient care [77] (Table 4.1).

TABLE 4.1 Key steps involved in integration of AI in anatomic pathologic workflow with an overview of challenges encountered in implementation.

Key steps in AI integration in anatomic pathology	Challenges
Correct identification of needs	• Lack of awareness of potential use • Incorrect assessment of end-user demands
Interdisciplinary collaboration	• Discordant goals among participants • Lack of clear communication
Concept and design	• Rationale and scientific background • Funding • Approval and regulation
Algorithm development	• IT personnel requirement • Preanalytic and analytic components • Lack of ground truth • Lack of interpretability • Lack of generalizability • "Black box" problem
Standardization and validation	• Lack of appropriate validation data • Overfitting
Data quality	• Lack of well-curated, annotated datasets
Regulation and reimbursement	• Lack of regulatory guidelines • Lack of CPT codes
Installation	• Overhead costs and IT infrastructure
Pathologists' attitude	• Resistance to change in "analog" workflow
Accreditation and clinical implementation	• Unclear audit process • Lack of FDA approval
Ethical issues	• Liability • Responsibility • Skepticism by pathologists, oncologists, and patients
Cloud storage	• High cost, high specification storage • Cost–benefit considerations

Role of artificial intelligence in education and research

AI tools can be implemented for training the next generation of pathologists, provide automated annotations and other interactive functions to create a dynamic teaching environment for trainees. WSIs are used for teaching at conferences, virtual workshops, presentations, and tumor boards [6,25]. The Ohio State University Wexner Medical Center has incorporated the use of a "digital cockpit" for fully digital sign-outs. Residents regularly preview digital slides using the Philips IMS. The annotation tools enable viewing, panning, and zooming enhanced digital slides, encircling regions of interest, including a single cell under question, thereby creating a more interactive learning interface. The university also leverages several of its add-on components to enable integration of WSI into the LIS. This underscores the state-of-the-art incorporation of DP tools in clinical workflow [78,79]. The use of Visopharm AI tools enables swift detection of isolated tumor cell metastases in lymph nodes in difficult cases. Integration of such AI tools in the reporting workflow can thus provide trainees with additional information, such as lists of differential diagnosis and potential auxiliary tests, that can be requested thereby honing their diagnostic skills. It also provides relevant educational resources which can potentially improve resident training. Such educational models can be complementary to the conventional educational processes provided by the pathologists and can be adopted by other institutions. Not only has it improved the in-house training and intersubspecialty consults, so has it made it a lot easier to collaborate with other institutions and provide efficient consults and second opinions on challenging cases [6,78,79].

Conclusions

Despite various challenges, such as difficulty in validating algorithms, lack of interpretability, overfitting, issues with regulation, reimbursement, clinical adoption, data storage, attitude of pathologists, and exorbitant costs of infrastructure, AI-assisted anatomic pathologic diagnosis is becoming a reality. While it is highly unlikely that AI applications will replace human pathologists, they can aid personalized cancer care after standardized usage recommendations, and harmonization with current information systems. In the beginning, in anatomical pathology workflows, these will serve as valuable decision support tools. Improved efficiency, high diagnostic accuracy, reproducible quantification, discovery of new patterns/associations, and pixel-level findings are geared to revolutionize the AP workflow. A multimodal approach using proteomics, genomics, and AI-based multiplexed biomarker quantifications, pathology reports can be enriched with personalized diagnostic, prognostic, and theranostic information, all tailored to individual patients to guide clinical management.

References

[1] McCarthy J, Minsky M, Rochester N. Artificial intelligence. Research Laboratory of Electronics (RLE) at the Massachusetts Institute of Technology, MIT; 1959.

[2] McCarthy J, Minsky ML, Rochester N, Shannon CE. A proposal for the dartmouth summer research project on artificial intelligence, august 31, 1955. AI Mag 2006;27(4):12.
[3] Prewitt JM, Mendelsohn ML. The analysis of cell images. Ann N Y Acad Sci 1966;128(3):1035–53.
[4] Prewitt JM. Parametric and nonparametric recognition by computer: an application to leukocyte image processing. Advances in computers, Vol. 12. Elsevier; 1972, p. 285–414.
[5] Prewitt JM. Intelligent microscopes: recent and near-future advances. Paper presented at: recent future dev med imaging II; 1979.
[6] Rakha EA, Toss M, Shiino S, et al. Current and future applications of artificial intelligence in pathology: a clinical perspective. J Clin Pathol 2021;74(7):409–14.
[7] Acs B, Rimm DL. Not just digital pathology, intelligent digital pathology. JAMA Oncol 2018;4(3):403–4.
[8] Yao X. Evolving artificial neural networks. Proc IEEE 1999;87(9):1423–47.
[9] Haykin S, Lippmann R. Neural networks, a comprehensive foundation. Int J Neural Syst 1994;5(4):363–4.
[10] Deng L, Yu D. Deep learning: methods and applications. Found Trends Signal Process 2014;7(3–4):197–387.
[11] LeCun Y, Bengio Y, Hinton G. Deep learning. Nature 2015;521(7553):436–44.
[12] Sinard JH, Castellani WJ, Wilkerson ML, Henricks WH. Stand-alone laboratory information systems versus laboratory modules incorporated in the electronic health record. Arch Pathol Lab Med 2015;139(3):311–18.
[13] Sepulveda JL, Young DS. The ideal laboratory information system. Arch Pathol Lab Med 2013;137(8):1129–40.
[14] Asa SL, Bodén AC, Treano D, Jarkman S, Lundström C, Pantanowitz L. 2020 vision of digital pathology in action. J Pathol Inform 2019;10.
[15] Fraggetta F, L'imperio V, Ameisen D, et al. Best practice recommendations for the implementation of a digital pathology workflow in the anatomic pathology laboratory by the European Society of digital and integrative pathology (ESDIP). Diagnostics. 2021;11(11):2167.
[16] Lujan G, Quigley JC, Hartman D, et al. Dissecting the business case for adoption and implementation of digital pathology: a white paper from the digital pathology association. J Pathol Inform 2021;12(1):17.
[17] Cheng JY, Abel JT, Balis UG, McClintock DS, Pantanowitz L. Challenges in the development, deployment, and regulation of artificial intelligence in anatomic pathology. Am J Pathol 2021;191(10):1684–92.
[18] Zarbo RJ. Creating and sustaining a lean culture of continuous process improvement, Vol. 138. Oxford, UK: Oxford University Press; 2012, p. 321–6.
[19] Dawson H. Digital pathology—rising to the challenge. Front Med 2022;9.
[20] Afework A, Beynon MD, Bustamante F, Cho S, Demarzo A, Ferreira R, et al. Digital dynamic telepathology—the virtual microscope. Paper presented at: proc AMIA symposium; 1998.
[21] Ferreira R, Moon B, Humphries J, et al. The virtual microscope. Paper presented at: proc AMIA annu fall symposium; 1997.
[22] Evans AJ, Bauer TW, Bui MM, et al. US Food and Drug Administration approval of whole slide imaging for primary diagnosis: a key milestone is reached and new questions are raised. Arch Pathol Lab Med 2018; 142(11):1383–7.
[23] Zarella MD, Bowman D, Aeffner F, et al. A practical guide to whole slide imaging: a white paper from the digital pathology association. Arch Pathol Lab Med 2019;143(2):222–34.
[24] Kohli MD, Summers RM, Geis JR. Medical image data and datasets in the era of machine learning—whitepaper from the 2016 C-MIMI meeting dataset session. J Dig Imaging 2017;30:392–9.
[25] Farahani N, Parwani AV, Pantanowitz L. Whole slide imaging in pathology: advantages, limitations, and emerging perspectives. Pathol Lab Med Int 2015;7(23–33):4321.
[26] Acs B, Rantalainen M, Hartman J. Artificial intelligence as the next step towards precision pathology. J Intern Med 2020;288(1):62–81.
[27] Acs B, Hartman J. Next generation pathology: artificial intelligence enhances histopathology practice. J Pathol 2020;250(1):7–8.
[28] Bejnordi BE, Veta M, Van Diest PJ, et al. Diagnostic assessment of deep learning algorithms for detection of lymph node metastases in women with breast cancer. JAMA 2017;318(22):2199–210.
[29] Chen J, Srinivas C. Automatic lymphocyte detection in H&E images with deep neural networks. ArXiv Prepr arXiv:161203217, 2016.
[30] Garcia E, Hermoza R, Castanon CB, Cano L, Castillo M, Castanneda C. Automatic lymphocyte detection on gastric cancer ihc images using deep learning. Paper presented at: 2017 IEEE 30th international symposium on computer-based medical systems (CBMS); 2017.

[31] Liu Y, Kohlberger T, Norouzi M, et al. Artificial intelligence–based breast cancer nodal metastasis detection: Insights into the black box for pathologists. Arch Pathol Lab Med 2019;143(7):859–68.
[32] NaikS, Doyle S, Agner S, Madabhushi A, Feldman M, Tomaszewski J. Automated gland and nuclei segmentation for grading of prostate and breast cancer histopathology. Paper presented at 2008 5th IEEE international symposium on biomedical imaging: from nano to macro; 2008.
[33] Zhu X, Yao J, Huang J. Deep convolutional neural network for survival analysis with pathological images. Paper presented at: 2016 IEEE international conference on bioinformatics and biomedicine (BIBM); 2016.
[34] Nagpal K, Foote. D, Tan F, et al. Development and validation of a deep learning algorithm for Gleason grading of prostate cancer from biopsy specimens. JAMA Oncol 2020;6(9):1372–80.
[35] Pantanowitz L, Quiroga-Garza GM, Bien L, et al. An artificial intelligence algorithm for prostate cancer diagnosis in whole slide images of core needle biopsies: a blinded clinical validation and deployment study. Lancet Digital Health 2020;2(8):e407–16.
[36] JenoskiR, Mayer S, Marks R, Salmon R. Color calibration for digital cytology scanner. J Pathol Inform. 2021;12:37.
[37] Raciti P, Sue J, Ceballos R, et al. Novel artificial intelligence system increases the detection of prostate cancer in whole slide images of core needle biopsies. Mod Pathol 2020;33(10):2058–66.
[38] Perincheri S, Levi AW, Celli R, et al. An independent assessment of an artificial intelligence system for prostate cancer detection shows strong diagnostic accuracy. Mod Pathol 2021;34(8):1588–95.
[39] da Silva LM, Pereira EM, Salles PG, et al. Independent real-world application of a clinical-grade automated prostate cancer detection system. J Pathol 2021;254(2):147–58.
[40] Bulten W, Kartasalo K, Chen P-HC, et al. Artificial intelligence for diagnosis and Gleason grading of prostate cancer: the PANDA challenge. Nat Med 2022;1–10.
[41] Vestjens J, Pepels M, de Boer M, et al. Relevant impact of central pathology review on nodal classification in individual breast cancer patients. Ann Oncol 2012;23(10):2561–6.
[42] Steiner DF, MacDonald R, Liu Y, et al. Impact of deep learning assistance on the histopathologic review of lymph nodes for metastatic breast cancer. Am J Surg Pathol 2018;42(12):1636.
[43] Osareh A, Shadgar B. Machine learning techniques to diagnose breast cancer. Paper presented at: 2010 5th international symposium on health informatics and bioinformatics; 2010.
[44] Wu S, Hong G, Xu A, et al. Artificial intelligence-based model for lymph node metastases detection on whole slide images in bladder cancer: a retrospective, multicentre, diagnostic study. Lancet Oncol 2023;24 (4):360–70.
[45] Lu MY, Chen TY, Williamson DF, et al. AI-based pathology predicts origins for cancers of unknown primary. Nature 2021;594(7861):106–10.
[46] Czyzewski T, Daniel N, Rochman M, et al. Machine learning approach for biopsy-based identification of eosinophilic esophagitis reveals importance of global features. IEEE Open J Eng Med Biol 2021;2:218–23.
[47] Lara H, Li Z, Abels E, et al. Quantitative image analysis for tissue biomarker use: a white paper from the digital pathology association. Appl Immunohistochem Mol Morphol 2021;29(7):479.
[48] Feng M, Deng Y, Yang L, et al. Automated quantitative analysis of Ki-67 staining and HE images recognition and registration based on whole tissue sections in breast carcinoma. Diagnostic Pathol 2020;15(1):1–12.
[49] Geread RS, Sivanandarajah A, Brouwer ER, et al. Pinet—an automated proliferation index calculator framework for Ki67 breast cancer images. Cancers. 2020;13(1):11.
[50] Liu Y, Li X, Zheng A, et al. Predict Ki-67 positive cells in H&E-stained images using deep learning independently from IHC-stained images. Front Mol Biosci 2020;7:183.
[51] Aubreville M, Stathonikos N, Bertram CA, et al. Mitosis domain generalization in histopathology images—the MIDOG challenge. Med Image Anal 2023;84:102699.
[52] Busby D, Grauer R, Pandav K, et al. Applications of artificial intelligence in prostate cancer histopathology. Paper presented at: urologic oncology: seminars and original investigations; 2023.
[53] Chung M, Lin W, Dong L, Li X. Tissue requirements and DNA quality control for clinical targeted next-generation sequencing of formalin-fixed, paraffin-embedded samples: a mini-review of practical issues. J Mol Genet Med 2017;11(262):1747-0862.
[54] Kim I, Kang K, Song Y, Kim T-J. Application of artificial intelligence in pathology: trends and challenges. Diagnostics. 2022;12(11):2794.

[55] Ferroni P, Zanzotto FM, Riondino S, Scarpato N, Guadagni F, Roselli M. Breast cancer prognosis using a machine learning approach. Cancers. 2019;11(3):328.
[56] Wulczyn E, Steiner DF, Xu Z, et al. Deep learning-based survival prediction for multiple cancer types using histopathology images. PLoS One 2020;15(6):e0233678.
[57] Bera K, Schalper KA, Rimm DL, Velcheti V, Madabhushi A. Artificial intelligence in digital pathology—new tools for diagnosis and precision oncology. Nat Rev Clin Oncol 2019;16(11):703–15.
[58] Mobadersany P, Yousefi S, Amgad M, et al. Predicting cancer outcomes from histology and genomics using convolutional networks. Proc Natl Acad Sci U S A 2018;115(13):E2970–9.
[59] Khoury T, Peng X, Yan L, Wang D, Nagrale V. Tumor-infiltrating lymphocytes in breast cancer: Evaluating interobserver variability, heterogeneity, and fidelity of scoring core biopsies. Am J Clin Pathol 2018; 150(5):441–50.
[60] Swisher SK, Castaneda Wu Y. CA, et al. Interobserver agreement between pathologists assessing tumor-infiltrating lymphocytes (TILs) in breast cancer using methodology proposed by the International TILs Working Group. Ann Surg Oncol 2016;23:2242–8.
[61] Gao G, Wang Z, Qu X, Zhang Z. Prognostic value of tumor-infiltrating lymphocytes in patients with triple-negative breast cancer: a systematic review and meta-analysis. BMC Cancer 2020;20:1–15.
[62] Lee K, Lockhart JH, Xie M, et al. Deep learning of histopathology images at the single cell level. Front Artif Intell 2021;137.
[63] Moorman A, Vink R, Heijmans H, Van Der Palen J, Kouwenhoven E. The prognostic value of tumour-stroma ratio in triple-negative breast cancer. Eur J Surg Oncol 2012;38(4):307–13.
[64] Roeke T, Sobral-Leite M, Dekker TJ, et al. The prognostic value of the tumour-stroma ratio in primary operable invasive cancer of the breast: a validation study. Breast Cancer Res Treat 2017;166:435–45.
[65] Kather JN, Krisam J, Charoentong P, et al. Predicting survival from colorectal cancer histology slides using deep learning: a retrospective multicenter study. PLoS Med 2019;16(1):e1002730.
[66] Geessink OG, Baidoshvili A, Klaase JM, et al. Computer aided quantification of intratumoral stroma yields an independent prognosticator in rectal cancer. Cell Oncol 2019;42(3):331–41.
[67] Zhang T, Xu J, Shen H, Dong W, Ni Y, Du J. Tumor-stroma ratio is an independent predictor for survival in NSCLC. Int J Clin Exp Pathol 2015;8(9):11348.
[68] Kasahara K, Katsumata K, Saito A, et al. Artificial intelligence predicts lymph node metastasis or risk of lymph node metastasis in T1 colorectal cancer. Int J Clin Oncol 2022;27(10):1570–9.
[69] Shen B, Saito A, Ueda A, et al. Development of multiple AI pipelines that predict neoadjuvant chemotherapy response of breast cancer using H&E-stained tissues. J Pathol Clin Res 2023;9:182–94.
[70] Donovan MJ, Khan FM, Fernandez G, et al. Personalized prediction of tumor response and cancer progression on prostate needle biopsy. J Urol 2009;182(1):125–32.
[71] Campanella G, Hanna MG, Geneslaw L, et al. Clinical-grade computational pathology using weakly supervised deep learning on whole slide images. Nat Med 2019;25(8):1301–9.
[72] Mun Y, Paik I, Shin S-J, Kwak T-Y, Chang H. Yet another automated Gleason grading system (YAAGGS) by weakly supervised deep learning. npj Digital Med 2021;4(1):99.
[73] Huang W, Randhawa R, Jain P, et al. Development and validation of an artificial intelligence–powered platform for prostate cancer grading and quantification. JAMA Netw Open 2021;4(11):e2132554.
[74] Arvaniti E, Fricker KS, Moret M, et al. Automated Gleason grading of prostate cancer tissue microarrays via deep learning. Sci Rep 2018;8(1):12054.
[75] Cheong J-H, Wang SC, Park S, et al. Development and validation of a prognostic and predictive 32-gene signature for gastric cancer. Nat Commun 2022;13(1):774.
[76] Park CH, Hong C, Lee A-r, Sung J, Hwang TH. Multi-omics reveals microbiome, host gene expression, and immune landscape in gastric carcinogenesis. Iscience 2022;25(3):103956.
[77] Zeng Q, Klein C, Caruso S, et al. Artificial intelligence predicts immune and inflammatory gene signatures directly from hepatocellular carcinoma histology. J Hepatol 2022;77(1):116–27.
[78] Lujan GM, Savage J, Shana'ah A, et al. Digital pathology initiatives and experience of a large academic institution during the coronavirus disease 2019 (COVID-19) pandemic. Arch Pathol Lab Med 2021; 145(9):1051–61.
[79] Lujan G, Parwani AV, Bui MM. Whole slide imaging: remote consultations/second opinions. Whole slide imaging. Springer; 2022, p. 153–62.

CHAPTER

5

Artificial intelligence in clinical microbiology

Kenneth Smith[1,2]

[1]Infectious Disease Diagnostics Laboratory, The Children's Hospital of Philadelphia, Philadelphia, PA, United States [2]Department of Pathology and Laboratory Medicine, Perelman School of Medicine at the University of Pennsylvania, Philadelphia, PA, United States

Faced with current and future shortages of trained technologists, many clinical laboratories are turning to automation and artificial intelligence (AI) to support higher test volumes with limited staffing [1]. In contrast to laboratories that primarily report numeric values derived from specimens in standard containers (i.e., blood collection tubes), many microbiology tests are reported qualitatively from a diversity of specimen types. This lack of standardization has hindered development of laboratory automation in this field, and most tests are still performed by visual inspection of slides or cultures. However, new technologies have enabled microbiology laboratory automation [2] allowing for digitization of slides or culture plates, data which have previously not been available. Increased automation combined with digitization of image data will ultimately serve as the foundation for training and implementation of AI-based diagnostics modalities in clinical microbiology.

Current-generation laboratory automation systems are primarily used for bacterial culture. These instruments can process specimens, incubate, and collect images of each culture plate at various time points without need for manual handling. Such images have been used to train AI models to interpret screening cultures for vancomycin-resistant enterococci [3], group A *Streptococcus* [4], and methicillin-resistant *Staphylococcus aureus* [5,6]. These cultures are particularly amenable to AI interpretation as they are grown using chromogenic media on which the organism of interest is easily identified by a distinct color and other flora is suppressed.

In addition to screening cultures, AI models have been developed to interpret urine cultures. This culture type is interpreted using quantitative rules, allowing relatively easier distinction between pathogens and normal flora [7,8]. Cultures from other specimen sources, notably wound or respiratory, are more difficult to interpret using AI due to

Artificial Intelligence in Clinical Practice
DOI: https://doi.org/10.1016/B978-0-443-15688-5.00008-5

qualitative interpretation and need to distinguish pathogens admixed with abundant and visually similar normal flora.

In addition to culture, microscopic observation of stained smears is a mainstay of the clinical microbiology laboratory and one of its most demanding and labor-intensive processes. The highest-volume stain is typically the Gram stain, which is performed on various specimen types (wound, respiratory, blood) and classifies bacteria into two major groups (Gram positive or Gram negative). A proof-of-concept AI model has been trained for blood culture Gram stains [9]. Blood culture Gram stains are typically monomicrobial with abundant organisms present, allowing for relatively simpler data collection and interpretation as most fields will contain objects of interest. In contrast, Gram stains from other sources (wound, respiratory) are much more likely to have multiple organisms, rare organisms, or extensive background staining, posing a challenge to automated microscopy-based data collection and therefore implementation of AI.

Other microscopy-based tests include blood parasite exams, the gold standard for identifying infection with *Plasmodium* and guiding malaria treatment. However, these smears require significant technical expertise to interpret, which may not always be available, especially in resource-limited settings where malaria is endemic. As such, significant effort has been devoted to development of AI methods to improve access to malaria diagnostics, some of which require only a basic microscope and a smartphone [10–12].

Microscopy is also used for the interpretation of ova and parasite exams, which consist of wet mounts of stool and fixed trichrome stains. This test takes significant technologist time to read, and typically has a very low positivity rate, especially in areas with low parasite burden. As such, AI has been leveraged to reduce technologist time spent on negative tests. Recently Mathison et al. developed a deep-learning method which achieved >98% positive and negative agreement with trained technologists and an increased sensitivity over manual microscopy [13]. This accuracy was achievable in part due to the exceptionally large training set of >50,000 unique slides.

Indeed, a major challenge of implementation of AI in clinical microbiology is availability of data for model training. Large amounts of image data are required to train models for slide or culture plate interpretation, but only a very small fraction of plates or slides are digitized in practice. Further, stained slides are typically read under oil immersion, which poses a significant challenge to current-generation automated microscopy platforms typically designed for anatomic pathology. Even when data are digitized, it remains unclear what fraction of it is retained or otherwise made available to train AI models. Finally, instrumentation for automation and digitization of plates or slides is often large and expensive, putting it out of reach for many space and resource constrained hospital-based laboratories.

Nevertheless, staffing shortages and expected retirement of experienced technologists are likely to persist and even increase in the future [1]. It is therefore likely that adoption of AI and automation will also increase, starting with high-volume tests with relatively simple interpretation including screening cultures and urine cultures. As the field develops, this technology is anticipated to be applied more broadly to more complex culture types as well as for slide interpretation. In the near term, benefits of automation will be unevenly distributed due to space or resource limitations, with adoption limited to those high-volume or otherwise well-resourced laboratories that can accommodate the instrumentation and technical expertise required. However, as existing laboratories upgrade their spaces, more and more will be designed with automation and AI in mind, allowing smaller laboratories to realize the benefits of these technologies.

Major takeaways

- Most clinical microbiology is currently performed using highly manual, labor-intensive methods.
- Laboratory automation is enabling AI-based plate reading in high-volume laboratories, especially for screening cultures and urine cultures.
- Interpretation of stained smears by microscopy is in its infancy but represents a promising use case for AI in the future.
- A major challenge to implementation of AI in microbiology is digitization of image data, a problem that is being addressed by introduction of laboratory automation systems.

References

[1] Garcia E, Kundu I, Kelly M, Soles R. The American Society for Clinical Pathology 2020 vacancy survey of medical laboratories in the United States. Am J Clin Pathol 2022;157(6):874–89. Available from: https://doi.org/10.1093/ajcp/aqab197.

[2] Antonios K, Croxatto A, Culbreath K. Current state of laboratory automation in clinical microbiology laboratory. Clin Chem 2021;68(1):99–114. Available from: https://doi.org/10.1093/clinchem/hvab242.

[3] Faron ML, Buchan BW, Coon C, Liebregts T, van Bree A, Jansz AR, et al. Automatic digital analysis of chromogenic media for vancomycin-resistant-enterococcus screens using Copan WASPLab. J Clin Microbiol 2016;54(10):2464–9. Available from: https://doi.org/10.1128/JCM.01040-16.

[4] Van TT, Mata K, Dien Bard J. Automated detection of streptococcus pyogenes pharyngitis by use of colorex Strep A CHROMagar and WASPLab artificial intelligence chromogenic detection module software. J Clin Microbiol 2019;57(11):e00811–19. Available from: https://doi.org/10.1128/JCM.00811-19.

[5] Faron ML, Buchan BW, Vismara C, Lacchini C, Bielli A, Gesu G, et al. Automated scoring of chromogenic media for detection of methicillin-resistant Staphylococcus aureus by use of WASPLab image analysis software. J Clin Microbiol 2016;54(3):620–4. Available from: https://doi.org/10.1128/JCM.02778-15.

[6] Gammel N, Ross TL, Lewis S, Olson M, Henciak S, Harris R, et al. Comparison of an automated plate assessment system (APAS Independence) and artificial intelligence (AI) to manual plate reading of methicillin-resistant and methicillin-susceptible Staphylococcus aureus CHROMagar surveillance cultures. J Clin Microbiol 2021;59(11):e0097121. Available from: https://doi.org/10.1128/JCM.00971-21.

[7] Alouani DJ, Ransom EM, Jani M, Burnham CA, Rhoads DD, Sadri N. Deep convolutional neural networks implementation for the analysis of urine culture. Clin Chem 2022;68(4):574–83. Available from: https://doi.org/10.1093/clinchem/hvab270.

[8] Brenton L, Waters MJ, Stanford T, Giglio S. Clinical evaluation of the APAS® Independence: automated imaging and interpretation of urine cultures using artificial intelligence with composite reference standard discrepant resolution. J Microbiol Methods 2020;177:106047. Available from: https://doi.org/10.1016/j.mimet.2020.106047.

[9] Smith KP, Kang AD, Kirby JE. Automated interpretation of blood culture Gram stains by use of a deep convolutional neural network. J Clin Microbiol 2018;56(3):e01521-17. Available from: https://doi.org/10.1128/JCM.01521-17.

[10] Poostchi M, Silamut K, Maude RJ, Jaeger S, Thoma G. Image analysis and machine learning for detecting malaria. Transl Res 2018;194:36–55. Available from: https://doi.org/10.1016/j.trsl.2017.12.004.

[11] Torres K, Bachman CM, Delahunt CB, Alarcon Baldeon J, Alava F, Gamboa Vilela D, et al. Automated microscopy for routine malaria diagnosis: a field comparison on Giemsa-stained blood films in Peru. Malar J 2018;17(1):339. Available from: https://doi.org/10.1186/s12936-018-2493-0.

[12] Das D, Vongpromek R, Assawariyathipat T, Srinamon K, Kennon K, Stepniewska K, et al. Field evaluation of the diagnostic performance of EasyScan GO: a digital malaria microscopy device based on machine-learning. Malar J 2022;21(1):122. Available from: https://doi.org/10.1186/s12936-022-04146-1.

[13] Mathison BA, Kohan JL, Walker JF, Smith RB, Ardon O, Couturier MR. Detection of intestinal protozoa in trichrome-stained stool specimens by use of a deep convolutional neural network. J Clin Microbiol 2020; 58(6):e02053-19. Available from: https://doi.org/10.1128/JCM.02053-19.

CHAPTER

6

Artificial intelligence on interventional cardiology

Chayakrit Krittanawong[1], *Scott Kaplin*[1] *and Samin K Sharma*[2]

[1]Cardiology Division, NYU Langone Health and NYU School of Medicine, New York, NY, United States [2]Cardiac Catheterization Laboratory of the Cardiovascular Institute, Mount Sinai Hospital, New York, NY, United States

Introduction

Over the past few decades the combination of percutaneous coronary intervention (PCI) and various emerging technologies [e.g., shockwave intravascular lithotripsy, optical coherence tomography (OCT)] has been a revolutionary innovation in the diagnosis and treatment of coronary artery disease (CAD). Within the field of interventional cardiology, risk stratification or outcome/prognosis prediction regarding revascularization strategies are crucial for the individualized management of patients. However, there are several areas of PCI that can still be improved upon (e.g., DAPT, PCI in TAVI, PCI in AF patients, revascularization in certain population—SCAD, pre-op assessment prior to transplantation). In general, decision-making regarding revascularization strategy is based on data obtained from clinical trials. However, there are complexities among individuals that require a more tailored approach beyond the data gleaned from clinical trials. Most importantly, traditional prognostic risk scores rely on phenotypic generalizations that cannot be applied for all phenotypes. For example, GRACE, TIMI, or ASCVD pooled cohort values can be underestimated or overestimated in certain populations (e.g., South Asian, Hispanic). With that said, current scoring systems for the prediction of the short-term prognosis of ACS patients which are primarily used in clinical practice, such as GRACE and TIMI, have areas for improvement. For example, integrating data regarding medication compliance and functional capacity provided by wearable technology could provide more accurate, individualized decision-making. Another example is the contradiction between the SYNTAX score (based on the SYNTAX trial) and the results from the NOBLE (Percutaneous Coronary Angioplasty

Artificial Intelligence in Clinical Practice
DOI: https://doi.org/10.1016/B978-0-443-15688-5.00040-1

Versus Coronary Artery Bypass Grafting in Treatment of Unprotected Left Main Stenosis) [1] and EXCEL (Evaluation of XIENCE Versus Coronary Artery Bypass Surgery for Effectiveness of Left Main Revascularization) trials [2]. More powerful tools and matrices are needed to supplement the data obtained from randomized clinical trials. For nonelective CAD patients who require PCI, a more accurate scoring system is needed to better assess long-term outcomes. It is crucial to integrate multidimensional variables into interventional cardiovascular scoring systems in order to provide a tailored, individualized decision-making approach. In the era of big data and rapid advancement of cardiac catherization technology, integration of clinical variables including EMR data, lifestyle habits, medication compliance, and angiographic data obtained from diagnostic and intravascular imaging could potentially be used to improve PCI outcomes and lead to the development of precision PCI.

Machine learning (ML) is a field of computer science that represents various techniques for solving complicated problems with big data by identifying interaction patterns among variables multidimensionally (e.g., anginal symptoms, equivalent stress test results, borderline physiologic or intravascular imaging measurements) [3]. Most importantly, ML is focused on building automated clinical decision systems (such as P2Y12 selection based on genetics and lifestyle variables) that help physicians make more accurate predictions, rather than those obtained through simplified estimated scoring systems.

ML can be used in robotic PCI, precatheterization planning (catheter/guide catheter selection based on prior chest X-ray/CT, BMI/BSA), preop strategy (whether revascularization is needed), and intraprocedural guidance [deep learning (DL)-guided angiography view selection, DL-guided intravascular ultrasound (IVUS)/OCT recommendation]. DL is a subset of ML, which utilizes dense neural networks to better imitate the way humans gain knowledge (more algorithmic layers than ML). DL networks can be applied to various imaging modalities by performing advanced pattern recognition to generate an output. Angiography and intravascular imaging are a great fit for DL integration. We previously proposed a quality assessment of ML research for clinical practice [4], as healthcare data can be challenging to work with due to disorganization and missing values.

Image quantification (deep learning and computer vision)

Artificial intelligence (AI) techniques, such as DL and computer vision, can be utilized for the purposes of image quantification (e.g., angiography quantify lesion, intravascular imaging assisted diagnosis) and the results thus far have been promising. For example, several studies have demonstrated that AI can quantify angiographic lesions with similar accuracy to human experts. In one preliminary study, DL approaches ResNet and U-Net were combined with traditional image processing to analyze the right coronary artery using an LAO view in a training set [5]. This was then validated against a testing set with reasonable results (precision of 98.4%, vessel segmentation F1 score of 0.891, and stenosis measurement 20.7% type I error rate). Although in this instance the technology was implemented solely for stenosis measurement, data obtained from this method could be used for reducing radiation exposure by optimizing angiographic views and radiation pad placement. Yang et al. used DL (fully convolutional networks) to segment the major vessels and assess stenotic lesions on coronary angiography and they demonstrated an average F1 score 0.917, and 93.7% of the images exhibited a high F1 score >0.8 [6]. Du et al. used a DL model to identify lesion morphology, stenosis, calcification, thrombosis, total occlusion, and dissection from coronary angiography and they demonstrated good accuracy (98.4%) and sensitivity (85.2%) [7]. In addition, the DL model demonstrated good F1 score for detecting lesion morphologies including stenosis, total occlusion, calcification, thrombosis, and dissection (0.829, 0.810, 0.802, 0.823, and 0.854, respectively). In preliminary data, Shen et al. proposed DBU-Net, a combination of U-Net and bidirectional convLSTM, tested in a private dataset for coronary angiography segmentation, which achieved an average accuracy, precision, recall values, and F1-score for coronary artery segmentation of 0.985, 0.913, 0.847, and 0.879 respectively [8].

Moreover, ML can be applied for physiologic measurements. Cho et al. used an XGBoost model to classify lesions as having fractional flow reserve (FFR) ≤0.80 versus >0.80 [9]. In 28 physiologic features, the ML model with 5-fold cross-validation in the 1204 training samples predicted FFR ≤0.80 with overall diagnostic accuracy of 78 ± 4% (averaged area under the curve: 0.84 ± 0.03). The final model included 12 features (segment; body surface area; distal lumen diameter; minimal lumen diameter; length of a lumen diameter <2.0, <1.5, and <1.25 mm; mean lumen diameter within the worst segment; sex; diameter stenosis; distal 5-mm reference lumen diameter; and length of diameter stenosis >70%), the ML predicted FFR ≤0.80 in the test set with sensitivity of 84%, specificity of 80%, and overall accuracy of 82% (area under the curve: 0.87). Gao et al. used a feature-based ensemble method (gradient-boosting decision tree) in analyzing coronary angiography and demonstrated that gradient-boosting decision tree outperformed common deep convolutional neural networks [10]. Tao et al. proposed Bottleneck Residual U-Net (BRU-Net), a modified U-Net model, to analyze coronary angiography and they demonstrated that the proposed model performed well for accurate coronary artery segmentation, achieving a sensitivity, specificity, accuracy, and area under the curve (AUC) of 0.8770, 0.9789, 0.9729, and 0.9910, respectively [11].

Schwalm et al. used ML models (e.g., logistic regression, random forests, gradient-boosted decision trees, and a multilayer perception deep neural network) to predict obstructive CAD from angiographic data, and they demonstrated that ML models outperformed existing clinical risk scores with a net reclassification index of 27.8% [95% confidence interval (CI): [24.9%–30.8%], *P* value <.01] and 44.7% (95% CI: [42.4%–47.0%], *P* value <.01), respectively [12].

Although these initial results seem promising, further studies are needed to compare various DL models to identify the most suitable model for assessing coronary angiography. DL models would then need to be validated in multiple cohorts using a variety of datasets. In order to implement DL models in clinical practice, randomized clinical trials may also be needed.

Intravascular imaging (intravascular ultrasound/optical coherence tomography)

Intravascular imaging technology, such as IVUS/OCT, can be used to accurately diagnose angiographic lesions, characterize plaque morphology, and identify microvascular dysfunction. Several studies showed that ML could be implemented in IVUS imaging.

Recently the IVUS-XPL (Impact of Intravascular Ultrasound Guidance on Outcomes of Xience Prime Stents in Long Lesions) trial demonstrated that use of IVUS guidance for stenting was associated with a reduction in target lesion revascularization at 12 months from 5.0% to 2.5% ($P = .02$), an effect that was sustained at the recently reported 5-year follow-up time point (ischemia-driven revascularization 8.4% vs 4.8%; $P = .007$) [13,14]. AI could be used to improve IVUS guidance for stenting performance. Min et al. [15] developed a DL model (convolutional neural network architecture) to predict postprocedural stent area and expansion from preprocedural IVUS imaging (accuracy of 94%). Yang et al. developed IVUS-Net (FCN architecture) to classify IVUS segmentation of lumen, vessel, and stent area and they demonstrated the results are comparable to manual segmentation by human experts [16]. Most importantly, they demonstrated that IVUS-Net model performed well on challenging images such as bifurcations, shadows, and side branches. Additional studies demonstrated the correlations of IVUS characteristics and significant physiologic measurements, potentially reducing the need for obtaining intraprocedural physiologic measurements. Lee et al. developed the IVUS-based supervised ML algorithms to identify lesions with an FFR ≤ 0.80. IVUS could potentially preemptively identify physiologically significant lesions [17] or as demonstrated in preliminary data, DeepIVUS, a convolutional neural network, could potentially be used to segment and classify the phenotype of each IVUS image (e.g., normal, fibrotic, fibroatheroma, calcified, or stented) [18]. This model can be improved to identify various plaque subtypes (lipidic, fibrofatty, fibrous) or calcified plaque with a certain degree arc of calcium. With sufficient training, the model could potentially recommend interventional strategies based on plaque characteristics and severity of calcification (e.g., angioplasty alone, shockwave lithotripsy, orbital atherectomy, or rotational atherectomy). Guo et al. demonstrated that ML-guided IVUS (FSI models) could be used to accurately predict plaque vulnerability changes based on morphological and biomechanical factors from multimodality image-based FSI models [19]. Shinohara et al. developed ML-guided IVUS image interpretation (U-Net model) to classify vessels that are likely to require treatment or special devices with a high accuracy [20]. ML-guided IVUS could also auto-generate MLA, distal reference diameter, proximal reference diameter, and lesion length. Postintervention, ML-guided IVUS could identify edge dissections and minimal stent area (MSA), ultimately predicting the risk of ISR. Based on the results of these prior studies, ML/DL-assisted IVUS interpretation is one of the most promising areas in interventional cardiology.

Several studies have also demonstrated the potential of ML/DL-guided OCT. Niioka et al. developed a DL model (DenseNet-121-based) to better identify vulnerable plaque. Similar to IVUS, Cha et al. developed ML-guided OCT and demonstrated that the model is comparable to the physiologic measurements by wire-based FFR ($r = 0.853$, $P < .001$) [21,22]. Athanasiou et al. designed a DL model (convolutional neural network) to define arterial wall area and plaques characteristics (e.g., calcium, lipid tissue, fibrous tissue, mixed tissue, nonpathological tissue or media, and no visible tissue) [23]. Abdolmanafi et al. created an ML model (encoder-decoder) with dilated convolutions to extract all atherosclerotic tissues [24]. Katagiri et al. compared AI to visual assessment of calcification visualized using OCT to determine whether to perform lesion modification prior to stenting and found the results comparable between the two cohorts [25]. Overall, both

IVUS and OCT can be improved through the implementation of ML/DL models to better characterize lesions and ultimately generate recommendations for intervention strategy.

Clinical decision support for precision percutaneous coronary intervention

AI-assisted precision PCI is an extremely promising area within interventional cardiology in order to optimize lesion preparation, obtain complete revascularization, and achieve a durable long-term result. AI can be implemented to optimize PCI strategy selection in the context of an anatomical and functional assessment or provide guidance for complex PCI based on data obtained from training sets comprised of millions of angiographic images. For example, AI could evaluate lesion characteristics such as angulation and tortuosity seen on diagnostic angiography and assist with guidewire selection, or recommend different modalities of atherectomy for calcified lesions, based on learning memory from multiple experts and data from CORE lab/clinical trials. In the TAILOR-PCI Randomized Clinical Trial, there was no statistically significant difference in a composite end point (e.g., cardiovascular death, myocardial infarction, stroke, stent thrombosis) between genotype-guided selection of an oral P2Y12 inhibitor and conventional clopidogrel therapy, among CYP2C19 LOF carriers with ACS and stable CAD undergoing PCI [26]. Further genetic variants should also be explored in this manner. AI could be used in guiding genotype-guided selection of an oral P2Y12 inhibitor based on an individual's genetics. In this way, AI could not only guide the precision PCI itself, but also guide pharmacogenomics post-PCI. For example, based on an individual's genetics, insurance carrier, and lifestyle characteristics obtained from wearable technology, AI could then generate recommendations lifestyle changes and appropriate P2Y12 agents. This includes monitoring therapy, and ensuring medication compliance and appropriate follow-up.

Another promising area for potential AI application is in CTO intervention. AI can be used to learn the thought process and approach to CTOs from experts in the field in order to generate recommendations for patients. For example, AI can create a CTO scoring system based on lesion characteristics and degree of collateralization, and provide a recommendation regarding whether to try an antegrade or retrograde approach. Li et al. compared DL-guided automated CTO reconstruction versus manual editing and found that automated segmentation and reconstruction of CTOs was successful in 95% of lesions (228 of 240) without manual editing and in 48% of lesions (116 of 240) with the conventional manual protocol ($P < .001$) [27]. Liu et al. demonstrated that a DL model could detect the entry point of a CTO and classify its morphology automatically based on coronary angiography images alone [28]. They found that the recall of CTO detection can reach up to 89.3%, and the sensitivity and specificity of CTO classification can reach up to 94.5% and 89.1%, respectively.

Risk stratification and outcome prediction

In clinical practice, cardiologists often utilize traditional cardiovascular risk scores such as TIMI or GRACE risk scores for AMI risk stratification, or SYNTAX and STS scores for revascularization strategy and outcome prediction. The TIMI risk score incorporates age,

CAD risk factors, prior coronary stenosis, ST deviation on ECG, severe anginal symptoms, aspirin use within the past week, and elevated serum cardiac markers [29], while the GRACE score uses Killip heart failure class, systolic blood pressure, heart rate, age, creatinine level, presence of cardiac arrest, ST-segment deviation, and elevated cardiac enzymes [30]. Several AI studies have been shown that AI models are comparable or even superior to traditional risk scores. In a single-center prospective study, Călburean et al. showed that an ML model outperformed risk scores such as GRACE, ACEF, SYNTAX II, and TIMI: 0.729 versus 0.474, 0.469, 0.365, and 0.389 for 3-year cardiovascular- and 0.718 versus 0.483, 0.466, 0.388 and 0.395 for 3-year all-cause mortality prediction, respectively (all $P \leq .001$) [31]. VanHouten et al. compared ML models to traditional risk scores for early risk stratification, and they found the random forest model (AUC = 0.848) significantly outperformed elastic net (AUC = 0.818), ridge regression (AUC = 0.810), and the TIMI (AUC = 0.745) and GRACE (AUC = 0.623) scores [32].

In one prospective study, Liu et al. showed that the random forest model exhibited the best performance for predicting all-cause mortality in CAD patients who underwent PCI, compared to other models [33]. In the Mayo Clinic PCI registry, Zack et al. found that random forest regression models were more predictive and discriminative than standard regression methods at identifying patients at risk for 180-day cardiovascular mortality and 30-day CHF rehospitalization after PCI [34]. For example, the American College of Cardiology CathPCI bleeding risk model has been developed to predict periprocedural bleeding based on candidate variables (<0.5% missing data) with 80% validation via training sets and the remaining 20% validation using logistic regression with backward selection model from the CathPCI Registry [35]. It has been known that some ML models can outperform logistic regression in certain circumstances. For example, in The New York PCI registry, Al'Aref et al. compared the performance of ML algorithms (AdaBoost and XGBoost) with Logistic Regression to predict in-hospital mortality in patients undergoing PCI and they found that a boosted ensemble algorithm (AdaBoost) had optimal discrimination with an AUC of 0.927 (95% CI 0.923–0.929) compared with an AUC of 0.908 for logistic regression (95% CI 0.907–0.910, $P < .01$) [36]. To date, there are no data comparing ML models versus logistic regression using the CathPCI Registry. Rayfield et al. found that the boosted classification tree algorithm model (AUC 0.873 vs 0.764; $P = .02$) accurately predicted bleeding post PCI and outperformed the American College of Cardiology CathPCI bleeding risk model in predicting the risk of bleeding in patients undergoing PCI [37]. In The Japan Cardiovascular Database-Keio Interhospital Cardiovascular Studies (JCD-KiCS), a large, ongoing, prospective multicenter PCI registry, Niimi et al. found that the XGBoost models modestly improved discrimination for AKI and bleeding (C-statistics of 0.84 in AKI and 0.79 in bleeding), compared with the NCDR-CathPCI risk scores [38]. In the NCDR CathPCI Registry, Mortazavi et al. demonstrated that XGBoost correctly identified an additional 3.7% of cases for in-hospital major bleeding within 72 h after PCI (C statistic of 0.82 with an F score of 0.31) [39]. Another study using NCDR CathPCI Registry, Huang et al. found that ML using 13 variables resulted in improved prediction of AKI risk after PCI [40]. In another study from five hospitals in the Barnes-Jewish Hospital system, Kulkarni et al. used neural network-based models to accurately predict AKI (77.9%), bleeding (86.5%), death (90.3%) and other adverse outcomes (80.6%) [41]. In the 2016–2018 National Inpatient Sample, Pushparaji et al. multivariable regression model performance was optimized using backward propagation neural networks to demonstrate that inpatient PCI is relatively safe among patients with active cancer [42]. ML has also been used to predict recurrent angina post PCI based on metabolic profile signatures with over 89% accuracy, sensitivity, and specificity across three independent cohorts [43]. In the national database, Monlezun et al. used ML-augmented propensity score to assess PCI mortality in cancer patients [44].

In nonatherosclerotic AMI, there is no presently available tool for risk stratification or outcome prediction. For example, the prediction of mortality/recurrent rate in SCAD patients or prediction of PPM requirement post-TAVI is unknown. We previously compared different AI models to predict SCAD mortality [45] and found that a deep neural network model was associated with higher predictive accuracy and discriminative power than logistic regression or ML models for the identification of SCAD patients.

Robotic-assisted procedures (percutaneous coronary intervention, TAVR, MITRA)

Although robotic-assisted procedures can provide physical assistance during intervention, they are not presently capable of decision-making in this area. DL could potentially process a combination of angiographic and TEE patterns by learning from millions of images to provide recommendations for mitraclip positioning. Similar training using DL could be used to optimize ASD or PFO closure device position.

Natural language processing—EHR extraction

NLP can be used to predict post-PCI outcomes using a combination of clinical variables and cardiac catheterization report results, leading to enhanced outcome prediction. For example, an integration of genetic data (P2Y12 or aspirin responding genes), bleeding risk, and medical comorbidities (e.g., CKD, autoimmune, PAD) could generate predictions for ISR rates and potentially generate recommendations for stent types based on millions of data points.

Several studies have used an NLP technique for evaluating cardiac catheterization text reports to identify the presence of obstructive CAD at the level of individual coronary arteries. Li et al. developed a DL model (a BERT-based model) for analyzing text content from catheterization reports [46]. In comparison to SYNTAX scores, they found that the DL model outperformed the residual SYNTAX score in predicting 5-year all-cause mortality (AUC, 0.867 [95% CI, 0.813–0.921] vs 0.590 [95% CI, 0.503–0.684]) and 5-year cardiovascular mortality (AUC, 0.880 [95% CI, 0.873–0.925] vs 0.649 [95% CI, 0.535–0.764]) respectively, after PCI among these patients. In a preliminary study, Levin et al. randomly selected 200 full-text catheterization reports and manually labeled them for the presence or absence of obstructive CAD in each of the left main, left anterior descending, left circumflex, and right coronary arteries [47]. The final random forest model was validated in a testing set. The model was then used to classify a set of 4226 unlabeled reports and identified obstructive CAD with high precision (0.951) and recall (0.940); the *F*1 statistic was 0.946.

In ACS patients, AI could generate strategic guidance based on clinical profiles from EMR data and ECG results. For example, an AI-integrated EMR system could generate warnings based on minute ECG changes from multiple outpatient visits. This would reduce the time required for emergency room physicians to search and identify ECGs from prior visits, particularly from different EMR systems or practice sites.

Augmented reality and artificial intelligence

AR or VR can be instituted in medical education and catheterization laboratory training in many ways including procedural training, procedural assistance, and patient or physician education. Several studies have demonstrated the benefit of AR/VR in providing procedural assistance. Google Glass for instance can assist with coronary angiography interpretation [48]. AR with holographic rendering could facilitate endovascular aortic repair, potentially shortening operating time, reducing contrast volume, and lowering the required radiation dose [49]. Chu et al. compared TEE alone versus AR-enhanced TEE in aiding mitral valve repair [50]. They found that AR-enhanced TEE could better navigate from left ventricular apex to MV leaflet ($P = .2$) compared to TEE alone. AR-enhanced TEE was also associated with fewer distance errors, shorter path lengths, and shorter navigation times compared with TEE alone. An AR enhanced 3D visual environment, identifying the 3D position of the catheter determined by processing two fluoroscopic images, could provide image guidance of transcatheter procedures for atrial septal defect and patent foramen ovale closure, valvular repair/replacement, and left atrial appendage closure [51]. The results appear promising in vitro as this method improves an interventionalist's visualization and potentially assists them in evaluating complex cases. However, further testing in this area is required. Bruno et al. demonstrated that VR interventions during TAVI to assist with conscious sedation are safe and reduce periprocedural anxiety (visual analog scales 2 [IQR 0–3.75] vs 5 [IQR 2–8], $P = .04$) compared to a control group [52]. During TAVI procedures, Arslan et al. demonstrated that remote proctoring by virtual support using smartglass (Rods & Cones) and a self-expandable TAVI system (Medtronic) is feasible [53]. Currie et al. compared AR guidance of TAVI deployment with fluoroscopic guidance. The precision of TAVI deployment using fluoroscopic guidance was 3.4 mm, whereas the precision of AR guidance was 2.9 mm, and its overall accuracy was 3.4 mm suggesting similar performance [54].

AR/VR can also be beneficial as a training tool for catheter engagement and various atherectomy modalities (Fig. 6.1). Today, the metaverse has the potential to revolutionize healthcare education. Abbott's new OCT training pilot program allows interventional cardiologists to train wearing Oculus Go VR goggles [55]. This could be expanded for the education of fellows in training.

Artificial intelligence diagnosis of challenging cases

AI has the potential to identify various challenging conditions such as coronary vasospasm, SCAD or MINOCA based on angiographic data by training on millions of angiographic and intravascular images both with and without provocative testing. In addition, AI could quickly process angiographic data during a case and generate the percentage and exact location of eccentric plaque or calcified nodules, allowing precise recommendations for balloon/stent selection.

AI also has the capability to promptly detect angiographic dissections, especially more challenging subtypes such as type 2 and type 3 SCAD, and provide treatment recommendations

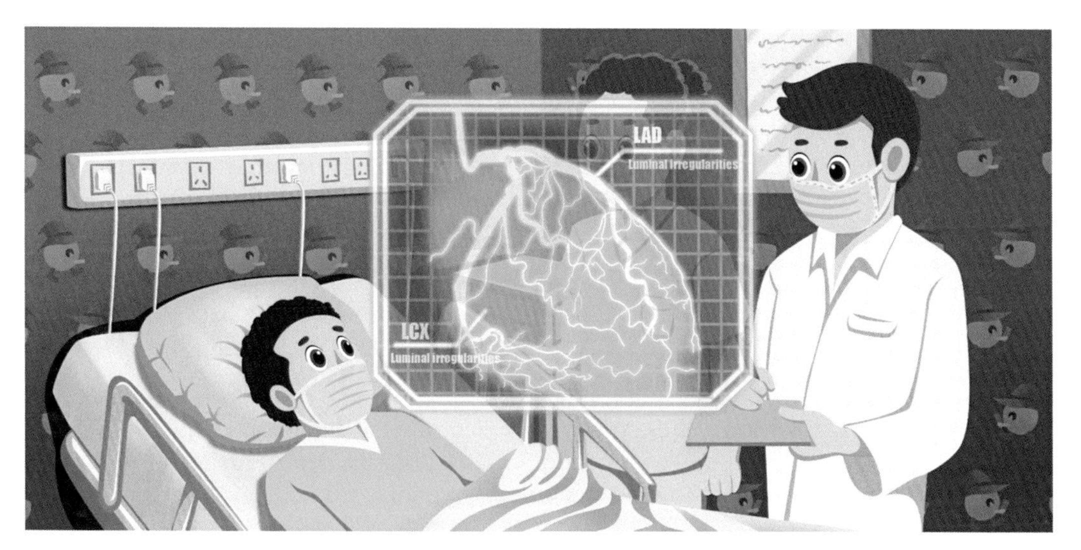

FIGURE 6.1 Example of artificial intelligence utilization in interventional cardiology.

based on clinical and angiographic variables. For example, AI could provide reommendations regarding revascularization options (PCI vs CABG) and ACS treatment (e.g., Heparin, DAPT, cardiac rehab) in LM SCAD patients [56]. Liu et al. analyzed 3786 patients who had triple vessel disease (TVD) with a SYNTAX score of $\geq$ 23 and classified them into 3 phenotypes by age and LVEDD [57]. They concluded that PCI may be a better option for relatively young patients with left ventricular dilation. However, this was a single-center trial without external validation. AI could assist with clinical decision-making to promptly diagnose early ACS using only ECG-based data, and differentiate from other conditions with similar presenting symptoms, such as PE [58]. ML could provide clinical decision-making to decide between heparin versus bivalirudin in specific PCI cases based on patient characteristics from clinical trial data. Similarly, ML could provide clinical decision-making to decide between mitraclip and traditional valve surgery based on patient characteristics from the COAPT or MITRA FR trials.

Another potential application is AI-guided zero-contrast PCI, as a technique for preventing contrast-induced nephropathy [59]. One prior case report demonstrated the feasibility of zero-contrast CTO-PCI with IVUS guidance [60]. AI has the potential to identify atherosclerotic lesions based on dataset traning on angraphic images using very minimal contrast. To reduce radiation exposure, AI could potentially guide PCI without fluoroscopy or contrast agents. Nair et al. performed PCI successfully using an electroanatomic mapping system (NavX; Abbot Inc., USA) with near zero use of fluoroscopy or contrast agents [61].

Challenges

First, randomized clinical trials regarding AI-assisted decision-making are required at this time. For example, a randomized clinical trial comparing human expert stent selection versus AI recommendation for minimizing ISR rates. Second, "mimicking" human tactile sensations could be challenging to reproduce. Third, robotic AI is currently unable to perform PCI independently but does have the capability to assist experts with optimal learning memory. Fourth, AI-guided hemodynamic interpretation and recommendations for MCS may be inefficient compared to human expert decision-making. Further randomized clinical trials are required to compare between human experts and AI in this regard. Lastly, optimal AI outcomes require integration from several moving parts including wearable technology, home lifestyle monitoring, and intraprocedural data. Integrating data from these various sources have proven to be the primary barrier for AI implementation.

Future directions

AI has the potential to advance the field of interventional cardiology by improving patient safety and clinical outcomes. When provided large volumes of training data from experienced centers, AI could be leveraged in several different critical areas. Box 1 demonstrates the potential of AI integration within interventional cardiology.

BOX 6.1 Summary of AI potentials in interventional cardiology.

Conditions/situations	AI potentials
Cath lab workflow	ML could learn simple pattern in the cath lab and generate automated recommendations (driving the tables, checking ACT, warning for radiation safety).
SCAD	DL could predict SCAD recurrence based on angiographic and IVUS characteristics.
CTO	ML could learn pattern how to tackle CTO from experts and generate recommendation for approaching CTO lesions.
SYNTAX	DL could predict outcomes that are more granular than SYNTAX score. For example, predicting outcomes between PCI and CABG in patients with or without diabetes.
IVUS measurement	DL could learn from millions of IVUS images and data regarding adverse outcomes in order to generate an optimal MLA cutoff. In addition, AI could generate an MSA recommendation.
Contrast utilization	DL could potentially be used for contrast recommendation in high-risk patients (CAD with CKD).
Cardiogenic shock	ML classification and management of cardiogenic shock or mixed shock. For example, it has been shown that cardiogenic shock can be classified into 3 phenotypes using ML.
Complication assessment	DL combined fluoroscopic and ultrasound imaging to generate recommendations for precise femoral site access to avoid RP bleed, pseudoaneurysm, and hematoma.
Physiologic measurements	ML could help with physiologic measurements (FFR, iFR, RFR) to precisely characterize lesion (e.g., location and length).
A combination of clinical and imaging	A combination of clinical and imaging variables could better predict outcomes using DL. For example, incorporating variables such as hemoglobin A1C, ASCVD score, and a h/o PAD with bifurcation LM, ML could learn from million patients' outcome and generate optimal recommendation.
Low contrast angiography	ML could identify dissection or perforation from low-contrast angiography.

In conclusion, AI has the potential to improve interventional cardiology in several aspects including risk stratification, AI-assisted IVUS/OCT-guided PCI, and outcome prediction post-PCI; ultimately leading to the facilitation of precision PCI.

Major takeaway points

- AI could improve catheterization lab workflow from patient triage to providing discharge recommendations.
- AI could facilitate precision PCI by combining clinical variables and imaging characteristics, navigating IVUS/OCT and physiologic measurements.
- AI could assist complex procedure such as high-risk PCI (high SYNTAX score, unprotected LM) or CTO intervention-based training data from experienced centers providing large volumes of data points.

References

[1] Mäkikallio T, et al. Percutaneous coronary angioplasty versus coronary artery bypass grafting in treatment of unprotected left main stenosis (NOBLE): a prospective, randomised, open-label, non-inferiority trial. Lancet 2016;388(10061):2743–52.
[2] Stone GW, et al. Everolimus-eluting stents or bypass surgery for left main coronary artery disease. N Engl J Med 2016;375(23):2223–35.
[3] Krittanawong C, et al. Artificial intelligence in precision cardiovascular medicine. J Am Coll Cardiol 2017;69 (21):2657–64.
[4] Krittanawong C, et al. Machine learning prediction in cardiovascular diseases: a meta-analysis. Sci Rep 2020;10(1):16057.
[5] Zhou C., et al. Automated deep learning analysis of angiography video sequences for coronary artery disease. 2021.
[6] Yang S, et al. Deep learning segmentation of major vessels in X-ray coronary angiography. Sci Rep 2019;9 (1):16897.
[7] Du T, et al. Training and validation of a deep learning architecture for the automatic analysis of coronary angiography. EuroIntervention 2021;17(1):32–40.
[8] Shen Y., et al. DBCU-Net: deep learning approach for segmentation of coronary angiography images, Research Square; 2022.
[9] Cho H, et al. Angiography-based machine learning for predicting fractional flow reserve in intermediate coronary artery lesions. J Am Heart Assoc 2019;8(4):e011685.
[10] Gao Z, et al. Vessel segmentation for X-ray coronary angiography using ensemble methods with deep learning and filter-based features. BMC Med Imaging 2022;22(1):10.
[11] Tao X, et al. A lightweight network for accurate coronary artery segmentation using X-ray angiograms. Front Public Health 2022;10:892418. Available from: https://doi.org/10.3389/fpubh.2022.892418.
[12] Schwalm JD, et al. A machine learning-based clinical decision support algorithm for reducing unnecessary coronary angiograms. Cardiovasc Digit Health J 2022;3(1):21–30.
[13] Hong SJ, et al. Effect of intravascular ultrasound-guided vs angiography-guided everolimus-eluting stent implantation: the IVUS-XPL randomized clinical trial. JAMA 2015;314(20):2155–63.
[14] Hong SJ, et al. Effect of intravascular ultrasound-guided drug-eluting stent implantation: 5-year follow-up of the IVUS-XPL Randomized Trial. JACC Cardiovasc Interv 2020;13(1):62–71.
[15] Min HS, et al. Prediction of coronary stent underexpansion by pre-procedural intravascular ultrasound-based deep learning. JACC Cardiovasc Interv 2021;14(9):1021–9.
[16] Yang J, et al. IVUS-Net: an intravascular ultrasound segmentation network. Smart multimedia. Cham: Springer International Publishing; 2018.

[17] Lee JG, et al. Intravascular ultrasound-based machine learning for predicting fractional flow reserve in intermediate coronary artery lesions. Atherosclerosis 2020;292:171–7.
[18] Molony D, Samady H. TCT-342 DeepIVUS: a machine learning platform for fully automatic IVUS segmentation and phenotyping. J Am Coll Cardiol 2019;74(13_Supplement) B339.
[19] Guo X, et al. Predicting plaque vulnerability change using intravascular ultrasound + optical coherence tomography image-based fluid-structure interaction models and machine learning methods with patient follow-up data: a feasibility study. Biomed Eng Online 2021;20(1):34.
[20] Shinohara H, et al. Automatic detection of vessel structure by deep learning using intravascular ultrasound images of the coronary arteries. PLoS One 2021;16(8):e0255577.
[21] Niioka H, et al. Automated diagnosis of optical coherence tomography imaging on plaque vulnerability and its relation to clinical outcomes in coronary artery disease. Sci Rep 2022;12(1):14067.
[22] Cha J-J, et al. Optical coherence tomography-based machine learning for predicting fractional flow reserve in intermediate coronary stenosis: a feasibility study. Sci Rep 2020;10(1):20421.
[23] Athanasiou L, et al. A deep learn approach classify atherosclerosis using intracoronary optical coherence tomography 2019;22.
[24] Abdolmanafi A, et al. A deep learning-based model for characterization of atherosclerotic plaque in coronary arteries using optical coherence tomography images. Med Phys 2021;48(7):3511–24.
[25] Katagiri Y, et al. Artificial intelligence vs visual assessment of calcified plaque in coronary artery using optical coherence tomography. JACC Adv 2022;1(4):1–3.
[26] Pereira NL, et al. Effect of genotype-guided oral P2Y12 inhibitor selection vs conventional clopidogrel therapy on ischemic outcomes after percutaneous coronary intervention: the TAILOR-PCI randomized clinical trial. JAMA 2020;324(8):761–71.
[27] Li M, et al. Deep learning segmentation and reconstruction for CT of chronic total coronary occlusion. Radiology 2022;221393.
[28] Liu X., et al. Detection and classification of chronic total occlusion lesions using deep learning. In: 2019 41st annual international conference of the IEEE Engineering in Medicine and Biology Society (EMBC); 2019.
[29] Antman EM, et al. The TIMI risk score for unstable angina/non-ST elevation MI: A method for prognostication and therapeutic decision making. JAMA 2000;284(7):835–42.
[30] Granger CB, et al. Predictors of hospital mortality in the global registry of acute coronary events. Arch Intern Med 2003;163(19):2345–53.
[31] Călburean P-A, et al. Prediction of 3-year all-cause and cardiovascular cause mortality in a prospective percutaneous coronary intervention registry: machine learning model outperforms conventional clinical risk scores. Atherosclerosis 2022;350:33–40.
[32] VanHouten JP, et al. Machine learning for risk prediction of acute coronary syndrome. AMIA Annu Symp Proc 2014;1940–9.
[33] Liu S, et al. Machine learning-based long-term outcome prediction in patients undergoing percutaneous coronary intervention. Cardiovasc Diagn Ther 2021;11(3):736–43.
[34] Zack CJ, et al. Leveraging machine learning techniques to forecast patient prognosis after percutaneous coronary intervention. JACC Cardiovasc Interv 2019;12(14):1304–11.
[35] Rao SV, et al. An updated bleeding model to predict the risk of post-procedure bleeding among patients undergoing percutaneous coronary intervention: a report using an expanded bleeding definition from the National Cardiovascular Data Registry CathPCI Registry. JACC Cardiovasc Interv 2013;6(9):897–904.
[36] Al'Aref SJ, et al. Determinants of in-hospital mortality after percutaneous coronary intervention: a machine learning approach. J Am Heart Assoc 2019;8(5):e011160.
[37] Rayfield C, et al. Machine learning on high-dimensional data to predict bleeding post percutaneous coronary intervention. J Invasive Cardiol 2020;32(5):E122–e129.
[38] Niimi N, et al. Machine learning models for prediction of adverse events after percutaneous coronary intervention. Sci Rep 2022;12(1):6262.
[39] Mortazavi BJ, et al. Comparison of machine learning methods with national cardiovascular data registry models for prediction of risk of bleeding after percutaneous coronary intervention. JAMA Netw Open 2019;2(7):e196835.
[40] Huang C, et al. Enhancing the prediction of acute kidney injury risk after percutaneous coronary intervention using machine learning techniques: a retrospective cohort study. PLoS Med 2018;15(11):e1002703.

[41] Kulkarni H, Amin AP. Artificial intelligence in percutaneous coronary intervention: improved risk prediction of PCI-related complications using an artificial neural network. BMJ Innov 2021;7(3):564.
[42] Pushparaji B, et al. Abstract 14712: PCI is safe and increasing among cancer patients: nationally representative propensity score adjusted and machine learning augmented study of 100 million hospitalizations. Circulation 2022;146(Suppl_1) A14712.
[43] Cui S, et al. Machine learning identifies metabolic signatures that predict the risk of recurrent angina in remitted patients after percutaneous coronary intervention: a multicenter prospective cohort study. Adv Sci (Weinh) 2021;8(10):2003893.
[44] Monlezun DJ, et al. Machine learning-augmented propensity score analysis of percutaneous coronary intervention in over 30 million cancer and non-cancer patients. Front Cardiovasc Med 2021;8:620857.
[45] Krittanawong C, et al. Machine learning and deep learning to predict mortality in patients with spontaneous coronary artery dissection. Sci Rep 2021;11(1):8992.
[46] Li Y-H, et al. Using text content from coronary catheterization reports to predict 5-year mortality among patients undergoing coronary angiography: a deep learning approach. Front Cardiovasc Med 2022;9:800864. Available from: https://doi.org/10.3389/fcvm.2022.800864.
[47] Levin MG, et al. Abstract 17565: high-throughput analysis of full-text cardiac catheterization reports using open-source natural language processing and machine learning tools. Circulation 2017;136(suppl_1) A17565.
[48] Duong T, et al. Interpretation of coronary angiograms recorded using google glass: a comparative analysis. J Invasive Cardiol 2015;27(10):443–6.
[49] Rynio P, et al. Holographically-Guided Endovascular Aneurysm Repair. J Endovasc Ther 2019;26(4):544–7.
[50] Chu MW, et al. Augmented reality image guidance improves navigation for beating heart mitral valve repair. Innov (Phila) 2012;7(4):274–81.
[51] Liu J, et al. An augmented reality system for image guidance of transcatheter procedures for structural heart disease. PLoS One 2019;14(7):e0219174.
[52] Bruno RR, et al. Virtual reality-assisted conscious sedation during transcatheter aortic valve implantation: a randomised pilot study. EuroIntervention 2020;16(12):e1014–20.
[53] Arslan F, Gerckens U. Virtual support for remote proctoring in TAVR during COVID-19. Catheter Cardiovasc Interv 2021;98(5):E733–6.
[54] Currie ME, et al. Augmented reality system for ultrasound guidance of transcatheter aortic valve implantation. Innov (Phila) 2016;11(1):31–9.
[55] West N. Bringing the cath lab to the doctor through virtual reality training for OCT. Available from: https://www.linkedin.com/pulse/bringing-cath-lab-doctor-through-virtual-reality-training-nick-west/?trk=read_related_article-card_title. [accessed 05.06.23].
[56] Khalaji A, et al. Machine learning algorithms for predicting mortality after coronary artery bypass grafting. Front Cardiovasc Med 2022;9:977747.
[57] Jie L, et al. Using machine learning to aid treatment decision and risk assessment for severe three-vessel coronary artery disease. J Geriatr Cardiol 2022;19(5):367–76.
[58] Al-Zaiti S, et al. Machine learning-based prediction of acute coronary syndrome using only the pre-hospital 12-lead electrocardiogram. Nat Commun 2020;11(1):3966.
[59] Ali ZA, et al. Imaging- and physiology-guided percutaneous coronary intervention without contrast administration in advanced renal failure: a feasibility, safety, and outcome study. Eur Heart J 2016;37(40):3090–5.
[60] Chen C-Y, et al. Zero-contrast percutaneous coronary intervention for chronic total occlusions guided by intravascular ultrasound with ChromaFlo mode: a case report. Eur Heart J - Case Rep 2020;4(6):1–6.
[61] Nair M, et al. First in man: percutaneous coronary angioplasty using non-fluoroscopic electro-anatomic mapping. Int J Cardiovasc Imaging 2020;36(7):1189–90.

CHAPTER

7

Artificial intelligence in heart failure and transplant

Kunaal Sarnaik[1] *and W. H. Wilson Tang*[2]

[1]Case Western Reserve University School of Medicine, Cleveland, OH, United States

[2]Kaufman Center for Heart Failure Treatment and Recovery, Heart, Vascular and Thoracic Institute, Cleveland Clinic, Cleveland, OH, United States

The use of artificial intelligence (AI) to gain insights into "big data"—an overarching term utilized to describe datasets too large or complex to be analyzed by traditional statistical methods—has grown exponentially in the field of healthcare over recent years. This paradigm shift has manifested primarily via the widespread adoption of clinical tools that capture vast amounts of data, such as electronic health records (EHRs), medical imaging modalities, genomic sequencing, wearable technologies, and medical devices [1]. Furthermore, with the increasing availability of machines possessing enormous computational capabilities, the development of AI systems as clinical decision support tools has skyrocketed to meet the growing need to interpret such extensive and heterogeneous data volumes. Consequently, use of AI in the clinical setting is in the process of transforming our everyday practice; these systems serve as potential means to fine tune decision-making, streamline workflows, and most importantly, improve patient outcomes.

The complex and often nonlinear interactions between various types of clinical data—the vast nature of each requiring tremendous computational power to analyze—simply cannot be adequately elucidated through use of traditional statistical methods. This is primarily why adoption of AI and machine learning (ML)—a subset of algorithms within the broader field of AI—has been such an appealing prospect for cardiovascular applications; chronic systemic disorders, such as heart failure (HF) and cardiomyopathy, are multifactorial and require the comprehensive integration of several components for proper management. However, several limitations continue to hinder the fruition of these idealistic algorithms into clinical settings as decision support tools for HF and transplantation, regardless of their perceived advantages in these spaces. The obstacles primarily include the lack of necessary training data quality and volume, the need for comprehensive external validation of those algorithms trained on single datasets, the "black-box" nature of

Artificial Intelligence in Clinical Practice
DOI: https://doi.org/10.1016/B978-0-443-15688-5.00036-X

these systems concurrent with the requirement for easy interpretability by clinicians, and their lack of interoperability with clinical technologies, such as the EHR.

This review hopes to shed further light on how this successful amalgamation may still take place by focusing on recent AI and ML developments in the diagnosis, management, and pre- and posttransplantation stages of HF and cardiomyopathy. Moreover, the review contextualizes this discussion to the principles of AI and ML and further explains limitations these systems face in their reliability and feasibility of implementation. Finally, it explores emerging technologies that may be utilized to iteratively refine and enhance the performance of these intelligent algorithms in HF and transplant management.

Overview of artificial intelligence and machine learning

In its most general form, AI is an umbrella term encompassing computer systems that employ human-like cognitive behaviors to perform a variety of tasks [2]. Examples of AI in daily life include personal digital assistants, self-driving vehicles, and facial recognition sensors. Although commonly mistaken as a synonym of AI, ML is instead a subclass of AI, specifically pertaining to those systems that learn—or "train"—from the data they are given. After training, ML algorithms are then validated on "test" data to gauge performance and external validity. Training of ML algorithms is classified into three categories: (1) supervised learning, (2) unsupervised learning, and (3) reinforcement learning (Fig. 7.1) [3]. The majority of ML methods

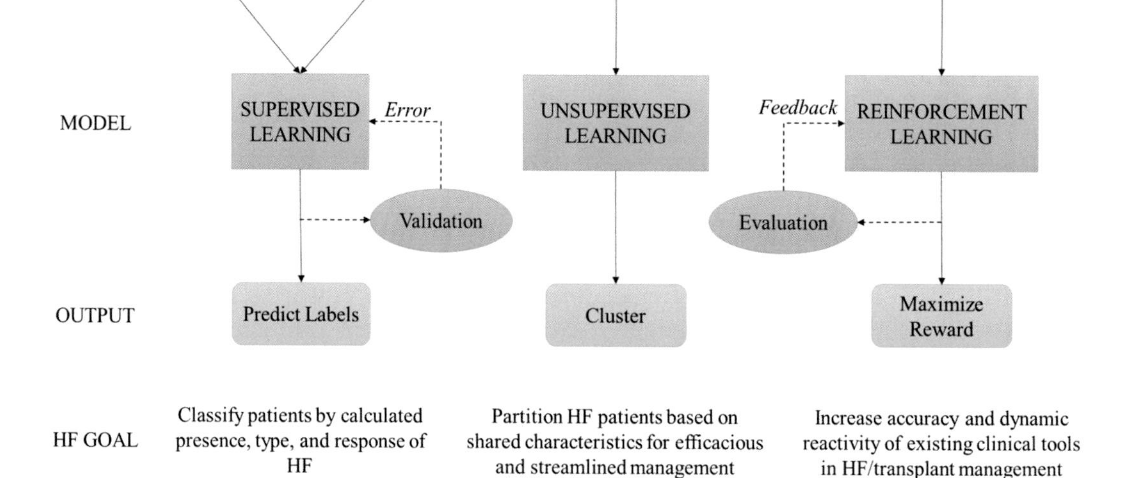

FIGURE 7.1 Types of ML training in HF. Supervised, unsupervised, and reinforcement learning methods comprise the three main categories of the ML training process. Supervised learning involves performing classification and regression tasks to predict outcomes from labeled training data during training. Unsupervised learning clusters unlabeled data into specific groups based on shared characteristics and values. Reinforcement learning is a dynamic process utilizing the calculated rewards (positive and negative) of possible model actions optimize behavior and performance.

developed for HF and cardiomyopathy employ supervised learning, in which labeled datasets are utilized for training [4]. For instance, a supervised algorithm may learn from data of patients classified by dilated cardiomyopathy (DCM) status (label) during training, and then, during testing, predict the cardiomyopathy status of another set of patients. Thus in supervised learning, both the input and output variables are given to the model during training, and the model attempts to predict output variables from only input variables given during testing based on interactions it captures from training. Examples of supervised algorithms include random forest classifiers (RFCs), support vector machines (SVMs), and naïve Bayes (Table 7.1).

In contrast, unsupervised ML algorithms utilize unlabeled data to partition data into groups based on shared characteristics. In other words, unsupervised algorithms are only given input variables during training. Furthermore, the testing process for unsupervised ML methods is relatively more complex than that for supervised methods due to the lack of labeling.

TABLE 7.1 Advantages and disadvantages of ML algorithms.

Algorithm	Description	Advantages	Disadvantages
Supervised			
Traditional logistic regression (LR)	Classic statistical model fitting logistic functions to relationships between inputs and outputs	– Easiest to implement and explain – No hyperparameter tuning – Lowest computational demand	– Poor performance with nonlinear data or missing data – Consistently outperformed
Naïve Bayes	Probabilistic algorithm based on Bayes' Theorem of conditional probabilities	– Multiclass prediction – Scalability – Large number of features for high-dimensional data	– Independence assumption between variables not often true in real world – Poor external validity due to vulnerability to nonrepresentation
K-nearest neighbors (KNN)	Nonparametric model generating classifications based on proximity of data points	– Easy to explain and implement – Readily adaptive to new data – No assumption of independence – Only tune 1 hyperparameter (K)	– Low scalability – Slow with large number of features – Poor performance with unbalanced data (outlier sensitivity)
Classification and regression tree (CART)	Performs recursive binary splits on data points and determines best features and values to split using various indices	– High interpretability – Easily handles missing data – Limited need for data preprocessing	– Highly prone to overfitting with poor external validity – Requires large amounts of data – Long training time
Random forest classification (RFC)	Builds numerous independent CART models and merges outputs for higher prediction accuracy and stability	– Low outlier sensitivity – Less prone to overfitting – Reduces overall error and variance between CART models	– Large "Black Box" nature – Low interpretability – Large computational demands – Long training time

(Continued)

TABLE 7.1 (Continued)

Algorithm	Description	Advantages	Disadvantages
Gradient boosting machines (GBM)	Successively builds dependent CART models that iteratively improve upon the preceding, merging outputs for higher prediction accuracy and stability	– Low outlier sensitivity – Outperforms other models – Fast training speed and efficiency – Flexibility and customizability	– Very difficult to tune due to large amount of hyperparameters – Tendency to overfit – Computationally expensive – Low interpretability
Support vector machines (SVM)	Attempts to identify a single hyperplane in a multidimensional space that best classifies data	– Minimal outlier sensitivity – High performance in higher dimensions (e.g., more features) – Highest performance with nonoverlapping binary classification	– Very slow to train on large datasets – Poor performance with overlapping classes – Requires careful hyperparameter tuning
Unsupervised			
K-nearest neighbors (KNN)	Nonparametric model clustering data points based on feature proximity to similar data and distance from distinct data	– Readily adaptive to new data – High interpretability – Easy implementation	– Low scalability – Slow with large number of features
Hierarchical clustering analysis (HCA)	Iterative process of grouping two similar clusters until none are left, ultimately generating a hierarchical dendrogram	– High interpretability – Easy implementation – High degree of visualization with dendrogram	– Large outlier sensitivity – Affected by ordering of data – Low scalability
K-means clustering (KMC)	Iteratively finds optimal centroid separating various data points into "K" clusters based on sum of squared distances between data points and centroid line	– Easy implementation – Readily adaptive to new data – Computationally fast with small class quantity – Highly specific clusters	– Affected by ordering of the data – Sensitive to data normalization – Dependent on user-given quantity of clusters/classes (i.e., presumptive) – Outlier sensitivity
Principal component analysis (PCA)	Dimensionality-reduction technique utilizing orthogonal transformations to decrease correlation among data features in new coordinate systems called principal components for clustering	– Easy implementation due to removal of correlated features – High interpretability with respect to principal components generated – Effective removal of noise – Scalable	– Requires data standardization – Low interpretability with respect to dataset's independent variables – Dimensionality reduction can result in information loss
Singular value decomposition (SVD)	Dimensionality-reduction technique that can be utilized with PCA that relies on the principle of factorizing matrices into three individual matrices	– High performance – Simplification of data points similar to PCA – Effective removal of noise – Scalable	– Low interpretability of transformed data – High complexity due to additional factorization processing step

(Continued)

TABLE 7.1 (Continued)

Algorithm	Description	Advantages	Disadvantages
Independent component analysis (ICA)	Seeks to find underlying factors affecting datasets by separating multivariate signals into subcomponents for clustering	– More powerful than PCA/SVD – Heterogenous data sources – High computational performance	– Computationally expensive – Low interpretability – Assumes independence of subcomponents
Deep learning			
Multilayer perceptron (MLP)	Artificial neural network model consisting fully connected feedforward input, hidden, and output layers	– Modest performance with nonlinear data – Real-time/adaptive learning – No probabilistic assumptions regarding data – Strong classification and regression accuracy – High interpretability relative to other DL methods	– Poor performance on image data relative to other DL methods – High sensitivity to feature scaling – Inefficiency with larger number of parameters in high dimensions
Convolutional neural network (CNN)	Artificial neural network model consisting of convolutional, pooling, and fully connected layers, with successive layers increasing complexity and focusing on finer details in data	– Strongest performance with image data – Not entirely fully connected relative to MLP, allowing for greater discriminative capacity	– Large number of parameters – Low interpretability without sensitivity analysis – Computationally expensive – Low interoperability
Recurrent neural network (RNN)	Artificial neural network model utilizing sequential data and distinguished by memory from earlier inputs affecting decisions made for future inputs and outputs	– Strongest performance for sequence prediction (e.g., natural language processing) – Does not assume independence between data	– Poor performance with tabular or image data – High complexity – Low interpretability – Specific for sequence prediction tasks – Difficult to train

Consequently, these algorithms are used as methods of gaining unique insights regarding complex disorders. For instance, patients with DCM may be partitioned into distinct phenotypes created by an unsupervised algorithm, and these classifications may then be used for subsequent management to determine if certain treatment options work better for patients within each phenotype. Unsupervised algorithms are characterized by the clustering process they use for partitioning; examples include hierarchical and k-means agglomerative or dendrogram clustering.

Reinforcement learning is a recent development, describing a dynamic training process rooted in positive and negative feedback based on accuracy of the algorithm's classifications [5]. Much of the current development in the field of AI is focused on reinforcement and deep learning

(DL). The latter describes a collection of advanced ML models composed of numerous processing layers that learn from data characterized by highly complex abstraction [6]. Examples of DL models include convolutional neural networks (CNNs), generative adversarial networks, and recurrent neural networks. DL models may use any of the training methods described previously, and applications of these algorithms include image classification and natural language processing (NLP). In fact, the analysis of cardiac MRI is among the most promising applications of DL techniques in cardiovascular medicine [7]. Finally, ensemble and hybrid methods combine outputs of various ML models applied to the same task to enhance performance [8].

The advantages of ML algorithms over traditional statistics (e.g., multivariate logistic regression) primarily lies in their rigorous training [9]. Regardless of the method, training of ML algorithms is an iterative process consisting of key cross-validation, dimensionality reduction, hyperparameter tuning, and error reduction steps, all of which serve to enhance performance. Due to this, ML algorithms have consistently outperformed traditional statistical models when evaluated on the same endpoints. In addition to this increased accuracy, advantages of ML techniques over traditional methods include the ability to analyze larger volumes of data, carry out unique tasks, such as image analysis and voice recognition, capture nonlinear interactions among numerous inputs, and provide clinical decision support. However, many of these advantages only manifest due to the tremendous computational power these algorithms require for training and execution. Supervised learning methods are the least computationally demanding, while unsupervised, reinforcement, and DL algorithms lie at the other end of the spectrum (Fig. 7.2). Complexity of ML algorithms follows a similar trend, and there exists a

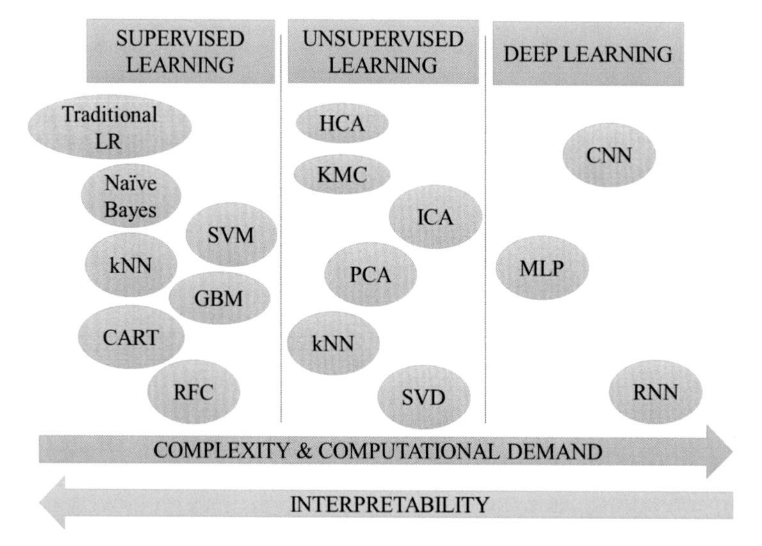

FIGURE 7.2 Tradeoffs between complexity and interpretability in ML algorithms. Inevitably, there exists a tradeoff between interpretability of ML models and both their complexity and computational demands. Supervised learning methods generally possess the least complexity and computational demand yet the highest interpretability. Deep learning methods lie on the other end of this spectrum, and unsupervised learning methods somewhere between the two. Note that specific models within each type of ML method also differ in their complexity, computational requirements, and interpretability.

tradeoff between model performance and computational requirements or complexity. Similarly, there is an inverse relationship between complexity and interpretability of ML algorithms, with higher complexity correlating to increased "black box" natures. Thus in clinical settings, performance of these algorithms should be optimized, but only to the extent that they are still able to be easily interpreted by providers using them. In essence, clinicians must be able to clearly understand why and how these algorithms draw conclusions, especially in the setting of HF and transplant when morbidity and mortality are frequently at stake.

Algorithms capitalizing on state-of-the-art ML and AI techniques can potentially make far-reaching impacts in the clinical management of HF and cardiomyopathy, from the diagnosis to the treatments of these disorders (Fig. 7.3). Moreover, the vast wealth of data provided via genomic, clinical, and patient-generated avenues can power the iterative advancement and refinement of these intelligent techniques for years to come [10]. For instance, genomic data obtained from DNA sequencing and microbiome analysis can be combined with clinical information present in EHRs to cluster patients into distinct subtypes based on characteristics underlying their cardiovascular disease. Then, patient-generated health data collected from wearable sensors and mobile health applications can be utilized to monitor progression and treatment efficacy, thereby providing a comprehensive pipeline of patient-specific diagnosis, treatment, and recovery monitoring for various HF conditions.

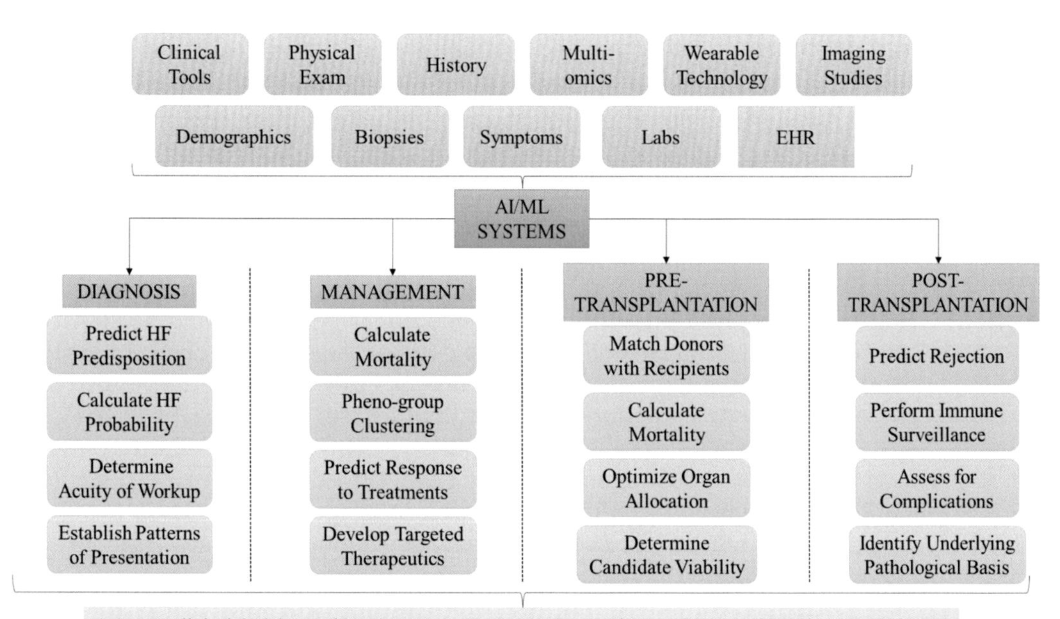

FIGURE 7.3 AI workflow in management of HF. Patient data reflecting the various inputs symbolized at the top of the schematic may be inputted into AI/ML models at each of the four general steps of HF management: (1) diagnosis, (2) management, (3) pretransplantation, and (4) posttransplantation. At each stage, various ML models may be utilized in the clinical setting to achieve various purposes, all of which ultimately serve to attain the end goal of acting as efficient, accurate, and informative decision support tools for the providers and patients using them.

Artificial intelligence in heart failure

Techniques developed using ML have been widely applied in the field of cardiovascular medicine, with HF and cardiomyopathy being among the most commonly studied topics [4]. The majority of these algorithms train via supervised learning before being extrapolated to perform a wide variety of tasks, including partitioning of data into categories (e.g., classifying cardiomyopathy patients into distinct phenotypes), making predictions of clinical outcomes (e.g., calculating mortality rates of patients with decompensated HF), and declaring diagnosis of HF at various stages (e.g., identifying decompensation based on electrocardiogram or echocardiogram findings) [11]. Similarly, ML strategies have been readily investigated in the transplantation of solid organs, including but not limited to that of the liver, kidney, and heart [12,13]. With regard to heart transplantation specifically, algorithms have primarily sought to identify optimal donor-recipient matches, determine mortality after transplantation, and assess posttransplantation rejection based on histologic characteristics of endomyocardial biopsy samples (EMBs) [14]. Furthermore, intriguing developments have been made that use ML methods to predict responses to immunotherapy and the development of complications, such as malignancy after cardiac transplantation.

Detection and diagnosis of heart failure

HF is a clinical syndrome with a variety of both cardiac and noncardiac etiologies, and characterizing the complex disorder relies on various imaging techniques and careful analysis of patient history, physical examination, and laboratory data. In other words, HF does not have a single noninvasive test that clinicians use for establishing a diagnosis. Consequently, one of the most challenging and pervasive issues in the detection of HF is the inconsistent method of defining the condition. Furthermore, due to the many nonspecific symptoms that may manifest (shortness of breath, edema, etc.) HF is repeatedly misdiagnosed or even undiagnosed, correlating to suboptimal sensitivity and specificity of current diagnostic methods [15]. Several physiologic measurements and clinical tools provide valuable and dynamic data regarding presence or absence of HF. Typically, these modalities present with complex patterns that may not be recognized easily by clinicians or traditional statistical methods. As such, diagnostics pertaining to HF and cardiomyopathy is an area in which AI and ML algorithms may be leveraged as helpful detection assistants, augmenting the abilities of clinicians to identify the presence of HF in any given patient at an early stage. On the other hand, recent developments suggest that imaging measurements can be performed much more efficiently if expert knowledge is supplemented with AI. In addition, studies have indicated that alternative tools not currently utilized for detection purposes may provide valuable information regarding presence of HF through use of AI. As such, AI can serve as an important tool in augmenting the abilities of clinicians to screen and diagnose HF using multimodal approaches. In this manner, unsuspected individuals may be flagged for further workup at earlier stages and complications may be reduced.

Echocardiography for diagnosing heart failure subtypes

Echocardiography is among the most important clinical tools providing valuable image-based data in the detection of HF. In recent years, several DL and ML algorithms have been developed to correlate both structured and unstructured echocardiogram data to HF status [16]. Commonly studied applications include functional left ventricle assessment, automatic segmentation, and diagnosis of various structural, valvular, and intracardiac mass disorders. For instance, in 2020, Choi et al. constructed and investigated the use of a novel supervised classification and regression tree (CART) ML analysis in the diagnosis of HF [17]. After training the CART model on 600 labeled patients, it initially achieved an overall diagnostic testing accuracy of 88.5% upon retrospective external validation ($n = 598$). However, an intriguing feature was an analysis investigating the use of an Artificial Intelligence-Clinical Decision Support System (AI-CDSS)—a merged algorithm which integrated components of the CART model with expert-driven knowledge rules obtained from physicians. Interestingly, AI-CDSS achieved a higher accuracy (98.3%) upon external validation of the same testing dataset. Furthermore, in a secondary analysis of 100 prospective patients presenting with dyspnea, AI-CDSS outperformed non-HF specialists in the diagnosis of HF (98% vs 76% concordance rate, respectively). This investigation not only bolsters evidence behind the clinical utility of AI in detecting HF, but also reinforces the theme of AI systems being utilized as decision support tools rather than autonomous decision-makers.

Many other ML algorithms have been successfully validated in screening for HF and related diseases using cardiac sonography [18–21]. Some of these investigations have elucidated intriguing insights regarding the ability of AI techniques to partition HF patients into subgroups at the diagnosis stage. In 2020, Mishra et al. utilized unsupervised clustering based on finite Gaussian mixture and expectation-maximization methods to classify 1,000 prospective patients by 15 transthoracic echocardiographic variables [22]. The algorithm identified four classes of these patients based on descending severity of cardiopulmonary structural and functional abnormalities, and left ventricular mass, left atrium volume, mitral regurgitation, velocity-time integral, and diastolic dysfunction indices were the strongest determinants of classification. Moreover, a follow-up analysis revealed that the rate of HF hospitalization was highest in the group determined to have the highest severity of abnormalities by the unsupervised model (86 per 1,000 person-years), while this rate was the lowest in the group determined to have the lowest severity (40 per 1,000 person-years). Furthermore, the investigators found a stepwise decrease in Kaplan-Meier survival from the group determined to have the lowest severity of abnormalities to the group determined to have the highest severity. Studies such as this suggest that AI can help classify severity of HF at the detection stage, allowing for more efficient subsequent workup.

Chest radiographs for acute heart failure exacerbations

Chest Radiographs are often utilized to characterize HF especially in the acute care settings, and similar ML algorithms have been constructed for their automated analysis [23,24]. Horng et al. recently developed a semi-supervised Bayesian model—trained on a small portion of labeled data and large portion of unlabeled data—using a variational autoencoder that achieved an area under the receiving operating characteristic curve (AUC) of 0.99 in diagnosing congestive HF (CHF) based on severity categories of pulmonary edema observed in radiographs [25]. Matsumoto et al. achieved similar success after constructing a DL model for the same task, also utilizing heatmap imaging to visualize decisions made by the algorithm and increase clinical interpretability [26]. Interpretability is an important feature for DL models in image analysis due to their high complexity, and Seah et al. highlighted this in 2018 by analyzing the utility of Generative Visual Rationales (GVRs) as a tool to visualize neural network learning of chest radiograph features in CHF [27]. They found that GVRs can enable detection of bias and overfitted models by identifying specific features of CHF on chest radiographs learned by neural networks. Sensitivity analysis and quality control techniques, such as GVRs and heatmap analysis are important requirements for clinical integration of AI and ML, as they help clinicians understand model decisions and minimize the "black-box" nature of these models that may lead to arbitrary decision-making.

Cardiac magnetic resonance imaging for myocardial characterization

Cardiac magnetic resonance imaging (MRI) is an area of rapid growth for DL development in cardiovascular diagnostics, with studies assessing structure, function, strain, motion, and tissue features that are highly relevant to identify HF and cardiomyopathy subtypes and disease severity [28]. Recently, Martin et al. analyzed the diagnostic accuracy of various DL and ML methods in the binary classification of cardiac amyloidosis based on late gadolinium enhancement (LGE) cardiac MRI [7]. After training, validation, hyperparameter tuning, and transfer learning, the "global CNN"—utilizing the average prediction score of three individual CNNs—achieved a testing diagnostic accuracy of 88% and AUC of 0.982 ($n = 42$). A gradient boosting machine (GBM) ML algorithm trained on manually extracted features additionally achieved a testing accuracy of 90% with an AUC of 0.952. The GBM model also found that circumferential subendocardial LGE pattern and blood-pool darkening were the strongest determinants of prediction. Although both DL and ML techniques displayed similar metrics, the inference time—time required to make a classification—was considerably smaller for the DL method. Since the ML model relied on manual extraction of cardiac MRI features by clinicians, the decreased inference time for the DL algorithm showcases the remarkable efficiency of AI in automated diagnostics. Despite the higher complexity inherent to the DL model, the investigators provided quality control via Gradient-weighted Class Action Mapping (Grad-CAM), which highlighted region(s) of the input image yielding the most diagnostic power. A similar study conducted by Böttcher et al. analyzed DL methods in the fully automated quantification of left ventricular volumes and function using cardiac MRI with expert correction [29]. A stronger correlation was found between fully automated and expert-corrected results relative to fully manual analysis by two clinicians. The analysis time for fully automated results with expert correction was also considerably smaller than manual analysis (2 minutes vs 3.5–9 minutes on average), again highlighting the remarkable efficiency of AI in clinical diagnostics.

Electrocardiographic signature for heart failure subtypes

An appealing prospect of ML systems in HF diagnostics lies in their inherent ability to elucidate unique insights and interactions among numerous heterogenous clinical studies. For instance, electrocardiograms (ECGs) are indicated for virtually every patient presenting to care facilities for nonspecific symptoms, yet they are not considered a reliable tool for HF detection. However, recent studies have demonstrated strong clinical utility of ECGs for this very purpose [30]. In 2020 Kwon et al. achieved AUCs of 0.866 and 0.869 in an ensemble neural network for the detection of preconfirmed HF with preserved ejection fraction (HFpEF) upon internal validation ($n = 1979$) and external validation ($n = 11{,}995$), respectively [31]. The study also found that among 1412 patients without HFpEF on initial ECG, a higher proportion of patients determined by the model to have a high risk of developing HFpEF actually developed HFpEF than patients in the low risk group. A sensitivity analysis elucidated that the neural network primarily focused on QRS complexes—specifically the R-wave—and T-waves when making classifications. A similar multicenter study recently conducted by Bachtiger et al. in 2022 analyzing single-lead ECGs obtained via stethoscopes with a CNN successfully screened patients for left ventricular ejection fraction of 40% or lower with an AUC of 0.91, sensitivity of 91.9%, and specificity of 80.2% ($n = 1050$) [32]. The findings of these studies suggest that AI can substantially diversify current methods of detecting HF, elucidating that even economical and widespread tools, such as ECGs, may provide useful diagnostic information.

Endomyocardial biopsy histological and genetic analyses

Along this same line, histological analysis is not currently utilized for detection of HF, but rather for characterization of phenotypes in predetermined HF patients or assessment of rejection after cardiac transplantation. However, Nirschl et al. recently developed a CNN classifier that achieved 99% sensitivity and 94% specificity during testing when identifying patients with HF based on H&E stained whole-slide images [33]. The classifier also outperformed two expert pathologists by approximately 20%, showcasing the keen ability of AI and ML methods in capturing visual intricacies that may be hidden to the human eye. Moreover, advanced techniques, such as DL and NLP, have been utilized to identify specific gene-associations in the detection of HF, and genome-wide association studies are currently a fast-growing investigative area for HF diagnostics with AI [34,35]. In 2020, Alimadadi et al. investigated the use of five supervised ML models in classifying patients with DCM, ischemic cardiomyopathy, and nonfailure control based on RNA-Seq data of their left ventricular tissues ($n = 137$) [36]. The models were trained using various highly contributing genes (HCGs) previously proposed as candidate biomarker genes in making the corresponding classifications, and they achieved remarkable accuracies ($>85\%$) in binary classification among the various phenotypes. Furthermore, a follow-up pathway analysis confirmed the involvement of various HCGs in each phenotypic classification, bolstering evidence behind the ability of AI and ML algorithms to model whole genome transcriptomic data in the diagnosis of various clinical cardiomyopathies.

Management of heart failure

Precision medicine describes the tailored application of medical therapeutics to unique characteristics of each patient [37]. Since this is a primarily data-driven field, precision medicine has contributed to rapid growth in the world of health-based AI systems, and applications for clinical management of HF and cardiomyopathy are no exception. There has been a tremendous growth of therapies designed to address HF, novel treatment strategies for management of hypertrophic and infiltrative diseases, and investigations seeking to understand the genetic basis of various phenotypes in recent years, all of which have contributed to advances in AI use for precision medicine in cardiovascular disease management [38]. These developments offer the ability to not only provide risk stratification for timely intervention(s), but also accurately match various conditions or phenotypes to the according therapeutic strategy.

Numerous ML and AI algorithms have been developed to identify high-risk subgroups among HF or cardiomyopathy patients, such that these individuals can be partitioned according to predicted urgency of further workup. Those models in which the primary endpoint is to classify patients into distinct subgroups based on risk of future complications primarily utilize unsupervised learning. As described previously, the investigation conducted by Mishra et al. suggested that clustering based on overall survival prediction can be performed as early as the detection stage using echocardiograms [22]. Similar investigations have also incorporated clinical characteristics obtained after the diagnosis of HF to partition patients. In 2021, Casebeer et al. utilized clustering techniques to classify patients ($n = 1515$) with HFpEF into three distinguishable subgroups [39]. The first cluster had the lowest prevalence of HF comorbidities, while the second cluster had a higher prevalence of metabolic syndrome and pulmonary disease, and the third cluster possessed the highest 1-year HF-related hospitalization rates. In a follow-up intracluster analysis, the model found that prior use of diuretics was directly associated with, and combination therapy was inversely associated with, these hospitalization rates. An investigation conducted by Uszko-Lencer et al. in 2021 also utilized unsupervised clustering methods to partition HF patients ($n = 603$) based on medical comorbidities, identifying five distinct clusters: (1) least morbidities, (2) cachective/implosive, (3) metabolic diabetes, (4) metabolic renal, and (5) psychologic [40]. Interestingly, this study found that characteristics, such as exercise performance, daily life activities, disease-specific health status, coping styles, and personality traits, in addition to number and type of comorbidities, were significantly different between clusters. Thus unsupervised methods provide a potential means to elucidate unique insights behind various classifications of HF phenotypes, augmenting clinician understanding of patients' unique needs and better informing therapeutic decisions based on future risk.

Supervised learning techniques have also been leveraged in ML models for the identification of high-risk HF subgroups, and these investigations primarily focus on calculating risk of morbidity and mortality. In 2019, Adler et al. developed a boosted decision tree algorithm to train and validate a mortality risk score using eight EHR variables—diastolic blood pressure, creatinine, blood urea nitrogen, hemoglobin, white blood cell count, platelets, albumin, and red blood cell distribution width—of both hospitalized and ambulatory HF patients [41]. The labels utilized by the supervised model to discriminate between high mortality risk and low mortality risk patients were mortality after 90 days and lack of recorded follow-up in the EHR after 800 days. After achieving an AUC of 0.88 during hold-out validation ($n = 5822$), the model also achieved AUCs of 0.84 and 0.81 upon external validation in two independent HF populations ($n = 1516$ and $n = 888$, respectively). Importantly, the supervised ML algorithm was found to be more accurate in discriminating mortality relative to previous calculators utilizing traditional statistical methods—Intermountain Risk Score, Get With the Guidelines-HF (GWTG-HF) Risk Score, and Acute Decompensated HF Registry Risk Score—as determined by C-statistic (0.88 vs 0.78, 0.74, and 0.63, respectively) [42–44]. This suggests superior discriminatory capacity of ML-derived mortality prediction compared with that of traditional multivariable logistic regression in identifying high-risk subgroups of HF patients. In another study, an artificial neural network (ANN) model developed by Kwon et al. in 2019 trained on demographic, ECG, echocardiography, and laboratory data of acute HF (AHF) patients in two hospitals ($n = 2165$) achieved a remarkable average AUC of 0.820 when externally validated on AHF patients in ten other hospitals ($n = 4759$) [45]. The endpoint of the model was ternary classification of AHF patients based on in-hospital, 12-month, or 36-month mortality. This model also outperformed the GWTG-HF Risk Score, in addition to the Meta-Analysis Global Group in Chronic HF (MAGGIC) score—developed using traditional multivariable piecewise Poisson regression with stepwise variable selection—when evaluated on the same endpoint [46]. Furthermore, in a 36-month follow-up, the group determined to be at a higher risk of mortality by the ANN had a significantly higher mortality rate than the low risk group ($P < 0.001$). Wang et al. took this analysis one step further and compared a DL-derived model, developed using sequential model architecture based on bi-directional long short-term memory layers, with both traditional ML-derived and logistic regression models in predicting hospitalizations, worsening HF events, and 30- and 90-day readmissions in HFrEF patients ($n = 47{,}498$) [47]. The data utilized to generate predictions for all models included sociodemographic factors, geographic region, insurance plan type, prior medical procedures, and medical comorbidities. The study found that the DL model outperformed all others for all outcomes assessed with regard to AUC, precision, and recall metrics. Although this suggests a hierarchy in ability to discriminate mortality risk between HF patients—with accuracy decreasing in a stepwise fashion from DL to ML to traditional statistics—it is important to remember the tradeoff between model complexity and interpretability in AI algorithms. Nevertheless, these studies highlight the unique ability of AI in incorporating various characteristics to identify HF patients that should be prioritized for further treatment.

The financial benefits of prioritization in a value-based payment model were analyzed by Jing et al. in 2020; they trained ML models to predict 1-year all-cause mortality of HF patients using EHR data reflecting clinical variables, diagnostic codes, ECG and echocardiogram findings, and uniquely, 8 evidence-based "care gaps" ($n = 26{,}971$) [48]. The lattermost represented actionable interventions, such as hemoglobin A1c and blood pressure levels within patient-specific goal ranges, along with active treatments, such as beta-blockers, angiotensin-converting enzyme (ACE) inhibitors, and aldosterone receptor antagonists. The best-performing ML model was an XGBoost algorithm which achieved an AUC of 0.77 during external validation. The investigators also simulated XGBoost predictions if there was complete closure of the care gap variables (i.e., imputing values reflecting execution of all actionable interventions regardless of actual status), and this provided an absolute mortality rate reduction of 1.7% (231 patients). Finally, the study compared simulating prioritization strategies based on the XGBoost model's risk score with strategies based on other metrics, such as random prioritization and the Seattle HF Risk score [49]. It was found that prioritization based on the XGBoost model's predicted risk had the steepest slope on a plot of the number of lives predicted to be saved versus the number of patients receiving an intervention, reflecting maximization of efficiency relative to all other metrics [48]. As such, this study showcases how AI and ML systems can go a step farther than simply partitioning HF patients into subgroups and also inform intervention strategies rooted in objective prioritization.

Clinical decision-making for durable mechanical circulatory support devices

Dynamic risk assessment and adaptive decision-making is especially important for HF patients with durable mechanical circulatory support (MCS) devices, as there has been an increasing push toward earlier implantation and lifetime support of these devices in patients with advanced HF [50]. Although MCS offers substantial benefits, there is a tremendous physical and financial burden associated with long-term MCS use [51]. Due to this, AI may offer a means to better identify not only viable candidates for MCS device implantation, but also long-term MCS users viable for device explantation. Models seeking to achieve the former purpose primarily predict morbidity and mortality of HF patients and have been described above. With regard to the latter purpose, Topkara et al. recently investigated the use of various ML methods in 2021 to predict myocardial recovery in patients with left ventricular assist device (LVAD) support in the INTERMACS (Interagency Registry for Mechanically Assisted Circulatory Support) registry ($n = 20{,}270$ with a 70:30 train-test split) [52]. After utilizing the least absolute shrinkage and selection operator (LASSO) to identify 28 unique clinical features with the highest predictive power—including HF etiology, psychosocial risk factors, laboratory values, and numerous electrocardiography and echocardiography indices—the supervised ML models achieved AUCs between 0.813–0.824 during external validation. Furthermore, these models outperformed scores of previous algorithms, such as the logistic regression-based new INTERMACS recovery risk score (AUC of 0.796), again highlighting the superior discriminatory capacity of ML relative to traditional statistics. Importantly, a higher proportion of patients that were predicted to have a greater likelihood of recovery by the ML models demonstrated a significantly higher incidence of myocardial recovery resulting in device explantation during validation relative to those predicted to have a lesser likelihood of recovery (18.8% versus 2.6%, respectively, at 4 years of pump support).

Therapeutic applications of phenogroups

ML algorithms have also been utilized to gain new insights that may be utilized to match various HF and cardiomyopathy phenotypes to corresponding therapeutic strategies. In 2021, Pandey et al. developed a DNN model integrating multidimensional echocardiographic data to identify distinct high-risk and low-risk subgroups of HFpEF patients ($n = 1242$), as HFpEF is a multifactorial and heterogenous clinical disorder [53]. Upon external validation of the model using two independent cohorts, the model possessed a higher AUC in predicting elevated left ventricular filling pressure relative to the 2016

American Society of Echocardiography guideline grades (0.88 vs 0.67; $P = 0.01$). Moreover, the high-risk subgroup, as determined by the model, demonstrated higher rates of HF hospitalization and/or death relative to the low-risk subgroup. The investigators further validated the clinical significance of the model by assessing the relationships of the subgroups with adverse clinical outcomes, cardiac biomarkers, and exercise parameters in three National Heart, Lung, and Blood Institute-funded trials. In addition to similar differences of actual hospitalization and death rates between high-risk and low-risk subgroups, it was found that the high-risk subgroup possessed higher rates of event-free survival with spironolactone therapy, burden of chronic myocardial injury, and neurohormonal activation, and lower exercise capacity relative to the low-risk subgroup. Thus findings of this study bolster the indications of AI in assessing prognosis of HF patients, while also providing evidence that DL methods can identify subgroups of these patients more likely to respond to certain therapies, such as spironolactone.

Molecular profiling of cardiomyopathies

Biomarker profiling

Some investigations have additionally incorporated molecular profile analysis into ML models to better inform therapeutic interventions. For instance, Woolley et al. initially utilized unsupervised learning methods in 2021 to identify four distinct clusters of HFpEF patients ($n = 429$) based on cardiac biomarker profiles [54]. However, the investigators also performed pathway overrepresentation—"enrichment"—analysis on each cluster; enrichment analysis is a statistical method determining whether certain components (e.g., genes or biomarkers) belonging to predefined pathways are present more or less than would be expected in a subset of data [55]. In this follow-up analysis, one model-derived cluster that possessed the highest prevalence of diabetes mellitus and renal clearance was demonstrated to be associated with protein biomarkers indicative of activation inflammatory pathways [54]. Another cluster that possessed the highest prevalence of ischemic etiology and smoking or chronic lung disease, most symptoms, and highest N-terminal pro B-type natriuretic peptide and troponin levels was found to be associated with protein biomarkers indicative of pathways implicated in cell proliferation regulation and cell survival. Importantly, both clusters possessed the highest occurrence of death or HF hospitalization relative to the others. As such, the unsupervised ML analysis elucidated specific high-risk subgroups of HFpEF patients characterized by distinct underlying pathophysiology that may better inform targeted therapeutic strategies. Hedmen et al. performed a similar analysis on HFpEF patients ($n = 320$) in 2020, identifying six distinct phenogroups with significant differences in prevalence of concomitant atrial fibrillation, anemia, and kidney disease [56]. In the follow-up inflammatory and cardiovascular proteomic analysis, it was found that 15 plasma proteins that included biomarkers of HF, atrial fibrillation, and kidney function differed among the phenogroups. These molecular correlates of each phenogroup may be useful when deciding between therapeutic interventions, such as ACE inhibitors, angiotensin receptor blockers, and aldosterone antagonists, as different phenogroups may be most responsive to different interventions.

Genomic integration

Some studies have delved deeper into patient-specific biomarker analysis by incorporating genomics. Cardiac imaging-genetics is a budding field of research seeking to accurately identify the genotypic basis underlying phenotypes determined from cardiovascular imaging [57]. In 2021, Zhou et al. investigated use of supervised ML for this purpose, analyzing the genomic data of HFpEF patients from the Framingham Heart Study cohort ($n = 149$) to identify individuals with higher 3-year mortality risks [58]. They found that a kernel partial least squares with the genetic algorithm (GA-KPLS) exhibited the best performance in predicting patient risk based on binary classification—poor-outcome or good-outcome. Importantly, the investigators identified 116 differentially expressed genes among both groups, highlighting the unique benefits AI may have in informing the development of novel therapeutics for HF management.

The heterogenous nature of HF and cardiomyopathy lends itself to poor prognosis, difficulty in selecting treatments, and heavy physical, mental, and financial burdens on the patient. However, AI and ML algorithms can improve these aspects of management by augmenting the abilities of clinicians to practice precision medicine. In other words, developments have indicated enhanced prioritization of HF patients by acuity of further workup and increased synchrony between HF phenotype and treatment strategy through use of AI. Supervised ML models can even be combined into a clinical decision support database to manage various HF patients at a granular level [59]. If nothing else, the use of unsupervised ML can uncover intriguing genomic, proteomic, and clinical relationships underlying HF phenotypes, which can in turn lead to rapid development of both novel therapeutics and clinical trials.

Artificial intelligence in cardiac transplantation

Pretransplantation risk stratification

Many of the recent AI and ML applications relevant to the pretransplantation stage of HF focus on predicting survival of waiting list patients using preoperative variables and optimizing donor-recipient matching [60]. These methods primarily utilize supervised learning techniques, such as RFC, to predict morbidity and mortality after transplantation. The biggest advantage of ML and AI systems in achieving this endpoint lies in their ability to comprehend the large quantities of variables that must be considered.

Several studies have utilized ML to predict pretransplantation survival using the UNOS (United Network for Organ Sharing) registry. In 2018, Medved et al. demonstrated that a DL model—International Heart Transplantation Survival Algorithm—trained on preoperative clinical characteristics of patients undergoing cardiac transplantation before 2009 ($n = 22{,}263$) possessed stronger discriminatory capacity in predicting 1-year survival after transplantation compared to a model based on multivariable logistic regression—Index for Mortality Prediction After Cardiac Transplantation (IMPACT)—when validated on patients transplanted between 2009–2011 ($n = 5597$) [61,62]. IHTSA achieved an AUC of 0.643, while IMPACT achieved an AUC of 0.608, demonstrating higher accuracy of short-term cardiac transplantation survival prediction using ML relative to traditional statistics.

Villela et al. achieved similar results after using auto-ML with stacking of GBM-derived algorithms to predict 1-year mortality and reimplantation (AUC of 0.66; $n = 18{,}612$), and this ML model demonstrated improvement in outcome prediction relative to logistic regression when evaluated on this endpoint [63].

Agrawal and Raman also performed a similar analysis with UNOS, using a variety of ML methods including SVMs, ANNs, decision tables, and random forest classification [64]. Although similar AUCs were achieved in predicting 1-year mortality after cardiac transplantation ($n = 19{,}429$), a follow-up information gain analysis revealed eight attributes possessing the highest predictive power across all ML methods. In decreasing order of importance, these features included estimated GFR at time of aortic cross clamp, intubation at transplant, hemodialysis at transplant, time spent on waiting list before transplantation, age of donor, MCS at transplant, ischemic time of donor organ, and extracorporeal membrane oxygenation at transplant. When evaluating the same endpoint with an ensemble ML method on UNOS registry patients, Ayers et al. achieved an AUC of 0.764 ($n = 33{,}657$), outperforming both singular ML models (random forest, Adaboost, and deep neural network) and logistic regression [65]. The most important features included total bilirubin, mechanical ventilation, ventricular assist device, and serum creatinine at time of transplantation. As such, ML algorithms not only have demonstrated high discriminatory capacity in predicting short-term survival of patients undergoing heart transplantation, but also have elucidated factors important in determining such survival.

Zhou et al. expanded upon this in 2021 by providing model interpretation via use of the Shapley Additive exPlanations (SHAP) method [66]. After utilizing random forest classification with bootstrapping to achieve an AUC of 0.801 ($n = 381$) in predicting 1-year mortality of heart transplantation, the SHAP method was leveraged to investigate feature importance to model output at the individual level. It was found that albumin level, left atrium diameter, and patient age at time of transplant were the most important features in affecting model output, and the SHAP analysis allowed the investigators to determine how the model generated classifications for each patient using these features. This interpretation is a necessary component, as it can help clinicians understand the most important factors determining any given patient's mortality classification such that according interventions may be made prior to transplantation to improve survival.

ML and AI methods have also displayed considerable accuracy in predicting long-term mortality of heart transplant recipients based on preoperative characteristics. Kampaktsis et al. developed ML models trained on UNOS registry data that achieved AUCs of 0.605 and 0.628 in predicting 3- and 5-year mortality, respectively ($n = 18{,}625$) [67]. Furthermore, Agasthi et al. developed a GBM model trained on heart transplant patients in the ISHLT (International Society of Heart and Lung Transplant) registry, achieving AUCs of 0.717 and 0.716 in predicting 5-year mortality and graft failure, respectively, based on preoperative variables ($n = 15{,}236$) [68]. Length of stay, recipient age, donor age, recipient body mass index, donor body mass index, and time of organ ischemia possessed the strongest predictive power.

Dolatsara et al. recently developed an intriguing two-stage framework using ML methods to predict personalized and monotonically-constrained probability curves based on preoperative data [69]. In the first stage of the study, an ML model was selected based on predictive performance of mortality, and variation of feature importance of variables utilized by the model was gauged over various time horizons (1−10 years posttransplantation). In the second stage, isotonic regression was utilized to generate an individualized survival probability curve for each heart transplant recipient. Interestingly, the results of the first stage demonstrated that logistic regression, with LASSO to determine feature importance, achieved a similar AUC to those achieved using ML methods. Based on ease of interpretability and interoperability with currently used clinical tools, the logistic regression model was then utilized in the second stage. Here, it was found that nonincreasing survival probabilities could be generated for each patient, resulting in personalized survival analysis functions that had not been achieved in previous investigations. Furthermore, the isotonic regression can be applied to any model selected from the first stage, which is important given that aforementioned investigations analyzing mortality prediction demonstrated increased discriminatory capacity of ML relative to logistic regression.

A study conducted by Yoon et al. also attempted to increase applicability of personalized medicine in ML models of pretransplantation mortality prediction [70]. Their Trees of Predictors (ToP) framework initially utilized unsupervised methods to discover clusters of patients registered for cardiac transplantation between 1985 and 2015, before leveraging supervised methods specific to each cluster that can provide the best mortality prediction. ToP framework achieved an AUC of 0.660 in determining 3-year mortality, and also showed improved predictive power of mortality at later time horizons when compared with existing risk scoring methods (e.g., IMPACT). The utilization of AI in the pretransplantation stage of HF management has demonstrated remarkable mortality prediction using preoperative variables derived from both donors and recipients. The applications of these developments primarily include optimal allocation of scarce organs to more suitable recipients, optimization of donor-recipient matching, and further direction into management of less suitable recipients such that they become more viable candidates. Finally, this work has opened avenues of research concerning the use of precision medicine in heart transplantation.

Posttransplantation rejection surveillance

Difficulties in managing HF patients after transplantation primarily involve allograft rejection and immune suppression. Clinicians must be careful to avoid such rejection, but not suppress the immune system so much that the patient is left vulnerable to dangerous infections. ML algorithms have demonstrated remarkable clinical utility in assessing rejection and immune surveillance of cardiac transplant recipients. They have repeatedly been shown to outperform traditional methods and may serve as more suitable prognostic indicators for personalized clinical decision-making, dose adjustment, and therapeutic strategies to improve posttransplantation outcomes. Combined with high generalizability, these findings support use of AI systems as decision support tools that augment clinician judgement with increased standardization, consistency, and objectivity in posttransplant care [60,71,72].

Histologic detection of transplant rejection and cardiac allograft vasculopathy

Peyster et al. first described the use of ML to assess rejection of cardiac transplant allografts from histologic EMB samples in 2018 [73]. Widely considered to be the gold standard of rejection diagnosis, EMB has several limitations, including interobserver variability, discordance with associated clinical prognosis after obtaining results, and multiple histologic criteria. However, these investigators outlined several advantages of ML in analyzing EMB samples, such as improved diagnostic accuracy, differentiation of histologic features that seem homogenously distributed to the human eye, and elucidation of unique insights regarding histological rejection features when sensitivity analysis is applied. In 2021 this group expanded upon this by developing and validating a data pipeline providing automated computational image analysis for histologic cardiac allograft rejection (CAR) grading [74]. The pipeline utilized K-means clustering to perform tissue compartment segmentation and lymphocyte detection, isolate lymphocyte clusters, identify foci of lymphocytes in concordance, and identify foci neighborhoods in digitized EMB samples, conforming to established ISHLT histologic grading guidelines. From there, 154 quantitative features pertaining to number of clusters and foci in different tissue compartments, size and density statistics for clusters, and spatial interactions of clusters and foci were utilized to train two predictive SVM models: (1) classifying between low-grade and high-grade rejection and (2) classifying between stages of the standard 4-grade classification. The final pipeline—Computer-Assisted Cardiac Histologic Evaluation (CACHE)-Grader—achieved an 84.5% agreement with the grade of record when discriminating high-grade rejection from low-grade rejection (0.92 AUC), and also achieved a 65.9% agreement with the grade of record when performing 4-grade classification, when validated on one internal ($n = 395$) and two external datasets ($n = 1052$). Furthermore, these accuracies were similar between internal validation and external validation (88.6% vs 82.9%, respectively, $P = 0.008$ for low-grade vs high-grade classification, and 66.1% vs 65.8%, respectively; $P = 0.91$ for 4-grade classification), demonstrating high generalizability of CACHE-Grader. Importantly, a noninferiority analysis demonstrated superiority of CACHE-Grader to pathologist performance when evaluated on a subset of slides ($n = 166$) re-graded by three independent pathologists, highlighting the potential of the ML-derived pipeline to minimize effects of interobserver variability in analyzing EMB samples for transplant rejection.

Peyster et al. also utilized ML to develop another pipeline analyzing digital pathology images from routine EMBs and clinical risk factors postcardiac transplantation to predict future cardiac allograft vasculopathy (CAV) development [75]. Relying on similar techniques to perform computational extraction of histological features, the Integrated Histological/Clinical Risk Factor Future Cardiac Allograft Vasculopathy Prediction Model (iCAV-Pr) achieved an AUC of 0.93 with an accuracy of 88.6% during validation ($n = 95$) after training ($n = 207$). Importantly, the analysis elucidated novel morphologic biomarkers—stromal proliferation, decreased overall vascular density, distribution of total nuclei within versus outside microvessels, and increased perivascular cellular density—providing strong predictive power. Furthermore, the investigation demonstrated increased discriminatory capacity when integrating conventional clinical risk factors with extracted histological features relative to using either input alone. The integrated model showcases how ML methods may utilize various data inputs to identify posttransplant patients who will develop aggressive CAV years prior to actual onset, enabling the development of further personalized screening and preventative treatment strategies to decrease cardiac allograft failure, and therefore morbidity and mortality associated with cardiac transplantation.

Another investigation conducted by Lipkova et al. utilized DL methods—specifically CNNs and transfer learning—to automatically assess whole-slide images obtained from EMB samples without any computational feature extraction [76]. The Cardiac Rejection Assessment Neural Estimator (CRANE) was trained, validated, and internally assessed on 5,054 whole-slide images from patient biopsies obtained at Brigham and Woman's Hospital, before being externally validated on two independent cohorts obtained from Turkey and Switzerland ($n = 1717$ and $n = 123$, respectively). CRANE achieved an AUC of 0.962 with an accuracy of 0.899 when validated on the hold-out testing subset of Brigham and Woman's Hospital whole-slide images. Similar AUCs and accuracies were achieved when performing independent tasks, such as identification of cellular rejection, antibody-mediated rejection, and Quilty B lesions. Furthermore, external validation of CRANE on the Turkey and Switzerland datasets only led to decreases in AUC and ACC of 0.02−0.13 and 0.02−0.14, respectively. Along with this high generalizability, strong interpretability was demonstrated with heatmap analysis to identify which histologic regions of the whole-slide images were utilized most when making a classification. Finally, follow-up analyses demonstrated noninferiority of CRANE to human reads and reduced human assessment time in analyzing EMB samples with use of CRANE as an assisting tool. This highlights the potential of ML to decrease interobserver variability and increase efficiency of EMB assessment when utilized as decision support tools after cardiac transplantation.

Risk stratification in transplant recipient and immunosuppression strategies

Bakir et al. also analyzed time-dependent clinical, molecular, and genetic phenomapping underlying death, allograft loss, retransplantation, and rejection events after cardiac transplantation [77]. Unsupervised K-means clustering analysis revealed high and low event rate groups, with the former being characterized by a steeper immunosuppression minimization. Furthermore, the high event rate group had increased HLA class I and II antibody titers, higher FLT3 gene expression, and lower March8 and WDNR40 gene expressions. An intramyocardial biomarker-related coexpression network analysis conducted on the FLT3 gene also demonstrated a well-differentiated inflammatory network underlying this genetic biomarker, with associated genes well known to be contributory to CAR. This study provides strong evidence behind the use of ML in elucidating hidden patterns underlying rejection pathophysiology, and the results may be used as the basis for development of personalized treatment strategies to reduce CAR.

Although posttransplantation immunosuppressive strategies can be better informed by these ML methods assessing rejection, increased immunosuppression can lead to infection, malignancy, and drug toxicity [78]. Consequently, other ML algorithms developed for patients in the postcardiac transplantation stage have focused on immune surveillance. In 2020, Woillard et al. utilized supervised ML to estimate the b.i.d. or q.d. area-under the blood concentration curve (AUBC) of Tacrolimus (TAC) in organ transplant patients [79]. XGBoost models were trained and cross-validated on interdose AUBCs obtained from high-performance liquid chromatography sampling of TAC blood levels in organ transplant recipients, before being externally validated on independent testing datasets. When evaluated on an independent dataset composed of the TAC pharmacokinetic profiles of 47 cardiac transplant recipients, XGBoost models achieved excellent interdose AUBC estimations, with relative bias $<5\%$ and relative root mean-squared error $<10\%$. Woillard et al. later replicated this investigation for mycophenolic acid (MPA) interdose AUBC prediction, achieving similar results [80]. Given that these models generated accurate interdose TAC and MPA AUBC predictions based on only 2 or 3 concentration blood profiles, they have the potential to be reliably utilized as immunosuppressive exposure markers for routine care of heart transplant recipients. This can better inform dose adjustment strategies, such that a patient-specific balance between immune suppression and protection against overutilization may be achieved.

Applications of ML and precision medicine to perform immune surveillance using molecular and genetic biomarkers have also been investigated. In 2020 Gim et al. utilized RFC and LASSO analysis to identify genetic variants that can affect exposure to TAC in Korean males ($n = 81$), inspired by TAC's relatively narrow therapeutic index [81]. Specific single nucleotide polymorphisms (SNPs) in the CYP3A5, CYP2A6, and SLC7A5 genes were found to be primary genotypic determinants of altered TAC exposures with respect to maximum plasma concentration and interdose AUBC. Although this study had a small sample size and was not conducted on heart transplant recipients, the results demonstrate that ML methods, such as RFC, may serve as useful clinical tools in predicting TAC exposure using individual genotypes.

ML models have also been developed to predict incidence of malignancy in cardiac transplant recipients, as this is an important sequela of pronounced immunosuppression strategies. In 2021 Chiu et al. utilized DNNs to identify UNOS heart transplant recipients more likely to develop squamous cell carcinoma (SCC) or basal cell carcinoma (BCC) based on donor and recipient characteristics [82]. 23 novel risk factors were identified with positive contribution to development of SCC or BCC skin cancer, and the DL models—DeepSurv and CoxTime—trained on these risk factors achieved excellent AUCs (>0.70) in predicting their incidence rate 1–10 years posttransplantation ($n = 152{,}095$ in an 80:20 train:test split). The ML-derived models also outperformed the traditional Cox proportional hazards model when evaluated on this same endpoint, highlighting their strong predictive power.

Current limitations of artificial intelligence in heart failure and transplant care

Remarkable developments in AI applications are not free from limitations plaguing AI and ML algorithms; these hinder the integration of such systems into the clinical management of HF and cardiomyopathy [83].

Data quality

One of the first aspects appraisers scrutinize regarding ML models is the training process, as almost every performance metric—external validity, applicability, and implementation—stems from this phase [84]. With regards to supervised models in HF and cardiomyopathy, data quality concerns primarily manifest due to improper labeling, lack of a gold standard, and dataset inadequacy [85]. Supervised EMB predictors are prime examples in which the first concern plays a role. For instance, interobserver variability in histological classification is alarmingly high, and ML models designed to predict EMB rejection or HF detection trained using one human observer's labels may entirely disagree with another's expert opinions.

Interobserver variability also relates to the lack of gold standards permeating management of HF. Since HF is a multifactorial disease without a single established noninvasive diagnostic test, incorrect labels utilized during training can severely limit predictive accuracy and power of these ML algorithms. Similarly, several models described above utilized data imputation techniques to fill in lapses in training datasets. This is a process wherein missing data values for a given patient are replaced by predicted values extrapolated from nonmissing data values of other patients [86]. Although certain imputation techniques are sophisticated, they still contribute to error inherent in replacing missing data, which can in turn propagate into overall model performance and bias.

Many datasets used for training of the previously described algorithms are also generated via retrospective means. Compounded by the constantly changing guidelines and definitions of HF detection and management, this can limit the applicability of such models when utilized prospectively. Thus there exists a need to supplement the training of these supervised ML models with prospective studies using reinforcement learning techniques, such that their predictive power can better match updated guidelines.

Finally, much of the intrigue behind the advent of ML is to identify hidden associations among large distributions of data, and datasets designed by humans are inherently flawed in this manner. This is because majority of collected data reflects categories already well-described to have an effect on the end result, thereby resulting in conclusions of described unsupervised ML algorithms that can be drawn simply via use of traditional statistical methods, such as logistic regression, with the benefit of increased interpretability and decreased computational requirements.

Generalizability

External validity of ML algorithms is another well-described concern, as they have a strong tendency to overfit the data they are trained on. Although overfitting is somewhat alleviated by techniques, such as cross-validation, it can extend into external validation of ML model development, since many of the testing phases described above were performed on a subset of the same dataset utilized to train the model. Furthermore, even those investigations which assessed external validation utilizing independent datasets suffered from low sample sizes and power. As a result, there exists a need to rigorously validate ML models on larger and more heterogenous independent external datasets to assess generalizability prior to their implementation for management of HF and transplant. Moreover, generalizability limitations can be compounded by the fact that many of the datasets utilized for training and testing arise from homogenous patient populations with regard to age, sex, and race.

Large temporal variation of HF is a primary contributing factor to suboptimal generalizability of previously described ML algorithms. For instance, 1-year survival after heart transplant is remarkable different relative to 10-year survival, since different periods of disease may have different determinants of morbidity and mortality [87]. The heterogenous and multifactorial nature of HF etiologies plays a similar role. Thus it is important for investigators to remember that one-size-fits-all models are dangerous, and there exists a need to develop models utilizing different techniques for different patient populations to improve overall generalizability and predictive power. As the ToP investigation demonstrated, the subsetting of patients through use of unsupervised ML methods, with subsequent supervised prediction algorithm development specific to each group, is one technique that can be leveraged to combat this problem and overcome limitations of ML surrounding external validity in clinical HF and transplant management [70].

Clinical interpretation

For proper and seamless clinical integration, ML algorithms also must be easily interpretable by clinicians using them for subsequent decision-making. Professionals must be able to efficiently understand not only what conclusions are generated by these models, but also why these classifications were made for each patient to better inform management. Interpretability is especially of concern about DL algorithms, which rely on complicated input–output mapping and highly abstract decision-making. The inherent "black-box" nature of such models must be minimized through sensitivity analysis techniques, such as heatmap imaging and weighting of predictive determinants.

The ability to clearly explain ML models on both the global and local scales is a prerequisite for clinical integration as decision support tools, especially given the aforementioned concerns regarding generalizability and training quality. As such, investigations utilizing such techniques should focus on providing probability calculations of each classification that could possibly be made for any given patient rather than simply producing an absolute classification without any descriptive context. Furthermore, an important limitation surrounding interpretation in ML algorithms utilized for HF and cardiomyopathy is that the predictions or classifications of these models do not inform clinicians exactly what should be done at the patient-specific level. In other words, future investigations must prospectively analyze the tangible efficacy of subsequent clinical decisions made for each patient informed by ML conclusions. This would enhance assessment of how they may be interpreted to improve patient outcomes.

Implementation

The use of ML models relies on high degrees of computational power and memory. Such machines are not readily available in all clinical settings, and this is compounded by interoperability concerns with currently utilized tools, such as EHRs and data storage systems. Furthermore, use of these algorithms in the clinical setting may result in tremendous computational strain placed on existing tools (e.g., EHRs and workflow management systems), undermining the benefits of ML systems in streamlining clinical efficiency. Due to the use of third-party software and programming languages in the development of these models, their implementation may also be incompatible with protected health information laws, such as HIPAA.

Tremendous monetary and time-related costs associated with training the professionals that will be interacting with such models also exist. Relating to the last limitation, professionals must know exactly how to utilize, interpret, and apply conclusions of these ML models for tangible impacts on patient care. However, there already exists a well-described hesitancy among the majority of healthcare professionals toward the adoption of new clinical technologies [88]. Furthermore, proper understanding of ML systems requires strong foundational knowledge in data science and statistics, and development of training modules, in addition to the time required to complete such modules, may be associated with tremendous expenses imposed on both hospitals and clinicians. Finally, the training, computational, and time expenses associated with clinical implementation of ML models may result in inequities between facilities that can afford integration and those that cannot, undermining the universal clinical benefits these systems are thought to provide.

The limitations reinforce an underlying theme regarding utilization of AI/ML systems for the clinical management of HF and transplant: they should be treated as guides that can supplement expert knowledge and decision-making, rather than absolute truths.

Future work

Several emerging technologies for HF and cardiomyopathy management are being rapidly developed, and AI/ML systems may be important in their innovation, interpretation, and analysis.

Network medicine

Network medicine is a broad field describing interdependent nonlinear molecular interactions between several nodes, each representing distinct biological entities, utilized to describe the underlying mechanisms or pathophysiology of various disease processes [89]. Systematic integration of large-scale multiomic data is an appealing prospect for HF, as such analysis can be utilized to precisely characterize endotypes and subset patients affected by the tremendous multifactorial nature of such diseases [90]. Consequently, the use of both supervised and unsupervised DL algorithms in network medicine is a large area of future research with respect to HF and transplant. Unsupervised methods can be utilized to classify patients by endotype utilizing data retrieved by network medicine approaches, and then supervised models may elucidate insights regarding efficacy of targeted therapeutics for management of these specific subtypes.

Sensors and devices

The field of AI has also seen rapid growth related to wearable technologies and implantable sensors with applications immediately relevant to HF detection and management [91–93]. These innovations allow for real-time data collection reflecting various aspects of cardiovascular surveillance, such as electrophysiology, metabolism, fitness, and environmental variables. Interconnectivity describes the pooling of massive amounts of data tracked via such noninvasive and invasive devices. AI/ML models analyzing the interactions present in these patient-specific interconnected networks may allow for elucidation of unique insights relevant to HF diagnostics, surveillance, risk-stratification, and triage as these technologies are iteratively improved. Furthermore, the use of AI-based virtual reality may allow for future adoption of mental cardiovascular stress tests, increasing diagnostic and surveillance accessibility for HF patients [90]. Thus use of AI in the development and analysis of sensors and devices may increase the cost-effectiveness and precision associated with HF management.

Biomarkers

Recent decades have also seen a paradigm shift toward noninvasive analysis of rejection after heart transplantation, with biomarkers playing a leading role [94]. Techniques, such as liquid biopsy, have allowed for noninvasive gene expression, donor-derived cell-free DNA, and even RNA-based analysis in heart transplant recipients. For example, Allomap is an innovative technique developed recently performing gene expression profiling of peripheral blood leukocytes and donor-derived circulating cell-free DNA that would generate large volumes of genomic data [95]. Similar to wearable technologies and noninvasive sensors, biomarker collection techniques will generate a vast wealth of multifactorial data as they continue to be implemented. As such, the leveraging of ML techniques in biomarker analysis is a strong future direction of AI in clinical management of HF and transplant.

Digital twins

An especially intriguing emerging technology central to AI and precision medicine in cardiovascular disease management is the "digital twin" [96]. The interconnected nature of several technologies described above—wearable devices, multiomics, clinical data, medical imaging—is being utilized in these computational models to build a inductive and deductive predictive framework that can dynamically follow a patient through the detection, management, and therapeutic stages of cardiovascular diseases, such as HF. At each stage, the patient's digital twin is utilized to predict disease prognosis and response to various treatments, allowing for patient-specific decision support. Due to the massive amounts of data being utilized to both generate digital twins and analyze potential clinical decisions, AI/ML is integral to the continued development and potential implementation of this technology into the clinical setting.

Major takeaways

The use of AI in HF and cardiac transplant medicine currently lies at a critical juncture. While the improper application of these advanced algorithms may result in this paradigm shift's failure, the complete avoidance of their integration into the clinical setting may leave potentially invaluable improvements pertaining to efficiency and accuracy of HF and transplant management unrealized. Regardless, for the foreseeable future, these systems are not likely to exist at a stage where they can be applied autonomously. Instead, their careful supervision, interpretation, and implementation by experienced clinicians with rigorous validation and refinements must take place for their successful integration into augmenting the clinical decision-making process in managing patients with HF and transplant.

Disclosure

Mr. Sarnaik has no relationships to disclose. Dr. Tang is a consultant for Sequana Medical A.V., Cardiol Therapeutics Inc, Genomics plc, and Zehna Therapeutics Inc. and has received honorarium from Springer Nature for authorship/editorship and American Board of Internal Medicine for exam writing committee participation.

References

[1] Catalyst N. Healthcare big data and the promise of value-based care. NEJM Catal 2022. Available from: https://catalyst.nejm.org/doi/full/10.1056/CAT.18.0290.

[2] Kersting K. Machine learning and artificial intelligence: two fellow travelers on the quest for intelligent behavior in machines. Front Big Data 2018;1. Available from: https://www.frontiersin.org/article/10.3389/fdata.2018.00006.

[3] Berry MW, Mohamed A, Yap BW, editors. Supervised and unsupervised learning for data science. Springer International Publishing; 2020. Available from: http://doi.org/10.1007/978-3-030-22475-2.

[4] Friedrich S, Groß S, König IR, et al. Applications of artificial intelligence/machine learning approaches in cardiovascular medicine: a systematic review with recommendations. Eur Heart J Digit Health 2021;2(3):424–36. Available from: https://doi.org/10.1093/ehjdh/ztab054.

[5] Sutton RS, Barto AG. *Reinforcement learning, second edition: an introduction.* MIT Press; 2018.

[6] LeCun Y, Bengio Y, Hinton G. Deep learning. Nature 2015;521(7553):436–44. Available from: https://doi.org/10.1038/nature14539.

[7] Martini N, Aimo A, Barison A, et al. Deep learning to diagnose cardiac amyloidosis from cardiovascular magnetic resonance. J Cardiovasc Magn Reson 2020;22(1):84. Available from: https://doi.org/10.1186/s12968-020-00690-4.

[8] Ardabili S, Mosavi A, Várkonyi-Kóczy AR. Advances in machine learning modeling reviewing hybrid and ensemble methods. In: Várkonyi-Kóczy AR, editor. Engineering for sustainable future. Lecture notes in networks and systems. Springer International Publishing; 2020. p. 215–27.

[9] Khanzode CA, Sarode RD. Advantages and disadvantages of artificial intelligence and machine learning: a literature review. IAEME Publ 2022. Available from: https://www.academia.edu/44895767/ADVANTAGES_AND_DISADVANTAGES_OF_ARTIFICIAL_INTELLIGENCE_AND_MACHINE_LEARNING_A_LITERATURE_REVIEW.

[10] Rashidi P, Bihorac A. Artificial intelligence approaches to improve kidney care. Nat Rev Nephrol 2020;16(2):71–2. Available from: https://doi.org/10.1038/s41581-019-0243-3.

[11] Seetharam K, Shrestha S, Sengupta PP. Artificial intelligence in cardiovascular medicine. Curr Treat Options Cardiovasc Med 2019;21(5):25. Available from: https://doi.org/10.1007/s11936-019-0728-1.

[12] Balch JA, Delitto D, Tighe PJ, et al. Machine learning applications in solid organ transplantation and related complications. Front Immunol 2021. Available from: https://doi.org/10.3389/fimmu.2021.739728 0.

[13] Spann A, Yasodhara A, Kang J, et al. Applying machine learning in liver disease and transplantation: a comprehensive review. Hepatology 2020;71(3):1093–105. Available from: https://doi.org/10.1002/hep.31103.

[14] Goswami R. The current state of artificial intelligence in cardiac transplantation. Curr Opin Organ Transpl 2021;26(3):296–301. Available from: https://doi.org/10.1097/MOT.0000000000000875.

[15] Wong CW, Tafuro J, Azam Z, et al. Misdiagnosis of heart failure: a systematic review of the literature. J Card Fail 2021;27(9):925–33. Available from: https://doi.org/10.1016/j.cardfail.2021.05.014.

[16] Zhou J, Du M, Chang S, Chen Z. Artificial intelligence in echocardiography: detection, functional evaluation, and disease diagnosis. Cardiovasc Ultrasound 2021;19(1):29. Available from: https://doi.org/10.1186/s12947-021-00261-2.

[17] Choi DJ, Park JJ, Ali T, Lee S. Artificial intelligence for the diagnosis of heart failure. Npj Digit Med 2020;3(1):1–6. Available from: https://doi.org/10.1038/s41746-020-0261-3.

[18] Kusunose K, Haga A, Abe T, Sata M. Utilization of artificial intelligence in echocardiography. Circ J 2019. Available from: https://doi.org/10.1253/circj.CJ-19-0420.

[19] Nedadur R, Wang B, Tsang W. Artificial intelligence for the echocardiographic assessment of valvular heart disease. Heart 2022. Available from: https://doi.org/10.1136/heartjnl-2021-319725.
[20] Alsharqi M, Woodward WJ, Mumith JA, Markham DC, Upton R, Leeson P. Artificial intelligence and echocardiography. Echo Res Pract 2018;5(4):R115–25. Available from: https://doi.org/10.1530/ERP-18-0056.
[21] de Siqueira VS, Borges MM, Furtado RG, Dourado CN, da Costa RM. Artificial intelligence applied to support medical decisions for the automatic analysis of echocardiogram images: a systematic review. Artif Intell Med 2021;120:102165. Available from: https://doi.org/10.1016/j.artmed.2021.102165.
[22] Mishra RK, Tison GH, Fang Q, Scherzer R, Whooley MA, Schiller NB. Association of machine learning-derived phenogroupings of echocardiographic variables with heart failure in stable coronary artery disease: the heart and soul study. J Am Soc Echocardiogr 2020;33(3):322–331.e1. Available from: https://doi.org/10.1016/j.echo.2019.09.010.
[23] Peterzan MA, Rider OJ, Anderson LJ. The role of cardiovascular magnetic resonance imaging in heart failure. Card Fail Rev 2016;2(2):115–22. Available from: https://doi.org/10.15420/cfr.2016.2.2.115.
[24] El Omary S., Lahrache S., El Ouazzani R. Detecting heart failure from chest X-ray images using deep learning algorithms. In: 2021 3rd IEEE Middle East and North Africa COMMunications conference (MENACOMM); 2021. p. 13–18. Available from: https://doi.org/10.1109/MENACOMM50742.2021.9678291.
[25] Horng S, Liao R, Wang X, Dalal S, Golland P, Berkowitz SJ. Deep learning to quantify pulmonary edema in chest radiographs. Radiol Artif Intell 2021;3(2):e190228. Available from: https://doi.org/10.1148/ryai.2021190228.
[26] Matsumoto T, Kodera S, Shinohara H, et al. Diagnosing heart failure from chest X-ray images using deep learning. Int Heart J 2020;61(4):781–6. Available from: https://doi.org/10.1536/ihj.19-714.
[27] Seah JCY, Tang JSN, Kitchen A, Gaillard F, Dixon AF. Chest radiographs in congestive heart failure: visualizing neural network learning. Radiology 2019;290(2):514–22. Available from: https://doi.org/10.1148/radiol.2018180887.
[28] Tao Q, Lelieveldt BPF, van der Geest RJ. Deep learning for quantitative cardiac MRI. Am J Roentgenol 2020;214(3):529–35. Available from: https://doi.org/10.2214/AJR.19.21927.
[29] Böttcher B, Beller E, Busse A, et al. Fully automated quantification of left ventricular volumes and function in cardiac MRI: clinical evaluation of a deep learning-based algorithm. Int J Cardiovasc Imaging 2020;36(11):2239–47. Available from: https://doi.org/10.1007/s10554-020-01935-0.
[30] Feeny AK, Chung MK, Madabhushi A, et al. Artificial intelligence and machine learning in arrhythmias and cardiac electrophysiology. Circ Arrhythm Electrophysiol 2020;13(8):e007952. Available from: https://doi.org/10.1161/CIRCEP.119.007952.
[31] Kwon Jm, Kim KH, Eisen HJ, et al. Artificial intelligence assessment for early detection of heart failure with preserved ejection fraction based on electrocardiographic features. Eur Heart J – Digit Health 2021;2(1):106–16. Available from: https://doi.org/10.1093/ehjdh/ztaa015.
[32] Bachtiger P, Petri CF, Scott FE, et al. Point-of-care screening for heart failure with reduced ejection fraction using artificial intelligence during ECG-enabled stethoscope examination in London, UK: a prospective, observational, multicentre study. Lancet Digit Health 2022;4(2):e117–25. Available from: https://doi.org/10.1016/S2589-7500(21)00256-9.
[33] Nirschl JJ, Janowczyk A, Peyster EG, et al. A deep-learning classifier identifies patients with clinical heart failure using whole-slide images of H&E tissue. PLoS One 2018;13(4):e0192726. Available from: https://doi.org/10.1371/journal.pone.0192726.
[34] Krittanawong C, Johnson KW, Choi E, et al. Artificial intelligence and cardiovascular genetics. Life 2022;12(2):279. Available from: https://doi.org/10.3390/life12020279.
[35] Lanzer JD, Leuschner F, Kramann R, Levinson RT, Saez-Rodriguez J. Big data approaches in heart failure research. Curr Heart Fail Rep 2020;17(5):213–24. Available from: https://doi.org/10.1007/s11897-020-00469-9.
[36] Alimadadi A, Manandhar I, Aryal S, Munroe PB, Joe B, Cheng X. Machine learning-based classification and diagnosis of clinical cardiomyopathies. Physiol Genomics 2020;52(9):391–400. Available from: https://doi.org/10.1152/physiolgenomics.00063.2020.
[37] Ginsburg GS, Phillips KA. Precision medicine: from science to value. Health Aff Proj Hope 2018;37(5):694–701. Available from: https://doi.org/10.1377/hlthaff.2017.1624.
[38] Weldy CS, Ashley EA. Towards precision medicine in heart failure. Nat Rev Cardiol 2021;18(11):745–62. Available from: https://doi.org/10.1038/s41569-021-00566-9.

[39] Casebeer A, Horter L, Hayden J, Simmons J, Evers T. Phenotypic clustering of heart failure with preserved ejection fraction reveals different rates of hospitalization. J Cardiovasc Med Hagerstown Md 2021;22 (1):45–52. Available from: https://doi.org/10.2459/JCM.0000000000001116.

[40] Uszko-Lencer NHMK, Janssen DJA, Gaffron S, et al. Clustering based on comorbidities in patients with chronic heart failure: an illustration of clinical diversity. ESC Heart Fail 2022;9(1):614–26. Available from: https://doi.org/10.1002/ehf2.13704.

[41] Adler ED, Voors AA, Klein L, et al. Improving risk prediction in heart failure using machine learning. Eur J Heart Fail 2020;22(1):139–47. Available from: https://doi.org/10.1002/ejhf.1628.

[42] Horne BD, May HT, Muhlestein JB, et al. Exceptional mortality prediction by risk scores from common laboratory tests. Am J Med 2009;122(6):550–8. Available from: https://doi.org/10.1016/j.amjmed.2008.10.043.

[43] Peterson PN, Rumsfeld JS, Liang L, et al. A validated risk score for in-hospital mortality in patients with heart failure from the American Heart Association get with the guidelines program. Circ Cardiovasc Qual Outcomes 2010;3(1):25–32. Available from: https://doi.org/10.1161/CIRCOUTCOMES.109.854877.

[44] Fonarow GC, Adams KF, Abraham WT, Yancy CW, Boscardin WJ. Risk stratification for in-hospital mortality in acutely decompensated heart failure classification and regression tree analysis ADHERE Scientific Advisory Committee SG and InvestigatorsJAMA. 2005;293(5):572–80. Available from: https://doi.org/10.1001/jama.293.5.572.

[45] Kwon Jm, Kim KH, Jeon KH, et al. Artificial intelligence algorithm for predicting mortality of patients with acute heart failure. PLoS One 2019;14(7):e0219302. Available from: https://doi.org/10.1371/journal.pone.0219302.

[46] Pocock SJ, Ariti CA, McMurray JJV, et al. Predicting survival in heart failure: a risk score based on 39 372 patients from 30 studies. Eur Heart J 2013;34(19):1404–13. Available from: https://doi.org/10.1093/eurheartj/ehs337.

[47] Wang Z, Chen X, Tan X, et al. Using deep learning to identify high-risk patients with heart failure with reduced ejection fraction. J Health Econ Outcomes Res 2021;8(2):6–13. Available from: https://doi.org/10.36469/jheor.2021.25753.

[48] Jing L, Ulloa CAE, Good CW, et al. A machine learning approach to management of heart failure populations. JACC Heart Fail 2020;8(7):578–87. Available from: https://doi.org/10.1016/j.jchf.2020.01.012.

[49] Levy WC, Mozaffarian D, Linker DT, et al. The Seattle Heart Failure Model: prediction of survival in heart failure. Circulation. 2006;113(11):1424–33. Available from: https://doi.org/10.1161/CIRCULATIONAHA.105.584102.

[50] Kanwar MK, Kilic A, Mehra MR. Machine learning, artificial intelligence and mechanical circulatory support: a primer for clinicians. J Heart Lung Transpl 2021;40(6):414–25. Available from: https://doi.org/10.1016/j.healun.2021.02.016.

[51] Magasi S, Buono S, Yancy CW, Ramirez RD, Grady KL. Preparedness and mutuality affect quality of life for patients with mechanical circulatory support and their caregivers. Circ Cardiovasc Qual Outcomes 2019;12 (1):e004414. Available from: https://doi.org/10.1161/CIRCOUTCOMES.117.004414.

[52] Topkara VK, Elias P, Jain R, Sayer G, Burkhoff D, Uriel N. Machine learning-based prediction of myocardial recovery in patients with left ventricular assist device support. Circ Heart Fail 2022;15(1):e008711. Available from: https://doi.org/10.1161/CIRCHEARTFAILURE.121.008711.

[53] Ambarish Pandey MD, Nobuyuki Kagiyama MD, Naveena, Yanamala MS, et al. Deep-learning models for the echocardiographic assessment of diastolic dysfunction. Cardiovasc Imaging 2021;. Available from: https://doi.org/10.1016/j.jcmg.2021.04.010.

[54] Woolley RJ, Ceelen D, Ouwerkerk W, et al. Machine learning based on biomarker profiles identifies distinct subgroups of heart failure with preserved ejection fraction. Eur J Heart Fail 2021;23(6):983–91. Available from: https://doi.org/10.1002/ejhf.2144.

[55] Gilbert N, Boyle S, Fiegler H, Woodfine K, Carter NP, Bickmore WA. Chromatin architecture of the human genome: gene-rich domains are enriched in open chromatin fibers. Cell 2004;118(5):555–66. Available from: https://doi.org/10.1016/j.cell.2004.08.011.

[56] Hedman ÅK, Hage C, Sharma A, et al. Identification of novel pheno-groups in heart failure with preserved ejection fraction using machine learning. Heart. 2020;106(5):342–9. Available from: https://doi.org/10.1136/heartjnl-2019-315481.

[57] de Marvao A, Dawes TJW, O'Regan DP. Artificial intelligence for cardiac imaging-genetics research. Front Cardiovasc Med 2020;6. Available from: https://www.frontiersin.org/article/10.3389/fcvm.2019.00195.

[58] Zhou L, Guo Z, Wang B, et al. Risk prediction in patients with heart failure with preserved ejection fraction using gene expression data and machine learning. Front Genet 2021;12. Available from: https://www.frontiersin.org/article/10.3389/fgene.2021.652315.
[59] Guidi G, Pettenati MC, Melillo P, Iadanza E. A machine learning system to improve heart failure patient assistance. IEEE J Biomed Health Inf 2014;18(6):1750–6. Available from: https://doi.org/10.1109/JBHI.2014.2337752.
[60] Connor KL, O'Sullivan ED, Marson LP, Wigmore SJ, Harrison EM. The future role of machine learning in clinical transplantation. Transplantation 2021;105(4):723–35. Available from: https://doi.org/10.1097/TP.0000000000003424.
[61] Medved D, Ohlsson M, Höglund P, Andersson B, Nugues P, Nilsson J. Improving prediction of heart transplantation outcome using deep learning techniques. Sci Rep 2018;8(1):3613. Available from: https://doi.org/10.1038/s41598-018-21417-7.
[62] Weiss ES, Allen JG, Arnaoutakis GJ, et al. Creation of a quantitative recipient risk index for mortality prediction after cardiac transplantation (IMPACT). Ann Thorac Surg 2011;92(3):914–21. Available from: https://doi.org/10.1016/j.athoracsur.2011.04.030 discussion 921-922.
[63] Villela MA, Bravo CA, Shah M, et al. Prediction of outcomes after heart transplantation using machine learning techniques. J Heart Lung Transpl 2020;39(4):S295–6. Available from: https://doi.org/10.1016/j.healun.2020.01.658.
[64] Agrawal A, Raman J, Russo MJ, Choudhary A. Heart transplant outcome prediction using UNOS data. Available from: https://chbrown.github.io/kdd-2013-usb/workshops/DMH/doc/dmh4167_Agrawal.pdf.
[65] Ayers B, Sandholm T, Gosev I, Prasad S, Kilic A. Using machine learning to improve survival prediction after heart transplantation. J Card Surg 2021;36(11):4113–20. Available from: https://doi.org/10.1111/jocs.15917.
[66] Zhou Y, Chen S, Rao Z, et al. Prediction of 1-year mortality after heart transplantation using machine learning approaches: A single-center study from China. Int J Cardiol 2021;339:21–7. Available from: https://doi.org/10.1016/j.ijcard.2021.07.024.
[67] Kampaktsis PN, Tzani A, Doulamis IP, et al. State-of-the-art machine learning algorithms for the prediction of outcomes after contemporary heart transplantation: results from the UNOS database. Clin Transpl 2021;35(8):e14388. Available from: https://doi.org/10.1111/ctr.14388.
[68] Agasthi P, Buras MR, Smith SD, et al. Machine learning helps predict long-term mortality and graft failure in patients undergoing heart transplant. Gen Thorac Cardiovasc Surg 2020;68(12):1369–76. Available from: https://doi.org/10.1007/s11748-020-01375-6.
[69] Ahady Dolatsara H, Chen YJ, Evans C, Gupta A, Megahed FM. A two-stage machine learning framework to predict heart transplantation survival probabilities over time with a monotonic probability constraint. Decis Support Syst 2020;137:113363. Available from: https://doi.org/10.1016/j.dss.2020.113363.
[70] Yoon J, Zame WR, Banerjee A, Cadeiras M, Alaa AM, van der Schaar M. Personalized survival predictions via Trees of Predictors: an application to cardiac transplantation. PLoS One 2018;13(3):e0194985. Available from: https://doi.org/10.1371/journal.pone.0194985.
[71] Glass C, Lafata KJ, Jeck W, et al. The role of machine learning in cardiovascular pathology. Can J Cardiol 2022;38(2):234–45. Available from: https://doi.org/10.1016/j.cjca.2021.11.008.
[72] Glass C, Davis R, Xiong B, Dov D, Glass M. The use of artificial intelligence (AI) machine learning to determine myocyte damage in cardiac transplant acute cellular rejection. J Heart Lung Transpl 2020;39(4, Supplement):S59. Available from: https://doi.org/10.1016/j.healun.2020.01.1250.
[73] Peyster EG, Madabhushi A, Margulies KB. Advanced morphologic analysis for diagnosing allograft rejection: the case of cardiac transplant rejection. Transplantation. 2018;102(8):1230–9. Available from: https://doi.org/10.1097/TP.0000000000002189.
[74] Peyster EG, Arabyarmohammadi S, Janowczyk A, et al. An automated computational image analysis pipeline for histological grading of cardiac allograft rejection. Eur Heart J 2021;42(24):2356–69. Available from: https://doi.org/10.1093/eurheartj/ehab241.
[75] Peyster EG, Janowczyk A, Swamidoss A, Kethireddy S, Feldman MD, Margulies KB. Computational analysis of routine biopsies improves diagnosis and prediction of cardiac allograft vasculopathy. Circulation. 2022;145(21):1563–77. Available from: https://doi.org/10.1161/CIRCULATIONAHA.121.058459.
[76] Lipkova J, Chen TY, Lu MY, et al. Deep learning-enabled assessment of cardiac allograft rejection from endomyocardial biopsies. Nat Med 2022;28(3):575–82. Available from: https://doi.org/10.1038/s41591-022-01709-2.

[77] Bakir M, Jackson NJ, Han SX, et al. Clinical phenomapping and outcomes after heart transplantation. J Heart Lung Transpl 2018;37(8):956–66. Available from: https://doi.org/10.1016/j.healun.2018.03.006.
[78] Patel JK. Blood-based immunological monitoring after heart transplant. Current status and future prospects. Indian J Thorac Cardiovasc Surg 2020;36(2):194–9. Available from: https://doi.org/10.1007/s12055-020-00928-x.
[79] Woillard JB, Labriffe M, Debord J, Marquet P. Tacrolimus exposure prediction using machine learning. Clin Pharmacol Ther 2021;110(2):361–9. Available from: https://doi.org/10.1002/cpt.2123.
[80] Woillard JB, Labriffe M, Debord J, Marquet P. Mycophenolic acid exposure prediction using machine learning. Clin Pharmacol Ther 2021;110(2):370–9. Available from: https://doi.org/10.1002/cpt.2216.
[81] Gim JA, Kwon Y, Lee HA, et al. A machine learning-based identification of genes affecting the pharmacokinetics of tacrolimus using the DMETTM plus platform. Int J Mol Sci 2020;21(7):E2517. Available from: https://doi.org/10.3390/ijms21072517.
[82] Chiu K.C., Du D., Nair N., Du Y. Deep neural network-based survival analysis for skin cancer prediction in heart transplant recipients. In: *2021 43rd annual international conference of the IEEE engineering in medicine biology society (EMBC)*; 2021. p. 2144–2147. doi:10.1109/EMBC46164.2021.9630234.
[83] Asselbergs FW, Fraser AG. Artificial intelligence in cardiology: the debate continues. Eur Heart J – Digit Health 2021;2(4):721–6. Available from: https://doi.org/10.1093/ehjdh/ztab090.
[84] van Smeden M, Heinze G, Van Calster B, et al. Critical appraisal of artificial intelligence-based prediction models for cardiovascular disease. Eur Heart J 2022;ehac238. Available from: https://doi.org/10.1093/eurheartj/ehac238.
[85] Miller PE, Pawar S, Vaccaro B, et al. Predictive abilities of machine learning techniques may be limited by dataset characteristics: insights from the UNOS database. J Card Fail 2019;25(6):479–83. Available from: https://doi.org/10.1016/j.cardfail.2019.01.018.
[86] Emmanuel T, Maupong T, Mpoeleng D, Semong T, Mphago B, Tabona O. A survey on missing data in machine learning. J Big Data 2021;8(1):140. Available from: https://doi.org/10.1186/s40537-021-00516-9.
[87] Miller RJ, Sabovčik F, Cauwenberghs N, et al. Temporal shift and predictive performance of machine learning for heart transplant outcomes. J Heart Lung Transpl 2022;. Available from: https://doi.org/10.1016/j.healun.2022.03.019.
[88] Khairat S, Marc D, Crosby W, Al, Sanousi A. Reasons for physicians not adopting clinical decision support systems: critical analysis. JMIR Med Inf 2018;6(2):e24. Available from: https://doi.org/10.2196/medinform.8912.
[89] Lee LYH, Loscalzo J. Network medicine in pathobiology. Am J Pathol 2019;189(7):1311–26. Available from: https://doi.org/10.1016/j.ajpath.2019.03.009.
[90] Gladding PA, Loader S, Smith K, et al. Multiomics, virtual reality and artificial intelligence in heart failure. Future Cardiol 2021;17(8):1335–47. Available from: https://doi.org/10.2217/fca-2020-0225.
[91] Bachtiger P, Plymen CM, Pabari PA, et al. Artificial intelligence, data sensors and interconnectivity: future opportunities for heart failure. Card Fail Rev 2020;6:e11. Available from: https://doi.org/10.15420/cfr.2019.14.
[92] Maurya MR, Riyaz NUSS, Reddy MSB, et al. A review of smart sensors coupled with Internet of Things and Artificial Intelligence approach for heart failure monitoring. Med Biol Eng Comput 2021;59(11):2185–203. Available from: https://doi.org/10.1007/s11517-021-02447-2.
[93] Kwon JM, Jo YY, Lee SY, et al. Artificial intelligence-enhanced smartwatch ECG for heart failure-reduced ejection fraction detection by generating 12-lead ECG. Diagn Basel Switz 2022;12(3):654. Available from: https://doi.org/10.3390/diagnostics12030654.
[94] Giarraputo A, Barison I, Fedrigo M, et al. A changing paradigm in heart transplantation: an integrative approach for invasive and non-invasive allograft rejection monitoring. Biomolecules. 2021;11(2):201. Available from: https://doi.org/10.3390/biom11020201.
[95] Khachatoorian Y, Khachadourian V, Chang E, et al. Noninvasive biomarkers for prediction and diagnosis of heart transplantation rejection. Transpl Rev 2021;35(1):100590. Available from: https://doi.org/10.1016/j.trre.2020.100590.
[96] Corral-Acero J, Margara F, Marciniak M, et al. The 'Digital Twin' to enable the vision of precision cardiology. Eur Heart J 2020;41(48):4556–64. Available from: https://doi.org/10.1093/eurheartj/ehaa159.

CHAPTER

8

Artificial intelligence in hematology

Joshua A. Fein[1] *and Roni Shouval*[1,2]

[1]Department of Medicine, Weill Cornell Medicine, New York, NY, United States
[2]Department of Medicine, Adult Bone Marrow Transplantation Service, Memorial Sloan Kettering Cancer Center, New York, NY, United States

Artificial intelligence (AI) remains a relative newcomer to the field of hematology. One can argue that machine learning (ML) in the hematologic cancers divides along two tracks—(1) diagnostic and histopathological and (2) prognostic or predictive. A cursory search of *PubMed* for "machine learning" and "leukemia," "lymphoma," or "myeloma" shows an exponential rise in these publications over the past decade. As a research tool, such as for clustering genomic and metagenomic alterations to define disease subtypes, ML has established a vital role in hematology [1–3]. In this chapter, we will point to several of the most important and promising studies using ML and then identify some of the key challenges the field has yet to overcome.

Though preceded by a number of demonstrations of ML applied to clinical hematology datasets [4–6], one of the earliest attempts to describe a large-scale clinically applicable ML tool in malignant hematology was the development of the AL (acute leukemia)-EBMT (European Society for Blood and Marrow Transplantation) score in 2015 [7]. AL-EBMT was developed on a dataset of over 25,000 patients with acute leukemia to predict 100-day mortality following allogeneic stem cell transplantation and uses an alternating decision tree methodology. Subsequent validation studies have suggested that its performance is on par with, though not better than, tools developed using traditional methods [8]. This work has been followed by a number of prognostic systems using different methods, though none has been adopted for widespread clinical use. The list in Table 8.1 is not exhaustive but highlights a number of interesting studies using ML for prognostication along these lines. A clinically promising case for ML in the hematology clinic is demonstrated in Agius et al., Chronic Lymphocytic Leukemia (CLL) Treatment Infection Model (CLL-TIM), which predicts risk of infection within 2 years of new CLL diagnosis [12]. Trained on over 4000 patients, this model is able to account for missing data when generating new predictions, provides bootstrap-derived confidence intervals, and it is interpretable by presenting an individual prediction-level metric of which features

Artificial Intelligence in Clinical Practice
DOI: https://doi.org/10.1016/B978-0-443-15688-5.00003-6

TABLE 8.1 Selected machine learning studies in hematology.

References	Year	Output	Input	Algorithm
[4]	1998	Prediction of chronic lymphocytic leukemia (Rai) stage	Immunophenotypic markers	Decision tree
[5]	2002	Prediction of cured versus refractory/fatal lymphoma	Gene expression	Weighted voting
[6]	2014	Complete response following different therapies	Gene expression	Several
[7]	2015	Day-100 mortality after allogenic stem cell transplantation in acute leukemia	Patient/disease characteristics	Alternating decision tree
[9]	2016	Prediction of clinical deterioration for patients hospitalized with hematologic malignancies	Laboratory values, vital signs	Neural network
[10]	2017	Relapse in childhood acute lymphoblastic leukemia	"Sociodemographic, clinical, immunological and cytogenetic variables"	Random forest
[11]	2019	Complete remission of acute myeloid leukemia	Gene expression	K-nearest neighbor and others
[12]	2020	Infection during 2 years following new CLL diagnosis	Patient/disease characteristics	Ensemble
[13]	2020	In-hospital mortality of patients with febrile neutropenia	Patient, hospital and diagnosis information	Several
[14]	2021	Recommendation of optimal tyrosine kinase inhibitor for individual patient with chronic myelogenous leukemia	Patient and comorbidity variables	Extreme gradient boosting
[15]	2021	Survival prediction (high versus low risk) in patients with follicular lymphoma	Patient/disease characteristics, clinical laboratory values	Random survival forests
[16]	2021	Survival of patients with multiple myeloma	Patient/disease characteristics, first-line treatment	Random survival forest
[17]	2021	Overall survival and time-to-leukemia transformation	Patient/disease characteristics, cytogenetic and molecular variables	Random survival forest
[18]	2021	Classification of bone marrow cells	Microscopic cytological images of bone marrow smears	Convolutional neural networks
[19]	2021	Prediction of leukemia subtype	Gene expression	Ensemble
[20]	2022	Graft-versus-host-free, relapse-free survival after allogeneic stem cell transplantation	Patient, disease, donor, and transplant information	Ensemble

contributed the greatest weight. Compellingly, a clinical trial (NCT03868722) is underway using CLL-TIM predictions to identify patients at highest risk of infection and randomly assigning them to early treatment versus observation, which is the current standard of care. Recent work by Nazha et al. led to a prognostic model in patients with myelodysplastic syndrome which uses a random survival forest algorithm to generate predictions of overall survival (OS) and the likelihood of transformation to acute leukemia [17]. Integrating both clinical/laboratory and genetic data, the authors find the model performs favorably when compared to the current standard, the IPSS-R. They also show that a meaningful proportion of patients would be reclassified to different IPSS risk categories based on the outcome of their model's predictions. The treatment of low- and high-risk MDS is radically different in intensity, making appropriate risk classification for these patients a significant clinical need. Notably, the Molecular International Prognosis Scoring System (IPSS-M), a concurrently developed tool similarly incorporating a wide array of genomic information, was built using traditional statistical techniques [21].

In the diagnostic arena, a particularly intriguing study is a gene expression-based ML diagnostic algorithm for identifying lymphoma subtype. Designed specifically for implementation in low/middle-income countries [19], Valvert and colleagues used an ensemble approach which integrates the results of a variety of base learners to assign probabilities to nine diagnostic bins of different lymphoma subtypes. In the absence of adequate access to pathologists in many treatment settings, this model would allow for rapid diagnosis and appropriate therapy selection. The authors report that a prospective trial using this tool in several countries in Latin America is underway. [19]

Critical challenges remain in bringing ML into the hematologic malignancy clinic. Perhaps chief among them is that tools so far developed for prognosis show incremental, or often no, improvement over traditional risk-prognostic scores. This lack of progress may be attributed to the similarity of input data upon which many ML and traditional scores are built. The advantages of ML are greatest in high-dimensional data, with complex structures of interaction between features. However, discovery in such a high-dimensional setting requires vastly larger datasets than currently exist. Individual hematologic diseases have low prevalence, making such datasets challenging to assemble. Collaboration across many institutions to construct multiomic knowledge banks with rich clinical annotation is imperative for the success of AI in this field.

Lack of interpretability remains a barrier to the adoption of ML tools, and approaches like the CLL-TIM which "show their work" have an inherent advantage. ML offers a unique platform for developing models which are updated and re-validated continually as new data reflecting new practice patterns emerge. The problem of attempting to generate useful predictions with outdated, static models is hardly new to AI. ML may overcome this barrier, however, if a reliable mechanism for continuously updating and validating models gains widespread acceptance. This would represent a paradigm-shift in how such tools are critically assessed by the medical community.

Arguably the most important direction for AI tools to move in for the hematology clinic is away from prognostic (outcome, irrespective of intervention) and toward predictive (outcome in the context of a modifiable intervention) modeling [22]. Tools that can predict response to different treatment strategies by integrating clinical and genomic information and guide therapy selection have an obvious role in precision medicine. This is particularly true as the number of new treatment options expands exponentially in many blood cancers.

We find several principal take-always from the work currently applying AI to hematologic cancers. First, data richness and quality are just as important as quantity in this setting; we are unlikely to make new discoveries using the same tried-and-true datasets and variables, just by applying a new algorithm. Second, translating ML approaches into broad clinical practice depends on developing tools where individual predictions have clear interpretability and measures of model confidence. Finally, models predicting short- and intermediate-term outcomes, such as toxicity or early treatment failure, and which might offer the opportunity for early interventions, should continue to be a focus of the field.

It is roughly a decade since AI established itself as an important tool in hematologic cancers. The complexity of clinical decision-making continues to grow, driven by new drug approvals and the discoveries of new and potentially actionable clinical and molecular risk factors. We anticipate that ML-powered decision aids will be essential to integrating increasingly complex patient data and to making the best treatment decisions in concert with our patients.

References

[1] Chapuy B, Stewart C, Dunford AJ, et al. Molecular subtypes of diffuse large B cell lymphoma are associated with distinct pathogenic mechanisms and outcomes. Nat Med 2018;24(5):679–90.

[2] Schmitz R, Wright GW, Huang DW, et al. Genetics and pathogenesis of diffuse large B-cell lymphoma. N Engl J Med 2018;378(15):1396–407.

[3] Janizek J.D., Dincer A.B., Celik S., et al. Uncovering expression signatures of synergistic drug responses via ensembles of explainable machine-learning models [published online ahead of print, 1 May 2023]. Nat Biomed Eng 2023.

[4] Mašić N, Gagro A, Rabatić S, et al. Decision-tree approach to the immunophenotype-based prognosis of the B-cell chronic lymphocytic leukemia. Am J Hematol 1998;59(2):143–8.

[5] Shipp MA, Ross KN, Tamayo P, et al. Diffuse large B-cell lymphoma outcome prediction by gene-expression profiling and supervised machine learning. Nat Med 2002;8(1):68–74.

[6] Amin SB, Yip W-K, Minvielle S, et al. Gene expression profile alone is inadequate in predicting complete response in multiple myeloma. Leukemia. 2014;28(11):2229–34.

[7] Shouval R, Labopin M, Bondi O, et al. Prediction of allogeneic hematopoietic stem-cell transplantation mortality 100 days after transplantation using a machine learning algorithm: a European Group for Blood and Marrow Transplantation Acute Leukemia Working Party Retrospective Data Mining Study. JCO 2015;33 (28):3144–51.

[8] Shouval R, Fein JA, Shouval A, et al. External validation and comparison of multiple prognostic scores in allogeneic hematopoietic stem cell transplantation. Blood Adv 2019;3(12):1881–90.

[9] Hu SB, Wong DJL, Correa A, Li N, Deng JC. Prediction of clinical deterioration in hospitalized adult patients with hematologic malignancies using a neural network model. PLoS One 2016;11(8):e0161401.

[10] Pan L, Liu G, Lin F, et al. Machine learning applications for prediction of relapse in childhood acute lymphoblastic leukemia. Sci Rep 2017;7(1):7402.

[11] Gal O, Auslander N, Fan Y, Meerzaman D. Predicting complete remission of acute myeloid leukemia: machine learning applied to gene expression. Cancer Inf 2019;18 1176935119835544.

[12] Agius R, Brieghel C, Andersen MA, et al. Machine learning can identify newly diagnosed patients with CLL at high risk of infection. Nat Commun 2020;11(1):363.

[13] Du X, Min J, Shah CP, et al. Predicting in-hospital mortality of patients with febrile neutropenia using machine learning models. Int J Med Inf 2020;139:104140.
[14] Sasaki K, Jabbour EJ, Ravandi F, et al. The LEukemia Artificial Intelligence Program (LEAP) in chronic myeloid leukemia in chronic phase: a model to improve patient outcomes. Am J Hematol 2021;96(2):241–50.
[15] Li C, Patil V, Rasmussen KM, et al. Predicting survival in veterans with follicular lymphoma using structured electronic health record information and machine learning. Int J Env Res Public Health 2021;18(5).
[16] Mosquera Orgueira A, González Pérez MS, Díaz Arias JÁ, et al. Survival prediction and treatment optimization of multiple myeloma patients using machine-learning models based on clinical and gene expression data. Leukemia. 2021;35(10):2924–35.
[17] Nazha A, Komrokji R, Meggendorfer M, et al. Personalized prediction model to risk stratify patients with myelodysplastic syndromes. J Clin Oncol 2021;39(33):3737–46.
[18] Matek C, Krappe S, Münzenmayer C, Haferlach T, Marr C. Highly accurate differentiation of bone marrow cell morphologies using deep neural networks on a large image data set. Blood. 2021;138(20):1917–27.
[19] Valvert F, Silva O, Solórzano-Ortiz E, et al. Low-cost transcriptional diagnostic to accurately categorize lymphomas in low- and middle-income countries. Blood Adv 2021;5(10):2447–55.
[20] Iwasaki M, Kanda J, Arai Y, et al. Establishment of a predictive model for GVHD-free, relapse-free survival after allogeneic HSCT using ensemble learning. Blood Adv 2022;6(8):2618–27.
[21] Bernard E. Molecular international prognosis scoring system for myelodysplastic syndromes; 2021.
[22] Ballman KV. Biomarker: predictive or prognostic? JCO 2015;33(33):3968–71.

CHAPTER

9

Artificial intelligence in oncology

Jirapat Likitlersuang[1,2] *and Benjamin H. Kann*[1,2]

[1]**Artificial Intelligence in Medicine (AIM) Program, Mass General Brigham, Harvard Medical School, Boston, MA, United States** [2]**Department of Radiation Oncology, Brigham and Women's Hospital, Dana-Farber Cancer Institute, Harvard Medical School, Boston, MA, United States**

Introduction

Recent advances in computer science and artificial intelligence (AI), particularly within the fields of computer vision, natural language processing, and machine learning, have resulted in multiple translations of knowledge to healthcare applications. Developed algorithms for solving multiple complex pattern recognition problem in images, such as object recognition, classification, and segmentation, are being adapted to biomedical imaging and clinical assessment. Given that the nature of oncology is highly associated with digital image and metadata processing, AI has enormous potential to improve the accuracy, precision, efficiency, and overall quality of care for patients with cancer [1–4]. In this chapter, we provide a brief discussion of the current work in AI within oncology, including examples of ground-breaking studies and challenges within the field.

Artificial intelligence research topics in oncology

The potential application of AI within the clinical patient pathway can be roughly broken down into seven stages (Table 9.1) [3]. At each of these AI "touchpoints" along the care pathway, there is an opportunity for AI, using a variety of potential algorithms, to guide cancer management.

Artificial Intelligence in Clinical Practice
DOI: https://doi.org/10.1016/B978-0-443-15688-5.00045-0

TABLE 9.1 AI clinical patient pathway model along with associated research topics and AI methods implemented.

Patient pathway	Research topic examples	AI methods
Prevention	- Automated risk prediction	- Random forest
Screening	- Automated nodule detection - Patient-reported outcome	- Support vector machine (SVM)
Diagnosis	- Automated diagnosis and grading	- XGBoost - Neural networks
Staging/risk stratification	- Cancer genomics	Convolutional neural networks (CNN)
Treatment	- Toxicity prediction and management	- Fully convolutional neural networks (FCN)
Response	- Automated tumor burden evaluation	- Variational auto-encoders (VAE)
Follow-up	- Predict patient recurrence risk	- Generative adversarial networks (GAN) - Reinforcement learning (RL) - Convolutional neural network (CNN)

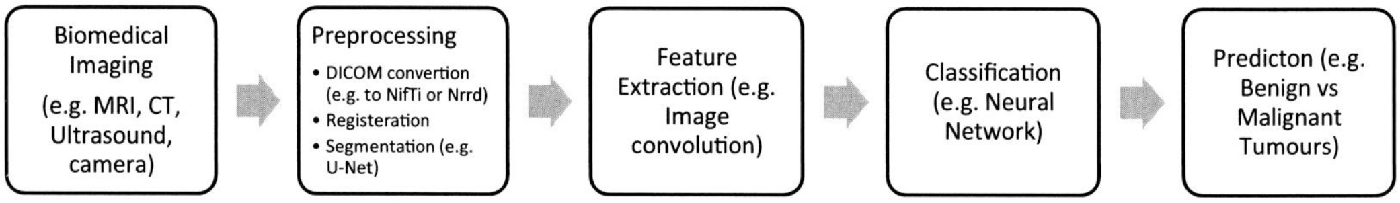

FIGURE 9.1 Representative workflow for utilizing artificial intelligence for biomedical imaging. Key steps include image preprocessing, feature extraction, machine learning classification, and prediction with performance evaluation.

Methodology in artificial intelligence oncology research

Artificial intelligence in cancer imaging

Research in AI for oncology has made great strides in qualitative interpretation and automation of cancer imaging, for example, tracking of volumetric information of tumors over time [5]. The goal of the AI systems is often to optimize and improve patient care, diagnosis, planning, or surgical treatments. In order to build robust AI models for cancer imaging, thoughtful, domain-informed, image preprocessing is essential.

Processing pipeline

The workflows of radiology AI research often involves preprocessing of biomedical images (e.g., MRI and CT). This includes but is not limited to registration (transforming multiple data into one coordinate system), segmentation (partitioning the region of interest within an image), and feature extraction (metrics extracted from the images or meta-data). On many occasions, the workflows pipeline will involve all these steps connected in sequential order (Fig. 9.1). Each of these steps, however, can either be built on AI methods or manually performed by a clinician. For example, a scan level system will consist of the whole workflows that are automated from image to prediction (i.e., automated preprocessing step), while a region of interest-based system will take manually segmented regions, often tumors or lymph nodes, for cancer classification. This workflow has been used in many biomedical imaging investigations.

Image classification

Early AI platform predictions were rule-based, whereby a computer system performed classification according to a set of steps and procedures defined or trained by human experts. Such a system required developers to define a set feature and train/test them to classifier (e.g., random forest, SVM). Unfortunately, such methods are often shown to be lacking in the generalizability and accuracy needed for clinical applications [1]. In recent years, advancements, such as CNN and similar deep learning methods (e.g., FCN, VAE, GAN, and RL), which utilize raw input data without engineering of features, as well as availability of large datasets have drastically increased our ability to discern complex, nonlinear relationships in oncology data. This includes, but not limited to, lung node detection/classification and tumor detection during mammography [6,7].

Artificial intelligence in genomic oncology

Increases in computer processing power and declining cost of genome sequencing have also resulted in translation of AI used in analysis of genomic data from cancer patients [3]. Research on the sequencing of associations of genetic mutations with clinical phenotypes is moving away from traditional lookup table. Work on several convolution filters is applied one-dimensionally for convolution of the genomic table. Traditional machine learning algorithms are often framed on detecting, classifying, optimizing, or solving single metrics. Given that cancer is a complex disease and often cause by multiple mutation points, research on genomic oncology has moved to integrate multilayered data through multimodal learning, that is, integration of several types of data (such as sequence and chromatin accessibility as inputs) [8]. In this process, AI automatically learns how to combine these various kinds of data, using variant analysis. In recent years, variant detection framework like Google's DeepVariant have also been explored in cancer research [8].

Cancer data: opportunities and challenges

The number of medical images available for training has significantly grown in recent years, with most studies able to train their classifier models using images from > 150 participants [8]. Furthermore, datasets available through online repositories have grown over the past few years, including data specific to radiographic images, digital pathology, medical record digitization, and genomics. This, in turn, has led to multiple research projects on data mining as well as model architecture and optimization [9]. In genomics, for example, multiple unsupervised learning algorithms have become popular, including clustering, data reduction, and network analysis [10,11].

Despite the increasing availability of datasets, the data are heterogeneous in nature, and often not suitable for machine learning in raw form. Often manual curation of data is necessary to improve documentation and annotation prior to machine learning analysis.

There are also barriers related to data transfer policy, privacy, and lack of standardization of repository remains a challenge [12].

To overcome these barriers, strategies like data augmentation and transfer learning which involves retrained the existing model for new purpose of medical imaging usually from a natural image dataset, such as ImageNet, have shown promising results [7]. Additionally, works related to auto-labeling and self-supervised learning have the potential to learns from unlabeled sample data. Such methods like Contrastive predictive coding (CPC) have also been used in oncology to solve the limited number of labeled data [13]. Perhaps most excitingly, multiple institutes have recently begun to implement federated learning and swarm learning, which would allow researchers to train their algorithm on local data at the institute holding the data (i.e., data do not have to be sent out or decentralized) [14].

Conclusion

Significant advances in applied AI technology in cancer medicine have resulted in 71 AI-associated devices that have received FDA approval [15]. Such advancement has shown promising results in increasing accuracy of cancer tumor detection, optimization of care and treatments as well as understanding factor in cancer genetics and pathology. Nevertheless, compared to natural image or language processing in other sectors, the sample sizes available for model training in healthcare, and particularly oncology are typically much smaller. However, over time, efforts are being made to increase dataset size, access, and quality. Coupled with novel techniques, such as transfer, self-supervised, and federated learning, these advances are helping developers and providers overcome the challenges of bringing AI to oncologic care.

References

[1] Huynh E, Hosny A, Guthier C, et al. Artificial intelligence in radiation oncology. Nat Rev Clin Oncol 2020;17 (12):771–81. Available from: https://doi.org/10.1038/s41571-020-0417-8.

[2] Kann BH, Hosny A, Aerts HJWL. Artificial intelligence for clinical oncology. Cancer Cell 2021;39(7):916–27. Available from: https://doi.org/10.1016/j.ccell.2021.04.002.

[3] Bera K, Schalper KA, Rimm DL, Velcheti V, Madabhushi A. Artificial intelligence in digital pathology - new tools for diagnosis and precision oncology. Nat Rev Clin Oncol 2019;16(11):703–15. Available from: https://doi.org/10.1038/s41571-019-0252-y.

[4] Chua IS, Gaziel-Yablowitz M, Korach ZT, et al. Artificial intelligence in oncology: Path to implementation. Cancer Med 2021;10(12):4138–49. Available from: https://doi.org/10.1002/cam4.3935.

[5] Xu Y, Hosny A, Zeleznik R, et al. Deep learning predicts lung cancer treatment response from serial medical imaging. Clin Cancer Res 2019;25(11):3266–75. Available from: https://doi.org/10.1158/1078-0432.CCR-18-2495.

[6] Le EPV, Wang Y, Huang Y, Hickman S, Gilbert FJ. Artificial intelligence in breast imaging. Clin Radiol 2019;74(5):357–66. Available from: https://doi.org/10.1016/j.crad.2019.02.006.

[7] Bitencourt A, Daimiel Naranjo I, Lo Gullo R, Rossi Saccarelli C, Pinker K. AI-enhanced breast imaging: where are we and where are we heading? Eur J Radiol 2021;142:109882. Available from: https://doi.org/10.1016/j.ejrad.2021.109882.

[8] Shimizu H, Nakayama KI. Artificial intelligence in oncology. Cancer Sci 2020;111(5):1452–60. Available from: https://doi.org/10.1111/cas.14377.

[9] Mayo CS, Kessler ML, Eisbruch A, et al. The big data effort in radiation oncology: data mining or data farming? Adv Radiat Oncol 2016;1(4):260–71. Available from: https://doi.org/10.1016/j.adro.2016.10.001.
[10] Nicora G, Vitali F, Dagliati A, Geifman N, Bellazzi R. Integrated multi-omics analyses in oncology: a review of machine learning methods and tools. Front Oncol 2020;10:1030. Available from: https://doi.org/10.3389/fonc.2020.01030.
[11] Sheikine Y, Kuo FC, Lindeman NI. Clinical and technical aspects of genomic diagnostics for precision oncology. J Clin Oncol 2017;35(9):929–33. Available from: https://doi.org/10.1200/JCO.2016.70.7539.
[12] Blasimme A, Fadda M, Schneider M, Vayena E. Data sharing for precision medicine: policy lessons and future directions. Health Aff (Millwood) 2018;37(5):702–9. Available from: https://doi.org/10.1377/hlthaff.2017.1558.
[13] Shurrab S, Duwairi R. Self-supervised learning methods and applications in medical imaging analysis: A survey. PeerJ Computer Science 2022;8:e1045. Available from: https://doi.org/10.48550/arXiv.2109.08685.
[14] Sheller MJ, Edwards B, Reina GA, et al. Federated learning in medicine: facilitating multi-institutional collaborations without sharing patient data. Sci Rep 2020;10(1):12598. Available from: https://doi.org/10.1038/s41598-020-69250-1.
[15] Luchini C, Pea A, Scarpa A. Artificial intelligence in oncology: current applications and future perspectives. Br J Cancer 2022;126(1):4–9. Available from: https://doi.org/10.1038/s41416-021-01633-1.

CHAPTER

10

Artificial intelligence in ophthalmology I: retinal diseases

Dawei Li[1], *Yingfeng Zheng*[2] *and Tien Yin Wong*[3,4]

[1]College of Future Technology, Peking University, Beijing, P.R. China [2]State Key Laboratory of Ophthalmology, Zhongshan Ophthalmic Center, Sun Yat-sen University, Guangzhou, P.R. China [3]School of Clinical Medicine, Tsinghua Medicine, Tsinghua University, Beijing, P.R. China [4]Singapore Eye Research Institute, Singapore National Eye Centre, Singapore

Many retinal diseases threaten the vision of hundreds of millions of people worldwide. Specifically, diabetic retinopathy (DR) affects one-third of people with diabetes and is the leading cause of vision loss in working adults [1]; age-related macular degeneration (AMD) is the leading cause of blindness in older people [2]. Retinopathy of prematurity (ROP) remains the leading cause of childhood blindness [3]. Retinal vein occlusion (RVO) is increasingly common with aging populations, and rising rates of hypertension and vascular diseases [4]. These retinal diseases impose significant demand on eye-care services and socio-economic costs.

Artificial intelligence (AI) can impact on management of retinal diseases in multiple points in the clinical journey (Fig. 10.1). For example, retinal diseases are often managed with by analyzing images, including those from fundus photography (FP) and optical coherence tomography (OCT). There has been progress in the application of AI in FP and OCT (Table 10.1).

Currently AI algorithms has been used for image enhancement, segmentation, classification for screening, and automated diagnosis [6,9,24]. In practice, AI systems have already demonstrated performances at or above expert levels in DR grading [11,16], AMD grading [12], and general retinal diagnosis [10]. In 2018, the U.S. Food and Drug Administration approved the first AI-based system for DR screening [8], opening the way for AI-based diagnostic systems.

Artificial intelligence in diabetic retinopathy

The major application of AI in DR is screening for early detection and intervention to change the outcomes of the disease, mainly in FP. In 2016 AI-based screening protocols

Artificial Intelligence in Clinical Practice
DOI: https://doi.org/10.1016/B978-0-443-15688-5.00024-3

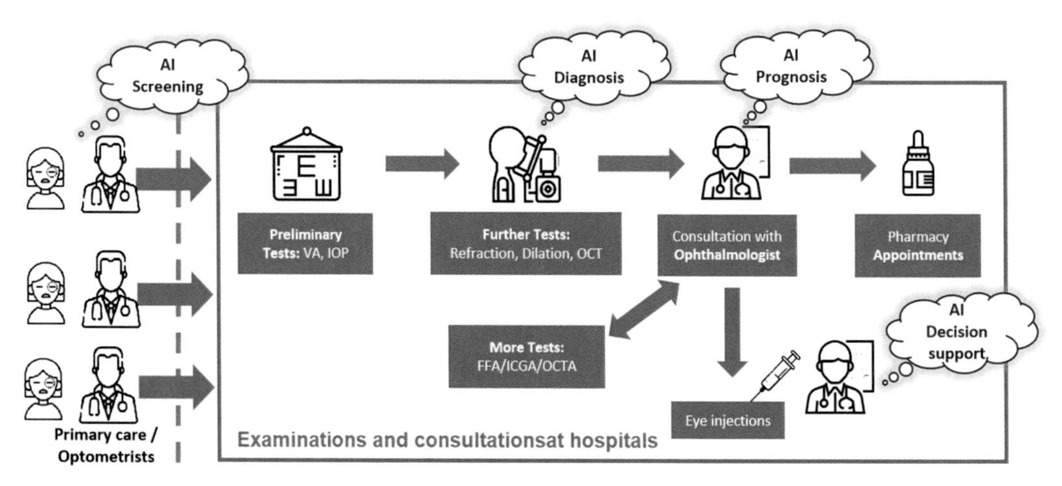

FIGURE 10.1 How artificial intelligence benefits ophthalmic clinical journey.

were reliable in distinguishing referable DR from nonreferable DR [3,25]. In 2017 a DR screening program in Singapore showed that the AI had a high sensitivity and specificity in identifying real-world DR when applying the system to the populations of different ethnicities, by using retinal images taken by different fundus cameras [5]. Since then, several DR screening systems have been officially approved, such as IDx-DR and EyeArt from the US, Selena + from Singapore, Retmarker DR from Europe, and Airdoc from China. AI-based DR screening system have the advantages of high efficiency and high accuracy, greatly reducing human resource workload and being more accuracy than manual screening by ophthalmologists.

AI also provides automated tools for the application of OCT angiography in DR, such as automatically segmenting and quantifying the capillary density in superficial foveal avascular zone [13], assessment of superficial and deep retinal capillary plexus [14], and higher classification accuracy in DR compared to FP [15].

Artificial intelligence in age-related macular degeneration

There is an increasing need for effective screening and follow up of macular status in the elderly population. Similar to DR screening, AI has demonstrated its ability to detect and refer patients with AMD [5] and to stage AMD [6] with FP.

Meanwhile, OCT has transformed the diagnosis and management of retinal/macular diseases by providing volumetric information with micrometer resolution. AI first proved its effectiveness for classifying normal versus AMD by using OCT images [7,17], and then demonstrated its ability to diagnose and differentiate AMD subtypes by using both segmentation and classification AI [11]. Such a system may help triage patients with macular diseases and enhance the referral system, thus reducing undue load [26]. AI has also shown its potential to predict AMD conversion by combining models based on

TABLE 10.1 Selected progress of AI in retinal diseases.

Authors	Input	Objective
Gulshan et al. [3]	FP	First published AI-based DR detection algorithm
Burlina et al. [5]	FP	Detect and refer patients with AMD
Ting et al. [6]	FP	Detect referable DR, vision-threatening DR, possible glaucoma, and AMD in multiethnic populations with diabetes
Lee et al. [7]	OCT	Classify AMD
Abràmoff et al. [8]	FP	First FDA approved DR diagnostic system
Grassmann et al. [9]	FP	Predict the stage of AMD
De Fauw et al. [10]	OCT	Segment features with expert performance, and classify OCT into 15 classes of diseases with expert performance on referral decisions
Sayre et al. [11]	FP	Improve the accuracy of, and confidence in, DR diagnosis in an assisted read setting
Peng et al. [12]	FP	Classify AMD severity
Guo et al. [13]	OCT angiography	Quantify superficial foveal avascular zone
Lo et al. [14]	OCT angiography	Microvasculature segmentation and intercapillary area quantification of the deep vascular complex
Heisler et al. [15]	OCT angiography	Detect DR
Varadarajan et al. [16]	FP, OCT	Predict OCT features from FP, Predict OCT-based DME features better than retinal specialists
Rim et al. [17]	OCT	Detect features associated with AMD in ethnically distinct data sets
Yim et al. [18]	OCT	Predict the progress to wet AMD
Moraes et al. [19]	OCT	Quantitative analysis for neovascular AMD
Brown et al. [20]	FP	Detect plus disease in ROP
Mulay et al. [21],	FP	Detect ROP in early stage
Waldstein et al. [22]	OCT	Evaluate the impact of vitreomacular adhesion for RVO treatment
Schlegl et al. [23]	OCT	Detect and quantify macular fluid for RVO

three-dimensional (3D) OCT images and corresponding automatic tissue maps [18]. Lastly, AI helps to predict treatment outcomes by quantifying segmented OCT features, such as neurosensory retina, drusen, intraretinal fluid, and subretinal fluid [19].

Artificial intelligence in retinopathy of prematurity

An ophthalmoscopic examination or digital FP can detect the stage and progression of ROP. However, diagnosis is often limited by a shortage of examiners and inconsistent examinations [27]. Recently AI systems have been able to diagnose plus disease, which is the most important feature of severe ROP [20]. AI has also shown its ability to grade ROP severity [21] or to classify zone of stage specifically [28]. Currently none of these systems is available for clinical use.

Artificial intelligence in retinal vein occlusion

AI models have shown good performance when recognizing different RVO types from normal and identifying RVO-related lesions using FP [29]. AI models also facilitate detection and quantification of RVO biomarkers by using OCT, such as macular fluid [23] and vitreomacular adhesion [22], which paves a way to evaluate the treatment of RVO.

Challenges and future

Despite the advances in ophthalmology, there are limitations that need to be overcome before AI can be fully applied to clinical practice [30].

- First, data collected with different devices, especially OCT/A, often come with different parameters and postprocessing procedures. While showing great accuracy during development, AI algorithms may produce inaccurate results when applied to other datasets.
- Second, the performance of AI model depends on the quality of the data that clinicians label to train the models. Data labeling is labor intensive, and inconsistent labeling can result in inaccurate AI models.
- Third, datasets often lack various features and are not representative of different ethnic groups. AI models trained on an ideal set of retinal images may still fail to produce accurate results for patients with different fundi pigmentation, poor dilation, and media opacity.
- Forth, data on rare diseases or conditions are not imaged regularly in clinical practice, which also results in a shortage of data sets when training AI models.
- Fifth, algorithms are often trained for a specific diseases and fail to incorporate the ambiguity and variability inherent in observation and decision making in clinical medicine.

In addition to these technical challenges, physicians' education is essential to the widespread adoption of AI in ophthalmology. For physicians, the path for AI to determine disease diagnosis and prognosis is unclear and is like a "black box." This also highlights the need for explainable AI. There is also a need to gain patient trust in the use of AI to make clinical decisions and to consider any ethical issues in terms of data privacy and safety.

Moving forward, the above considerations must be overcome to apply AI to real-world practice. Furthermore, AI will become even more powerful in advanced research and large scale clinical practice, if multimodal data are available as input, such as volumetric OCTA.

Conclusion

AI can help provide safe, personalized, efficient, and cost-effective ophthalmological practices. Recent regulatory guidelines have helped to ensure that thoughtful research are conducted and that they are commercialized. Commercial AI systems have been developed and validated for DR screening by using fundus photographs. Researches have demonstrated the ability of AI to diagnose and manage of eye diseases in OCT and OCTA. However, standardized collection and postprocessing procedures are necessary for OCT scans collected from different devices, and strong leadership is needed to develop guidelines for collecting data, training and adopting AI models.

References

[1] Lee R, Wong TY, Sabanayagam C. Epidemiology of diabetic retinopathy, diabetic macular edema and related vision loss. Eye Vis (Lond) 2015;2:17.
[2] Wong WL, Su X, Li X, et al. Global prevalence of age-related macular degeneration and disease burden projection for 2020 and 2040: a systematic review and meta-analysis. Lancet Glob Health 2014;2(2):e106–16.
[3] Gulshan V, Peng L, Coram M, et al. Development and validation of a deep learning algorithm for detection of diabetic retinopathy in retinal fundus photographs. JAMA 2016;316(22):2402–10.
[4] Wong TY, Scott IU. Clinical practice. Retinal-vein occlusion. N Engl J Med 2010;363:2135–44.
[5] Burlina PM, Joshi N, Pekala M, Pacheco KD, Freund DE, Bressler NM. Automated grading of age-related macular degeneration from color fundus images using deep convolutional neural networks. JAMA Ophthalmol 2017;135(11):1170–6.
[6] Ting DSW, Cheung CY-L, Lim G, et al. Development and validation of a deep learning system for diabetic retinopathy and related eye diseases using retinal images from multiethnic populations with diabetes. JAMA 2017;318(22):2211–23.
[7] Lee CS, Baughman DM, Lee AY. Deep learning is effective for the classification of OCT images of normal versus age-related macular degeneration. Ophthalmol Retina 2017;1:322–7.
[8] Abràmoff MD, Lavin PT, Birch M, Shah N, Folk JC. Pivotal trial of an autonomous AI-based diagnostic system for detection of diabetic retinopathy in primary care offices. npj Digit Med 2018;1:1.
[9] Grassmann F, Mengelkamp J, Brandl C, et al. A deep learning algorithm for prediction of age-related eye disease study severity scale for age-related macular degeneration from color fundus photography. Ophthalmology 2018;125(9):1410–20.
[10] De Fauw J, Ledsam JR, Romera-Paredes B, et al. Clinically applicable deep learning for diagnosis and referral in retinal disease. Nat Med 2018;24(9):1342–50.
[11] Sayres R, Taly A, Rahimy E, et al. Using a deep learning algorithm and integrated gradients explanation to assist grading for diabetic retinopathy. Ophthalmology 2019;126(4):552–64.
[12] Peng Y, Dharssi S, Chen Q, et al. DeepSeeNet: a deep learning model for automated classification of patient-based age-related macular degeneration severity from color fundus photographs. Ophthalmology 2019;126(4):565–75.
[13] Guo M, Zhao M, Cheong AMY, Dai H, Lam AKC, Zhou Y. Automatic quantification of superficial foveal avascular zone in optical coherence tomography angiography implemented with deep learning. Vis Comput Ind Biomed Art 2019;2(1):21.
[14] Lo J, Heisler M, Vanzan V, et al. Microvasculature segmentation and intercapillary area quantification of the deep vascular complex using transfer learning. Transl Vis Sci Technol 2020;9(2):38.
[15] Heisler M, Karst S, Lo J, et al. Ensemble deep learning for diabetic retinopathy detection using optical coherence tomography angiography. Transl Vis Sci Technol 2020;9(2):20.
[16] Varadarajan AV, Bavishi P, Ruamviboonsuk P, et al. Predicting optical coherence tomography-derived diabetic macular edema grades from fundus photographs using deep learning. Nat Commun 2020;11(1):1–8.

[17] Rim TH, Lee AY, Ting DS, Teo K, Betzler BK, Teo ZL, et al. Detection of features associated with neovascular age-related macular degeneration in ethnically distinct data sets by an optical coherence tomography: trained deep learning algorithm. Br J Ophthalmol 2020;78:7321–40.

[18] Yim J, Chopra R, Spitz T, et al. Predicting conversion to wet age-related macular degeneration using deep learning. Nat Med 2020;26:892–9.

[19] Moraes G, Fu DJ, Wilson M, et al. Quantitative analysis of optical coherence tomography for neovascular age-related macular degeneration using deep learning. Ophthalmology 2021;128(5):693–705.

[20] Brown JM, Campbell JP, Beers A, et al. Automated diagnosis of plus disease in retinopathy of prematurity using deep convolutional neural networks. JAMA Ophthalmol 2018;136:803–10.

[21] Mulay S., Ram K., Sivaprakasam M., Vinekar A. Early detection of retinopathy of prematurity stage using deep learning approach. Paper presented at SPIE medical imaging; 2019.

[22] Waldstein SM, Montuoro A, Podkowinski D, Philip AM, Gerendas BS, Bogunovic H, et al. Evaluating the impact of vitreomacular adhesion on anti-VEGF therapy for retinal vein occlusion using machine learning. Sci Rep 2017;7:2928.

[23] Schlegl T, et al. Fully automated detection and quantification of macular fluid in OCT using deep learning. Ophthalmology 2018;125:549–55.

[24] Raman R, Srinivasan S, Virmani S, Sivaprasad S, Rao C, Rajalakshmi R. Fundus photograph-based deep learning algorithms in detecting diabetic retinopathy. Eye (Lond) 2019;33(1):97–109.

[25] Abràmoff MD, Lou Y, Erginay A, et al. Improved automated detection of diabetic retinopathy on a publicly available dataset through integration of deep learning. Invest Ophthalmol Vis Sci 2016;57(13):5200–6.

[26] Buchan JC, Amoaku W, Barnes B, et al. How to defuse a demographic time bomb: the way forward? Eye (Lond) 2017;31(11):1519–22.

[27] Gilbert C. Retinopathy of prematurity: a global perspective of the epidemics, population of babies at risk and implications for control. Early Hum Dev 2008;84:77–82.

[28] Hu J, Chen Y, Zhong J, Ju R, Yi Z. Automated analysis for retinopathy of prematurity by deep neural networks. IEEE Trans Med Imaging 2019;38:269–79.

[29] Chen Q, Yu W, Lin S, Liu B, Wang Y, Wei Q, et al. Artificial intelligence can assist with diagnosing retinal vein occlusion. Int J Ophthalmol 2021;14(12):1895–902.

[30] Ting DSW, Pasquale LR, Peng L, Campbell JP, Lee AY, Raman R, et al. Artificial intelligence and deep learning in ophthalmology. Br J Ophthalmol 2019;103(2):167–75.

CHAPTER

11

Artificial intelligence in ophthalmology II: glaucoma

Lavanya Raghavan[1], *Ching-Yu Cheng*[1,2] *and Tien Yin Wong*[1,3]

[1]Singapore Eye Research Institute, Singapore National Eye Centre, Singapore [2]Department of Ophthalmology, Yong Loo Lin school of Medicine, National University of Singapore, Singapore [3]School of Clinical Medicine, Tsinghua Medicine, Tsinghua University, Beijing, P.R. China

Glaucoma is the leading cause of irreversible blindness worldwide [1], characterized by loss of retinal ganglion cell (RGC) axons, leading to progressive damage to the visual field (VF). Glaucoma is presently an incurable condition, but early detection and prompt treatment can slow the progression and prevent loss of vision. In clinical practice, glaucomatous damage can be detected and monitored by a complete eye examination and a plethora of structural and functional tests, repeated over several visits. These are subjective and can be challenging to interpret and largely depend on the clinician's expertise and experience [2]. Structural damage from glaucoma manifests itself as thinning of the retinal nerve fiber layer (RNFL, where the ganglion cell axons reside) and changes in the optic nerve head (ONH) parameters including optic cup-to-disk ratio (CDR). Higher CDR value indicates a higher probability of glaucoma. These changes are evaluated on stereoscopic fundus photographs and optical coherence tomography (SD-OCT) images (a 3D imaging modality that allows fast, high-resolution and noninvasive visualization of the retina and optic nerve head). Functional changes are assessed by VF analysis performed on standard automated perimetry. Rate of VF loss helps to guide how aggressively to treat glaucoma and also strongly correlates with quality of life measures.

Role of artificial intelligence in glaucoma

The last decade has seen rapid advances in artificial intelligence (AI) systems in glaucoma, with focus mainly on screening, detection and progression of the disease, using structural or functional tests or a combination of both [3–14]. These studies have utilized

Artificial Intelligence in Clinical Practice
DOI: https://doi.org/10.1016/B978-0-443-15688-5.00052-8

(1) traditional machine learning (ML) and/or deep learning (DL) approaches; (2) segmentation based and nonsegmentation-based techniques; (3) ML and DL classifiers, such as vector machines or convolutional neural networks (CNN); (4) public and private types of data sets; (5) based on full ONH image or optic disk (OD) or OD and optic cup or RNFL thickness and macular RGC plus inner plexiform layer thickness as a region-of-interest selection. Performance metrics were reported as the area under the receiver operating characteristic curve (AUC) or accuracy with sensitivity and specificity.

Using structural data as input

A recent systematic review and meta-analysis of the studies using fundus photographs and OCT images demonstrated that AI algorithms perform well for diagnosing glaucoma, with both modalities having similar accuracy (96.2% AUC and 96.0% AUC, respectively) [15]. Nonsegmentation-based approaches seem to have superior diagnostic performance compared to segmentation-based methods, which are easily affected by noise and low image quality [15]. RNFL segmentation errors are common, especially in myopic or glaucoma eyes with thin RNFL and are impacted by blood vessels and media opacity [16]. To overcome this, feature agnostic approaches directly from raw OCT volumes of ONH, using a 3D DL network (focusing on neuroretinal rim, OD area, the lamina cribrosa and its surrounding regions) were also explored to classify glaucomatous and healthy eyes (AUC: 0.94) [17].

Studies have demonstrated that the CNN-based architectures including Visual Geometry Group 19 (VGG19), Inception, and ResNet, which were pretrained on a large data set (over a million images with 1000 categories) and then fine-tuned using transfer learning [13,18], showed excellent performance. The ResNet-152 achieved a 3.57% error rate on ImageNet test data [19] and was the most frequently used DL architecture for transfer learning with high diagnostic performance in glaucoma [3–6,20–22].

Using functional data as input

Many studies analyzing VF data inputs for AI algorithms to detect glaucoma showed variable results, probably due to differing levels of test reliability and patient performance [23–25]. To predict progression and predict future VFs, studies have used deep variational auto-encoder [26], recurrent neural networks [27], to take into account the time component within the prediction, or archetypal analysis [28], to extract characteristics VF features before making a prediction.

There is now strong evidence that training an ML classifier model with combined structural and functional features could enhance discriminatory power compared to models trained with either structure or function alone [29]. Some researchers have suggested that VF loss progression could potentially be predicted from structural imaging, such as OCT [22,30] or fundus imaging [31]. Thompson et al. [32] presented a deep learning program that uses minimum optic nerve rim width relative to Bruch's membrane opening (BMO-MRW), when assessing photos of optic nerves of glaucomatous eyes, glaucoma suspects, and healthy eyes and showed that the predicted structural data correlated with VF loss (AUROC 0.945).

Current challenges and limitations

Glaucoma represents a spectrum of disease ranging from glaucoma suspects to preperimetric glaucoma to early manifest disease with VF changes and later advanced glaucoma. The definitions are not very clear and there is a lack of a true clear cut reference standard even among experienced ophthalmologists, that is objective, validated, and standardized globally to serve as a ground truth. This has remained a formidable challenge in applying AI to glaucoma. Also, glaucoma is a multifactorial disease, with several risk factors, such as ethnicity, age, family history, use of steroid medications, intraocular pressure, ocular factors, such as myopia, and associated systemic conditions, such as diabetes and hypertension. With AI assessment solutions, the diagnosis of glaucoma is solely based on the appearance of the ONH and RNFL, without considering others risk factors, and may lead to oversimplification of data.

Currently, the datasets used in many studies are small and collected from homogeneous populations. Modern AI systems require very large, diverse training and validation datasets with different population characteristics and phenotypes, encompassing the normal anatomical variations of ONH and RNFL values. This will require global collaborations to create programs that yield unbiased, reproducible, and accurate results, using standardized definitions. Finally, for clinical acceptance, AI models need to move from "black boxes" toward "explainable AI," and concerns about deskilling of clinicians needs to be overcome.

Future considerations

For translation into clinical practice, future studies should focus on the optimal imaging parameters with the best output for classification into various glaucoma stages and disease trajectories. Development of algorithms to evaluate the risk of progression may serve as an adjunct to assist clinical decision making, thereby also enhancing the efficiency, productivity, and quality of glaucoma care. Unsupervised AI techniques offer the potential of uncovering currently unrecognized patterns of disease. AI has potential to be used in virtual review clinics and telemedicine, thereby providing expert-level evaluation even in remote areas with limited access to healthcare.

Conclusion

Despite the challenges ahead, AI will have positive impact on research and clinical practice in glaucoma. Accurate automated diagnosis and prognosis of glaucoma may assist clinicians to personalize and improve overall quality of glaucoma treatments. By extracting meaningful information from high dimensional and complex multimodal data, AI may help to discover new biomarkers, patterns, or knowledge to improve the current understanding of glaucoma, which could promote new paradigms in treatment.

References

[1] Tham YC, Li X, Wong TY, Quigley HA, Aung T, Cheng CY. Global prevalence of glaucoma and projections of glaucoma burden through 2040: a systematic review and meta-analysis. Ophthalmology 2014;121 (11):2081–90.

[2] Schuman JS. Detection and diagnosis of glaucoma: ocular imaging. Invest Ophthalmol Vis Sci 2012;53:2488–90.

[3] Liu H, Li L, Wormstone IM, et al. Development and validation of a deep learning system to detect glaucomatous optic neuropathy using fundus photographs. JAMA Ophthalmol 2019;137:1353–60.

[4] Li F, Yan L, Wang Y, et al. Deep learning-based automated detection of glaucomatous optic neuropathy on color fundus photographs. Graefes Arch Clin Exp Ophthalmol 2020;258:4.

[5] Medeiros FA, Jammal AA, Thompson AC. From machine to machine: an OCT-trained deep learning algorithm for objective quantification of glaucomatous damage in fundus photographs. Ophthalmology 2019;126:513–21.

[6] Wang P, Shen J, Chang R, et al. Machine learning models for diagnosing glaucoma from retinal nerve fiber layer thickness maps. Ophthalmol Glaucoma 2019;2:6.

[7] Murtagh P, Greene G, O'Brien C. Current applications of machine learning in the screening and diagnosis of glaucoma: a systematic review and meta-analysis. Int J Ophthalmol 2020;13:149.

[8] Asaoka R, Murata H, Hirasawa K, et al. Using deep learning and transfer learning to accurately diagnose early-onset glaucoma from macular optical coherence tomography images. Am J Ophthalmol 2019;198:136–45.

[9] Camara J, Neto A, Pires IM, Villasana MV, Zdravevski E, Cunha A. Literature review on artificial intelligence methods for glaucoma screening, segmentation, and classification. J Imaging 2022;8(2):19.

[10] Shuldiner SR, Boland MV, Ramulu PY, De Moraes CG, et al. Predicting eyes at risk for rapid glaucoma progression based on an initial visual field test using machine learning. PLoS One 2021;16(4):e0249856.

[11] Akter N, Fletcher J, Perry S, et al. Glaucoma diagnosis using multi-feature analysis and a deep learning technique. Sci Rep 2022;12(1):8064.

[12] Schuman JS, Cadena M, McGee R, Al-Aswad LA, Medeiros FACollaborative Community on Ophthalmic Imaging Executive Committee and Glaucoma Workgroup. A case for the use of artificial intelligence in glaucoma assessment. Ophthalmol Glaucoma 2022;5(3):e3–e13.

[13] Gómez-Valverde J.J., et al. Automatic glaucoma classifcation using color fundus images based on convolutional neural networks and transfer learning. Biomed Opt Express 10:892–913.

[14] Asaoka R, et al. Validation of a deep learning model to screen for glaucoma using images from different fundus cameras and data augmentation. Ophthalmol Glaucoma 2019;2:224–31.

[15] Chaurasia AK, Greatbatch CJ, Hewitt AW. Diagnostic accuracy of artificial intelligence in glaucoma screening and clinical practice. J Glaucoma 2022;31(5):285–99.

[16] Zhixi L, He Y, Keel S, Meng W, Chang R, He M. Effcacy of a deep learning system for detecting glaucomatous optic neuropathy based on color fundus photographs. Ophthalmology. 2018;125(8):1199–206.

[17] Maetschke S, Antony BJ, Ishikawa H, et al. A feature agnostic approach for glaucoma detection in OCT volumes. PLoS One 2019;14:e0219126.

[18] Simonyan K., Zisserman A. Very deep convolutional networks for large-scale image recognition. <http://arxiv.org/abs/1409.1556>; 2014.

[19] Zhao B, Feng J, Wu X, et al. A survey on deep learning-based fine-grained object classification and semantic segmentation. Int J Autom Comput 2017;14:119–35.

[20] Thompson AC, Jammal AA, Berchuck SI, et al. Assessment of a segmentation-free deep learning algorithm for diagnosing glaucoma from optical coherence tomography scans. JAMA Ophthalmol 2020;138:333–9.

[21] Christopher M, Bowd C, Belghith A, et al. Deep learning approaches predict glaucomatous visual field damage from OCT optic nerve head en face images and retinal nerve fiber layer thickness maps. Ophthalmology. 2020;127:346–56.

[22] Medeiros FA, Jammal AA, Mariottoni EB, et al. Detection of progressive glaucomatous optic nerve damage on fundus photographs with deep learning. Ophthalmology. 2020;128:83–392.

[23] Li F, Wang Z, Qu G, et al. Automatic differentiation of glaucoma visual field from non-glaucoma visual field using deep convolutional neural network. BMC Med Imaging 2019;19(1):40.

[24] Yousefi S, Kiwaki T, Zheng Y, et al. Detection of longitudinal visual field progression in glaucoma using machine learning. Am J Ophthalmol 2018;193:71–9.

[25] Kucur ŞS, Holló G, Sznitman R. A deep learning approach to automatic detection of early glaucoma from visual felds. PLoS One 2018;13:e0206081.
[26] Berchuck SI, Mukherjee S, Medeiros FA. Estimating rates of progression and predicting future visual fields in glaucoma using a deep variational autoencoder. Sci Rep 2019;9:18113.
[27] Park K, Kim J, Lee J. Visual field prediction using recurrent neural network. Sci Rep 2019;9:8385.
[28] Wang M, Tichelaar J, Pasquale LR, et al. Characterization of central visual field loss in end-stage glaucoma by unsupervised artificial intelligence. JAMA Ophthalmol 2020;138:190–8.
[29] Thompson AC, Jammal AA, Medeiros FA. A review of deep learning for screening, diagnosis, and detection of glaucoma progression. Transl Vis Sci Technol 2020;9:42.
[30] Xu L, Asaoka R, Kiwaki T, Murata H, et al. Predicting the glaucomatous central 10-degree visual field from optical coherence tomography using deep learning and tensor regression. Am J Ophthalmol 2020;218:304–13.
[31] Thakur A, Goldbaum M, Yousefi S. Predicting glaucoma before onset using deep learning. Ophthalmol Glaucoma 2020;3(4):262–8.
[32] Thompson AC, Jammal AA, Medeiros FA. A deep learning algorithm to quantify neuroretinal rim loss from optic disc photographs. Am J Ophthalmol 2019;201:9–18.

CHAPTER

12

Artificial intelligence in ophthalmology III: systemic disease prediction

An Ran Ran[1], *Herbert Y.H. Hui*[1], *Carol Y. Cheung*[1] *and Tien Yin Wong*[2,3]

[1]Department of Ophthalmology and Visual Sciences, The Chinese University of Hong Kong, Hong Kong, P.R. China [2]School of Clinical Medicine, Tsinghua Medicine, Tsinghua University, Beijing, P.R. China [3]Singapore Eye Research Institute, Singapore National Eye Centre, Singapore

Introduction

Many systemic diseases exhibit pathological changes that manifest in the eye, particularly in the retina which is accessible to direct noninvasive visualization and shares similar anatomical and physiological characteristics with microcirculation in the body. For over decades, researchers have sought to measure retinal changes in identifying ocular biomarkers for systemic diseases, newly termed *oculomics* [1]. Retinal imaging modalities, such as fundus photography, optical coherence tomography (OCT), and OCT angiography (OCT-A), are unique tools for studying "eye–body" research, based on a widely accepted concept that retinal vascular health mirrors other vascular beds (e.g., cerebral and coronary) and retinal neuronal cells display typical properties of central nervous system (CNS) neurons with its axons (optic nerve) joining the CNS.

As big multimodal ocular image datasets are increasingly available and complex, the use of artificial intelligence (AI), especially deep learning (DL), is useful to be integrated and scrutinized for associations between the eye and systemic diseases, aiming to improve disease prediction and risk stratification. This chapter aims to summarize the application of AI in studying the eye–body relationship which may provide novel insights on potential retinal biomarkers for detecting common systemic diseases.

DOI: https://doi.org/10.1016/B978-0-443-15688-5.00041-3

Artificial intelligence-based ocular image analysis for cardiovascular disease

Studies have demonstrated parallel pathological changes morphologically between the retinal vasculature and the coronary-cerebral vasculatures [2–5]. Since changes in the microvascular circulation precede clinical manifestation, retinal vascular signs have been shown to act as an independent predictor for cardiovascular vascular disease (CVD) events such as myocardial infarction, stroke, and CVD mortality [6–19]. Current AI-based retinal photograph analyses are promising for predicting CVD with different kinds of outcomes, including CVD risk factors [20–23] (e.g., age, sex, smoking status, blood pressure, diabetes), direct CVD event [20,21,24] (e.g., atherosclerosis and CVD mortality), retinal features [25–27] (e.g., retinal vessel caliber and retinopathy), and CVD biomarkers [28–30] (e.g., coronary artery calcium score). For instance, Poplin et al. have developed a DL algorithm that predicts multiple CVD risk factors, such as age, sex, smoking status, HbA_{1c} level, systolic blood pressure, and body mass index, as well as major adverse cardiac events solely from retinal photographs [20]. Besides the prediction of CVD variables, systems such as the RetiAGE were also developed to stratify morbidity and mortality risk directly [21]. In terms of retinal features, the development of the Singapore I Vessel Assessment DL system provides automated computation of retinal vessel calibers from retinal photographs, showing that such caliber measurements are correlated with CVD risk factors and associated with incident CVD events [31]. Other applications are currently under development, such as the StrokeSave platform which combines retinal photographs, heart rate, blood pressure, blood oxygen, facial images, and vocal recordings for a comprehensive, mobile-phone-based stroke evaluation [32].

Artificial intelligence-based ocular image analysis for renal diseases

Retinal vascular changes are also associated with chronic kidney disease (CKD), that is, renal dysfunction and reduced estimated glomerular filtration rate (eGFR) [33,34]. Traditional community-based screening remains ineffective, as eGFR can only be estimated through the serum creatinine level obtained with invasive measures. Therefore, Sabanayagam et al. developed a DL algorithm to estimate renal function through retinal vascular signs, such as retinopathy, arteriolar narrowing, and venular dilatation in community-based populations [35]. AI-based retinal photograph analysis can detect CKD solely with knowledge of specific retinal signs (e.g., retinopathy) or simple risk factor information, which supports the potential of its utilization in specific settings (e.g., community-based, primary care, or high-risk groups). It performed similarly in subgroups of participants with diabetes and hypertension [35]. Another study found that among patients with diabetes or elevated HbA_{1c} level, an AI algorithm for early CKD detection achieved superior performance [36], suggesting that the concurrent existence of other systemic diseases may affect CKD detection from retinal photographs. Besides binary classification of CKD, the AI algorithm can also predict the continuous value of eGRF from retinal photographs and CKD development in longitudinal cohorts. Zhang et al. utilized a

DL model to predict eGFR solely from retinal photographs. Prediction of CKD progression was also performed by integrating retinal photographs with clinical metadata, which achieved significantly better performance than the model using only metadata [23]. These studies illustrate the feasibility of AI-based retinal image analysis as a non-invasive tool for CKD screening and prognosis prediction.

Artificial intelligence-based ocular image analysis for neurological diseases

The retina has long been considered a platform to study neurological disorders, due to its embryological, anatomical, and physiological similarities with the brain [37,38]. Alzheimer's disease (AD) [39], the main cause of dementia, can be exhibited in neural structures of the retina, as studies demonstrated thinning of the retinal nerve fiber layer and macular ganglion cell-inner plexiform layer from OCT images [40], mirroring neuronal lost in the brain. Retinal vascular changes are also observed, with narrowing of vessel caliber and an increase in tortuosity on the retinal photographs in AD patients [41], along with alterations in the foveal avascular zone on the OCT-A images in preclinical AD patients [42]. Based on these associations, Tian et al. have developed a highly modular AI algorithm that enables automated image selection, vessel segmentation, and classification of AD, achieving an accuracy of over 80% [43]. Wisely et al. recently proposed a DL system to predict AD using images and measurements from multiple ocular imaging modalities (OCT, OCT-A, ultra-widefield retinal photography, and retinal autofluorescence) [44]. These studies provided evidence for using AI-based ocular image analysis for AD detection. Neurological diseases can also cause papilledema which is detectable with high accuracy by AI-based retinal photographs analysis [45].

Artificial intelligence-based ocular image analysis for other systemic diseases

Although conspicuous eye—body relations are observed in some systemic diseases, systemic biomarkers remain obscure in ocular images. For example, Rim et al. found that the DL algorithm had a suboptimal predictive performance of two diabetes-related biomarkers (i.e., HbA_{1c} and fasting glucose) from retinal photographs. Lipid-related biomarkers such as high-density lipoprotein, cholesterol, and triglyceride also showed unsatisfactory results, showing that more work is needed to investigate the relationship of metabolic-related biomarkers with the retina [46]. Despite the inadequacy of AI-based ocular image analysis in predicting systemic biomarkers, Zhang et al. demonstrated the potential of AI in identifying and predicting the progression of type 2 diabetes mellitus (T2DM). Risk scores were extracted from retinal photographs and analyzed simultaneously with clinical metadata for identifying T2DM, as well as predicting its progression [23]. Besides metabolic-related diseases, AI-based ocular image analysis may be able to identify specific ocular features related to other systemic diseases. For example, hepatobiliary disease has been traditionally linked with ocular changes, such as jaundice, gaze palsy, nystagmus, or xanthelasma. To identify clinical independent diagnostic features in the eye, Xiao et al., have studied the application of DL on two common ocular imaging modalities, slit lamp and retinal photographs, for the diagnosis of seven categories of hepatobiliary diseases [47]. The ability of DL to identify abnormal features that is not perceivable even by experienced clinicians can potentially enhance our understanding of the pathophysiological mechanisms of systemic diseases.

Current challenges

Although the future of AI-based systemic disease detection seems promising, there are still several challenges that have yet to be addressed. First, the training dataset's inherent selection bias can affect the DL models' generalizability. For example, in the study by Lim et al., their model did not perform well in the European population for predicting body composition from retinal photographs, as the model was trained on the Asian population [46]. Second, the training and validation of DL algorithms for disease prediction generally require extensive datasets with high-quality images, which is labor-intensive and costly if built from scratch, while sharing of existing datasets is limited by legal barriers [48]. Third, concomitant retinal pathologies may diminish the accuracy of systemic disease prediction. In fact, many studies often exclude patients with pathologies that severely disrupt the retinal structure, such as diabetic retinopathy and macular degeneration, thus limiting its transition into clinical use where many patients suffer from comorbidities. Fourth, the "black box" issue of DL leaves uncertainties in understanding the eye–body relations and diminishes the credibility of identified ocular biomarkers in systemic diseases.

Future directions

To accelerate the implementation of AI into clinical settings, a few directions should be emphasized. First, local and international collaboration is encouraged to improve the availability of extensive and accessible datasets, as well as to increase the generalizability of AI algorithms and facilitate the comparison of algorithms across studies in an objective manner [49]. Second, advanced techniques such as federated learning [50] and block-chain [51] should be used to ensure privacy-preserving collaborations. Third, the multitask technique is potentially useful in screening as it can provide multiple outcomes when subjects are affected with concurrent systemic diseases or retinal pathologies. Fourth, more research on the visualization of ocular biomarkers identified by the AI algorithms should be directed, thus enhancing the interpretability and understanding of the eye–body relationship, ultimately providing more incentive for clinicians to adopt AI-based ocular image analysis for systemic disease detection in their daily work. Fifth, to ensure the true utility of AI-based ocular image analysis, more prospective studies combined with portable devices such as smartphone-based retinal imaging [23] should be conducted, such that improvements can be made by encountering real-world data and improve clinical feasibility.

Major takeaway points

- AI-based ocular image analysis is promising in providing a noninvasive tool for the prediction of various systemic diseases, through the evaluation of risk factors, retinal features and biomarkers.
- Most existing AI algorithms focus on retinal photography-based systemic disease detection. Other retinal imaging modalities, such as OCT and OCT-A, hold great potential for further research.
- It is important to visualize the specific ocular features identified by AI-based ocular image analysis for systemic diseases and provide a more comprehensive understanding of eye–body relations.
- Multitask and multimodal AI algorithms should be established to improve the accuracy of systemic disease identification (e.g., concurrent systemic diseases and eye pathologies).
- Combining the use of portable devices and telemedicine, AI-based ocular image analysis can potentially facilitate systemic disease screening in the general population and less-developed areas, enabling more patients to be diagnosed and treated at an earlier stage.

References

[1] Wagner SK, Fu DJ, Faes L, Liu XX, Huemer J, Khalid H, et al. Insights into systemic disease through retinal imaging-based oculomics. Transl Vis Sci Technol 2021;10(8).
[2] Flammer J, Konieczka K, Bruno RM, Virdis A, Flammer AJ, Taddei AS, et al. The eye and the heart. Eur Heart J 2013;34(17):1270–8.
[3] Liew G, Wang JJ. Retinal vascular signs: a window to the heart? Rev Esp Cardiol 2011;64(6):515–21.
[4] Taylor R. AI and the retina: finding patterns of systemic disease. EyeNet magazine. American Academy of Ophthalmology; 2021.
[5] Wagner SK, Fu DJ, Faes L, Liu X, Huemer J, Khalid H, et al. Insights into systemic disease through retinal imaging-based oculomics. Transl Vis Sci Technol 2020;9(2):6.
[6] Wong TY, Mohamed Q, Klein R, Couper DJ. Do retinopathy signs in non-diabetic individuals predict the subsequent risk of diabetes? Br J Ophthalmol 2006;90(3):301–3.
[7] Günthner R, Hanssen H, Hauser C, Angermann S, Lorenz G, Kemmner S, et al. Impaired retinal vessel dilation predicts mortality in end-stage renal disease. Circ Res 2019;.
[8] McGeechan K, Liew G, Macaskill P, Irwig L, Klein R, Klein BEK, et al. Meta-analysis: retinal vessel caliber and risk for coronary heart disease. Ann Intern Med 2009;151(6):404–13.
[9] Wong TY, Mitchell P. Hypertensive retinopathy. N Engl J Med 2004;351(22):2310–17.
[10] Wong TY, Klein R, Couper DJ, Cooper LS, Shahar E, Hubbard LD, et al. Retinal microvascular abnormalities and incident stroke: the atherosclerosis risk in communities study. Lancet 2001;358(9288):1134–40.
[11] Kawasaki R, Xie J, Cheung N, Lamoureux E, Klein R, Klein BE, et al. Retinal microvascular signs and risk of stroke: the Multi-Ethnic Study of Atherosclerosis (MESA). Stroke 2012;43(12):3245–51.
[12] Cheung CY, Tay WT, Ikram MK, Ong YT, De Silva DA, Chow KY, et al. Retinal microvascular changes and risk of stroke: the Singapore Malay Eye Study. Stroke 2013;44(9):2402–8.
[13] Yatsuya H, Folsom AR, Wong TY, Klein R, Klein BE, Sharrett AR, et al. Retinal microvascular abnormalities and risk of lacunar stroke: atherosclerosis risk in communities study. Stroke 2010;41(7):1349–55.
[14] Seidelmann SB, Claggett B, Bravo PE, Gupta A, Farhad H, Klein BE, et al. Retinal vessel calibers in predicting long-term cardiovascular outcomes: the atherosclerosis risk in communities study. Circulation 2016;134 (18):1328–38.

[15] Wang J, Leng F, Li Z, Tang X, Qian H, Li X, et al. Retinal vascular abnormalities and their associations with cardiovascular and cerebrovascular diseases: a Study in rural southwestern Harbin, China. BMC Ophthalmol 2020;20(1):136.
[16] Wieberdink RG, Ikram MK, Koudstaal PJ, Hofman A, Vingerling JR, Breteler MM, et al. Retinal vascular calibers and the risk of intracerebral hemorrhage and cerebral infarction: the Rotterdam Study. Stroke 2010;41 (12):2757–61.
[17] Ikram MK, de Jong FJ, Bos MJ, Vingerling R, Hofman A, Koudstaal PJ, et al. Retinal vessel diameters and risk of stroke: the Rotterdam Study. Neurology 2006;66(9):1339–43.
[18] Lindley RI, Wang JJ, Wong MC, Mitchell P, Liew G, Hand P, et al. Retinal microvasculature in acute lacunar stroke: a cross-sectional study. Lancet Neurol 2009;8(7):628–34.
[19] Betzler BK, Sabanayagam C, Tham YC, Cheung CY, Cheng CY, Wong TY, et al. Retinal vascular profile in predicting incident cardiometabolic diseases among individuals with diabetes. Microcirculation 2022;e12772.
[20] Poplin R, Varadarajan AV, Blumer K, Liu Y, McConnell MV, Corrado GS, et al. Prediction of cardiovascular risk factors from retinal fundus photographs via deep learning. Nat Biomed Eng 2018;2(3):158–64.
[21] Nusinovici S, Rim TH, Yu M, Lee G, Tham YC, Cheung N, et al. Retinal photograph-based deep learning predicts biological age, and stratifies morbidity and mortality risk. Age Ageing 2022;51:4.
[22] Vaghefi E, Yang S, Hill S, Humphrey G, Walker N, Squirrell D, et al. Detection of smoking status from retinal images; a Convolutional Neural Network study. Sci Rep 2019;9(1):7180.
[23] Zhang K, Liu X, Xu J, Yuan J, Cai W, Chen T, et al. Deep-learning models for the detection and incidence prediction of chronic kidney disease and type 2 diabetes from retinal fundus images. Nat Biomed Eng 2021;5 (6):533–45.
[24] Chang J, Xu D, Cheng CY, Sabanayagam C, Tham YC, Yu M, et al. Association of cardiovascular mortality and deep learning-funduscopic atherosclerosis score derived from retinal fundus images. Am J Ophthalmol 2020;217:121–30.
[25] Cheung CY, Xu D, Cheng CY, Sabanayagam C, Tham YC, Yu M, et al. A deep-learning system for the assessment of cardiovascular disease risk via the measurement of retinal-vessel calibre. Nat Biomed Eng 2021;5 (6):498–508.
[26] Zekavat SM, Raghu VK, Trinder M, Ye Y, Koyama S, Honigberg MC, et al. Deep learning of the retina enables phenome- and genome-wide analyses of the microvasculature. Circulation 2022;145(2):134–50.
[27] Lim, G, Lim ZW, Xu D, Ting DSW, Wong TY, Lee ML, et al. Feature isolation for hypothesis testing in retinal imaging: an ischemic stroke prediction case study. In: Proceedings of the AAAI conference on artificial intelligence; 2019. p. 9510–15.
[28] Son J, Shin JY, Chun EJ, Jung KH, Park KH, Park SJ, et al. Predicting high coronary artery calcium score from retinal fundus images with deep learning algorithms. Transl Vis Sci Technol 2020;9(6):28.
[29] Rim TH, Lee CJ, Tham Yc, Cheung N, Yu M, Lee G, et al. Deep-learning-based cardiovascular risk stratification using coronary artery calcium scores predicted from retinal photographs. Lancet Digit Health 2021;3(5):e306–16.
[30] Diaz-Pinto A, Ravikumar N, Attar R, Suinesiaputra A, Zhao Y, Levelt E, et al. Predicting myocardial infarction through retinal scans and minimal personal information. Nat Mach Intell 2022;4(1):55–61.
[31] Huang QF, Wei FF, Zhang ZY, Raaijmakers A, Asayama K, Thijs L, et al. Reproducibility of retinal microvascular traits decoded by the Singapore I Vessel Assessment Software across the human age range. Am J Hypertens 2018;31(4):438–49.
[32] Gupta A. StrokeSave: a novel, high-performance mobile application for stroke diagnosis using deep learning and computer vision; 2019.
[33] Lim LS, Cheung CY, Sabanayagam C, Lim SC, Tai ES, Huang L, et al. Structural changes in the retinal microvasculature and renal function. Invest Ophthalmol Vis Sci 2013;54(4):2970–6.
[34] Ooi QL, Tow FK, Deva R, Alias MA, Kawasaki R, Wong TY, et al. The microvasculature in chronic kidney disease. Clin J Am Soc Nephrol 2011;6(8):1872–8.
[35] Sabanayagam C, Xu D, Ting DSW, Nusinovici S, Banu R, Hamzah H, et al. A deep learning algorithm to detect chronic kidney disease from retinal photographs in community-based populations. Lancet Digital Health 2020;2(6):e295–302.
[36] Kang EY-C, Hsieh Y-T, Li C-H, Huang Y-J, Kou C-F, Kang J-H, et al. Deep learning-based detection of early renal function impairment using retinal fundus images: model development and validation. JMIR Med Inform 2020;8(11):e23472.

[37] London A, Benhar I, Schwartz M. The retina as a window to the brain—from eye research to CNS disorders. Nat Rev Neurol 2013;9(1):44–53.
[38] Cheung CY, Mok V, Foster PJ, Trucco E, Chen C, Wong TY, et al. Retinal imaging in Alzheimer's disease. J Neurol Neurosurg Psychiatry 2021;92(9):983–94.
[39] Scheltens P, De Strooper B, Kivipelto M, Holstege H, Chetelat G, Teunissen CE, et al. Alzheimer's disease. Lancet 2021;397(10284):1577–90.
[40] Chan VTT, Sun Z, Tang S, Chen LJ, Wong A, Tham CC, et al. Spectral-domain OCT measurements in Alzheimer's disease: a systematic review and meta-analysis. Ophthalmology 2019;126(4):497–510.
[41] Cheung CY, Ong YT, Ikram MK, Ong SY, Li X, Hilal S, et al. Microvascular network alterations in the retina of patients with Alzheimer's disease. Alzheimers Dement 2014;10(2):135–42.
[42] O'Bryhim BE, Apte RS, Kung N, Coble D, Van Stavern GP, et al. Association of preclinical alzheimer disease with optical coherence tomographic angiography findings. JAMA Ophthalmol 2018;136(11):1242–8.
[43] Tian J, Smith G, Guo H, Liu B, Pan Z, Wang Z, et al. Modular machine learning for Alzheimer's disease classification from retinal vasculature. Sci Rep 2021;11(1):238.
[44] Wisely CE, Najjar RP, Zhubo J, Ting D, Vasseneix C, Xu X, et al. Convolutional neural network to identify symptomatic Alzheimer's disease using multimodal retinal imaging. Br J Ophthalmol 2022;106(3):388–95.
[45] Milea D, Najjar RP, Zhubo J, Ting D, Vasseneix C, Xu X, et al. Artificial intelligence to detect papilledema from ocular fundus photographs. N Engl J Med 2020;382(18):1687–95.
[46] Rim TH, Lee G, Kim Y, Tham YC, Lee CJ, Baik SJ, et al. Prediction of systemic biomarkers from retinal photographs: development and validation of deep-learning algorithms. Lancet Digit Health 2020;2(10): e526–36.
[47] Xiao W, Huang X, Wang JH, Lin DR, Zhu Y, Chen C, et al. Screening and identifying hepatobiliary diseases through deep learning using ocular images: a prospective, multicentre study. Lancet Digital Health 2021;3(2): e88–97.
[48] Maurovich-Horvat P. Current trends in the use of machine learning for diagnostics and/or risk stratification in cardiovascular disease. Cardiovascular Res 2021;117(5):e67–9.
[49] Kelly CJ, Karthikesalingam A, Suleyman M, Corrado G, King D. Key challenges for delivering clinical impact with artificial intelligence. BMC Med 2019;17(1):195.
[50] Xu J, Glicksberg BS, Su C, Walker P, Bian J, Wang F. Federated learning for healthcare informatics. J Healthc Inform Res 2021;5(1):1–19.
[51] Tan TE, Anees A, Chen C, Li S, Xu X, Li Z, et al. Retinal photograph-based deep learning algorithms for myopia and a blockchain platform to facilitate artificial intelligence medical research: a retrospective multicohort study. Lancet Digit Health 2021;3(5):e317–29.

CHAPTER

13

Artificial intelligence in respiratory medicine

Sherif Gonem[1,2]

[1]Department of Respiratory Medicine, Nottingham University Hospitals NHS Trust, Nottingham, United Kingdom [2]Division of Respiratory Medicine, University of Nottingham, Nottingham, United Kingdom

Introduction discussing the role of artificial intelligence, including a review of the landmark trials, within your field

Advances in artificial intelligence (AI) within respiratory medicine over the past decade have focused on achieving more accurate diagnosis and phenotyping of respiratory conditions, as well as on predicting the risk of future events occurring over both long and short timescales. In common with many other medical specialties, the advent of deep neural networks (DNNs) coupled with the availability of large training datasets has had a transformative effect on AI research within respiratory medicine. A wide variety of inputs have been used to develop AI models including radiological imaging, molecular and histopathology, physiological measurements, and clinical/demographic data. Several AI models are now being deployed clinically within interventional trials, showing that the implementation gap between model development and clinical utilization is starting to be bridged. Key recent advances in the use of AI within respiratory medicine include:

(1) Chest radiograph (CXR) interpretation

The CXR is one of the most commonly requested radiological investigations, and one which provides a wealth of diagnostic information within all fields of medicine. Therefore it is not surprising that a large amount of research has been devoted to developing AI tools that can help to interpret CXRs. Hwang et al. developed an all-purpose AI algorithm for the diagnosis of a range of common CXR abnormalities, and demonstrated superior diagnostic accuracy compared to expert radiologists [1]. More specific CXR interpretation applications that have been successfully developed include differentiating COVID-19 pneumonia from other viral and nonviral

Artificial Intelligence in Clinical Practice
DOI: https://doi.org/10.1016/B978-0-443-15688-5.00027-9

pneumonias [2], diagnosing acute respiratory distress syndrome [3], and detecting malignant pulmonary nodules [4]. Two recent studies evaluated commercially available AI algorithms for diagnosing tuberculosis (TB) on CXRs, in 23,954 individuals attending a TB screening clinic in Dhaka, Bangladesh [5], and in 2198 adults attending the TB clinic at the Indus Hospital, Karachi, Pakistan [6]. Two algorithms (qXR and CAD4TB) were found to meet, or be statistically noninferior to, the World Health Organisation target standards for triage tests (sensitivity of $\geq 90\%$ together with specificity $\geq 70\%$).

(2) Computed tomography (CT) scan interpretation

CT scans are less widely available than CXRs but provide a more detailed assessment of a variety of thoracic conditions including lung cancer, interstitial lung disease, airways disease and pulmonary infection. Ardila et al. developed an AI system which out-performed radiologists in accurately diagnosing lung cancer in CT scans taken as part of a large lung cancer screening trial [7]. An important problem in the field of thoracic radiology is to determine the malignant potential of indeterminate pulmonary nodules (IPNs). Massion et al. developed a convolutional neural network model which classified IPNs into low or high-risk categories with greater accuracy than current risk prediction models [8]. This model was externally validated using data from three UK centers, once again showing an advantage in accuracy compared to the currently used Brock model [9].

Walsh et al. reported a DNN which could, with human-level accuracy, distinguish idiopathic pulmonary fibrosis from other interstitial lung diseases on chest CT scans [10], while Huang et al. developed a DNN which could detect pulmonary emboli with moderate accuracy [11]. A number of groups have demonstrated that DNNs can detect and stage the severity of chronic obstructive pulmonary disease [12–14]. The COVID-19 pandemic triggered a large research effort to develop improved diagnostic tools, and a number of research groups developed AI models that could distinguish between chest CT scans of patients with COVID-19 pneumonia and non-COVID-19 pneumonia, with equal or greater accuracy than expert thoracic radiologists [15–19].

(3) Histological diagnosis of lung cancer

Lung biopsies are most commonly used to diagnose lung cancer subtypes. Coudray et al. trained a DNN to distinguish between adenocarcinoma and squamous cell carcinoma on digital histopathology images with high accuracy, comparable to that of human pathologists [20]. The model could also detect common mutations that have therapeutic significance, with moderate accuracy.

(4) Pulmonary function test (PFT) interpretation

The standard battery of PFTs comprises spirometry, static lung volumes and diffusing capacity of the lungs. These tests produce a large number of numerical outputs which require considerable expertise to interpret. Topalovic et al. developed and validated an AI model which classified PFT results into one of nine broad diagnostic categories, and demonstrated that the model significantly out-performed even expert pulmonologists in this task [21].

(5) Predicting lung cancer risk

The early detection of future events or occult diagnoses is an important application of AI, particularly in cases where early intervention leads to improved clinical outcomes. Gould et al. developed a machine learning model using routinely collected clinical and demographic data to predict the risk of developing non-small-cell lung cancer (NSCLC) [22]. They demonstrated that their model is more accurate for predicting a future diagnosis of NSCLC than the standard eligibility criteria for lung cancer screening, as well as previously published cancer risk models.

(6) Risk stratification of patients with COVID-19 infection

The emergence of the COVID-19 pandemic led to a number of machine learning models being developed to predict severe disease in patients presenting to hospital with COVID-19 [23–29]. Dayan et al. used a federated learning approach, allowing them to train their AI model using data from 20 institutions worldwide, but without the requirement for data sharing [23]. They developed a model that accurately predicted the oxygen requirements of patients with COVID-19 at 24 and 72 hours after initial presentation, using inputs of vital signs, laboratory data and chest radiographs. The model maintained its accuracy when it was externally validated in three further test sites.

Current examples of artificial intelligence implementation in clinical practice (published works to date using artificial intelligence in your area of specialty)

Clinical implementation of AI models within respiratory medicine has so far focused on radiological imaging applications. Seah et al. assessed the accuracy of radiologist diagnosis of 127 CXR abnormalities with and without the assistance of an AI model, and found significant improvements in accuracy for the majority of findings when the AI model was used [30]. Jones et al. integrated the same AI model into the real-life clinical workflow of 11 radiologists [31]. Out of 2972 cases reviewed with the model, 3.1% of reports were materially changed due to the AI interpretation, and patient management was potentially altered as a result in 1.4% of cases. However, 13% of cases had one or more findings of the AI model rejected by the reporting radiologist, highlighting the importance of combining AI model outputs with human expertise. Indeed, it was shown by Patel et al. that a human-AI partnership achieved higher accuracy for the diagnosis of pneumonia than either acting alone [32].

Schmuelling et al. coupled an AI-based system for detecting pulmonary embolism on CT pulmonary angiograms with an electronic notification system to relay this information in real time to clinicians in the Emergency Department [33]. The investigators confirmed the technical feasibility of implementing their system, but there were no significant improvements in clinical performance measures, such as radiology report reading time, radiology report communication time, time to anticoagulation, and patient turnaround time in the Emergency Department. This showed that merely introducing an AI tool into the clinical workflow is not sufficient to improve clinical outcomes without accompanying changes in clinical processes and procedures.

Pierce et al. integrated an AI algorithm for detecting pneumothorax into a mobile CXR scanner, and published their experience of implementing this into their clinical workflows [34]. Studies that were flagged by the AI as suspicious for pneumothorax were prioritized for review by the reporting radiologist. In a number of cases this led to expedited patient management, particularly when studies were performed outside usual working hours and would not otherwise have been reported until the following day. Similarly, Hong et al. integrated an AI tool for detecting pneumothorax into their clinical workflows for patients undergoing CT-guided lung biopsy procedures [35]. They found that accuracy for detecting postprocedure pneumothorax improved following introduction of the AI-tool compared to the period prior to its introduction. Hwang et al. compared the diagnostic yield for pulmonary metastases on CXR before and after the implementation of a computer-aided detection tool, and found a significantly greater rate of detection postimplementation, without a significant increase in the false referral rate [36].

Current challenges preventing artificial intelligence application

Implementation of AI within respiratory medicine faces a number of challenges, many of which are in common with other medical specialties. Most AI algorithms that are developed in a single center do not progress to external validation, and those that do are often found to be less accurate when tested in new settings, or alternatively their accuracy can wane over time due to changes in patient characteristics or administrative coding [37]. Algorithmic bias is a further important problem that needs to be addressed to ensure the confidence of the medical community and general public in AI models. Seyyed-Kalantari et al. found that AI models trained on three public CXR datasets systematically underdiagnosed patients who were female, Black or Hispanic, young, and of lower socioeconomic status [38]. Such biases raise particular ethical concerns if AI algorithms do not have the capacity to explain their outputs, or if they are being used to make important medical decisions without human input, such as selecting which patients will be entered into cancer screening programs. It has been suggested that AI algorithms should undergo a rigorous and systematic audit prior to and during their deployment in a new setting, including review of false positive and false negative cases, subgroup testing, and adversarial testing [39].

Practical deployment of AI within real-life clinical workflows comes with both technical and socio-cultural challenges. Successful implementation requires careful planning and collaboration with multiple stakeholders including clinicians, information technology specialists and managers [40]. Clinicians need to be convinced that AI algorithms are reliable and that they will add value to patient care, without increasing their workload [41]. This may be easier to achieve if AI algorithms are able to provide a human-understandable explanation for their output, as clinicians are understandably reluctant to accept recommendations from an opaque "black box." Producing AI models that are interpretable or explainable is an active field of current research [42].

Future directions with an emphasis on how artificial intelligence can help solve specific problems within your field

Developing AI models requires large training datasets which are often not available from a single center, so future progress is likely to require large multicenter collaborations to assemble datasets of the required size. Examples of how this can work in practice include the establishment of the UK National COVID-19 Chest Imaging Database [43,44]. This is a national repository of CXR, CT, and MRI scans of patients with COVID-19 infection, together with a COVID-negative control population, assembled for the purpose of research. This initiative was facilitated by emergency legislation to temporarily ease data sharing restrictions and support scientific collaboration in response to the COVID-19 pandemic. However, under normal circumstances, such a project would have taken months or even years to set up, due to the regulatory barriers to data sharing. There is increasing interest in training AI models using federated or swarm learning, in which the training of AI models is distributed among multiple sites, without the need for data to be shared between centers. This approach has been used to train AI models to recognize COVID-19 changes on chest CT scans [45] and to identify TB using blood transcriptomes or CXRs [46]. A further barrier to AI model training is the need for labor-intensive annotation of cases by medical experts, such as radiologists or histopathologists. This may be mitigated by automated methods using natural language processing of radiology reports. Bressem et al. developed an AI model to extract diagnostic labels from free-text CXR reports with high accuracy [47]. Similarly, Weikert et al. reported an AI model that accurately classified CT pulmonary angiogram reports as indicating the presence or absence of a pulmonary embolism [48].

A number of novel AI applications within respiratory medicine are still at an early stage of development but show promise for the future. These include the prediction of respiratory decompensation events, both for patients at home and those in hospital. Zhang et al. developed AI models for predicting asthma exacerbations using home monitoring of asthma symptoms and peak expiratory flow [49]. Similar systems have been developed for predicting exacerbations of chronic obstructive pulmonary disease using home telemonitoring [50,51]. Prospective studies are needed to assess whether using predictive strategies such as these will result in improved clinical outcomes, through earlier detection and treatment of exacerbations. Richards et al. used an AI model based on wearable sensor data, measuring parameters including heart rate, respiratory rate and step count, to predict hospital admission in patients with COVID-19 diagnosed in the community [52]. Un et al. used a similar wearable sensor to predict clinical deterioration events and length of stay in patients already hospitalized with COVID-19 [53]. Further promising applications of AI within respiratory medicine include automated interpretation of cardiopulmonary exercise tests [54], and the diagnosis of obstructive sleep apnea using overnight pulse oximetry [55].

While the current explosion of interest in medical applications of AI owes much to the development of DNNs with often superhuman pattern recognition capabilities, other approaches to AI should not be forgotten. A limitation of DNNs is that they do not employ causal reasoning in the same way that humans do, and instead rely on learning complex patterns or associations using large training datasets. Incorporating causal reasoning into AI models may lead to more explainable model outputs, and models that can learn from fewer examples. Jiao et al. recently reported a system for the diagnosis of dyspnea using dynamic uncertain causality graphs, which achieved 96.5% accuracy in the validation dataset [56]. An emerging area is the use of reinforcement learning based on retrospective data to derive optimal treatment policies. This approach came to prominence following the study by Komorowski et al. who developed an AI agent that learned a policy for administering intravenous fluid and vasopressor doses in order to minimize mortality, based on a large retrospective intensive care dataset [57]. Peine et al. similarly developed a reinforcement learning algorithm to optimize mechanical ventilation settings with the goal of minimizing 90-day mortality, based on retrospective intensive care data [58]. However, algorithms developed in this way must be treated with caution given the high risk of confounding from unmeasured variables in retrospective datasets [59]. Nevertheless such studies may provide useful hypothesis-generating insights that can be tested in prospective interventional studies.

Major takeaways (3–5, one sentence each)

- There are a wide variety of current and potential future applications of AI within respiratory medicine, with CXR and CT interpretation algorithms being the closest to clinical deployment.
- Promising areas of future development include the analysis of continuous physiological data for diagnosis and event prediction, and the incorporation of causal inference or reinforcement learning into AI models.
- Clinical deployment of AI models will require not only technical solutions but close collaboration with patients and clinicians to ensure that the products are acceptable and provide clinical benefit in a real-world setting.
- AI models should be rigorously assessed both before and after clinical deployment, to ensure that they remain accurate in new clinical settings, and that they do not systematically disadvantage particular groups of people.

References

[1] Hwang EJ, Park S, Jin KN, Kim JI, Choi SY, Lee JH, et al. DLAD Development and Evaluation Group Development and validation of a deep learning-based automated detection algorithm for major thoracic diseases on chest radiographs. JAMA Netw Open 2019;2(3):e191095.

[2] Wang G, Liu X, Shen J, Wang C, Li Z, Ye L, et al. A deep-learning pipeline for the diagnosis and discrimination of viral, non-viral and COVID-19 pneumonia from chest X-ray images. Nat Biomed Eng 2021;5(6):509–21.

[3] Sjoding MW, Taylor D, Motyka J, Lee E, Co I, Claar D, et al. Deep learning to detect acute respiratory distress syndrome on chest radiographs: a retrospective study with external validation. Lancet Digit Health 2021;3(6):e340–8.

[4] Nam JG, Park S, Hwang EJ, Lee JH, Jin KN, Lim KY, et al. Development and validation of deep learning-based automatic detection algorithm for malignant pulmonary nodules on chest radiographs. Radiology. 2019;290(1):218–28.

[5] Qin ZZ, Ahmed S, Sarker MS, Paul K, Adel ASS, Naheyan T, et al. Tuberculosis detection from chest X-rays for triaging in a high tuberculosis-burden setting: an evaluation of five artificial intelligence algorithms. Lancet Digit Health 2021;3(9):e543–54.

[6] Khan FA, Majidulla A, Tavaziva G, Nazish A, Abidi SK, Benedetti A, et al. Chest x-ray analysis with deep learning-based software as a triage test for pulmonary tuberculosis: a prospective study of diagnostic accuracy for culture-confirmed disease. Lancet Digit Health 2020;2(11):e573–81.

[7] Ardila D, Kiraly AP, Bharadwaj S, Choi B, Reicher JJ, Peng L, et al. End-to-end lung cancer screening with three-dimensional deep learning on low-dose chest computed tomography. Nat Med 2019;25(6):954–61.

[8] Massion PP, Antic S, Ather S, Arteta C, Brabec J, Chen H, et al. Assessing the accuracy of a deep learning method to risk stratify indeterminate pulmonary nodules. Am J Respir Crit Care Med 2020;202(2):241–9.

[9] Baldwin DR, Gustafson J, Pickup L, Arteta C, Novotny P, Declerck J, et al. External validation of a convolutional neural network artificial intelligence tool to predict malignancy in pulmonary nodules. Thorax. 2020;75 (4):306–12.

[10] Walsh SLF, Calandriello L, Silva M, Sverzellati N. Deep learning for classifying fibrotic lung disease on high-resolution computed tomography: a case-cohort study. Lancet Respir Med 2018;6(11):837–45.
[11] Huang SC, Kothari T, Banerjee I, Chute C, Ball RL, Borus N, et al. PENet-a scalable deep-learning model for automated diagnosis of pulmonary embolism using volumetric CT imaging. NPJ Digit Med 2020;3:61.
[12] Tang LYW, Coxson HO, Lam S, Leipsic J, Tam RC, Sin DD. Towards large-scale case-finding: training and validation of residual networks for detection of chronic obstructive pulmonary disease using low-dose CT. Lancet Digit Health 2020;2(5):e259–67.
[13] Humphries SM, Notary AM, Centeno JP, Strand MJ, Crapo JD, Silverman EK, , et al. Genetic Epidemiology of COPD (COPDGene) Investigators Deep learning enables automatic classification of emphysema pattern at CT. Radiology. 2020;294(2):434–44.
[14] González G, Ash SY, Vegas-Sánchez-Ferrero G, Onieva Onieva J, Rahaghi FN, Ross JC, et al. COPDGene and ECLIPSE Investigators Disease staging and prognosis in smokers using deep learning in chest computed tomography. Am J Respir Crit Care Med 2018;197(2):193–203.
[15] Bai HX, Wang R, Xiong Z, Hsieh B, Chang K, Halsey K, et al. Artificial intelligence augmentation of radiologist performance in distinguishing COVID-19 from pneumonia of other origin at chest CT. Radiology. 2021;299(1):E225.
[16] Lee EH, Zheng J, Colak E, Mohammadzadeh M, Houshmand G, Bevins N, et al. Deep COVID DeteCT: an international experience on COVID-19 lung detection and prognosis using chest CT. NPJ Digit Med 2021;4(1):11.
[17] Jin C, Chen W, Cao Y, Xu Z, Tan Z, Zhang X, et al. Development and evaluation of an artificial intelligence system for COVID-19 diagnosis. Nat Commun 2020;11(1):5088.
[18] Harmon SA, Sanford TH, Xu S, Turkbey EB, Roth H, Xu Z, et al. Artificial intelligence for the detection of COVID-19 pneumonia on chest CT using multinational datasets. Nat Commun 2020;11(1):4080.
[19] Mei X, Lee HC, Diao KY, Huang M, Lin B, Liu C, et al. Artificial intelligence-enabled rapid diagnosis of patients with COVID-19. Nat Med 2020;26(8):1224–8.
[20] Coudray N, Ocampo PS, Sakellaropoulos T, Narula N, Snuderl M, Fenyö D, et al. Classification and mutation prediction from non-small cell lung cancer histopathology images using deep learning. Nat Med 2018;24 (10):1559–67.
[21] Topalovic M, Das N, Burgel PR, Daenen M, Derom E, Haenebalcke C, et al. Pulmonary Function Study Investigators Artificial intelligence outperforms pulmonologists in the interpretation of pulmonary function tests. Eur Respir J 2019;53(4):1801660.
[22] Gould MK, Huang BZ, Tammemagi MC, Kinar Y, Shiff R. Machine learning for early lung cancer identification using routine clinical and laboratory data. Am J Respir Crit Care Med 2021;204(4):445–53.
[23] Dayan I, Roth HR, Zhong A, Harouni A, Gentili A, Abidin AZ, et al. Federated learning for predicting clinical outcomes in patients with COVID-19. Nat Med 2021;27(10):1735–43.
[24] Kamran F, Tang S, Otles E, McEvoy DS, Saleh SN, Gong J, et al. Early identification of patients admitted to hospital for covid-19 at risk of clinical deterioration: model development and multisite external validation study. BMJ 2022;376:e068576.
[25] Shamout FE, Shen Y, Wu N, Kaku A, Park J, Makino T, et al. An artificial intelligence system for predicting the deterioration of COVID-19 patients in the emergency department. NPJ Digit Med 2021;4(1):80.
[26] Jiao Z, Choi JW, Halsey K, Tran TML, Hsieh B, Wang D, et al. Prognostication of patients with COVID-19 using artificial intelligence based on chest X-rays and clinical data: a retrospective study. Lancet Digit Health 2021;3(5):e286–94.
[27] Soltan AAS, Kouchaki S, Zhu T, Kiyasseh D, Taylor T, Hussain ZB, et al. Rapid triage for COVID-19 using routine clinical data for patients attending hospital: development and prospective validation of an artificial intelligence screening test. Lancet Digit Health 2021;3(2):e78–87.
[28] Liang W, Yao J, Chen A, Lv Q, Zanin M, Liu J, et al. Early triage of critically ill COVID-19 patients using deep learning. Nat Commun 2020;11(1):3543.
[29] Shashikumar SP, Wardi G, Paul P, Carlile M, Brenner LN, Hibbert KA, et al. Development and prospective validation of a deep learning algorithm for predicting need for mechanical ventilation. Chest. 2021;159 (6):2264–73.
[30] Seah JCY, Tang CHM, Buchlak QD, Holt XG, Wardman JB, Aimoldin A, et al. Effect of a comprehensive deep-learning model on the accuracy of chest X-ray interpretation by radiologists: a retrospective, multireader multicase study. Lancet Digit Health 2021;3(8):e496–506.

[31] Jones CM, Danaher L, Milne MR, Tang C, Seah J, Oakden-Rayner L, et al. Assessment of the effect of a comprehensive chest radiograph deep learning model on radiologist reports and patient outcomes: a real-world observational study. BMJ Open 2021;11(12):e052902.
[32] Patel BN, Rosenberg L, Willcox G, Baltaxe D, Lyons M, Irvin J, et al. Human-machine partnership with artificial intelligence for chest radiograph diagnosis. NPJ Digit Med 2019;2:111.
[33] Schmuelling L, Franzeck FC, Nickel CH, Mansella G, Bingisser R, Schmidt N, et al. Deep learning-based automated detection of pulmonary embolism on CT pulmonary angiograms: no significant effects on report communication times and patient turnaround in the emergency department nine months after technical implementation. Eur J Radiol 2021;141:109816.
[34] Pierce JD, Rosipko B, Youngblood L, Gilkeson RC, Gupta A, Bittencourt LK. Seamless integration of artificial intelligence into the clinical environment: our experience with a novel pneumothorax detection artificial intelligence algorithm. J Am Coll Radiol 2021;18(11):1497–505.
[35] Hong W, Hwang EJ, Lee JH, Park J, Goo JM, Park CM. Deep learning for detecting pneumothorax on chest radiographs after needle biopsy: clinical implementation. Radiology 2022;211706. Available from: https://doi.org/10.1148/radiol.211706.
[36] Hwang EJ, Lee JS, Lee JH, Lim WH, Kim JH, Choi KS, et al. Deep learning for detection of pulmonary metastasis on chest radiographs. Radiology. 2021;301(2):455–63.
[37] Kelly CJ, Karthikesalingam A, Suleyman M, Corrado G, King D. Key challenges for delivering clinical impact with artificial intelligence. BMC Med 2019;17(1):195.
[38] Seyyed-Kalantari L, Zhang H, McDermott MBA, Chen IY, Ghassemi M. Underdiagnosis bias of artificial intelligence algorithms applied to chest radiographs in under-served patient populations. Nat Med 2021;27(12):2176–82.
[39] Liu X, Glocker B, McCradden MM, Ghassemi M, Denniston AK, Oakden-Rayner L. The medical algorithmic audit. Lancet Digit Health 2022;. Available from: https://doi.org/10.1016/S2589-7500(22)00003-6.
[40] Strohm L, Hehakaya C, Ranschaert ER, Boon WPC, Moors EHM. Implementation of artificial intelligence (AI) applications in radiology: hindering and facilitating factors. Eur Radiol 2020;30(10):5525–32.
[41] Scott IA, Carter SM, Coiera E. Exploring stakeholder attitudes towards AI in clinical practice. BMJ Health Care Inf 2021;28(1):e100450.
[42] Markus AF, Kors JA, Rijnbeek PR. The role of explainability in creating trustworthy artificial intelligence for health care: a comprehensive survey of the terminology, design choices, and evaluation strategies. J Biomed Inf 2021;113:103655.
[43] Cushnan D, Berka R, Bertolli O, Williams P, Schofield D, Joshi I, et al. Towards nationally curated data archives for clinical radiology image analysis at scale: learnings from national data collection in response to a pandemic. Digit Health 2021;7.
[44] Jacob J, Alexander D, Baillie JK, Berka R, Bertolli O, Blackwood J, et al. Using imaging to combat a pandemic: rationale for developing the UK National COVID-19 Chest Imaging Database. Eur Respir J 2020;56(2):2001809.
[45] Dou Q, So TY, Jiang M, Liu Q, Vardhanabhuti V, Kaissis G, et al. Federated deep learning for detecting COVID-19 lung abnormalities in CT: a privacy-preserving multinational validation study. NPJ Digit Med 2021;4(1):60.
[46] Warnat-Herresthal S, Schultze H, Shastry KL, Manamohan S, Mukherjee S, Garg V, et al. Swarm Learning for decentralized and confidential clinical machine learning. Nature 2021;594(7862):265–70.
[47] Bressem KK, Adams LC, Gaudin RA, Tröltzsch D, Hamm B, Makowski MR, et al. Highly accurate classification of chest radiographic reports using a deep learning natural language model pre-trained on 3.8 million text reports. Bioinformatics 2021;36(21):5255–61.
[48] Weikert T, Nesic I, Cyriac J, Bremerich J, Sauter AW, Sommer G, et al. Towards automated generation of curated datasets in radiology: application of natural language processing to unstructured reports exemplified on CT for pulmonary embolism. Eur J Radiol 2020;125:108862.
[49] Zhang O, Minku LL, Gonem S. Detecting asthma exacerbations using daily home monitoring and machine learning. J Asthma 2021;58(11):1518–27.
[50] Shah SA, Velardo C, Farmer A, Tarassenko L. Exacerbations in chronic obstructive pulmonary disease: identification and prediction using a digital health system. J Med Internet Res 2017;19(3):e69.

[51] Orchard P, Agakova A, Pinnock H, Burton CD, Sarran C, Agakov F, et al. Improving prediction of risk of hospital admission in chronic obstructive pulmonary disease: application of machine learning to telemonitoring data. J Med Internet Res 2018;20(9):e263.
[52] Richards DM, Tweardy MJ, Steinhubl SR, Chestek DW, Hoek TLV, Larimer KA, et al. Wearable sensor derived decompensation index for continuous remote monitoring of COVID-19 diagnosed patients. NPJ Digit Med 2021;4(1):155.
[53] Un KC, Wong CK, Lau YM, Lee JC, Tam FC, Lai WH, et al. Observational study on wearable biosensors and machine learning-based remote monitoring of COVID-19 patients. Sci Rep 2021;11(1):4388.
[54] Inbar O, Inbar O, Reuveny R, Segel MJ, Greenspan H, Scheinowitz M. A machine learning approach to the interpretation of cardiopulmonary exercise tests: development and validation. Pulm Med 2021;2021:5516248.
[55] Nikkonen S, Afara IO, Leppänen T, Töyräs J. Artificial neural network analysis of the oxygen saturation signal enables accurate diagnostics of sleep apnea. Sci Rep 2019;9(1):13200.
[56] Jiao Y, Zhang Z, Zhang T, Shi W, Zhu Y, Hu J, et al. Development of an artificial intelligence diagnostic model based on dynamic uncertain causality graph for the differential diagnosis of dyspnea. Front Med 2020;14(4):488–97.
[57] Komorowski M, Celi LA, Badawi O, Gordon AC, Faisal AA. The Artificial Intelligence Clinician learns optimal treatment strategies for sepsis in intensive care. Nat Med 2018;24(11):1716–20.
[58] Peine A, Hallawa A, Bickenbach J, Dartmann G, Fazlic LB, Schmeink A, et al. Development and validation of a reinforcement learning algorithm to dynamically optimize mechanical ventilation in critical care. NPJ Digit Med 2021;4(1):32.
[59] Gottesman O, Johansson F, Komorowski M, Faisal A, Sontag D, Doshi-Velez F, et al. Guidelines for reinforcement learning in healthcare. Nat Med 2019;25(1):16–18.

CHAPTER

14

Artificial intelligence in critical care

Chao-Ping Wu[1] *and Piyush Mathur*[2]

[1]Department of Critical Care, Respiratory Institute, Cleveland Clinic, Cleveland, OH, United States [2]Department of Intensive Care and Resuscitation, Anesthesiology Institute, Cleveland Clinic, Cleveland, OH, United States

Introduction

Critical care medicine is a specialty focused on managing patients with life-threatening illnesses. With increasing adoption of sensors, digital devices, and the electronic health records (EHRs), patients in intensive care units (ICUs) generate a sheer amount of data which paves the way for application of artificial intelligence (AI). Critical care has been at the forefront of adapting AI in clinical research among diseases, such as sepsis, acute kidney injury (AKI), acute respiratory distress syndrome (ARDS), and COVID-19 [1,2].

Artificial intelligence applications in critical care

Predictive models and phenotypes

The most common AI studies in critical care have been focused on developing predictive and prognostic models using machine learning (ML) algorithms. There are limited "positive" randomized control trials in critical care, in part due to inadequate sample size and heterogeneous patient population with different biological mechanisms [3,4]. With a large dataset, AI can analyze this complex patient population in real-time and predict clinically relevant outcomes of interests to support physicians decision-making.

Data quality is central to data-driven AI research. In critical care, several high quality databases are publicly available: Medical Information Mart for Intensive Care (MIMIC-3) [5], de-identified data from single center ICU patients at Beth Israel Deaconess Medical Center in Boston, Massachusetts between 2001 and 2012, eICU collaborative Research Database (eICU-CRU) consist many ICUs throughout the US with over 200,000 admissions

DOI: https://doi.org/10.1016/B978-0-443-15688-5.00006-1

from 2014 to 2015 [6], and Amsterdam University Center Database (AmsterdamUMCdb) which includes 20,109 ICU patients from 2003 to 2016 [7].

Sepsis remains one of the major causes of mortality and morbidity in the ICU and has been a research focus area due to the detrimental outcome of delayed diagnosis. One of the major challenges is to identify and recognize patients at-risk of sepsis and impending organ damages [8]. A retrospective study of sepsis warning using traditional penalized logistic regression models embedded in EHR showed low accuracy and created alert fatigue in action [9]. Novel ML models have been developed for early sepsis identification from 1 to 24 or 48 hours prior to clinical onset with great accuracy with area under receiver operating characteristic curve (AU-ROC) 0.83–0.9 [10–12] and consistently outperformed traditional predictive tools, such as quick Sepsis Related Organ Failure Assessment, systemic inflammatory response syndrome, or Modified early warning score. Many studies used supervised ML algorithms to predict mortality and yield high accuracy of 0.8–0.9 seconds, especially with gradient-boosted decision tree and convolutional neural network [13–15]. These studies have demonstrated the advantage of higher predictive ability with novel ML algorithms. In recent years an increasing number of institutions have piloted ML models into clinical practice with the aim to reduce patient length of stay and improve mortality [16,17]. However, most of the studies are retrospective by designed and lack external validation. Thus further prospective randomized controlled trials are necessary to validate and generalize.

Mechanical ventilation is a common procedure in the ICU and early extubation can lead to shorter length of stay and lead to lower rate of complications, such as ventilator-associated pneumonia, delirium, and pneumothorax. Several studies investigated using ML to predict successful extubation [18,19], at-risk for prolonged intubation requiring tracheostomy [20,21], duration of intubation [22] or difficult intubation [23] with promising results. ML is also applied for predicting AKI in the ICU patients with excellent accuracy [24,25] with the aim to early identification of at-risk population and consider prospective interventional trials.

To address heterogeneity in patients with critical illness, ML is utilized for disease phenotyping. In the era of "omics," these studies hold the promise of leading toward precision care for patients in the ICU. For example, in patients with ARDS, prognosis and treatment responses are different in hyperinflammation and hyperinflammatory phenotypes. Accurate predictive models have developed for disease phenotyping of such heterogeneous disorders using supervised ML [26] and unsupervised ML [27].

Imaging and waveform monitoring

There is a trend of increasing utilization of point-of-care ultrasound in ICU for assessing patient's hemodynamics, such as cardiac output, regional wall abnormalities, and volume status. Deep learning, a subset of ML, models are developed to interpret echocardiography images and can accurately identify the presence of structural heart disease, atrial abnormality and left ventricular ejection fraction [28–30]. In addition, deep learning algorithms are used to interpret chest-X-ray and computed tomographic scans and have reported accurate diagnosis. Seah et al. [31] demonstrated deep-learning models can improve performance in interpreting chest X-ray in a study that compared the diagnostic accuracy between radiologists assisted with deep learning model (AUC: 0.81, 95% CI: 0.65–0.79) and unassisted radiologists (AUC: 0.71, 95% CI: 0.76–0.84). In point-of-care lung ultrasounds images, deep learning models also demonstrated comparable accuracy to expert interpretations in classifying lung pathologies [32]. In addition, using ML for waveform analysis from ventilator data can accurately identify ventilator asynchrony, such as delayed termination and premature termination [33], double triggers and ineffective inspiratory effort during expiration [34]. With a novel AI method, one study uses noninvasive blood pressure cuff, continuous electrocardiogram and pulse oximeter waveform to impute continuous arterial waveform without arterial catheterization [35]. Other examples of ML derived waveform research includes monitoring including prediction of hypotension from arterial waveform based device data [36], automated seizure detection from continuous electroencephalography [37], and tachy- or brady-arrhythmia detection from continuous electrocardiographic tracing [38]. With advances in the AI field, newly developed technologies promise better clinician decision supporting tools and minimize the burden of alarm fatigue.

Novel machine learning algorithms in critical care

Natural language processing (NLP) is a subset of ML which unlocks the capacity of analyzing one of the most underutilized yet essential data—clinical notes. Traditionally, ML studies focused on tabular data including vital signs, lab values, diagnostic codes and demographics. With the assistance of NLP, we could understand more granular details about a patient's progress in the ICU. In one retrospective study, investigators used clinical notes within 24 hours of ICU admission from the MIMIC3 database to predict short, mid and long-term mortality and yield AUC-ROC scores of 0.82–0.87 [5,39]. Another study demonstrated that using NLP can augment in-hospital mortality prediction models from AUC-ROC 0.83 to 0.92 [40]. In addition to improving model prediction, NLP was also evaluated to be used as a tool to improve the efficiency of identifying eligible critically ill patients for clinical trials. The study can screen patients much faster with high sensitivity of 90% which can potentially reduce the burden of manual reviewing medical records, especially in enrolling patients into time-sensitive critical care trials [41]. With the capacity of handling free-text, NLP can also be used to automate generating problem list [42] and CPT billing codes [43].

Reinforcement learning (RL) is another subset of ML which is about learning optimal decisions with predefined goals. RL is especially applicable in the ICU because patients' outcomes are affected by countless complex decisions made by providers. The role of RL is to predict an optimal strategy based on cumulative feedback (positive or negative) in similar situations from the database. In sepsis patients, the decision of prescribing intravenous fluid and vasopressors is one of the most common decisions made by providers in the ICUs. To address this dilemma, a study uses RL-derived models that provide optimal treatment policy dynamically for septic patients to achieve the lowest mortality [44]. Peine et al. [45] developed a RL model that can dynamically support providers in choosing optimal positive end-expiratory pressors (PEEP), fraction of inspired oxygen (FiO_2), tidal volume (TV) with best survival outcome. The studies provide the potential benefit of constantly monitoring and optimizing ventilator setting and allow personalized management. RL were also researched in minimizing ordering unnecessary laboratory tests [46], optimal dosage of heparin [47,48], and oxygen delivery in severe COVID-19 patients [49].

Administrative

Integration of AI into healthcare also holds the promise of improving value based care. ML could improve patient flow in ICU by triaging patients in the ED and regular nursing floor [50] via identifying patients at-risk for decompensation. With AI, hospitals can automatically extract essential quality metrics, such as mortality, length of stay, ICU acquired complications, and adherence to best practice of care [51].

Challenges of artificial intelligence in critical care

Although AI holds the promise for improving healthcare delivery, there are challenges in implementation and real-life adaptation. First, data quality and collection remain a major challenge in healthcare in general. While data quality is of the utmost importance in ML research, a vast amount of healthcare data is considered unstructured data and is not stored within a predefined and organized structure. Other real-world data quality issues, such as missing data, improper features selection, bias, inconsistent documentation, and definition, could significantly affect model performance. These factors make it difficult for extracting real-time meaningful data without specialized programs or established data pipelines. It is also important to recognize the algorithmic bias could affect ML prediction results due to the underrepresented group in the dataset that generates the model. Second, most of the ML studies are retrospective by design and only a limited number of prospective studies demonstrated clinical benefit. Third, lack of interpretability of some of the AI models could be a barrier for providers to fully adapt in real life practice. However, novel ML models and other solutions are being developed to address all these gaps that exist limiting application of AI to critical care and healthcare in general.

Conclusion

In the future, AI generated models derived from real-time data will no doubt integrate into critical care clinician's daily practice to augment clinical decision-making and facilitate personalized care delivery while decreasing the burden of data overload. Hence, it is important for clinicians to understand, properly interpret and lead the fast paced AI-based critical care research.

References

[1] Mathur P, Burns ML. Artificial Intelligence in Critical Care. Int Anesthesiol Clin 2019;57:89–102.
[2] Mathur P., Mishra S., Awasthi R., Khanna A., Maheshwari K., Papay F., et al. Artificial Intelligence in Healthcare: 2021 Year in Review. 2022.

[3] Laffey JG, Kavanagh BP. Negative trials in critical care: why most research is probably wrong. Lancet Respir Med 2018;6:659–60.

[4] Harhay MO, Wagner J, Ratcliffe SJ, Bronheim RS, Gopal A, Green S, et al. Outcomes and statistical power in adult critical care randomized trials. Am J Respir Crit Care Med 2014;189:1469–78.

[5] Johnson AEW, Pollard TJ, Shen L, Lehman L-WH, Feng M, Ghassemi M, et al. MIMIC-III, a freely accessible critical care database. Sci Data 2016;3:160035.

[6] Pollard TJ, Johnson AEW, Raffa JD, Celi LA, Mark RG, Badawi O. The eICU Collaborative Research Database, a freely available multi-center database for critical care research. Sci Data 2018;5:180178.

[7] Thoral PJ, Peppink JM, Driessen RH, Sijbrands EJG, Kompanje EJO, Kaplan L, et al. Sharing ICU Patient Data Responsibly Under the Society of Critical Care Medicine/European Society of Intensive Care Medicine Joint Data Science Collaboration: The Amsterdam University Medical Centers Database (AmsterdamUMCdb) Example*. Critical Care Medicine 2021;e563–77. Available from: https://doi.org/10.1097/ccm.0000000000004916.

[8] Yealy DM, Huang DT, Delaney A, Knight M, Randolph AG, Daniels R, et al. Recognizing and managing sepsis: what needs to be done? BMC Med 2015;13. Available from: https://doi.org/10.1186/s12916-015-0335-2.

[9] Wong A, Otles E, Donnelly JP, Krumm A, McCullough J, DeTroyer-Cooley O, et al. External Validation of a Widely Implemented Proprietary Sepsis Prediction Model in Hospitalized Patients. JAMA Intern Med 2021;181:1065–70.

[10] Barton C, Chettipally U, Zhou Y, Jiang Z, Lynn-Palevsky A, Le S, et al. Evaluation of a machine learning algorithm for up to 48-hour advance prediction of sepsis using six vital signs. Comput Biol Med 2019;109:79–84.

[11] Delahanty RJ, Alvarez J, Flynn LM, Sherwin RL, Jones SS. Development and Evaluation of a Machine Learning Model for the Early Identification of Patients at Risk for Sepsis. Ann Emerg Med 2019;73:334–44.

[12] Bedoya AD, Futoma J, Clement ME, Corey K, Brajer N, Lin A, et al. Machine learning for early detection of sepsis: an internal and temporal validation study. JAMIA Open 2020;3:252–60.

[13] Li K, Shi Q, Liu S, Xie Y, Liu J. Predicting in-hospital mortality in ICU patients with sepsis using gradient boosting decision tree. Medicine 2021;e25813. Available from: https://doi.org/10.1097/md.0000000000025813.

[14] Perng J-W, Kao I-H, Kung C-T, Hung S-C, Lai Y-H, Su C-M. Mortality Prediction of Septic Patients in the Emergency Department Based on Machine Learning. J Clin Med Res 2019;8. Available from: https://doi.org/10.3390/jcm8111906.

[15] Kong G, Lin K, Hu Y. Using machine learning methods to predict in-hospital mortality of sepsis patients in the ICU. BMC Med Inform Decis Mak 2020;20:251.

[16] Shimabukuro DW, Barton CW, Feldman MD, Mataraso SJ, Das R. Effect of a machine learning-based severe sepsis prediction algorithm on patient survival and hospital length of stay: a randomised clinical trial. BMJ Open Respir Res 2017;4:e000234.

[17] Sendak MP, Ratliff W, Sarro D, Alderton E, Futoma J, Gao M, et al. Real-World Integration of a Sepsis Deep Learning Technology Into Routine Clinical Care: Implementation Study. JMIR Med Inform 2020;8:e15182.

[18] Lin M-Y, Li C-C, Lin P-H, Wang J-L, Chan M-C, Wu C-L, et al. Explainable Machine Learning to Predict Successful Weaning Among Patients Requiring Prolonged Mechanical Ventilation: A Retrospective Cohort Study in Central Taiwan. Front Med 2021;8:663739.

[19] Hsieh M-H, Hsieh M-J, Chen C-M, Hsieh C-C, Chao C-M, Lai C-C. An Artificial Neural Network Model for Predicting Successful Extubation in Intensive Care Units. J Clin Med Res 2018;7. Available from: https://doi.org/10.3390/jcm7090240.

[20] Parreco J, Hidalgo A, Parks JJ, Kozol R, Rattan R. Using artificial intelligence to predict prolonged mechanical ventilation and tracheostomy placement. J Surg Res 2018;228:179–87.

[21] Castiñeira D, Schlosser KR, Geva A, Rahmani AR, Fiore G, Walsh BK, et al. Adding Continuous Vital Sign Information to Static Clinical Data Improves the Prediction of Length of Stay After Intubation: A Data-Driven Machine Learning Approach. Respir Care 2020;65:1367–77.

[22] Sayed M, Riaño D, Villar J. Predicting Duration of Mechanical Ventilation in Acute Respiratory Distress Syndrome Using Supervised Machine Learning. J Clin Med Res 2021;10. Available from: https://doi.org/10.3390/jcm10173824.

[23] Yamanaka S, Goto T, Morikawa K, Watase H, Okamoto H, Hagiwara Y, et al. Machine Learning Approaches for Predicting Difficult Airway and First-Pass Success in the Emergency Department: Multicenter Prospective Observational Study. Interact J Med Res 2022;11:e28366.
[24] Flechet M, Güiza F, Schetz M, Wouters P, Vanhorebeek I, Derese I, et al. AKIpredictor, an online prognostic calculator for acute kidney injury in adult critically ill patients: development, validation and comparison to serum neutrophil gelatinase-associated lipocalin. Intensive Care Med 2017;43:764–73.
[25] Chiofolo C, Chbat N, Ghosh E, Eshelman L, Kashani K. Automated Continuous Acute Kidney Injury Prediction and Surveillance: A Random Forest Model. Mayo Clin Proc 2019;94:783–92.
[26] Sinha P, Churpek MM, Calfee CS. Machine Learning Classifier Models Can Identify Acute Respiratory Distress Syndrome Phenotypes Using Readily Available Clinical Data. Am J Respir Crit Care Med 2020;202:996–1004.
[27] Duggal A, Kast R, Van Ark E, Bulgarelli L, Siuba MT, Osborn J, et al. Identification of acute respiratory distress syndrome subphenotypes de novo using routine clinical data: a retrospective analysis of ARDS clinical trials. BMJ Open 2022;e053297. Available from: https://doi.org/10.1136/bmjopen-2021-053297.
[28] Tromp J, Seekings PJ, Hung C-L, Iversen MB, Frost MJ, Ouwerkerk W, et al. Automated interpretation of systolic and diastolic function on the echocardiogram: a multicohort study. Lancet Digit Health 2022;4:e46–54.
[29] Asch FM, Mor-Avi V, Rubenson D, Goldstein S, Saric M, Mikati I, et al. Deep Learning-Based Automated Echocardiographic Quantification of Left Ventricular Ejection Fraction: A Point-of-Care Solution. Circ Cardiovasc Imaging 2021;14:e012293.
[30] Wegner FK, Benesch Vidal ML, Niehues P, Willy K, Radke RM, Garthe PD, et al. Accuracy of Deep Learning Echocardiographic View Classification in Patients with Congenital or Structural Heart Disease: Importance of Specific Datasets. J Clin Med Res 2022;11. Available from: https://doi.org/10.3390/jcm11030690.
[31] Seah JCY, Tang CHM, Buchlak QD, Holt XG, Wardman JB, Aimoldin A, et al. Effect of a comprehensive deep-learning model on the accuracy of chest x-ray interpretation by radiologists: a retrospective, multireader multicase study. Lancet Digit Health 2021;3:e496–506.
[32] Tsai C-H, van der Burgt J, Vukovic D, Kaur N, Demi L, Canty D, et al. Automatic deep learning-based pleural effusion classification in lung ultrasound images for respiratory pathology diagnosis. Phys Med 2021;83:38–45.
[33] Gholami B, Phan TS, Haddad WM, Cason A, Mullis J, Price L, et al. Replicating human expertise of mechanical ventilation waveform analysis in detecting patient-ventilator cycling asynchrony using machine learning. Comput Biol Med 2018;97:137–44.
[34] Zhang L, Mao K, Duan K, Fang S, Lu Y, Gong Q, et al. Detection of patient-ventilator asynchrony from mechanical ventilation waveforms using a two-layer long short-term memory neural network. Comput Biol Med 2020;120:103721.
[35] Hill BL, Rakocz N, Rudas Á, Chiang JN, Wang S, Hofer I, et al. Imputation of the continuous arterial line blood pressure waveform from non-invasive measurements using deep learning. Sci Rep 2021;11:15755.
[36] van der Ven WH, Terwindt LE, Risvanoglu N, Ie ELK, Wijnberge M, Veelo DP, et al. Performance of a machine-learning algorithm to predict hypotension in mechanically ventilated patients with COVID-19 admitted to the intensive care unit: a cohort study. J Clin Monit Comput 2021;1–9.
[37] Saab K, Dunnmon J, Ré C, Rubin D, Lee-Messer C. Weak supervision as an efficient approach for automated seizure detection in electroencephalography. NPJ Digit Med 2020;3:59.
[38] Au-Yeung W-TM, Sevakula RK, Sahani AK, Kassab M, Boyer R, Isselbacher EM, et al. Real-time machine learning-based intensive care unit alarm classification without prior knowledge of the underlying rhythm. Eur Heart J Digit Health 2021;2:437–45.
[39] Mahbub M, Srinivasan S, Danciu I, Peluso A, Begoli E, Tamang S, et al. Unstructured clinical notes within the 24 hours since admission predict short, mid & long-term mortality in adult ICU patients. PLoS One 2022;17:e0262182.
[40] Marafino BJ, Park M, Davies JM, Thombley R, Luft HS, Sing DC, et al. Validation of Prediction Models for Critical Care Outcomes Using Natural Language Processing of Electronic Health Record Data. JAMA Netw Open 2018;1:e185097.
[41] Tissot HC, Shah AD, Brealey D, Harris S, Agbakoba R, Folarin A, et al. Natural Language Processing for Mimicking Clinical Trial Recruitment in Critical Care: A Semi-Automated Simulation Based on the LeoPARDS Trial. IEEE J Biomed Health Inform 2020;24:2950–9.

[42] Spyns P. Natural Language Processing in Medicine: Design, Implementation and Evaluation of an Analyser for Dutch. Leuven University Press; 2000.
[43] Kim JS, Vivas A, Arvind V, Lombardi J, Reidler J, Zuckerman SL, et al. Can Natural Language Processing and Artificial Intelligence Automate The Generation of Billing Codes From Operative Note Dictations? Global Spine J 2022; 21925682211062831.
[44] Komorowski M, Celi LA, Badawi O, Gordon AC, Faisal AA. The Artificial Intelligence Clinician learns optimal treatment strategies for sepsis in intensive care. Nat Med 2018;24:1716–20.
[45] Peine A, Hallawa A, Bickenbach J, Dartmann G, Fazlic LB, Schmeink A, et al. Development and validation of a reinforcement learning algorithm to dynamically optimize mechanical ventilation in critical care. NPJ Digit Med 2021;4:32.
[46] Cheng L-F, Prasad N, Engelhardt BE. An Optimal Policy for Patient Laboratory Tests in Intensive Care Units. Pac Symp Biocomput 2019;24:320–31.
[47] Nemati S, Ghassemi MM, Clifford GD. Optimal medication dosing from suboptimal clinical examples: a deep reinforcement learning approach. Conf Proc IEEE Eng Med Biol Soc 2016;2016:2978–81.
[48] Lin R, Stanley MD, Ghassemi MM, Nemati S. A Deep Deterministic Policy Gradient Approach to Medication Dosing and Surveillance in the ICU. Conf Proc IEEE Eng Med Biol Soc 2018;2018:4927–31.
[49] Zheng H, Zhu J, Xie W, Zhong J. Reinforcement learning assisted oxygen therapy for COVID-19 patients under intensive care. BMC Med Inform Decis Mak 2021;21:350.
[50] McManus S, Almuqati R, Khatib R, Khanna A, Cywinski J, Papay F, et al. 1214: Machine learning-based early mortality prediction at the time of ICU admission. Crit Care Med 2022;50 607–607.
[51] Robles Arévalo A, Maley JH, Baker L, da Silva Vieira SM, da Costa Sousa JM, Finkelstein S, et al. Data-driven curation process for describing the blood glucose management in the intensive care unit. Sci Data 2021;8:80.

CHAPTER

15

Artificial intelligence in dermatopathology

Puneet K. Bhullar[1], Dennis Murphree[2,3], Anirudh Choudhary[4], Margot S. Peters[2,5], Olayemi Sokumbi[6,7] and Nneka I. Comfere[2,3,5]

[1]Mayo Clinic Alix School of Medicine, Scottsdale, AZ, United States [2]Department of Dermatology, Mayo Clinic, Rochester, MN, United States [3]Mayo Clinic Dermatology Digital Health, Artificial Intelligence and Innovations Program, Rochester, MN, United States [4]Department of Computer Science, University of Illinois, Urbana-Champain, IL, United States [5]Department of Laboratory Medicine and Pathology, Mayo Clinic, Rochester, MN, United States [6]Department of Dermatology, Mayo Clinic, Jacksonville, FL, United States [7]Department of Laboratory Medicine and Pathology, Mayo Clinic, Jacksonville, FL, United States

Introduction

Dermatology has taken a spotlight for artificial intelligence (AI) application in the medical field due to the specialty's vast clinical, dermatoscopic, and dermatopathologic image database.

Within dermatopathology, applications of AI have revolved around whole slide imaging (WSI) also known as virtual microscopy, which is the process of digitizing whole slides into images by scanners [1]. Unlike traditional microscopes, digital image viewers have unique practical features, such as displaying multiple images (multiple "slides") at once and allowing multiple viewers in different locations to view an image at the same time [2]. The clinical applications for WSI have been broad: telepathology for primary diagnosis, consultations, tumor boards, and more. Multiple studies have investigated the validity and performance of WSI in dermatology diagnoses and confirmed diagnostic concordance between WSI and conventional glass slides [3–6]. Furthermore, quantitative applications of digital pathology driven by AI have been shown to reduce error rates and improve efficiency in pathology [7]. Within AI, various models have been explored to analyze skin histopathology images [8]. Cruz-Roa et al. compared the deep learning method

Artificial Intelligence in Clinical Practice
DOI: https://doi.org/10.1016/B978-0-443-15688-5.00050-4

to traditional machine learning with feature descriptors and concluded that the deep learning architecture was superior to traditional approaches (89.4% in F-measure and 91.4% in balanced accuracy) [9]. In image analysis, the most frequently used form of deep neural network is known as a convolutional neural network (CNN). CNNs can input complex visual stimuli and break them down into individual pixels that are analyzed by the deep neural network, a model that is ideal for dermatopathology.

Summary of literature

In subspecialties, such as dermatopathology, studies have focused on applications of AI for diagnosis to highlight the important role telepathology can play in mitigating the problem of limited access to experts [1]. Table 15.1 outlines the literature on AI in dermatopathology. Olson et al. utilized WSI of previously established diagnoses of nodular basal cell carcinomas (BCCs), dermal nevi, and seborrheic keratoses [10]. Distinct morphological areas were annotated for these diagnostic slides. Unannotated WSIs with five distractor diagnoses of common neoplastic and inflammatory diagnoses were also included in the

TABLE 15.1 Summary of current literature on artificial intelligence in dermatopathology.

Authors	Year	Summary of findings
Olson et al. [10]	2018	Olson et al. utilized WSI of previously established diagnoses of nodular basal cell carcinomas (BCCs), dermal nevi, and seborrheic keratoses [10]. Distinct morphological areas were annotated. A proprietary CNN was developed and accurately classified 123/124 (99.45%) nodular BCCs, 113/114 (99.4%), dermal nevi, and 123/123 (100%) seborrheic keratoses.
Hekler et al. [12]	2019	695 lesions were studied and classified by a board-certified dermatopathologist (350 nevi and 345 melanomas) [12]. Of the 695 total lesions, 595 were used to train a CNN, and 100 were used to assess the outcome of the CNN. The discordance between CNN and DP was 20% for nevi, 18% for melanoma, and 19% overall.
Andres et al. [13]	2017	A model identifying mitotic cells within tumor regions was trained using 59 WSIs, and yielded a diagnostic accuracy of 83% [13]. Highlighting regions of increased mitotic figures may decrease the need for immunohistochemical stains for mitosis and augment the dermatopathologist's diagnostic evaluation [14]
Ianni et al. [15]	2020	Three independently trained CNNs were used sequentially to categorize WSI into four classes (basaloid, melanocytic, squamous, and other) with a measure of confidence: overall accuracy was 78%. As the confidence level of the model increased, fewer images were included in the analysis, and poor predictions were discarded.
Hart et al. [16]	2019	A CNN was utilized for binary classification of Spitz nevi vs conventional nevi [16]. The algorithm developed showcased classification accuracy at the patch level of up to 99.0% on application to WSI. An additional arm of the study utilized images that were not curated by dermatopathologists; this model learned more slowly and its validation accuracy was only 52%. This study emphasized the importance of pathologist input to select representative image patches for CNN training efficiency and accuracy [16].

(Continued)

TABLE 15.1 (Continued)

Authors	Year	Summary of findings
Ba et al. [11]	2021	A deep learning algorithm was developed to differentiate melanoma from nevus using whole slide pathological images and was tested on 104 WSIs (29 melanomas, 75 nevi) [11]. Performance of the deep learning algorithm showcased equivalent accuracy to that of 7 expert dermatopathologists.
Xie et al. [17]	2019	2241 histopathological images from 2008 to 2019 were used to test deep learning architecture classifications of melanoma and nevi at different magnification scales, and achieved high accuracy: average F1 (0.89), sensitivity (0.92), specificity (0.94), and AUC (0.98) [17].
Crowley et al. [18]	2005	ReportTutor was created, an intelligent tutoring system which provides students with feedback regarding style and nonstandard ordering of features in a dermatopathology diagnostic report [18].
Crowley et al. [19]	2006	Crowley et al. developed SlideTutor, an intelligent tutoring system which teaches algorithmic problem-solving via visual classification of inflammatory dermatoses [19]. Learners begin by identifying morphologic features and progress through step-by-step feedback to form a differential diagnosis and then final diagnosis. To assess the effects of this AI-based tutoring, a prospective study was conducted among residents who preferred the knowledge focused interface of SlideTutor and expressed general approval of this type of interactive AI-based learning [20].

training set. A proprietary CNN was developed and accurately classified 123/124 (99.45%) nodular BCCs, 113/114 (99.4%), dermal nevi, and 123/123 (100%) seborrheic keratoses. Similar studies have reinforced the success of CNNs to classify images [11]. Nevi can present challenges to diagnostic classification due to overlapping features between benign nevi and malignant nevi (melanomas). One of the landmark studies exemplifying histopathologic application in this area was conducted by Hekler et al. A total of 695 lesions were studied and classified by a board-certified dermatopathologist (350 nevi and 345 melanomas) [12]. Of the 695 total lesions, 595 were used to train a CNN and 100 were used to assess the outcome of the CNN. The performance was assessed on randomly cropped 10x magnification sections. Over the course of 11 test runs, the CNN achieved higher mean sensitivity, specificity, and accuracy than the 11 dermatopathologists [12]. Of note, these results should be interpreted with caution as the dermatopathologists were limited to randomly cropped segments. This is not representative of the normal clinical setting where they would have the ability to evaluate the entire slide at varied magnification for diagnosis.

While artificial intelligence has been primarily investigated for use in medical decision support, applications to educate health professionals also have great potential. Crowley et al. created SlideTutor, an intelligent tutoring system which teaches algorithmic problem-solving via visual classification of inflammatory dermatoses [19]. Learners begin by identifying morphologic features in pathology and progress through step-by-step feedback to form a differential diagnosis and then final diagnosis. To assess the effects of this AI-based tutoring, a prospective study was conducted among residents who preferred the knowledge focused interface of SlideTutor and expressed general approval of this type of interactive AI-based learning [20]. Furthermore, WSI has garnered high levels of satisfaction among dermatopathologists and residents for its use in training workshops and as an e-learning tool [21–23].

Challenges and opportunities

AI tools should be viewed as methods to augment the physician's practice rather than any sort of replacement. The invaluable physician skillset of empathy, trust, trained history taking, physical examination, and judgement based on these factors is fundamental to medicine and is not threatened by the integration of deep learning. Instead, these advances will serve to improve efficiency and minimize repetitive and mechanical tasks, thus allowing physicians to expend their energy on patient-centered care [24].

It is important to note that in a field, such as dermatopathology, atypical lesions may not be perfectly framed in defined criteria and have been associated with well-documented diagnostic discordance among even expert dermatopathologists. Therefore one can imagine the difficulty in training a CNN on the diagnosis of such lesions. Finally, we must not forget that any AI algorithm is ultimately programmed and trained by a human being, introducing some form of participation from a pathologist [25,26].

One of the additional challenges with deep learning is the limited transparency of the process. The input and output are known to the user but the methodology to form these predictions is unknown and indiscernible. Some clinicians may argue that transparent justifications for reaching a diagnosis are necessary, and deep learning does not allow for this transparency. However, there are multiple examples of scenarios in medical practice where treatment plans are made despite having unclear mechanisms, such as naltrexone for the management of Hailey–Hailey disease [27].

As discussed in this chapter, there are many studies which show CNN classification tasks may perform equivalent to or even better than dermatologists [28]. Although, one should interpret these findings with caution, since research to date has been primarily retrospective and utilized in silico (computer-based) validation. Further prospective, randomized controlled trials still need to be conducted to vigorously evaluate deep learning prior to integration into clinical practice.

Because the infrastructure for complex computer algorithms, in terms of storage and computational power, remains naïve to the current clinical environment, incorporation of AI into routine clinical practice is still lacking. Additionally, AI algorithms often suffer from unbalanced training datasets which overweigh benign lesions on fair skinned adults [29]. In addition, AI algorithms are trained on limited classes of skin lesions within datasets and may be unreliable in distinguishing skin cancers from close mimics [30]. We have not yet reached a point where AI can completely replace any aspect of dermatopathology practice. However, there is enough literature to demonstrate its ability to augment diagnosis and monitoring of skin lesions.

Future directions

It is clear that alongside these advancements, harmony and a positive perception of AI in dermatology must exist for this technology to reach its full potential. In a survey of

90 dermatologists, Wei et al. found that respondents had an overall positive attitude toward AI integration into their clinical practice [31]. Nelson et al. further researched perceptions of AI in dermatology and found that of 46% of 121 dermatologists who completed a survey thought AI would positively influence their practice. However, when asked about augmented intelligence, the concept that deep learning is made to assist and not replace human intelligence, that opinion rose to 64%, indicating that while dermatologists welcome AI tools in clinical practice, they value the human physician-patient relationship and accuracy of physician judgement [32].

Artificial intelligence research has shown significantly optimistic results for dermatologic applications in histopathology—at times, reportedly better results than dermatopathologists can provide. However, AI is far from becoming the new reality, as there are many limitations that have yet to be addressed before achieving real-world clinical diagnostic and prognostic utility. AI algorithms require further refinement and validation in prospective, randomized controlled trials in order to augment the dermatopathologist's practice.

References

[1] Onega T, Reisch LM, Frederick PD, Geller BM, Nelson HD, Lott JP, et al. Use of digital whole slide imaging in dermatopathology. J Digit Imaging 2016;29(2):243–53.
[2] Al-Janabi S, Huisman A, Vink A, Leguit R, Offerhaus GJ, Ten Kate F, et al. Whole slide images for primary diagnostics in dermatopathology: a feasibility study. J Clin Pathol 2012;65(2):152–8.
[3] Al Habeeb A, Evans A, Ghazarian D. Virtual microscopy using whole-slide imaging as an enabler for teledermatopathology: a paired consultant validation study. J Pathol Inf 2012;3:2.
[4] Nielsen PS, Lindebjerg J, Rasmussen J, Starklint H, Waldstrøm M, Nielsen B. Virtual microscopy: an evaluation of its validity and diagnostic performance in routine histologic diagnosis of skin tumors. Hum Pathol 2010;41(12):1770–6.
[5] Massone C, Brunasso AMG, Soyer HP. Teledermatopathology: current status and perspectives. In: Kumar S, Dunn BE, editors. Telepathology. Berlin/Heidelberg: Springer; 2009. p. 163–78.
[6] Kent MN, Olsen TG, Feeser TA, Tesno KC, Moad JC, Conroy MP, et al. Diagnostic accuracy of virtual pathology vs traditional microscopy in a large dermatopathology study. JAMA Dermatol 2017;153(12):1285–91.
[7] Zhang X, Zou J, He K, Sun J. Accelerating very deep convolutional networks for classification and detection. IEEE Trans Pattern Anal Mach Intell 2016;38(10):1943–55.
[8] Cazzato G, Colagrande A, Cimmino A, Arezzo F, Loizzi V, Caporusso C, et al. Artificial intelligence in dermatopathology: new insights and perspectives. Dermatopathology (Basel) 2021;8(3):418–25.
[9] Cruz-Roa AA, Arevalo Ovalle JE, Madabhushi A, González, Osorio FA. A deep learning architecture for image representation, visual interpretability and automated basal-cell carcinoma cancer detection. Med Image Comput Comput Assist Interv 2013;16(Pt 2):403–10.
[10] Olsen TG, Jackson BH, Feeser TA, Kent M, Moad J, Krishnamurthy S, et al. Diagnostic performance of deep learning algorithms applied to three common diagnoses in dermatopathology. J Pathol Inf 2018;9:32.
[11] Ba W, Wang R, Yin G, Zhigang S, Zou J, Zhong C, et al. Diagnostic assessment of deep learning for melanocytic lesions using whole-slide pathological images. Transl Oncol 2021;14(9):101161.
[12] Hekler A, Utikal JS, Enk AH, Berking C, Klode J, Schadendorf D, et al. Pathologist-level classification of histopathological melanoma images with deep neural networks. Eur J Cancer 2019;115:79–83.
[13] Andres C, Andres-Belloni B, Hein R, Biedermann T, Schape A, Brieu N, et al. iDermatoPath—a novel software tool for mitosis detection in H&E-stained tissue sections of malignant melanoma. J Eur Acad Dermatology Venereology 2017;31(7):1137–47.
[14] Puri P, Comfere N, Drage LA, Shamim H, Bezalel S, Pittelkow M, et al. Deep learning for dermatologists: part II. Current applications. J Am Acad Dermatology 2020.

[15] Ianni JD, Soans RE, Sankarapandian S, Chamarthi RV, Ayyagari D, Olsen TG, et al. Tailored for real-world: a whole slide image classification system validated on uncurated multi-site data emulating the prospective pathology workload. Sci Rep 2020;10(1):3217.

[16] Hart SN, Flotte W, Norgan AP, Shah K, Buchan Z, Mounajjed T, et al. Classification of melanocytic lesions in selected and whole-slide images via convolutional neural networks. J Pathol Inf 2019;10 5.

[17] Xie P, Zuo K, Zhang Y, Li F, Yin M, Lu K. Interpretable classification from skin cancer histology slides using deep learning: a retrospective multicenter study. arXiv preprint arXiv:190406156, 2019.

[18] Crowley RS, Tseytlin E, Jukic D. ReportTutor—an intelligent tutoring system that uses a natural language interface. AMIA Annu Symp Proc 2005;2005:171–5.

[19] Crowley RS, Medvedeva O. An intelligent tutoring system for visual classification problem solving. Artif Intell Med 2006;36(1):85–117.

[20] Crowley RS, Medvedeva O. A general architecture for intelligent tutoring of diagnostic classification problem solving. AMIA Annu Symp Proc 2003;2003:185–9.

[21] Brick KE, Comfere NI, Broeren MD, Gibson LE, Wieland CN. The application of virtual microscopy in a dermatopathology educational setting: assessment of attitudes among dermatopathologists. Int J Dermatol 2014;53(2):224–7.

[22] Foster K. Medical education in the digital age: digital whole slide imaging as an e-learning tool. J Pathol Inf 2010;1.

[23] Pantanowitz L, Szymas J, Yagi Y, Wilbur D. Whole slide imaging for educational purposes. J Pathol Inf 2012;3:46.

[24] Dermatology AAo. Position statement on augmented intelligence. Dermatology AAo 2019.

[25] Esteva A, Kuprel B, Novoa RA, Ko J, Swetter S, Blau H, et al. Dermatologist-level classification of skin cancer with deep neural networks. Nature. 2017;542(7639):115–18.

[26] Winkler JK, Fink C, Toberer F, Enk A, Deinlein T, Hofmann-Wellenhof R, et al. Association between surgical skin markings in dermoscopic images and diagnostic performance of a deep learning convolutional neural network for melanoma recognition. JAMA Dermatol 2019;155(10):1135–41.

[27] Albers LN, Arbiser JL, Feldman RJ. Treatment of Hailey-Hailey disease with low-dose naltrexone. JAMA Dermatol 2017;153(10):1018–20.

[28] Höhn J, Hekler A, Krieghoff-Henning E, Nikolas Kather J, Sven Utikal J, Friedegund M, et al. Integrating patient data into skin cancer classification using convolutional neural networks: systematic review. J Med Internet Res 2021;23(7):e20708.

[29] Krizhevsky A, Sutskever I, Hinton GE. Imagenet classification with deep convolutional neural networks. Adv neural Inf Process Syst 2012;25.

[30] Goyal M, Knackstedt T, Yan S, Hassanpour S. Artificial intelligence-based image classification methods for diagnosis of skin cancer: challenges and opportunities. Comput Biol Med 2020;127:104065.

[31] Wei C, Adusumilli N, Friedman A, Patel V. Perceptions of artificial intelligence integration into dermatology clinical practice: a cross-sectional survey study. J Drugs Dermatol 2022;21(2):135–40.

[32] Nelson CA, Pachauri S, Balk R, Miller J, Theunis R, Ko J, et al. Dermatologists' perspectives on artificial intelligence and augmented intelligence—a cross-sectional survey. JAMA Dermatol 2021;157(7):871–4.

CHAPTER

16

Artificial intelligence in infectious diseases

Yousra Kherabi[1] *and Nathan Peiffer-Smadja*[1,2,3]

[1]Infectious Diseases Department, Bichat-Claude Bernard Hospital, Assistance-Publique Hôpitaux de Paris, Paris, France [2]Université Paris Cité, INSERM, IAME, Paris, France [3]National Institute for Health Research Health Protection Research Unit in Healthcare Associated Infections and Antimicrobial Resistance, Imperial College London, London, United Kingdom

Introduction

Taking care of patients with infectious diseases requires the real-time analysis of an important number of dynamic variables, such as causal microorganism, host factors, anti-infectious drugs pharmacodynamics and pharmacokinetics parameters, and local epidemiology. Thus artificial intelligence (AI) and machine learning (ML) have the potential to become increasingly important in the management of infectious diseases. In this chapter, we present an overview of current applications of AI in infectious diseases, from the research laboratory to the bedside of infected patients.

Artificial intelligence for clinical microbiology

The identification of causal microorganisms is a crucial step in the management of infectious diseases. ML techniques may accelerate the analysis of traditional microbiology methods, such as microscopic imaging and bacterial culture [1,2] and open the way to the clinical use of new data. Within the past 15 years, new clinical microbiology tools such as matrix-assisted laser desorption ionization—time-of-flight mass spectrometry or whole-genome sequencing have become a turning point in microbiological diagnosis [3]. These new technologies brought large quantity of microbiological data to support clinical

Artificial Intelligence in Clinical Practice
DOI: https://doi.org/10.1016/B978-0-443-15688-5.00020-6

decision making in the diagnosis of infectious diseases. ML applications in the microbiology laboratory use these data to improve the detection, identification and quantification of microorganisms. For instance, researchers recently reported the use of Raman optical spectroscopy of a single bacterial colony in suspension coupled with ML to identify a microorganism among 30 bacterial and yeast species [4]. Multiple applications of AI are also developed to help assessing antimicrobial susceptibility and resistance of bacteria by inferring the phenotypic antibiotic susceptibility pattern from genomic data [5]. AI is particularly useful when knowledge of the precise mutational event associated with antibiotic resistance is not complete. Researchers recently used whole genome sequencing data to predict susceptibility of 12 bacterial species to 56 antibiotics [6]. Overall, a recent review described 97 ML systems aiming to assist clinical microbiologists using various data sources: 82 ML systems (85%) targeted bacterial infections, 11 (11%) parasitic infections, nine (9%) viral infections and three (3%) fungal infections. Forty ML systems (41%) focused on microorganism detection, identification and quantification, 36 (37%) evaluated antimicrobial susceptibility, and 21 (22%) targeted the diagnosis, disease classification and prediction of clinical outcomes.

Artificial intelligence for the clinical diagnosis and management of infected patients

A review by Peiffer-Smadja et al. identified sixty different AI tools designed to support decision making for clinical diagnosis and management of infection [7]. AI was used to support the diagnosis of infection (20/60; 33%), the early detection or stratification of sepsis (18/60; 30%), the prediction of response to antimicrobial therapy (13/60; 22%), the presence of antibiotic resistance (4/60; 7%), and the choice of antibiotic regimen (3/60; 5%). AI input in the diagnosis or exclusion of infection has also been explored in a broad range of infectious conditions. These range from the diagnosis of tuberculosis in outpatient settings, to the diagnosis of bacterial infection in hospitalized patients, or the distinction between benign, viral and potentially lethal bacterial meningitis. However, the evaluation of AI methods in the management of patients with infectious diseases was very often limited to measures of technical performances (e.g., sensitivity, specificity) with no data in clinical practice or routine use. This review highlighted the importance of well-conducted clinical trials assessing clinical outcomes to evaluate the potential of ML systems in real life. However, in a recent study, researchers showed that an ML system using individualized 10-year history of urine cultures and antibiotic purchases could predict antimicrobial resistance and optimize empirical therapy in urinary tract infections [8]. As the COVID-19 pandemic has recently shown, surveillance and prevention are two crucial activities in infectious diseases. ML systems have been developed to predict the risk of transmission, for example, deep-learning image recognition has been applied to the analysis of radiological examinations to evaluate the risk of spreading tuberculosis. This application of AI could be particularly useful in low resources settings where there is limited access to specialists.

Challenges for artificial intelligence in infectious diseases

Current challenges in the application of AI to surveillance systems and infectious diseases include the lack of standardized data and large openly available databases. The unequal spreading of ML initiatives across healthcare settings reflects this heterogeneous availability of databases. Vulnerable populations tend to have a lower access to the healthcare system with a more fragmented care leading to lesser quality data that could undermine the development of ML tools and increase healthcare inequalities. Moreover, most AI applications in the literature were developed for secondary and tertiary care whereas most infections are managed in primary care. There is thus a need to encourage the systematic collection of data in primary care, especially in low- and middle-income countries.

Conclusion

Though mostly used at an experimental phase, AI has great potential to help better decision making in the management of infectious diseases. Implementation of AI technology must be supported by robust, randomized-control trial evidence. These have also to take into account AI potential application in endemic, low resources settings.

References

[1] Peiffer-Smadja N, Dellière S, Rodriguez C, Birgand G, Lescure F-X, Fourati S, et al. Machine learning in the clinical microbiology laboratory: has the time come for routine practice? Clin Microbiol Infect 2020;26(10):1300–9. Available from: https://doi.org/10.1016/j.cmi.2020.02.006.

[2] Weis CV, Jutzeler CR, Borgwardt K. Machine learning for microbial identification and antimicrobial susceptibility testing on MALDI-TOF mass spectra: a systematic review. Clin Microbiol Infect 2020;26:1310–17.

[3] Buchan BW, Ledeboer NA. Emerging technologies for the clinical microbiology laboratory. Clin Microbiol Rev 2014;27:783–822.

[4] Ho C-S, Jean N, Hogan CA, Blackmon L, Jeffrey SS, Holodniy M, et al. Rapid identification of pathogenic bacteria using raman spectroscopy and deep learning. Nat Commun 2019;10(1):4927. Available from: https://doi.org/10.1038/s41467-019-12898-9.

[5] Ruppé E, Cherkaoui A, Lazarevic V, Emonet S, Schrenzel J. Establishing genotype-to-phenotype relationships in bacteria causing hospital-acquired pneumonia: a prelude to the application of clinicalmetagenomics. Antibiotics 2017;6:1–15.

[6] Drouin A, Letarte G, Raymond F, Marchand M, Corbeil J, Laviolette F. Interpretable genotype-to-phenotype classifiers with performance guarantees. Sci Rep 2019;9(1):4071. Available from: https://doi.org/10.1038/s41598-019-40561-2.

[7] Peiffer-Smadja N, Rawson TM, Ahmad R, Buchard A, Georgiou P, Lescure F-X, et al. Machine learning for clinical decision support in infectious diseases: a narrative review of current applications. Clin Microbiol Infect 2020;26(5):584–95. Available from: https://doi.org/10.1016/j.cmi.2019.09.009.

[8] Yelin I, Snitser O, Novich G, Katz R, Tal O, Parizade M, et al. Personal clinical history predicts antibiotic resistance of urinary tract infections. Nat Med 2019;25(7):1143–52. Available from: https://doi.org/10.1038/s41591-019-0503-6.

CHAPTER

17

Artificial intelligence in neglected tropical diseases

Girish Thunga[1], Sohil Khan[2], Pooja Gopal Poojari[1], Asha K. Rajan[1], Muhammed Rashid[1], Harsimran Kaur[1] and Viji Pulikkel Chandran[1]

[1]Department of Pharmacy Practice, Manipal College of Pharmaceutical Sciences, Manipal Academy of Higher Education, Manipal, Karnataka, India [2]School of Pharmacy and Medical Sciences, Griffith University, Gold Coast Campus, Queensland, Australia

Present problems in neglected tropical diseases

Infectious diseases are responsible for majority of the deaths, illnesses, and loss of productive years of life. Although chemotherapeutic options play a major role in the management, high cost, relative lack of efficacy, and emergence of drug resistance, has led to unsatisfactory progress for better patient outcomes [1]. Neglected tropical diseases (NTDs) are an assembly of communicable diseases spread due to bacterial, parasitical or viral infestations. NTDs mainly affect rural populations with low socioeconomic status and people who live outside the reach of the health sector. These illnesses are now regarded as one of the biggest public health issues since they affect vulnerable and marginalized demographic groups [2]. They are neglected due to their low prevalence in western countries and are commonly found in areas with poor sanitation, unhygienic water, scarcity in healthcare facilities and poverty [3].

In current times when universal health coverage is a top priority, NTDs have very limited financial resources and are mostly disregarded by international funding organizations. NTDs being diseases of neglected groups results in social stigma and marginalization, poor educational outcomes, and few professional opportunities [4]. Many endemic nations have made progress toward integrating the prevention and control of NTDs into their healthcare policies, but there are several obstacles. As the burden of NTDs is increasing especially in lower- and middle-income countries (LMICs), there is a need to monitor and

Artificial Intelligence in Clinical Practice
DOI: https://doi.org/10.1016/B978-0-443-15688-5.00029-2

evaluate changes in disease epidemiology, transmission, and treatment compliance, which remains a challenge.

Vector-borne diseases have suffered for decades due to lack of investment in medication development, training, education and novel diagnostics. The majority of available medications for NTDs are outdated, due to the growing drug resistance all over the world. There are 20 NTDs as per the recent World Health Organization update and management of these NTDs by nations as separate vertical programs, each with a unique set of diagnostic and surveillance instruments, is neither practical nor effective. Finding similarities and creating a unified approach for the diagnosis and monitoring of NTDs are necessary [5].

Some of the present challenges in NTD are:

- Overlapping clinical presentations: Overlapping clinical presentation of different NTDs may lead to the misdiagnosis and improper management.
- Lack of resources: The resources needed on a global scale to tackle NTDs are not adequate.
- Population, poverty, and malnutrition: LMICs enormous populations, pervasive poverty, and malnutrition continue to pose problems in the management of neglected diseases.
- Delay in diagnosis: Rapid diagnostic tests that can be performed in primary care are lacking for most of NTDs.
- Treatment gaps: Lack of vaccines and development of novel drugs for the management.
- Global climate change: Unpredictable climate changes are altering pathogen reproduction patterns and spread, thus adding another complication to prevention and control of NTDs.
- Lack of research and development (R&D) initiatives: Stagnant knowledge reduces the ability to fill significant gaps in NTD R&D.

Dilemma in diagnosis of neglected tropical disease

The symptoms of many NTD patients are quite vague, making it difficult to diagnose NTDs accurately from their clinical manifestations alone. For proper patient management, a precise laboratory diagnosis is crucial because the preferred treatment differs greatly among various NTDs. Despite of variety of laboratory resources available, the conventional light microscopy continues to be the gold standard for many of these conditions, reflecting the widespread underutilization and underappreciation of diagnostic methods for NTDs. There are not many rapid diagnostic tests that explicitly target NTDs. In addition, very sensitive diagnostic techniques like polymerase chain reaction tests are still making their way into clinical practice [6].

From asymptomatic infections to fulminant, life-threatening, or chronically incapacitating illnesses, there is a wide range of potentially overlapping clinical manifestations even in presence of varied pathogenesis of the infectious agent. Since clinical symptoms and patient complaints are frequently difficult to distinguish, a definitive diagnosis based solely on the history and clinical examination is rarely possible without knowledge of epidemiology and additional diagnostic testing [7]. Insufficient and nonintegrated diagnostic tests which lead to empirical misdiagnosis are the culprit for the unending stories of NTDs though there are treatments available in this modern era of medicine [8]. Innovative technical interventions that make the diagnosis quicker, affordable and flexible for multi-disease diagnosis are the need of hour. These technologies also should have an ability to reciprocate to different clinical and geographical settings [9]. Incorporating intelligent technologies like artificial intelligence (AI) may provide promising solutions for existing dilemma in diagnosis of NTDs.

Applications of artificial intelligence in neglected tropical disease diagnosis

Diagnosis of disease is essential for formulating an effective treatment strategy and assuring patients' wellbeing. Due to the complexity and cognitive difficulty of comprehending medical data, human error makes accurate diagnosis difficult. Thus, more systems are dependent upon information technology in medical field. The use of AI can raise the efficiency and accuracy of diagnostic procedures [10].

AI is a technology that can analyze and learn from data to generate a certain outcome, has made significant progress in the field of NTD diagnosis and could help develop clinical and public health services in LMICs. The application of AI to the diagnosis of NTDs may facilitate clinical decision-making at the point of care, detect outbreaks before they spread, and map these diseases to direct public health surveillance and control initiatives [11]. This facilitates a speedy and precise diagnosis while saving consultation time and brainwork.

Machine learning, convolutional neural networks (CNNs), support vector machines (SVM), decision tree, classification and regression trees (CART), and artificial neural networks (ANNs) are the available AI technologies introduced in NTD diagnosis. AI applications in NTD diagnosis are broadly focused on two attributes: (1) image based and (2) symptomology and laboratory parameters based. AI tools based on image-based diagnosis in NTDs are based on large databases of images obtained from several geographical locations from a heterogenous population of patients. FecalNet, and MobileNet V2, are such examples of fecal databases of microscopic images used to detect and classify fecal constituents using deep learning approach [12,13]. A notable example is the prototype AI-digital pathology tool developed by Ward et al., for soil transmitted helminth infections [14]. Image based-diagnostic tools integrated with machine learning and CNN technology, have also been introduced for diagnosing schistosomiasis and chagas disease, respectively [15–17]. These image-based AI diagnostic tools carry the advantage of reduction in human error by manual microscopic analysis and also facilitate rapid diagnosis. Additionally, they have even facilitated remote diagnosis and telemedicine [18]. In addition to parasitic NTDs, computer vision algorithm and crowdsourcing have lead to better identification and management of snake bite envenomation [19].

AI based diagnostic tools based on patient's symptomology (e.g., fever, malaise, metallic taste, joint pain, abdominal pain, myalgia, and headache) and laboratory parameters (e.g., white blood cell counts, platelets, monocytes, hematocrit, and urine protein) are also emerging concepts in NTDs. These approaches have been used for NTDs spread through arboviruses which cause dengue, chikungunya and congenital zika syndrome.

The most common examples include diagnosis of dengue through decision tree models [9], CART models [20], SVM models [21], and machine learning [22]. In chikungunya, a symptomology based AI tool has also been developed using a comparative analysis between ANN, SVM, and a Fuzzy Logic Based Expert System (FLBES) [23]. For diagnosing children with congenital zika syndrome, laboratory parameter and symptomology attributes-based AI tools have shown promising results [24]. Another notable example is the machine learning diagnostic prototype tool developed by Shenoy et al., for differential diagnosis between dengue, malaria, leptospirosis and scrub typhus using clinical symptomology and laboratory parameters. Thus, extracting the maximum benefit through image-based and clinical data-based AI technologies is the way forward in NTD diagnosis [25].

Table 17.1 provides an insight into the available literature evidence on the application of AI in diagnosis of NTDs.

TABLE 17.1 Application of AI in diagnosis of neglected tropical diseases.

Author, year and country	Target NTDs	Sources and types of data	Model developed	Software used	Performance metrics
Shenoy et al. [25], India	Dengue, malaria, leptospirosis and scrub typhus	Clinical presentation (myalgia, arthralgia, abdominal pain, urine output), laboratory parameters (total bilirubin, sodium level, albumin level, red blood cell, lymphocytes, hematocrit, platelets, and erythrocyte sedimentation rate)	Binary classification analysis (one vs one; one vs rest strategy), multiclassification analysis (neural networks, decision tree, random forest, and multinomial regression)	WEKA	Predictability, accuracy, true positive rate/ sensitivity/recall, false positive rate, precision/positive predictive value, *F*-measure, and ROC area
Ojeda-Pat et al. [15], Mexico	Chagas disease (*Trypanosoma cruzi*)	Images of the parasite and their morphology in RGB color space	Residual convolutional neural network (Res2Unet)	Not reported	Dice coefficient, precision, recall, true positives, false positives, false negatives, active contour loss
Dacal et al. [18], Spain	Soil-transmitted helminths (*Trichuris trichiura*)	Helminth eggs in stool prepared with the Kato-Katz technique	Deep learning-based object detection model; convolutional neural network	GPU Nvidia Tesla K80 12GB	Precision, recall, and *F*-score
Veiga et al. [24], Brazil	Zika syndrome	Clinical data, and laboratorial tests	k-nearest neighbor, classification and regression trees, random forest, and gradient boosting	Not reported	Accuracy, precision, recall and F_1 score

(*Continued*)

TABLE 17.1 (Continued)

Author, year and country	Target NTDs	Sources and types of data	Model developed	Software used	Performance metrics
Tallam et al. [16], United States	Schistosoma	Images of snails and images of cercaria	Convolutional neural network	TensorFlow.js (Python based)	True positive, true negative, false positive, false negative, sensitivity, specificity, and F_1 score
Sajana et al. [22], India	Dengue	Laboratory parameters (WBC, hemoglobin, platelets) and clinical presentations (temperature, pain behind eyes, abdominal pain, vomiting, diarrhea, severe headache, metallic taste, joint/muscle pain)	C-4.5 decision tree, multilayer perception, classification and regression tree	Not reported	Accuracy, precision, recall, *F*-measure class values of positive, negative, true positive, true negative, false positive and false negative
Phakhounthong et al. [20], Thailand	Dengue prediction	Laboratory parameters (hematocrit, urine protein, creatinine, and platelet count) and clinical examination (Glasgow Coma Score)	Classification and regression tree	WEKA	Sensitivity, specificity, and accuracy
Ward et al. [14], Sweden	Soil-transmitted helminths (*Ascaris lumbricoides*, *Trichuris trichiura*, hookworms, and SCH) and *Schistosoma mansoni* eggs	Helminth eggs in stool prepared with the Kato-Katz technique	Artificial intelligence-based digital pathology device (deep learning-based object detection model); convolutional neural network	NVIDIA GeForce RTX 2070 GPU, and Jetson AGX Xavier Developer Kit	True positives, false negatives, and false positives, the recall, precision and F_1-score
Fathima and Hundewale [21], Saudi Arabia	Dengue	Initial clinical manifestation, and clinical investigation	Support vector machine algorithm and Naive bayes	R interface	Accuracy, sensitivity, specificity, and risk rate
Tanner et al. [26], Singapore	Dengue	Clinical, hematological, and virological data	C4.5 decision tree classifier	Inforsense (InforSense Ltd., London, UK)	Accuracy, sensitivity, specificity ROC curve, and AUC

AUC: area under the curve; NTD: neglected tropical diseases; ROC: receiver-operating characteristic; WBC: white blood cells; WEKA: Waikato Environment for Knowledge Analysis.

Challenges of artificial intelligence in neglected tropical disease diagnosis

The development and implementation of AI based diagnostic tools for NTDs in clinical practice is a significant challenge. Development of AI integrated tool in NTDs requires collaborative research with multidisciplinary team including data scientists, physicians and policy makers. These collaborative efforts are lacking in most of the LMICs where NTDs are predominant. As NTDs are diseases of neglected groups, adequate interest in R&D, funding efforts and expertise is another challenge.

AI algorithms are based on datasets derived from the clinical settings. The quality of datasets has a major role in differentiating the clinical condition and the model performance. AI in general, presents with number of moral and ethical issues such as guarantee of patient privacy and autonomy, patient explicability, and degree of trust by healthcare professionals on AI diagnostic tools. Availability of the sophisticated technology and internet facility are the challenges for the implementation of AI assisted tools.

Future implications of artificial intelligence in neglected tropical disease diagnosis

The use of AI to support practitioners and researchers in the diagnostics and therapeutic domain could prove beneficial for the healthcare industry and optimal health outcomes. By incorporating AI into current technical infrastructure, important medical data from various sources that are suited to the needs of the patient and the course of therapy can be quickly identified. Integration of AI in the diagnostic process creates possibilities for innovative digital health while also being able to guarantee improved patient outcomes.

References

[1] Winkler DA. Use of artificial intelligence and machine learning for discovery of drugs for neglected tropical diseases. Front Chem 2021;9:614073.

[2] Shrivastava SR, Shrivastava PS. Neglected tropical diseases: the present status and the planning for the future. J Curr Res Sci Med 2019;5(2):134.

[3] Falcone M. Neglected tropical diseases in Europe: an emerging problem for health professionals. Intern Emerg Med 2017;12(4):423–4.

[4] World Health Organization. Neglected tropical diseases, World Health organization Bulletin. <https://www.who.int/news-room/questions-and-answers/item/neglected-tropical-diseases>; 2021.

[5] Taylor EM. NTD diagnostics for disease elimination: a review. Diagnostics 2020;10(6):375.

[6] Becker SL. Diagnosis and treatment of neglected tropical diseases in europe: laboratory infrastructure, diagnostic techniques, disease notification, and surveillance systems. In: Steinmann P, Utzinger J, editors. Neglected tropical diseases—Europe and Central Asia, Cham: Springer; 2021. p. 157–83.

[7] Utzinger J, Becker SL, Knopp S, Blum J, Neumayr AL, Keiser J, et al. Neglected tropical diseases: diagnosis, clinical management, treatment and control. Swiss Med Wkly 2012;142:w13727.

[8] Bharadwaj M, Bengtson M, Golverdingen M, Waling L, Dekker C. Diagnosing point-of-care diagnostics for neglected tropical diseases. PLoS Negl Trop Dis 2021;15(6):e0009405.

[9] Souza AA, Ducker C, Argaw D, King JD, Solomon AW, Biamonte MA, et al. Diagnostics and the neglected tropical diseases roadmap: setting the agenda for 2030. Trans R Soc Trop Med Hyg 2021;115(2):129–35.

[10] Mirbabaie M, Stieglitz S, Frick NR. Artificial intelligence in disease diagnostics: a critical review and classification on the current state of research guiding future direction. Health Technol 2021;11(4):693–731.

[11] Vaisman A, Linder N, Lundin J, Orchanian-Cheff A, Coulibaly JT, Ephraim RK, et al. Artificial intelligence, diagnostic imaging and neglected tropical diseases: ethical implications. Bull World Health Organ 2020;98(4):288.
[12] Lin L, Bermejo-Peláez D, Capellán-Martín D, Cuadrado D, Rodríguez C, García L, et al. Combining collective and artificial intelligence for global health diseases diagnosis using crowdsourced annotated medical images. In: 2021 43rd annual international conference of the IEEE Engineering in Medicine & Biology Society (EMBC). IEEE; 2021. p. 3344–3348.
[13] Li Q, Li S, Liu X, He Z, Wang T, Xu Y, et al. FecalNet: automated detection of visible components in human feces using deep learning. Med Phys 2020;47(9):4212–22.
[14] Ward P, Dahlberg P, Lagatie O, Larsson J, Tynong A, Vlaminck J, et al. Affordable artificial intelligence-based digital pathology for neglected tropical diseases: a proof-of-concept for the detection of soil-transmitted helminths and Schistosoma mansoni eggs in Kato-Katz stool thick smears. PLoS Negl Trop Dis 2022;16(6):e0010500.
[15] Ojeda-Pat A, Martin-Gonzalez A, Brito-Loeza C, Ruiz-Piña H, Ruz-Suarez D. Effective residual convolutional neural network for Chagas disease parasite segmentation. Med Biol Eng Comput 2022;60(4):1099–110.
[16] Tallam K, Liu ZYC, Chamberlin AJ, Jones IJ, Shome P, Riveau G, et al. Identification of snails and schistosoma of medical importance via convolutional neural networks: a proof-of-concept application for human schistosomiasis. Front Public Health 2021;9:642895.
[17] Garcia-Vidal C, Sanjuan G, Puerta-Alcalde P, Moreno-García E, Soriano A. Artificial intelligence to support clinical decision-making processes. EBioMedicine 2019;46:27–9.
[18] Dacal E, Bermejo-Peláez D, Lin L, Álamo E, Cuadrado D, Martínez Á, et al. Mobile microscopy and telemedicine platform assisted by deep learning for the quantification of Trichuris trichiura infection. PLoS Negl Trop Dis 2021;15(9):e0009677.
[19] de Castañeda RR, Durso AM, Ray N, Fernández JL, Williams DJ, Alcoba G, et al. Snakebite and snake identification: empowering neglected communities and health-care providers with AI. Lancet Digit Health 2019;1(5):e202–3.
[20] Phakhounthong K, Chaovalit P, Jittamala P, Blacksell SD, Carter MJ, Turner P, et al. Predicting the severity of dengue fever in children on admission based on clinical features and laboratory indicators: application of classification tree analysis. BMC Pediatrics 2018;18(1):1–9.
[21] Fathima SA, Hundewale N. Comparitive analysis of machine learning techniques for classification of arbovirus. In: Proceedings of 2012 IEEE-EMBS international conference on biomedical and health informatics. IEEE; 2012. p. 376–379.
[22] Sajana T, Navya M, Gayathri YVSSV, Reshma N. Classification of dengue using machine learning techniques. Int J Eng Technol 2018;7(2.32):212–18.
[23] Hossain MS, Sultana Z, Nahar L, Andersson K. An intelligent system to diagnose chikungunya under uncertainty. J Wirel Mob Netw Ubiquitous Comput Dependable Appl 2019;10(2):37–54.
[24] Veiga RV, Schuler-Faccini L, França GV, Andrade RF, Teixeira MG, Costa LC, et al. Classification algorithm for congenital Zika Syndrome: characterizations, diagnosis and validation. Sci Rep 2021;11(1):1–7.
[25] Shenoy S, Rajan AK, Rashid M, Chandran VP, Poojari PG, Kunhikatta V, et al. Artificial intelligence in differentiating tropical infections: a step ahead. PLoS Negl Trop Dis 2022;16(6):e0010455.
[26] Tanner L, Schreiber M, Low JG, Ong A, Tolfvenstam T, Lai YL, et al. Decision tree algorithms predict the diagnosis and outcome of dengue fever in the early phase of illness. PLoS Negl Trop Dis 2008;2(3):e196.

CHAPTER

18

Artificial intelligence in psychiatry: current practice and major challenges

Ali Amer Hazime and Marc Fakhoury

Department of Natural Sciences, School of Arts and Sciences, Lebanese American University (LAU), Beirut, Lebanon

Introduction

As intelligence resembles the major asset of human's existence, artificial intelligence (AI) tends to mimic this ability and reflect it through machines and programs that process data and perform various actions only faster and with higher specificity than humans [1]. Despite the limited progress in research, the field of AI has already reached relatively advanced levels of efficiency with respect to modern medicine. AI has contributed to various enhancements in medical practice that includes diagnosis, symptom management, and healthcare cost especially in mental illness control [2]. However, the establishment of these effects within the area of psychiatry and neuroscience constitutes a greater challenge for researchers. This is mainly caused by the fact that most psychiatric disorders lack the biological markers that would allow a direct and thorough detection of the condition [3]. Moreover, one crucial issue regarding the identification of such psychiatric disorders lies within the practitioners themselves, where most of the time they are unable to provide their patients with objective diagnosis of their symptoms [4]. In an era of leading AI systems, such problems can be eliminated by implementing advanced programs that use machine learning (ML) as a basis for diagnosis with better accessibility, lower costs, and accurate results. Even in the past few years, the urgent need of AI was evident due to the COVID-19 pandemic during which many people sought psychiatric help due to the harsh counter measures that the governments had to take for the sake of human lives [5]. Furthermore, improving AI integration into psychiatric and neurological studies along with determining its limitations will provide researchers with a profound understanding of the full extent of what this field is able to provide to patients with mental disorders or patients at high risk of developing one.

Artificial Intelligence in Clinical Practice
DOI: https://doi.org/10.1016/B978-0-443-15688-5.00015-2

Examples of artificial intelligence implementation in clinical practice

In the last decade AI has revolutionized the clinical practices used for diagnosing various mental conditions [6]. As part of this development, ML techniques have provided practitioners with several superior methods that paved the way toward an improved ability to identify numerous mental issues [7]. ML systems are able to learn patterns of data consisting of patient's medical history (such as laboratory and imaging tests) and assessments performed by these systems, eventually leading to sets of numerical and categorical variables whose analysis could lay out explicit results for diagnoses in computational psychiatry [6,8–10]. Converging lines of evidence indicated that the use of ML in this context relies on the underlying relationship between psychopathology of psychiatric disorders and their corresponding neurophysiology [11,12]. Thus ML programs tend to analyze the data input (neuroimaging) and detect complex abnormal patterns in the brain that are possibly related to psychiatric disorders [13]. Moreover, the implementation of computer-based techniques has widened the scope of AI applications within the field of psychiatry. For example, latent semantic analysis has granted clinicians the ability to obtain explicit diagnosis by analyzing patients' speech transcripts [14]. Accordingly, a diagnosis can be established based on how the words used in similar context are semantically related [14]. When applied on patients with schizophrenia and healthy control group, this approach successfully determined whether a certain transcript belongs to a patient or healthy individual [15]. Following diagnosis, AI has been able to surpass traditional treatment methods in psychiatry by using computer-assisted therapy (CAT) to provide patients with some aspects of psychotherapy or behavioral treatment through videos and questionnaires [16]. For instance, the National institute for Health and Clinical Excellence (NICE) has recommended a CAT program called "Beating the Blues" due to its high-yield result in reducing symptoms of depression and anxiety [17–19]. Moreover, AI has even exceeded expectations by unraveling an extremely important feature which is the prediction of the possibility of having a mental health disorder [20]. It has been shown that upon the usage of automated speech analysis combined with ML, the prediction of psychosis's development in high-risk youths was greatly enhanced compared to clinical interviews [21]. As suicide rates dramatically increased, similar methods were relatively successful in predicting the intent of suicide in high-risk individuals [22]. Finally, AI has undoubtedly enhanced the efficiency of all fields within psychiatry and provided beneficial approaches that shortens time to diagnosis, thus saving more lives in return.

Current limitations and challenges preventing artificial intelligence application

Despite the immense improvements AI has introduced to the clinical practices of psychiatry, plenty of challenges remain intact, hindering the complete implementation of AI in this domain. Up until now, AI application in psychiatry raises some ethical concerns regarding its explicability [23]. Furthermore, AI systems may yield unfair outcomes based on biased training data fed to their corresponding algorithm [24]. Another crucial issue is that numerous studies that promote AI application in psychiatry are investigated by their own developers. These developers present the effectiveness of the methods demonstrated in favor of their own financial purposes. Thus many CATs are poorly programmed and randomly published, which can discourage individuals from pursuing the needed treatment since they are often met with an inadequate outcome [16]. It is important to note that these CATs should be clinically supervised to reduce the risk of misusage or nonadherence to the program. Individuals with severe conditions should have professionals guide them in choosing the appropriate programs and methods. For instance, unmonitored techniques can result in the formulation of suicidal and dangerous thoughts in patients [16]. In spite of the huge beneficial aspect of computational psychiatry, most of them focus on biological rather than psychological factors [25]. This means that most mental disorders are not taken into consideration during the diagnosis and treatment process with the psychological aspect being the core of most mental health diseases [26]. Overcoming these limitations and challenges is crucial as this would allow researchers and clinicians to save more lives through an advanced mental healthcare system.

Future directions

Although AI aims at mimicking human intelligence to overcome current challenges in the field of psychiatry, it still lacks the ability to comprehend emotions, diversity, and ethics [27]. Therefore future plans for AI should focus on introducing the ability to experience emotional intelligence, morality, and empathy [28–30]. Achieving this goal will generate a much efficient AI that portrays the key to guarantee individual and societal wellbeing [31–33]. Several attempts to study the association of neural activity with mental health disorders have been limited by AI's inability to understand latent neural functions and how they contribute to pathophysiology [34]. The fact that mental health disorders are directly related to brain activity indicates that any attempt to upgrade AI systems to aid in diagnosis and therapy must discuss aspects of neurostimulation based on imaging and clinical tests [31]. For this matter, explainable AI (XAI) has been introduced as a possible resolution to relate inputs and outputs of AI systems and unveil new treatment methods that can be performed by modern ML-integrated programs [35]. Furthermore, to achieve such abilities, some technical and conceptual advances are necessary. XAI has to quantify behavioral information in a way that allows interpreting it. Moreover, it should able to handle multimodal data using nonlinear computational algorithms granting them the capacity to understand behavior as numbers and codes [35]. If XAI is to achieve such heights, we can start to look up to a generation where psychiatry has created its own synthetic bio-markers and restructured mental healthcare.

Conclusion

All things considered, the integration of AI systems in psychiatry has utilized plenty of methods that enhanced physicians' ability to diagnose, treat and predict mental health conditions. The wide spectrum of AI and ML techniques such as CAT made psychiatric help even more accessible to overcome symptoms of depression and anxiety during pandemics and other compelling circumstances. However, leading research should focus on providing AI systems with greater efficacy in understanding the pathophysiology of psychiatric disorders in order to provide a better quality of life for patients.

References

[1] Macpherson T, et al. Natural and artificial intelligence: a brief introduction to the interplay between AI and neuroscience research. Neural Networks 2021;144:603–13. Available from: https://doi.org/10.1016/j.neunet.2021.09.018.

[2] Trautmann S, et al. The economic costs of mental disorders: do our societies react appropriately to the burden of mental disorders? EMBO Rep 2016;17(9):1245–9. Available from: https://doi.org/10.15252/embr.201642951.

[3] Krystal JH, Matthew WS. Psychiatric disorders: diagnosis to therapy. Cell 2014;157(1):201–14. Available from: https://doi.org/10.1016/j.cell.2014.02.042.

[4] Graham S, et al. Artificial intelligence for mental health and mental illnesses: an overview. Curr Psychiatry Rep 2019;21(11):116. Available from: https://doi.org/10.1007/s11920-019-1094-0.

[5] Di Carlo F, et al. Telepsychiatry and other cutting-edge technologies in COVID-19 pandemic: Bridging the distance in mental health assistance. Int J Clin Pract 2021;75(1). Available from: https://doi.org/10.1111/ijcp.13716.

[6] Bzdok D, Meyer-Lindenberg A. Machine learning for precision psychiatry: opportunities and challenges. Biol Psychiatry Cognit Neurosci Neuroimaging 2018;3(3):223–30. Available from: https://doi.org/10.1016/j.bpsc.2017.11.007.

[7] Shatte ABR, et al. Machine learning in mental health: a scoping review of methods and applications. Psychol Med 2019;49(9):1426–48. Available from: https://doi.org/10.1017/S0033291719000151.

[8] Jordan MI, Mitchell TM. Machine learning: trends, perspectives, and prospects. Science (New York, NY) 2015;349(6245):255–60. Available from: https://doi.org/10.1126/science.aaa8415.

[9] Nevin L. PLOS Medicine Editors. Advancing the beneficial use of machine learning in health care and medicine: toward a community understanding. PLoS Med 2018;15(11):e1002708. Available from: https://doi.org/10.1371/journal.pmed.1002708.

[10] Wiens J, Erica SS. Machine learning for healthcare: on the verge of a major shift in healthcare epidemiology. Clin Infect Dis 2018;66(1):149–53. Available from: https://doi.org/10.1093/cid/cix731.

[11] Bray S, et al. Applications of multivariate pattern classification analyses in developmental neuroimaging of healthy and clinical populations. Front Hum Neurosci 2009;3:32. Available from: https://doi.org/10.3389/neuro.09.032.2009.

[12] Lessov-Schlaggar CN, et al. The fallacy of univariate solutions to complex systems problems. Front Neurosci 2016;10(267):8. Available from: https://doi.org/10.3389/fnins.2016.00267.

[13] Nielsen AN, et al. Machine learning with neuroimaging: evaluating its applications in psychiatry. Biol Psychiatry Cognit Neurosci Neuroimaging 2020;5(8):791–8. Available from: https://doi.org/10.1016/j.bpsc.2019.11.007.

[14] Elvevåg B, et al. Quantifying incoherence in speech: an automated methodology and novel application to schizophrenia. Schizophr Res 2007;93(1–3):304–16. Available from: https://doi.org/10.1016/j.schres.2007.03.001.

[15] Elvevåg Brita, et al. An automated method to analyze language use in patients with schizophrenia and their first-degree relatives. J Neurolinguistics 2010;23(3):270–84. Available from: https://doi.org/10.1016/j.jneuroling.2009.05.002.

[16] Carroll KM, Bruce JR. Computer-assisted therapy in psychiatry: be brave-it's a new world. Curr Psychiatry Rep 2010;12(5):426–32. Available from: https://doi.org/10.1007/s11920-010-0146-2.

[17] Szanton SL, et al. Beat the Blues decreases depression in financially strained older African-American adults. Am J Geriatric Psychiatry 2014;22(7):692–7. Available from: https://doi.org/10.1016/j.jagp.2013.05.008.

[18] Proudfoot J, et al. Computerized, interactive, multimedia cognitive-behavioural program for anxiety and depression in general practice. Psychol Med 2003;33(2):217–27. Available from: https://doi.org/10.1017/s0033291702007225.

[19] Proudfoot J, et al. Clinical efficacy of computerised cognitive-behavioural therapy for anxiety and depression in primary care: randomised controlled trial. Br J Psychiatry J Ment Sci 2004;185:46–54. Available from: https://doi.org/10.1192/bjp.185.1.46.

[20] Stevens JR, et al. Psychotic disorders in children and adolescents: a primer on contemporary evaluation and management. Prim Care Companion CNS Disord 2014;16(2). Available from: https://doi.org/10.4088/PCC.13f01514.

[21] Bedi G, et al. Automated analysis of free speech predicts psychosis onset in high-risk youths. NPJ Schizophr 2015;1:15030. Available from: https://doi.org/10.1038/npjschz.2015.30.

[22] Walsh CG, et al. Predicting risk of suicide attempts over time through machine learning. Clin Psychol Sci 2017;5(3):457–69. Available from: https://doi.org/10.1177/2167702617691560.

[23] Floridi L, et al. AI4People—an ethical framework for a good AI society: opportunities, risks, principles, and recommendations. Minds Mach 2018;28(4):689–707. Available from: https://doi.org/10.1007/s11023-018-9482-5.

[24] Binns R. Fairness in machine learning: lessons from political philosophy. J Mach Learn Res 2018.

[25] Uusitalo S, et al. Mapping out the philosophical questions of AI and clinical practice in diagnosing and treating mental disorders. J Eval ClPract 2021;27(3):478–84. Available from: https://doi.org/10.1111/jep.13485.

[26] Wiese W, Karl JF. AI ethics in computational psychiatry: From the neuroscience of consciousness to the ethics of consciousness. Behav Brain Res 2022;420:113704. Available from: https://doi.org/10.1016/j.bbr.2021.113704.

[27] Lee EE, et al. Artificial intelligence for mental health care: clinical applications, barriers, facilitators, and artificial wisdom. Biol Psychiatry Cognit Neurosci Neuroimaging 2021;6(9):856–64. Available from: https://doi.org/10.1016/j.bpsc.2021.02.001.

[28] Fan L, Scheutz M, Lohani M, McCoy M, Stokes C. Do we need emotionally intelligent artificial agents? First results of human perceptions of emotional intelligence in humans compared to robots. In: Beskow J, Peters C, Castellano G, O'Sullivan C, Leite I, Kopp S, editors. Intelligent virtual agents. IVA 2017. Lecture notes in computer science, vol. 10498. Cham: Springer; 2017. Available from: https://doi.org/10.1007/978-3-319–67401-8_15.

[29] Conitzer V., Sinnott-Armstrong W., Schaich Borg J., Deng Y., Kramer M. Moral decision making frameworks for artificial intelligence. In: Proceedings of the AAAI conference on artificial intelligence, vol. 31, no. 1; February 2017. Available from: https://doi.org/10.1609/aaai.v31i1.11140.

[30] Banerjee S. A framework for designing compassionate and ethical artificial intelligence and artificial consciousness. Interdiscip Description Complex Syst 2018;18. Available from: https://doi.org/10.7906/indecs.18.2.2.

[31] Torous J, et al. Smartphones, sensors, and machine learning to advance real-time prediction and interventions for suicide prevention: a review of current progress and next steps. Curr Psychiatry Rep 2018;20(7):51. Available from: https://doi.org/10.1007/s11920-018-0914-y.

[32] Torous J, Firth J. Bridging the dichotomy of actual versus aspirational digital health. World Psychiatry 2018;17(1):108–9. Available from: https://doi.org/10.1002/wps.20464.

[33] Nebeker C, et al. Building the case for actionable ethics in digital health research supported by artificial intelligence. BMC Med 2019;17(1):137. Available from: https://doi.org/10.1186/s12916-019-1377-7.

[34] Fellous JM, et al. Explainable artificial intelligence for neuroscience: behavioral neurostimulation. Front Neurosci 2019;13:1346. Available from: https://doi.org/10.3389/fnins.2019.01346.

[35] Core M.G. Building explainable artificial intelligence systems. In: Proceedings of the 21st national conference on artificial intelligence and the 18th innovative applications of artificial intelligence conference 2; 2006. p. 1766–1773.

CHAPTER

19

Application of artificial intelligence frameworks in the clinical practice of neurology: recent advances and future directions

Nick Corriveau-Lecavalier[1], *Filip Mivalt*[1] *and David T. Jones*[1,2]

[1]Department of Neurology, Mayo Clinic, Rochester, MN, United States [2]Department of Radiology, Mayo Clinic, Rochester, MN, United States

Introduction

The practice of clinical neurology requires the collection, integration, and modeling of relevant clinical information within the context of accumulated knowledge to guide decision-making about a single patient. These steps are often performed focally by a single or a handful of clinicians and are informed by static or slowly growing knowledge. The expanding era of open and large datasets, the increased use of computational technologies and data-driven methods for healthcare-related purposes, and the very nature of neurological disorders make the practice of clinical neurology poised for an artificial intelligence (AI)-driven transformation. In this chapter, we first highlight recent advances at the intersection of clinical neurology and computational neuroscience. We then list the major challenges that pose the deployment of AI frameworks into healthcare, and concomitantly describe an integrated model aimed at the successful implementation of AI-driven frameworks into the practice clinical neurology, the *ABC* cycle [1].

Current state of artificial intelligence in clinical neurology

The recent years have seen an exponential application of AI-related techniques to healthcare-related purposes. However, there are only a few instances where AI algorithms have either equated, surpassed, or improved expert-level performance in a rigorous and

Artificial Intelligence in Clinical Practice
DOI: https://doi.org/10.1016/B978-0-443-15688-5.00032-2

clinically meaningful manner. These examples are can be found in the disciplines of dermatology, cardiology, radiology and oncology and include classification of skin cancer types [2], prediction of all-cause mortality based on echocardiographic videos [3,4], analysis of electrocardiography recordings [5,6], fetal heart rate assessment [7], tumor segmentation [8], microscopy image analysis [9,10], and analysis of genomes [11].

In the field of neurology, AI algorithms have so far largely been used for research purposes and have yet to be implemented in clinical practice. We below review recent advances that exemplify how AI could eventually shape the practice of clinical neurology, with a particular emphasis on aging and dementia and epilepsy.

Brain imaging is essential to assess abnormalities in brain structure and function relevant to neurological disorders. Many recent advances using AI algorithms have been made in the field of aging and dementia. For instance, studies have leveraged unsupervised machine learning (ML) to delineate patterns of macro-scale anatomy associated with mental functions selectively degraded across a wide range of dementia syndromes. A recent study by Jones et al. [12] showed that up to 50% of covariance in fluorodeoxyglucose-positron emission tomography (FDG-PET) images from a large cohort of individuals standing along the biological spectrum of Alzheimer's disease (AD) could be explained by a small number of global spatial patterns that also describe functional brain networks. Importantly, these latent patterns could also index seven clinical dementia syndromes, highlighting the generalizability of such method. Other studies have used similar techniques to decipher the clinico-radiological heterogeneity within relatively circumscribed dementia syndromes [13,14]. For instance, Townley et al. [14] showed that eight latent factors could explain 50% of covariance in patterns of FDG-PET hypometabolism in a large cohort of patients with posterior cortical atrophy, and that these patterns related to differential clinical and cognitive symptomatology. Data-driven techniques including Subtype and Stage Inference (SuStaIn) [15], Bayesian clustering [16], and hierarchical clustering [17] algorithms have also been used to derive biological subtypes of AD in large cohorts of patients with available multimodal imaging and characterize their clinical presentation and trajectories. These studies have expanded the clinical and biological understanding of AD by quantifying the patterns associated with the well-known heterogeneity in AD associated with distinct clinical presentations and patterns of progression [18].

A different category of studies leveraging supervised ML and deep learning (DL) methods probed biological questions of direct relevance for clinical purposes. For instance, a few groups of researchers used biological, genetic and imaging data either alone or in combination to predict cross-sectional and future tau accumulation in AD [19,20]. One of these studies used a convolutional neural network architecture to generate tau-synthesized images based on FDG-PET images with impressive accuracy in large cohorts including individuals in the normal aging spectrum as well as patients with AD and non-AD dementia syndromes [20]. Others have used a variety of DL models based on MRI and FDG-PET to yield "brain age gap" metrics across aging and dementia syndromes [21–24], which have the potential to identify individuals at risk of pathology and future clinical decline. Overall, these studies not only provide a deeper understanding of the relationships and interactions between different biological processes across neurodegenerative diseases of the mind, but also have high clinical practicality. For example, such algorithms could be used to differentiate between various degenerative etiologies and diseases staging in clinical practice and clinical trials with minimal cost while keeping excellent accuracy.

Important advances have also been made in the field of epilepsy in recent years. A particular area of progress pertains to the development, deployment and utilization of systems based on implantable neurostimulators that allow the recording of long-term and continuous intracranial electroencephalography (iEEG) data in patients with refractory temporal lobe epilepsy while they live in their natural environments [25,26]. Using ML and DL models, these studies were able to develop a system for long-term monitoring of patients with epilepsy [25,26] tracking abnormal brain activity including seizures [27] and interictal epileptiform discharges [28] but also sleep [29,30]. These studies aim to provide a better understanding of the intricate associations between seizures, interictal epileptiform discharges, sleep, psychiatric comorbidities of epilepsy [31], their cycles [32,33] and how they can be modified by deep brain stimulation [34,35]. The longer-term aim would be to use such a system not only to track behavioral brain states but also to forecast seizures using brain signals and wearable devices [36,37]. This has potentially positive implications for patients with complex epilepsy disorders in terms of seizure monitoring and management (e.g., avoiding sensitive activities during high-risk periods) as well as the optimization of pharmacological and nonpharmacological therapies including deep brain stimulation. However, the implementation of such systems bears challenges to reach the desired aims. This is mainly due to the difficulty in patient enrollment, high within- and between-patient variability and consequently limited availability of training datasets to develop AI models [30]. Recent initiatives have been pursued to address these caveats, including the release of a publicly available multicenter iEEG dataset [38].

While these recent developments in the field of behavioral neurology and epilepsy highlight the potential of AI-related technologies to enhance healthcare-related decision-making, disease monitoring and advance our understanding of complex diseases, their incorporation into clinical practice remains minimal. We below describe the *ABC* cycle, an integrated framework aimed at addressing the challenges associated with the deployment of the integration of AI algorithms into healthcare systems.

Deployment of artificial intelligence frameworks in healthcare: the *ABC* cycle

Before taking a deep dive into the *ABC* cycle, it is important to define what is meant by "artificial intelligence" and the broader term "intelligence." While definitions of AI-related terms are notorious for varying, the field has generally referred to "AI" as machines being capable of generalizable and flexible human-like reasoning and rationalizing capacities, and other terms, such as augmented intelligence or augmented human intelligence, to describe how computational machines and/or algorithms can enhance human intelligence and decision-making. These definitions do not explicitly define intelligence, but only implicitly defines it as "human-like." The definition of intelligence currently used by the Mayo Clinic Department of Neurology Artificial Intelligence Program refers to "*the collection, processing, and modeling of data and information about environments that is needed by an agent to perform beneficial actions within those, or related, environments*" [1]. This definition emphasizes three core components: the agent, the product, and the process. When applied to healthcare systems, these components respectively refer to a given healthcare delivery organization, the services provided by this organization, and the elements of the *ABC* cycle (see Fig. 19.1). This implies that the tripartite components included in this definition are not intelligent in isolation but

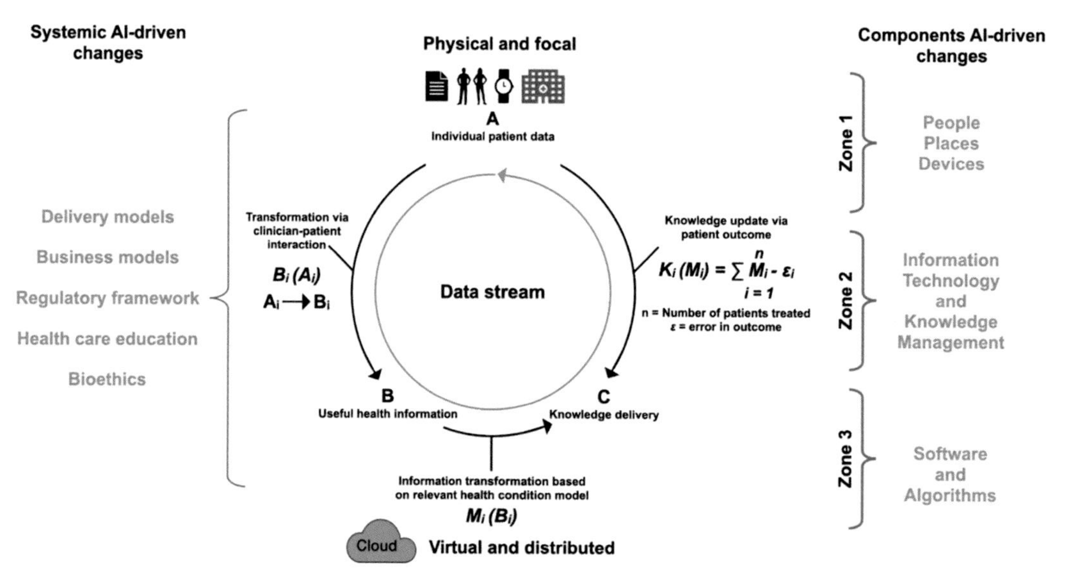

FIGURE 19.1 The ABC cycle model. This model positions at its core the data stream flow originating and ending with the individual patient (circular blue line) and the transformations applied to data and knowledge (black arrows). (A) Raw data are collected focally and physically from patients, places and/or devices and is then transformed into useful health information (B), which is most often done through the current clinician-patient interaction, that is, $B_i(A_i)$. This information is then contextualized within a model relevant to the patient's condition, that is, M. to update knowledge delivery in health care settings (C). Healthcare knowledge (C) can be updated in a feedback fashion through patient outcome (K) in real-world clinical settings. This whole data-centered process should organically drive systemic changes, such as delivery models and delivery models, regulatory framework, health care education and bioethics (left-hand side of the figure) and components of the ABC cycle spanning physical and focal components (zone 1) to virtual and distributed components (zone 3), which are tied by information and technology and knowledge management (zone 2).

can serve intelligence by fostering a dynamic synergy between high quality patient care, continuously informed disease models, and technological innovation.

The deployment of AI algorithms into healthcare delivery systems and clinical practice will require proper standards and integration. Such standards imply a fundamental transformation of how data streams flow within and even potential across healthcare institutions [1,39]. The actual model state in the vast majority of healthcare systems involves the focal acquisition of patient data that is transformed into relevant information mainly through patient-physician dyads ($Ai \rightarrow Bi$ in Fig. 19.1). The *ABC* cycle proposes the extension of this model for data streams to become virtually stored distributed in a systematic manner, with or without the implementation of federated learning infrastructure depending on the centralized or decentralized nature of the dataset [39]. This virtually distributed health information will allow for the development and implementation of AI algorithms that are naturally grounded in clinical purposes, leading to optimized physician-level knowledge in delivering high quality patient care. Importantly, these AI-driven models will be directly tested through clinical practice, allowing for their validation and refinement at every step of the way.

This data-centered transformation will inevitably rely on extensive information technology infrastructure that is scalable, robust, and adapted to every component of the *ABC* cycle. This will be critical for various purposes, which include but are not restricted to the automated processing of raw and heterogeneous data into curated information that is readily available to be modeled by AI algorithms, addressing data shifts and ensuring the storing, monitoring, labeling and securing of incoming and outgoing data. Ensuring success in the implementation of such infrastructure, especially at its early stages, will require the adoption of industry-wide standards, and even perhaps partnerships with big technology companies (or Big Tech) for cloud storage, computational infrastructure and expertise. It is however essential that offers made to healthcare institutions by such companies must be tailored to the *ABC* cycle core components (product, process, agent) to foster the development and growth of a dynamic data-centered system that begins and ends at the point of patient care. To put this another way, the healthcare sector must lead the technology sector and not the other way around.

The AI-driven extension of the practice of neurology and medicine in general will eventually and organically lead to the emergence of systemic changes surrounding the core components of the *ABC* cycle (left-hand side of Fig. 19.1). This includes regulatory frameworks concerning the use of computational software and patient data in the context of AI-related activities. Large-scale examples of this already exist, including the European Union's General Data Protection Regulation [40] or the US and the Food and Drug Administration's (FDA) AI/ML-Based Software as a Medical Device Action Plan (https://www.fda.gov/media/145022/download). Revamped bioethics guidelines and delivery models will also be needed to ensure that AI algorithms are in line with the rigorous tenets of research and clinical practice and respect patients' needs and rights. This includes, among other things, protecting patient privacy and safety, representativeness in terms of demographics and clinical diversity, and ensuring the accountability of AI algorithms (e.g., medical errors). Such a framework will also be essential to build public trust toward AI-enhanced health care delivery. Health education will also need to undergo changes to account for the upcoming AI-driven changes in clinical practice. This is tremendous to ensure physician-patient dyads, communication between healthcare providers and between the medical field and policy makers keep track with the technological advances aimed at the modernization of the practice of medicine [41].

Conclusions

The recent advances at the intersection of medicine and computational sciences have highlighted the potential of AI algorithms to aid decision-making in healthcare, improve patient outcomes, probe into biological questions and inform clinical trials. However, the successful integration of AI-driven technologies into the practice of clinical neurology is yet to come and is dependent on the implementation of integrated data-centered frameworks relying on extensive information technology infrastructure and industry-wide standards, as depicted by the *ABC cycle*.

Major takeaways

- The growing influence of AI on the practice of medicine is ubiquitous and has the potential to enhance high quality clinical care and inform disease models.
- Despite promising recent advances, AI algorithms have yet to be concretely implemented in the practice of clinical neurology.
- Integrated frameworks beginning and ending at the point of care centered on the needs of the patient and relying on extensive information technology infrastructure and industry-wide standards are required for the successful deployment of AI algorithms into health systems.
- Systemic changes should emerge from and track with the technological advances aimed at the modernization of the practice of medicine.

References

[1] Jones DT, Kerber KA. Artificial intelligence and the practice of neurology in 2035: the neurology future forecasting series. Neurology 2022;98(6):238–45. Available from: https://doi.org/10.1212/WNL.0000000000013200.

[2] Esteva A, Kuprel B, Novoa RA, Ko J, Swetter SM, Blau HM, et al. Dermatologist-level classification of skin cancer with deep neural networks. Nature 2017;542(7639):115–18.

[3] Ulloa Cerna AE, Jing L, Good CW, vanMaanen DP, Raghunath S, Suever JD, et al. Deep-learning-assisted analysis of echocardiographic videos improves predictions of all-cause mortality. Nat Biomed Eng 2021;5 (6):546–54.

[4] di Ruffano LF, Takwoingi Y, Dinnes J, Chuchu N, Bayliss SE, Davenport C, et al. Computer-assisted diagnosis techniques (dermoscopy and spectroscopy-based) for diagnosing skin cancer in adults. Cochrane Database Syst Rev 2018;12.

[5] Ivora A, Viscor I, Nejedly P, Smisek R, Koscova Z, Bulkova V, et al. QRS detection and classification in Holter ECG data in one inference step. Sci Rep 2022;12(1):1–9.

[6] Nejedly P, Ivora A, Viscor I, Koscova Z, Smisek R, Jurak P, et al. Classification of ECG using ensemble of residual CNNs with or without attention mechanism. Physiol Meas 2022;43(4):044001.

[7] Lutomski JE, Meaney S, Greene RA, Ryan AC, Devane D. Expert systems for fetal assessment in labour. Cochrane Database Syst Rev 2015;4.

[8] Veiga-Canuto D, Cerdà-Alberich L, Sangüesa Nebot C, Martínez de las Heras B, Pötschger U, Gabelloni M, et al. Comparative multicentric evaluation of inter-observer variability in manual and automatic segmentation of neuroblastic tumors in magnetic resonance images. Cancers 2022;14(15):3648.

[9] Kromp F, Fischer L, Bozsaky E, Ambros IM, Dörr W, Beiske K, et al. Evaluation of Deep Learning architectures for complex immunofluorescence nuclear image segmentation. IEEE Trans Med Imaging 2021; 40(7):1934–49.

[10] Lazic D, Kromp F, Rifatbegovic F, Repiscak P, Kirr M, Mivalt F, et al. Landscape of bone marrow metastasis in human neuroblastoma unraveled by transcriptomics and deep multiplex imaging. Cancers 2021; 13(17):4311.

[11] Hess L, Moos V, Lauber AA, Reiter W, Schuster M, Hartl N, et al. A toolbox for class I HDACs reveals isoform specific roles in gene regulation and protein acetylation. PLoS Genet 2022;18(8):e1010376.

[12] Jones D, Lowe V, Graff-Radford J, Botha H, Barnard L, Wiepert D, et al. A computational model of neurodegeneration in Alzheimer's disease. Nat Commun 2022;13(1):1–13.

[13] Groot C, Yeo BT, Vogel JW, Zhang X, Sun N, Mormino EC, et al. Latent atrophy factors related to phenotypical variants of posterior cortical atrophy. Neurology 2020;95(12):e1672–85.

[14] Townley RA, Botha H, Graff-Radford J, Whitwell J, Boeve BF, Machulda MM, et al. Posterior cortical atrophy phenotypic heterogeneity revealed by decoding 18F-FDG-PET. Brain Commun 2021;3(4):fcab182.

[15] Vogel JW, Young AL, Oxtoby NP, Smith R, Ossenkoppele R, Strandberg OT, et al. Four distinct trajectories of tau deposition identified in Alzheimer's disease. Nat Med 2021;27(5):871–81.

[16] Poulakis K, Pereira JB, Muehlboeck J, Wahlund LO, Smedby Ö, Volpe G, et al. Multi-cohort and longitudinal Bayesian clustering study of stage and subtype in Alzheimer's disease. Nat Commun 2022;13(1):1–15.

[17] Levin F, Ferreira D, Lange C, Dyrba M, Westman E, Buchert R, et al. Data-driven FDG-PET subtypes of Alzheimer's disease-related neurodegeneration. Alzheimers Res Ther 2021;13(1):1–14.

[18] Graff-Radford J, Yong KX, Apostolova LG, Bouwman FH, Carrillo M, Dickerson BC, et al. New insights into atypical Alzheimer's disease in the era of biomarkers. Lancet Neurol 2021;20(3):222–34.

[19] Giorgio J, Jagust WJ, Baker S, Landau SM, Tino P, Kourtzi Z. A robust and interpretable machine learning approach using multimodal biological data to predict future pathological tau accumulation. Nat Commun 2022;13(1):1–14.

[20] Lee J, Burkett BJ, Min HK, Senjem ML, Lundt ES, Botha H, et al. Deep learning-based brain age prediction in normal aging and dementia. Nat Aging 2022;2(5):412–24.

[21] Abrol A, Fu Z, Salman M, Silva R, Du Y, Plis S, et al. Deep learning encodes robust discriminative neuroimaging representations to outperform standard machine learning. Nat Commun 2021;12(1):1–17.

[22] Jones D, Lee J, Topol E. Digitising brain age. Lancet 2022;400(10357):988.

[23] Lee J, Burkett BJ, Min HK, Senjem ML, Dicks E, Corriveau-Lecavalier N, et al. Synthesizing images of tau pathology from cross-model neuroimaging using deep learning. BioRxiv 2022.

[24] Bashyam VM, Erus G, Doshi J, Habes M, Nasrallah IM, Truelove-Hill M, et al. MRI signatures of brain age and disease over the lifespan based on a deep brain network and 14 468 individuals worldwide. Brain 2020;143(7):2312–24.

[25] Kremen V, Brinkmann BH, Kim I, Guragain H, Nasseri M, Magee AL, et al. Integrating brain implants with local and distributed computing devices: a next generation epilepsy management system. IEEE J Transl Eng Health Med 2018;6:1–12.

[26] Pal Attia T, Crepeau D, Kremen V, Nasseri M, Guragain H, Steele SW, et al. Epilepsy personal assistant device—a mobile platform for brain state, dense behavioral and physiology tracking and controlling adaptive stimulation. Front Neurol 2021;1195.

[27] Sladky V, Nejedly P, Mivalt F, Brinkmann BH, Kim I, St. Louis EK, et al. Distributed brain co-processor for tracking spikes, seizures and behaviour during electrical brain stimulation. Brain Commun 2022;4(3): fcac115.

[28] Janca R, Jezdik P, Cmejla R, Tomasek M, Worrell GA, Stead M, et al. Detection of interictal epileptiform discharges using signal envelope distribution modelling: application to epileptic and non-epileptic intracranial recordings. Brain Topogr 2015;28(1):172–83.

[29] Dell KL, Payne DE, Kremen V, Maturana MI, Gerla V, Nejedly P, et al. Seizure likelihood varies with day-to-day variations in sleep duration in patients with refractory focal epilepsy: a longitudinal electroencephalography investigation. EClinicalMedicine 2021;37:100934.

[30] Mivalt F, Kremen V, Sladky V, Balzekas I, Nejedly P, Gregg NM, et al. Electrical brain stimulation and continuous behavioral state tracking in ambulatory humans. J Neural Eng 2022;19(1):016019.

[31] Balzekas I, Sladky V, Nejedly P, Brinkmann BH, Crepeau D, Mivalt F, et al. Invasive electrophysiology for circuit discovery and study of comorbid psychiatric disorders in patients with epilepsy: Challenges, opportunities, and novel technologies. Front Hum Neurosci 2021;416.

[32] Karoly PJ, Rao VR, Gregg NM, Worrell GA, Bernard C, Cook MJ, et al. Cycles in epilepsy. Nat Rev Neurol 2021;17(5):267–84.

[33] Baud MO, Kleen JK, Mirro EA, Andrechak JC, King-Stephens D, Chang EF, et al. Multi-day rhythms modulate seizure risk in epilepsy. Nat Commun 2018;9(1):1–10.

[34] Gregg NM, Sladky V, Nejedly P, Mivalt F, Kim I, Balzekas I, et al. Thalamic deep brain stimulation modulates cycles of seizure risk in epilepsy. Sci Rep 2021;11(1):1–12.

[35] Marks VS, Richner TJ, Gregg NM, Sladky V, Dolezal J, Kremen V, et al. Deep brain stimulation of anterior nuclei of the thalamus and hippocampal seizure rate modulate verbal memory performance. 2022 IEEE international conference on electro information technology (eIT). IEEE; 2022. p. 1–4.

[36] Pal Attia T, Viana PF, Nasseri M, Duun-Henriksen J, Biondi A, Winston JS, et al. Seizure forecasting using minimally invasive, ultra-long-term subcutaneous EEG: generalizable cross-patient models. Epilepsia 2022.

[37] Baud MO, Proix T, Gregg NM, Brinkmann BH, Nurse ES, Cook MJ, et al. Seizure forecasting: bifurcations in the long and winding road. Epilepsia 2022.

[38] Nejedly P, Kremen V, Sladky V, Cimbalnik J, Klimes P, Plesinger F, et al. Multicenter intracranial EEG dataset for classification of graphoelements and artifactual signals. Sci Data 2020;7(1):1–7.

[39] Zhang A, Xing L, Zou J, Wu JC. Shifting machine learning for healthcare from development to deployment and from models to data. Nat Biomed Eng 2022;1–16.

[40] Das AK. European Union's general data protectionregulation, 2018: a brief overview. Ann Libr Inf Stud 2018;65(2):139–40.

[41] Kolyshkina I, Simoff S. Interpretability of machine learning solutions in public healthcare: the CRISP-ML approach. Front Big Data 2021;4:660206.

CHAPTER

20

Artificial intelligence in rheumatology

Junjie Peng[1], *George Robinson*[1,2], *Elizabeth C. Jury*[1,2], *Pierre Dönnes*[3] *and Coziana Ciurtin*[1]

[1]Department of Medicine, Centre for Adolescent Rheumatology Versus Arthritis, University College London, London, United Kingdom [2]Department of Medicine, Centre for Rheumatology Research, University College London, London, United Kingdom [3]Scicross AB, Skövde, Sweden

Introduction

Autoimmune rheumatic diseases (ARDs) are characterized by chronic inflammation driven by abnormal immune responses reflected in a significant heterogeneity in clinical presentation and a multitude of disease and organ specific biomarkers. Recent advances in research try to make use of various tests and patient and disease related outcomes to improve disease recognition, and support personalized therapeutic interventions, ultimately aiming for improved patient outcomes. Artificial intelligence (AI) is a broad terminology comprising various analytical models that aim to mimic the function of the human brain by learning from existing knowledge. Machine learning (ML) is one of the most important AI methods, which has the ability to handle high-dimensional data, automatically without human intervention, allowing for easy pattern identification in a wide range of applications. The clinical need to identify patterns of response to treatment or predict disease flares as well as damage accrual in ARDs has led to the development of various AI techniques in rheumatology. AI excels at solving complex data made available by advances in high throughput biological techniques which are increasingly used in rheumatology research (Fig. 20.1). This chapter showcases some of key ML applications in rheumatology, emphasizing how ML can speed up the study of ARDs toward the goal of personalized medicine. Major challenges and limitations of applying ML in clinical study will also be discussed.

Artificial Intelligence in Clinical Practice
DOI: https://doi.org/10.1016/B978-0-443-15688-5.00033-4

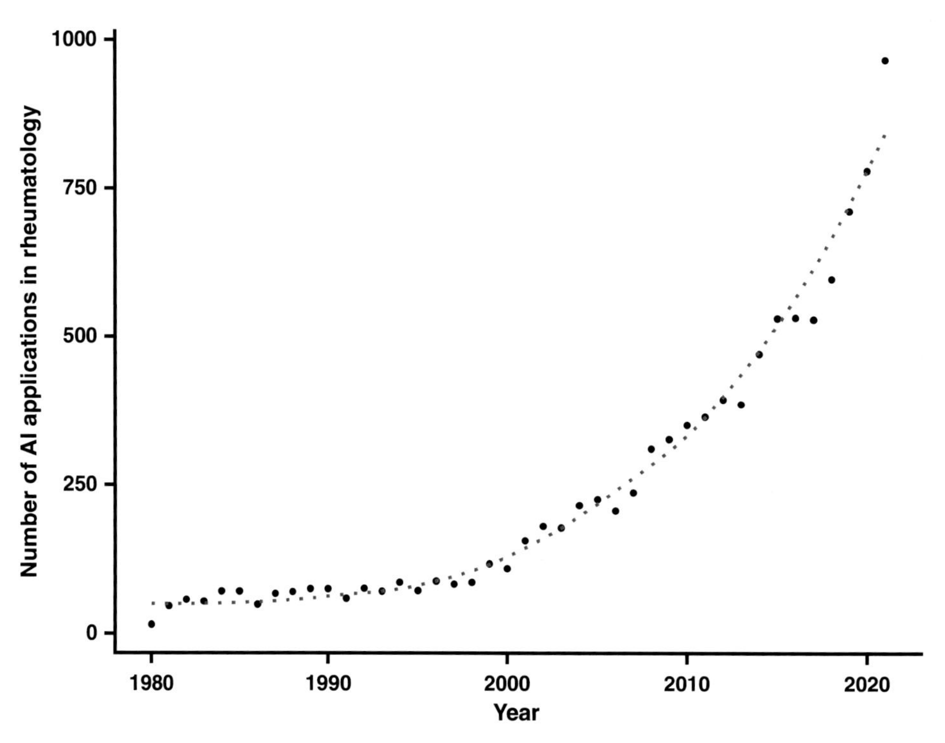

FIGURE 20.1 The rapid increase of AI application in rheumatology. Number of results appeared in PubMed from 1980 to 2021; under search term ("rheumatology" or "rheumatoid arthritis" or "Lupus") AND ("AI" OR "machine learning" OR "deep learning").

Current examples of artificial intelligence implementation in rheumatology

Machine learning applications in diagnosis

In recent years substantial breakthroughs using AI/ML applications have been made in the field of rheumatology, including disease prevention, diagnosis, prognosis, drug discovery and clinical trial design (Fig. 20.2).

Diagnostic biomarkers are useful in supporting early disease recognition and patient stratification. Some diagnostic biomarkers have significant value in providing valuable insight into the disease pathogenesis which can inform therapeutic approaches.

A number of studies have examined the application of ML techniques to improve diagnosis of ARDs using routinely collected medical record data, such as electronic health records (EHRs) and electronic medical records (EMRs) [1–3]. ML algorithms were used to identify patients with systemic lupus erythematosus (SLE) with good performance (area under curve—AUC = 0.909, with a 92% positive prediction rate), a complex disease whose diagnosis requires multiple criteria, including clinical presentation, history of symptoms, and laboratory data using a combination of combined the rule-based and natural language processing algorithms [3].

Another study [2] used an ensemble algorithm (AdaBoost learners, EasyEnsemble) [4] applied to an imbalanced dataset (derived from 583 SLE, and 16174 non-SLE individual

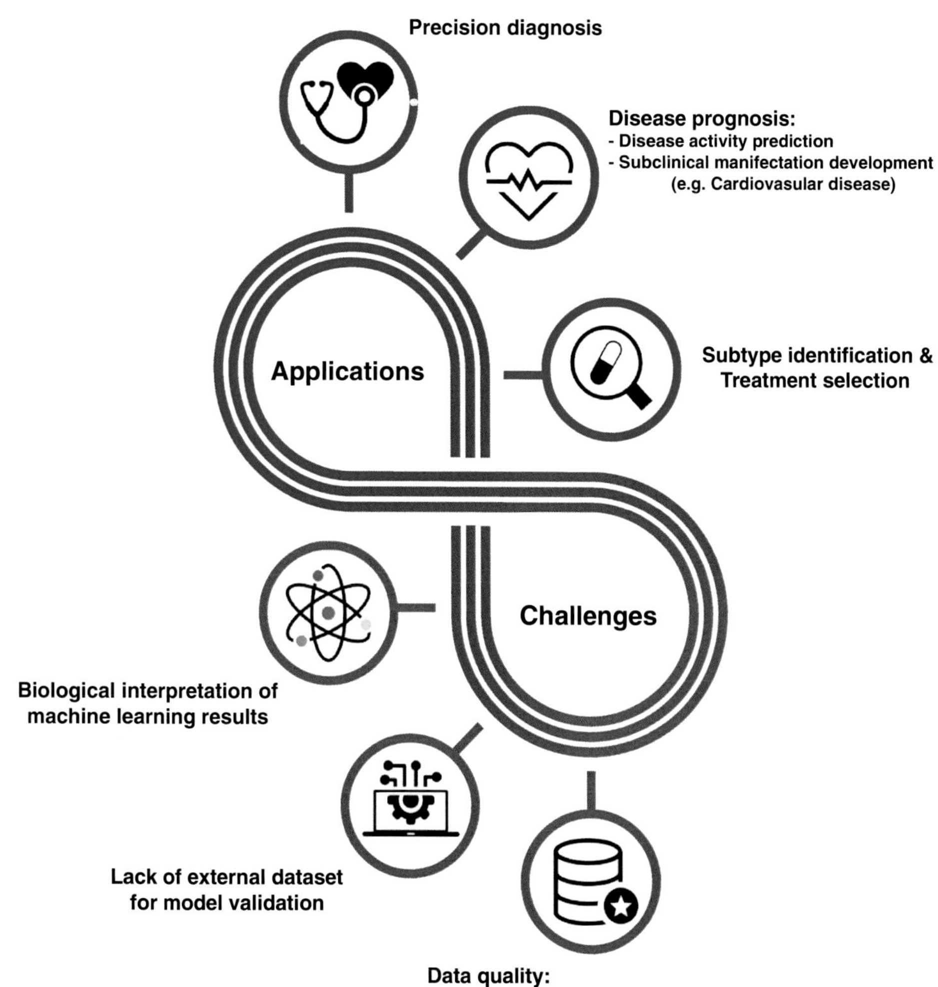

FIGURE 20.2 Applications and challenges of AI/ML in rheumatology. *ARDs*, Autoimmune rheumatic diseases.

patient EHRs) to identify patients with SLE. A high model performance was achieved (AUC = 0.97) and maintained in the testing dataset (AUC = 0.94), when compared to the rheumatologist expert classification. Similar studies have applied EMR data to classify patients with rheumatoid arthritis (RA) [1].

In a study of neuropsychiatric SLE [5], Simos et al. applied ML algorithms to enhance current neuropsychiatric SLE diagnosis approaches based on resting-state functional connectivity MRI (fMRI) imaging data of the brain, and achieved the best performance in identifying neuropsychiatric SLE patients with AUC = 0.75, validated by cross-validation, using a support vector machine model, which was used in addition to other classifiers, such as random forest, naïve Bayes and k-nearest neighbors, which were trained by the fMRI connectivity matrix derived from fMRI images of the brain network of 41 neuropsychiatric SLE patients and 31 healthy controls.

Van Nieuwenhove et al. [6] applied ML algorithms to identify diagnostic biomarkers in juvenile idiopathic arthritis (JIA), derived from the immunophenotyping of 72 JIA patients and 43 age-matched healthy controls, and found that iNKT cell subtype was the variable that contributed most to the random forest model used to discriminate JIA patients from healthy controls (AUC = 0.90).

Machine learning application in disease prognosis

ARDs are chronic conditions with no known cure; therefore being able to predict the disease course, treatment response or and future outcomes relevant to patients is key to personalized management strategies.

ML classification models can also be applied in disease activity prediction of complex ARDs. Disease activity in SLE has been predicted using whole blood gene expression data [7]. Supervised ML classifiers including random forest, k-nearest neighbors and generalized linear models were used to separate SLE patients with active versus inactive disease, with the random forest classifier scoring the highest performance with a peak accuracy of 83% when using raw gene expression data as a predictor. The mean decrease in Gini impurity from the random forest model indicated an important role for CD14 + monocytes in the pathogenesis of SLE patients with active disease, providing a valuable insight into the cellular processes potentially driving inflammation in SLE.

Another study [8] attempted to identify SLE patients with high disease activity (HDA) using ML algorithms including measurements of HDA, defined as Systemic Lupus Erythematosus Disease Activity Index (SLEDAI) $-2K \geq 10$, 16 laboratory and three demographic parameters (age, sex, and ethnicity) were used to build a multinomial logistic regression model. After screening a total of 216 models, the final model including seven laboratory variables and three demographic variables identified with 88.6% accuracy whether a certain SLE patient had HDA or not. The study highlights the possibility of using a limited amount of routinely available laboratory measurements and demographics to select SLE patients with HDA, which could help the early identification of SLE patients likely to require treatment escalation after testing the model in a clinical setting.

In a recent study of SLE [9], researchers attempted to predict subclinical atherosclerosis in SLE patients using serum metabolomics data quantified by nuclear magnetic resonance spectroscopy. The presence of subclinical atherosclerosis was assessed by femoral and carotid artery ultrasound scans. Five supervised classification models using metabolomic data were applied to predict subclinical atherosclerosis. The logistic regression with interactions model achieved the highest classification accuracy (80%). As SLE patients are known to be at higher risk of developing cardiovascular disease compared to age and sex-matched healthy individuals, this study revealed the possibility of applying serum biomarkers to identify SLE patients at risk early and support the use of early cardio-vascular risk preventative strategies.

Machine learning application in disease phenotype identification and treatment selection

Personalized treatment is a fundamental aim of precision medicine, aiming to support tailored treatment strategies instead of the "one-size-fits-all" approach. An effective way of delivering personalized treatment is by performing precise patient classification based on clinical manifestations and/or distinct pathogenetic signatures. Signatures can be identified from clinical and routine serological data, as well as genomes, metabolomics, immunophenotyping and other types of omic data. Supervised ML is an ideal tool, specialized in the identification of unique signatures, while clustering approaches from both supervised and unsupervised ML are designed for partitioning complex high dimensional data. An increasing number of studies have applied ML models to identify subgroups of patients and show promising results toward more personalized treatment [10–13].

Figgett et al. [14] applied ML clustering approaches to whole-blood RNA-sequencing data in patients with SLE. SLE patients were stratified into four clusters by k-means clustering based on their gene expression, which also had distinct clinical manifestations, which improves the understanding of SLE heterogeneity and emphasizes the need for individualized therapeutic strategies.

In the recent study of primary Sjögren's syndrome (pSS) and SLE [15], researchers applied supervised ML models to identify shared immunological characteristics between pSS and SLE. These two diseases share some clinical and laboratory features, despite differences in disease pathogenesis and overall clinical presentation [16]. Various analyses including supervised ML models (balanced random forest and sparse partial least squares discriminant analysis), univariate logistic regression and multiple *t*-tests were used to confirm the immunological similarity between pSS and SLE. The balanced random forest model identified a T-cell signature in the subsets that differentiated between the two groups of patients with high performance (AUC = 0.99). The study suggests the potential of differentiating pSS and SLE patients based on their immunological profile and could provide the opportunity for more accurate targeted treatments across diagnostic boundaries.

In a study of JIA [11], ML algorithms were employed to predict the efficiency of biological therapy (etanercept) in JIA patients using EMR data. A wide range of supervised ML approaches including extreme gradient boosting (XGBoost), random forest, gradient boosting decision tree, extremely random trees and logistic regression were tested as potential predictive models after optimization. The XGBoost strategy outperformed the other models with an AUC = 0.79 indicating a good predictive performance for prediction of treatment response. A recent ML application for personalized treatment response in RA investigated with success molecular signatures predictive of response to adalimumab and etanercept using differential gene expression in peripheral blood mononuclear cells (PBMCs), monocytes and CD4 + T cells and methylation analysis in PBMCs [17]. The random forest algorithms implemented to exploit the transcriptome signatures had an overall accuracy of 85.9% and 79% for response to adalimumab and etanercept and they have been validated in a partial dataset (a follow-up study).

Current challenges

Although AI/ML applications have shown great potential in addressing unmet clinical needs in rheumatology, through better disease characterization and improved understanding of factors that underly biological heterogeneity, researchers should be aware of their limitations, such as lack of sample size and poor data quality, poor reproducibility, difficulty in model implementation, and several ethical concerns.

Robust models require sufficient high-quality data

The performance of AI applications depends heavily on the quality of the data, including adequate sample size and adequate representability of target population for specific clinical applications. ML models trained by small sample sizes often suffer from the problem of "overfitting," where the model over relies on characteristics from the underrepresented training data and loses the ability to effectively perform in practice.

One way to improve the model reliability in the context of a small sample size is by reducing the model variance, as low variance algorithms are less influenced by the specificity of the training data. However, model variance reduction often results in an increase in model biased error, leading to a weakened predictive performance of models [18]. Meanwhile, obtaining a larger sample size often requires more resources (time, funding, access to large patient populations and computer power etc.).

The study by Robinson et al. [19] applied ML models to stratify a small number of patients with juvenile SLE (which is a rare ARD), based on their immune profile [19]. A balanced random forest algorithm was selected to characterize 67 JSLE patients and 39 healthy controls with 28 immune cell predictors, as it was less likely to overfit the data due to an implanted bagging method and random feature selection in the model ensembled by a large number of decision trees [20,21]. The results of the ML model were combined with additional analysis, such as the sparse-PLS-DA and univariate logistic regression, and were further validated by 10-fold cross-validation, showing the potential for applying a ML-based pipeline to other rare and heterogeneous autoimmune conditions [22]. This analysis identified a disease signature useful for patient stratification in relation to long-term disease trajectories highlighting the potential application of ML techniques in translational research in rare diseases.

External validation with independent datasets

Another challenge in the development of ML models is the access to high quality and well-defined datasets, needed for algorithm training and evaluation. In recent years a greater emphasis has been made on the need to make research data Findable, Accessible, Interoperable, and Reusable [23]. Ideally, clear rules for data access and use should be available, as well as use of domain-specific ontologies to describe the data. There should also be enough information available describing how the acquisition of data was carried out, enabling the reuse of data.

To achieve the highest model performance, many clinical studies tend to avoid data splitting for model development. Resampling methods, such as bootstrapping and *k*-fold cross-validation, are economical internal validation; therefore they are often applied to prevent model overfitting. On the other hand, external validation using an independent cohort is not often performed, potentially due to limited access to similar cohorts. Less than 10% of autoimmune studies combine cross-validation with an independent test dataset for validating model performance [24]. However, external validation remains a crucial step for model implementation in real-world clinical practice and the absence of external validation will raise several concerns for the model integrity including bias of the model, lack of reproducibility and lack of model generalizability [25].

Future direction

Advances in the use of ML techniques in research in rheumatology have highlighted the power of computers in generating information previously not accessible through traditional research analysis, in addition to improving the speed and efficiency. However, future investment in the standardization of ML applications, improvement of research study design to facilitate granular and relevant data collection, as well as use of an adequate sample size in relation to data multidimensionality, are all required to minimize the risk of significant data redundancy which can hamper the relevance of findings [26]. Identification of reproducible biomarkers with prognostic value is one of the key requirements for personalized medicine approaches and we advocate for the use of truly independent data sets for validation. Although in theory, personalized medicine could be advanced by the use of ML algorithms for individual disease risk identification and prognostic, as well as therapy selection, its implementation in large health systems poses the ethical challenges of reconciling health risk inequalities with finite health care resources [27]. Future research should provide answers regarding the advantages of ML-driven personalized medicine strategies for improved patient outcomes in real-life.

Future research should be focused on developing predictive ML models with outstanding biomarker selection capability for diagnosis and prognosis. Patient stratification by unsupervised models and advanced drug development strategies for personalized treatment selection are especially relevant for patients with rheumatic diseases, because of heterogeneity in clinical presentation, disease course and response to therapy. Future research in rheumatology should aim to make use of the advantages of AI methods both in lab and clinical research, as well as routine practice, while using unbiased, reliable and high-quality data sources to promote better diagnosis, advanced disease understanding and characterization, as well as individualized management for improved health care outcomes in rheumatology.

Key messages

1. ML techniques have improved the analysis of high dimensionality data relevant to autoimmune processes underlying rheumatic diseases.
2. AI applications in rheumatology include diagnosis, phenotype identification, prognosis, and treatment response.
3. ML can facilitate improved predictions for clinically relevant outcomes and personalized medicine approaches in rheumatic diseases characterized by high heterogeneity.
4. Future research is required to address the need for improved data quality, AI algorithm optimization and external validation before ML techniques can be implemented in clinical practice.

References

[1] Liao KP, Cai T, Gainer V, Goryachev S, Zeng-treitler Q, Raychaudhuri S, et al. Electronic medical records for discovery research in rheumatoid arthritis. Arthritis Care Res 2010;62(8):1120–7.

[2] Murray SG, Avati A, Schmajuk G, Yazdany J. Automated and flexible identification of complex disease: building a model for systemic lupus erythematosus using noisy labeling. J Am Med Inf Assoc 2019;26 (1):61–5.

[3] Jorge A, Castro VM, Barnado A, Gainer V, Hong C, Cai T, et al. Identifying lupus patients in electronic health records: development and validation of machine learning algorithms and application of rule-based algorithms. Semin Arthritis Rheum 2019;49(1):84–90. Available from: https://doi.org/10.1016/j.semarthrit.2019.01.002.

[4] Liu X-Y, Wu J, Zhou Z-H. Exploratory undersampling for class-imbalance learning. IEEE Trans Syst Man Cybern B Cybern 2008;39(2):539–50.

[5] Simos NJ, Manikis GC, Papadaki E, Kavroulakis E, Bertsias G, Marias K. Machine learning classification of neuropsychiatric systemic lupus erythematosus patients using resting-state fMRI functional connectivity. In: 2019 IEEE international conference on imaging systems and techniques (IST); 2019.

[6] Van Nieuwenhove E, Lagou V, Van Eyck L, Dooley J, Bodenhofer U, Roca C, et al. Machine learning identifies an immunological pattern associated with multiple juvenile idiopathic arthritis subtypes. Ann Rheum Dis 2019;78(5):617–28.

[7] Kegerreis B, Catalina MD, Bachali P, Geraci NS, Labonte AC, Zeng C, et al. Machine learning approaches to predict lupus disease activity from gene expression data. Sci Rep 2019;9(1):1–12.

[8] Hoi A, Nim HT, Koelmeyer R, Sun Y, Kao A, Gunther O, et al. Algorithm for calculating high disease activity in SLE. Rheumatology 2021;60(9):4291–7.

[9] Coelewij L, Waddington KE, Robinson GA, Chocano E, McDonnell T, Farinha F, et al. Serum metabolomic signatures can predict subclinical atherosclerosis in patients with systemic lupus erythematosus. Arterioscler Thromb Vasc Biol 2021;41(4):1446–58.

[10] McKinney EF, Lyons PA, Carr EJ, Hollis JL, Jayne DR, Willcocks LC, et al. A CD8+ T cell transcription signature predicts prognosis in autoimmune disease. Nat Med 2010;16(5):586–91.

[11] Mo X, Chen X, Ieong C, Zhang S, Li H, Li J, et al. Early prediction of clinical response to etanercept treatment in juvenile idiopathic arthritis using machine learning. Front Pharmacol 2020;11:1164.

[12] Rehberg M, Giegerich C, Praestgaard A, van Hoogstraten H, Iglesias-Rodriguez M, Curtis JR, et al. Identification of a rule to predict response to sarilumab in patients with rheumatoid arthritis using machine learning and clinical trial data. Rheumatol Ther 2021;8(4):1661–75. Available from: https://doi.org/10.1007/s40744-021-00361-5.

[13] Waljee AK, Wallace BI, Cohen-Mekelburg S, Liu Y, Liu B, Sauder K, et al. Development and validation of machine learning models in prediction of remission in patients with moderate to severe Crohn disease. JAMA network Open 2019;2(5):e193721.

[14] Figgett WA, Monaghan K, Ng M, Alhamdoosh M, Maraskovsky E, Wilson NJ, et al. Machine learning applied to whole-blood RNA-sequencing data uncovers distinct subsets of patients with systemic lupus erythematosus. Clin Transl Immunol 2019;8(12):e01093.
[15] Martin-Gutierrez L, Peng J, Thompson N, Robinson G, Naja M, Peckham H, et al. Two shared immune cell signatures stratify patients with Sjögren's syndrome and systemic lupus erythematosus with potential therapeutic implications. Arthritis Rheumatol 2021;.
[16] Pasoto SG, de Oliveira Martins VA, Bonfa E. Sjögren's syndrome and systemic lupus erythematosus: links and risks. Open Access Rheumatol, 2019;11:33.
[17] Tao W, Concepcion AN, Vianen M, Marijnissen AC, Lafeber FP, Radstake TR, et al. Multiomics and machine learning accurately predict clinical response to adalimumab and etanercept therapy in patients with rheumatoid arthritis. Arthritis Rheumatol 2021;73(2):212–22.
[18] Kohavi R, Wolpert DH. Bias plus variance decomposition for zero-one loss functions. ICML; 1996.
[19] Robinson GA, Peng J, Dönnes P, Coelewij L, Naja M, Radziszewska A, et al. Disease-associated and patient-specific immune cell signatures in juvenile-onset systemic lupus erythematosus: patient stratification using a machine-learning approach. Lancet Rheumatol 2020;2(8):e485–96.
[20] Breiman L. Random forests. Mach Learn 2001;45(1):5–32.
[21] Tin Kam H. Random decision forests. In: Proceedings of 3rd international conference on document analysis and recognition; 1995.
[22] Choi MY, Ma C. Making a big impact with small datasets using machine-learning approaches. Lancet Rheumatol 2020;2(8):e451–2.
[23] Wilkinson MD, Dumontier M, Aalbersberg IJ, Appleton G, Axton M, Baak A, et al. The FAIR Guiding Principles for scientific data management and stewardship. Sci Data 2016;3(1):1–9.
[24] Stafford I, Kellermann M, Mossotto E, Beattie RM, MacArthur BD, Ennis S. A systematic review of the applications of artificial intelligence and machine learning in autoimmune diseases. NPJ Digit Med 2020; 3(1):1–11.
[25] Ho SY, Phua K, Wong L, Goh WWB. Extensions of the external validation for checking learned model interpretability and generalizability. Patterns 2020;1(8):100129.
[26] Plant D, Barton A. Machine learning in precision medicine: lessons to learn. Nat Rev Rheumatol 2021; 17(1):5–6.
[27] Rose N. Personalized medicine: promises, problems and perils of a new paradigm for healthcare. Proc Soc Behav Sci 2013;77:341–52.

CHAPTER

21

Artificial intelligence in endocrinology

Ethan D.L. Brown[1], *Fady Hannah-Shmouni*[2] and *Skand Shekhar*[1]

[1]National Institute of Environmental Health Sciences, National Institutes of Health (NIH), Research Triangle Park, NC, United States [2]National Institute of Child Health and Human Development, National Institutes of Health (NIH), Bethesda, MD, United States

Introduction

The past several decades have produced an explosion of technological advancement in the sphere of digital automation and data exchange in what has been popularly termed the Fourth Industrial Revolution [1]. Artificial intelligence (AI), a preeminent driver of these technological gains, is defined as the use of intelligent agents by a system to perceive its environment and subsequently institute actions to realize the system's objectives [2]. While in medical literature, the term AI is frequently used interchangeably with machine learning (ML), the two processes are distinct, with ML being a subfield of AI focused on learning from data and applying these intelligence gains to make predictions. ML can be further subdivided into two general types of learning: (1) supervised, where labeled inputs and outputs are mapped to create an approximation of their relationship and (2) unsupervised, where hidden patterns in an unlabeled dataset can be identified and explored [3].

For clinicians, the rise of AI and ML bears the most weight for how it will facilitate the processing of mass data in research, improve diagnostic accuracy, or facilitate personalized medicine, and for how it might improve health system efficiency through the automation of organizational workflow and data transmission [4–6]. These shifts hold important implications for the field of endocrinology too, at the levels of research, clinical practice, and health systems.

Both AI and ML have been widely applied in endocrinology research, with the primary objective being the optimization of diagnostics and therapeutics. Such applications of ML

DOI: https://doi.org/10.1016/B978-0-443-15688-5.00022-X

TABLE 21.1 Definitions and sample applications of artificial intelligence in endocrinology.

Type of machine learning	Description of technique	Applications in endocrinology research
Unsupervised learning	Structures within unlabeled datasets are identified through clustering or association (C-means, K-means, etc.)	Screening and stratification of patients into low-, medium-, and high-risk cohorts for bone fracture risk and subsequent osteoporosis intervention based on DXA scans and patient comorbidities [8].
Reinforcement learning	An algorithm learns and optimizes itself to respond predictively through sequential feedback to data (Q-learning, SARSA, etc.)	Prediction of type 1 diabetes patients optimal insulin dosing regimen using biomarker and covariate data [9].
Semisupervised learning	A small amount of labeled data is used to initially train a model before employing it to identify structures within a larger unlabeled dataset (Generative model, semisupervised SVM, etc.)	Data mining of patient records and incorporation of biomarkers to produce a prognostic model for type 2 diabetes [10].
Supervised Learning	The best approximation of a relationship between labeled inputs and outputs is generated (i.e., linear regression, logistic regression, SVM, KNN, etc.)	Estimation of postthyroidectomy levothyroxine dosing for maintenance of euthyroidism [11].

in endocrinology research are not novel, with neural networks being applied to polycystic ovarian syndrome (PCOS) disease status using endocrine measurements as early as 1997 [7] (Table 21.1).

The increasing adoption of AI has wide-scale impacts on clinical practice and research in endocrinology. For instance, AI- and ML-based endocrinology publications are already expanding our understanding of many diseases. One review from 2020 found that among 611 ML-based endocrinology studies published from 2015 to 2020, 52% focused on diabetes, 14% on retinopathy, 14% on thyroid dysfunction, 8% on endocrine-related carcinoma, 7% on osteoporosis, and 5% on other disease states [6]. Subsequent AI and ML contributions will improve patient outcomes and reduce healthcare costs by enabling earlier disease detection and precise targeting of high-risk individuals. Such advancements have already been brought to light in the screening of gestational diabetes mellitus in Israel, where Artzi et al. achieved higher than NIH-standard disease prediction accuracy using electronic health records (EHR) in an ML model [12]. Another study, Silva et al. generated a hierarchy of risk factors beyond traditional clinical criteria to diagnose and stratify PCOS patients using biomarkers which marked a major development for patients who do not present with clear evidence of polycystic ovaries, or those who have a differential diagnosis relating to hyperandrogenism [13].

AI has also dramatically accelerated our interrogation of the human genome for the genetic causes of known phenotypes, leading to a renaissance in genetics research [14]. This development has immediate applications for clinical endocrinologists and researchers. Already, such genetics research has advanced our understanding of several endocrine disorders, such as PCOS, a disorder of polygenic origin with constantly evolving pathophysiology. One ML-based study identified 233 candidate genes that might act proximally to the PCOS phenotype, some of which have now been independently confirmed [15]. Better defining the genetic causes of endocrine disorders could consequently yield relatively rapid improvements to patient care through improved genetic counseling of disorders, such as Von Hippel-Lindau syndrome or sulphonylurea-responsive diabetes [16].

The relative efficiency of AI in processing large datasets and its consequent use in synthesizing genetics, biomarkers, and environmental factors is now facilitating developments in *Personalized Medicine* that are revolutionizing the management of public health problems like type 2 diabetes (T2D) [17]. A shining example is the integration of genetic risk factors into ML algorithms leading to improved hazard ratio predictions in T2D [18]. The expanding adoption of the "internet of things," which refers to multisource data capture, is providing additional avenues for employment of AI in various clinical settings, including endocrine clinics. This is readily visible in the management of type 1 diabetes (T1D), where continuous glucose monitoring (CGM) and "on the fly" ML is already leading to better patient outcomes [19].

Challenges to adoption

The past five years have seen a dramatic increase in the number of PubMed endocrinology-related publications referencing AI and there are now several annual academic and industry conferences, such as AI Med, dedicated to advancing the adoption of AI in clinical practice. However, despite the deep interest in AI among biomedical researchers and leaders of all specialties, including endocrinology, clinical trials and randomized controlled trials involving AI-based medical devices continue to lag [20]. As described by Goldfarb et al., the most significant obstacles preventing faster translation of the AI advances in biomedical research from bench to bedside are (1) algorithmic limitations involving interpretability; (2) limitations on large scale data capture in healthcare; and (3) novel regulatory barriers to entry for AI-based medical devices [21] (Fig. 21.1).

Primary technological insufficiencies, such as interpretability and discrimination bias, are one factor preventing AI's widespread application in clinical practice. With the greater training of ML models on clinical data, tradeoffs between model performance and the interpretability of results can emerge [20]. Consequently, the methodology of predictions is often poorly defined in the highest performing AI models which carry the greatest predictive strength. This creates questions regarding medico-legal liability in the case of erroneous clinical decision-making based on "black box" AI medical devices. Areas of unreliability and underperformance have been identified in AI medical device algorithms, particularly with regard to their application on racially and socioeconomically disadvantaged groups, raising questions about discriminatory bias that may be introduced following greater adoption of AI in clinics [22]. Future improvements in performance and interpretabilitiy should heighten clinician confidence toward the use of AI in clinical decision-making [23].

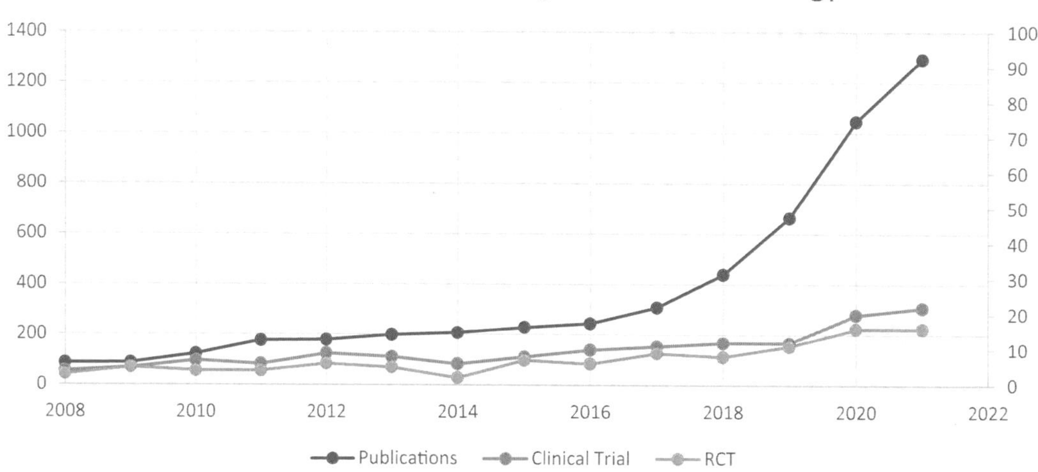

FIGURE 21.1 PubMed query results for publications, clinical trials, and randomized controlled trials which reference the terms related to artificial intelligence and endocrinology are shown.[a] Numerical tally of all publications can be made using the left axis, while numerical tally of clinical trials and randomized controlled trials (RCTs) can be made using the right axis. Quantification of publications from each year was generated using: (artificial intelligence[MeSH Terms] or "artificial intelligence" or "decision support systems, clinical"[MeSH Terms] or "clinical decision support") AND ("endocrinology" or "diabetes" or "PCOS" or "retinopathy" or "thyroid" or "osteoporosis" or "endocrine" or "gnrh" or "HPA" or "HPG"). Quantification of clinical trials using: (artificial intelligence[MeSH Terms] or "artificial intelligence" or "decision support systems, clinical"[MeSH Terms] or "clinical decision support") AND ("endocrinology" or "diabetes" or "PCOS" or "retinopathy" or "thyroid" or "osteoporosis" or "endocrine" or "gnrh" or "HPA" or "HPG"). Quantification of RCTs using: (artificial intelligence[MeSH Terms] or "artificial intelligence" or "decision support systems, clinical"[MeSH Terms] or "clinical decision support") AND ("randomized controlled trial"[Publication Type]) AND ("endocrinology" or "diabetes" or "PCOS" or "retinopathy" or "thyroid" or "osteoporosis" or "endocrine" or "gnrh" or "HPA" or "HPG").

Regulatory challenges

AI algorithms that seek to influence clinical decision-making must receive regulatory approval through the FDA or an equivalent regulatory body. While the time-consuming and resource-intensive nature of this process is justified, there is a lag between laboratory innovations and their clinic application. Novel regulatory questions regarding the use of "actively learning" AI in the clinic have already been addressed by the FDA, which waived reapproval requirements for devices that use clinical data to continually optimize following their initial approval [24]. Over the medium-long term, accelerating the creation or optimization of AI medical devices requires large, high-quality datasets of health information. This is challenging due to existing privacy regulation and the silos of noncompatible EHR systems held by different medical facilities, both of which hinder the ability of AI innovators to leverage data [25]. Further innovations in the deidentification of data to allow for its rapid sharing may help reduce regulatory oversight and accelerate the introduction of AI algorithms into the endocrine clinic.

The future of artificial intelligence in endocrinologist daily practice

Some subfields of endocrinology, such as diabetes mellitus, have rapidly adopted AI-based technologies, including glucose monitoring systems and insulin pumps. The recently released DreaMed decision support software is one such approved device which applies AI to the management of T1D and T2D through the analysis of CGM data to produce insulin titration recommendations and personalized behavioral suggestions [26]. Numerous studies suggest that DreaMed can provide titration recommendations at the level of a practicing endocrinologist and do so across age cohorts [27–29]. The EndoTool SubQ Insulin Dosing Calculator is a more modest application, resembling traditional insulin dosing calculators except for its use of a feedback algorithm to personalize insulin dosing based on previous blood glucose readings [30]. These advances are role models for adoption of AI devices by other subfields of endocrinology, such as women's health, thyroid disorders, bone disease, and genetic endocrine syndromes [6,7,11,13,31].

Advances and improvements in AI utilizing medical devices and clinical support systems will lead to their increased use in clinical endocrinology practice. Greater flexibility in regulatory requirements will further promote the use of AI in clinical settings, but this requires continued public and private enterprise engagement. To ensure that incentives between medical practitioners and stakeholders remain aligned, it is vital that the interface between AI and practitioners be collaborative, not competitive. Notwithstanding all challenges, AI has a very bright future in the endocrine clinic for the management of all metabolic, reproductive, and thyroid disorders.

Major takeaway points

- There has been an exponential growth in the use of AI and ML in the field of endocrinology in the past decade.
- Subfields of endocrinology, such as T1D and T2D, have already incorporated AI into clinical care using continuos glucose monitors and pumps.
- Some challenges facing clinical adoption of AI/ML include paucity of data on from underprivileged groups and the lack of clinically viable outcomes in AI studies.
- Improvements in regulatory requirements and use of her based data will hasten the application of AI/ML into endocrine practice.
- There is a growing need for public and private enterprises to partner in developing AI-based solutions for complex endocrine disorders.

Acknowledgments

We acknowledge Erin Knight and Kelly Miller from the NIEHS library for their assistance in the literature review.

Funding

This work was funded by the intramural research program (IRP) of the National Institute of Environmental Health Sciences (NIEHS), National Institutes of Health (NIH).

References

[1] Melo CE, JAGd ME, Faria Araújo NM. Impact of the fourth industrial revolution on the health sector: a qualitative study. Healthc Inform Res 2020;26(4):328–34.
[2] Poole D, Mackworth AK, Goebel R. Computational intelligence: a logical approach. New York: Oxford University Press; 1998.
[3] Rashidi HH, et al. Machine learning in health care and laboratory medicine: general overview of supervised learning and auto-ML. Int J Laboratory Hematol 2021;43(S1):15–22.
[4] Amisha, et al. Overview of artificial intelligence in medicine. J Family Med Prim Care 2019;8(7):2328–31.
[5] Gubbi S, et al. Artificial intelligence and machine learning in endocrinology and metabolism: the dawn of a new era. Front Endocrinol (Lausanne) 2019;10:185.
[6] Hong N, Park H, Rhee Y. Machine learning applications in endocrinology and metabolism research: an overview. Endocrinol Metab (Seoul) 2020;35(1):71–84.
[7] Lehtinen JC, et al. Visualization of clinical data with neural networks, case study: polycystic ovary syndrome. Int J Med Inf 1997;44(2):145–55.
[8] Kruse C, Eiken P, Vestergaard P. Clinical fracture risk evaluated by hierarchical agglomerative clustering. Osteoporos Int 2017;28(3):819–32.
[9] Oroojeni Mohammad Javad M, et al. A reinforcement learning-based method for management of type 1 diabetes: exploratory study. JMIR Diabetes 2019;4(3):e12905.
[10] Sumathi A, Meganathan S. Semi supervised data mining model for the prognosis of pre-diabetic conditions in type 2 diabetes mellitus. Bioinformation 2019;15(12):875–82.
[11] Zaborek NA, et al. The optimal dosing scheme for levothyroxine after thyroidectomy: a comprehensive comparison and evaluation. Surgery 2019;165(1):92–8.
[12] Artzi NS, et al. Prediction of gestational diabetes based on nationwide electronic health records. Nat Med 2020;26(1):71–6.
[13] Silva IS, et al. Polycystic ovary syndrome: clinical and laboratory variables related to new phenotypes using machine-learning models. J Endocrinol Invest 2022;45(3):497–505.
[14] Dias R, Torkamani A. Artificial intelligence in clinical and genomic diagnostics. Genome Med 2019;11(1):70.
[15] Zhang XZ, et al. Computational characterization and identification of human polycystic ovary syndrome genes. Sci Rep 2018;8(1):12949.
[16] De Sousa SM, et al. Genetic testing in endocrinology. Clin Biochem Rev 2018;39(1):17–28.
[17] Azizi F. Precision medicine for endocrinology. Int J Endocrinol Metab 2016;14(3):e40283.
[18] Wang Y, et al. Genetic risk score increased discriminant efficiency of predictive models for type 2 diabetes mellitus using machine learning: cohort study. Front Public Health 2021;9:606711.
[19] Rodríguez-Rodríguez I, et al. On the possibility of predicting glycaemia 'on the fly' with constrained IoT devices in type 1 diabetes mellitus patients. Sens (Basel) 2019;19(20).
[20] Varghese J. Artificial intelligence in medicine: chances and challenges for wide clinical adoption. Visc Med 2020;36(6):443–9.
[21] Avi Goldfarb FT. Why is AI adoption in health care lagging? In: Seamans R, editor. The economics and regulation of artificial intelligence and emerging technologies. Brookings Institute; 2022. Available from: Available from: https://www.brookings.edu/research/why-is-ai-adoption-in-health-care-lagging/.
[22] Kelly CJ, et al. Key challenges for delivering clinical impact with artificial intelligence. BMC Med 2019;17(1):195.
[23] Asan O, Bayrak AE, Choudhury A. Artificial intelligence and human trust in healthcare: focus on clinicians. J Med Internet Res 2020;22(6):e15154.
[24] Stern AD, Price II WN. Regulatory oversight, causal inference, and safe and effective health care machine learning. Biostatistics 2020;21(2):363–7.

[25] Braghin S, et al. An extensible de-identification framework for privacy protection of unstructured health information: creating sustainable privacy infrastructures. Stud Health Technol Inf 2019;264:1140–4.

[26] Benjamens S, Dhunnoo P, Meskó B. The state of artificial intelligence-based FDA-approved medical devices and algorithms: an online database. npj Digital Med 2020;3(1):118.

[27] Nimri R, et al. Adjustment of insulin pump settings in type 1 diabetes management: advisor pro device compared to physicians' recommendations. J Diabetes Sci Technol 2020;16(2):364–72.

[28] Nimri R. Decision support systems for insulin treatment adjustment in people with type 1 diabetes. Pediatr Endocrinol Rev 2020;17(Suppl 1):170–82.

[29] Nimri R, et al. Insulin dose optimization using an automated artificial intelligence-based decision support system in youths with type 1 diabetes. Nat Med 2020;26(9):1380–4.

[30] Fogel SL, Baker CC. Effects of computerized decision support systems on blood glucose regulation in critically ill surgical patients. J Am Coll Surg 2013;216(4):828–33.

[31] Kasuki L, Wildemberg LE, Gadelha MR. Management of endocrine disease: personalized medicine in the treatment of acromegaly. Eur J Endocrinol 2018;178(3):R89–100.

CHAPTER

22

Artificial intelligence in sleep medicine

Anuja Bandyopadhyay[1] *and Cathy Goldstein*[2]

[1]Indiana University School of Medicine, Indianapolis, IN, United States [2]University of Michigan Sleep Disorders Center, Ann Arbor, MI, United States

Introduction

The cornerstone of diagnostic testing in sleep medicine is the polysomnogram r(PSG), which results in the overnight collection of multidimensional, complex physiological signals that require manual annotation for clinical use. Therefore artificial intelligence (AI) emerged in sleep medicine as method to automate and streamline the sleep laboratory process.

However, over the last decade, in addition to marked improvements in AI sleep scoring accuracy, other uses cases for AI have emerged [1,2]. The practice of reducing PSG data into summarized parameters, including the apnea hypopnea index that defines obstructive sleep apnea (OSA), has resulted in a loss of valuable complexity. Sleep disorders with heterogeneous presentations, impact, and underlying mechanisms are distilled into a few disorders that are treated imprecisely [3]. Therefore the analysis of PSG with AI algorithms, particularly if combined with ancillary clinical data, is expected to identify sleep disorder phenotypes, endotypes, and precision treatments. Furthermore, collection of massive amounts of patient-generated sleep-related data through positive airway pressure devices, mobile applications, and ubiquitous consumer geared wearable devices will provide another substrate for AI algorithms to reveal yet unknown contributions of sleep to health and disease at scale.

Here, we will briefly review examples of AI applications in the screening, diagnosis, and treatment of multiple sleep disorders, opportunities to improve public health related to sleep, challenges, and future directions.

Current examples of artificial intelligence implementation in clinical practice

Polysomnography records electroencephalogram (EEG), electro-oculogram, electromyogram, electrocardiogram, pulse oximetry, as well as airflow and respiratory effort.

Artificial Intelligence in Clinical Practice
DOI: https://doi.org/10.1016/B978-0-443-15688-5.00001-2

To quantify and summarize these signals over the course of the night, technologists manually view PSG data in 30-second intervals and assign sleep stages and annotate respiratory and movement events. Therefore the most immediate use of AI in sleep medicine is the automated scoring of PSG, which has gained increasing popularity due to the promise of reducing workload associated with scoring multiple hours of physiological data [4–7]. Most of the early commercially available PSG scoring software utilized some form of supervised machine learning, which involves feature extraction and use of classifiers.

Studies utilizing AI-assisted PSG scoring demonstrate performance comparable to interrater reliability of human scorers (82%) [8]. There is a wide variability in the type of signals, filters and classifiers used in published studies. Moreover, the feature-based approach is not well suited to the large, complex, and heterogenous data obtained from a PSG. Therefore as machine learning evolved in sleep scoring, deep learning was increasingly adopted. Over the last few years, several studies applied deep learning algorithms (recurrent neural network, convoluted neural network) directly on raw PSG signals with excellent performance. Additionally, AI analysis of non-EEG signals incorporated in PSG that are not traditionally used in sleep staging (i.e., photoplethysmography, respiratory effort) may one day allow the derivation of sleep stages from the minimal parameters recorded during home testing [9].

In addition to the automation of sleep scoring, AI has the potential to offer novel methods to diagnose and understand several sleep disorders. For example, apart from utilizing the traditional airflow and respiratory effort signals, AI-assisted algorithms applied to alternative signals, such as EEG, ECG, photoplethysmography, and even pupil size can diagnose OSA [9]. Additionally, OSA diagnosis completely independent from PSG may be realized by AI analysis of cranial X-rays or large datasets of demographic and clinical attributes, decreasing patient burden [10,11]. Importantly, unsupervised learning has revealed subtypes of OSA that associate with impact on daytime function and predict sequela, such as incident cardiovascular disease [12]. AI has also identified OSA endotypes that contribute to severity and may even predict treatment efficacy [13,14].

Apart from sleep disordered breathing, AI-assisted algorithms have been used to diagnose narcolepsy from a single night of PSG by retaining sleep-stage probabilities as opposed to discrete classifications to construct a hypnodensity graph. In addition to improved depiction of sleep state stability and sleep architecture over the course of the night, features derived from the hypnodensity graph distinguished narcolepsy patients from controls comparable to the traditional two-day PSG and multiple sleep latency test protocol [15]. Insomnia, encountered frequently across medical specialties, demonstrates significant heterogeneity and AI-assisted algorithms have also been used to identify insomnia phenotypes through EEG analysis [16]. Our internal body clock, or circadian rhythm, regulates our timing and quality of sleep but also plays a role in nearly all physiological processes. However, measurement of circadian phase requires complex, in laboratory protocols. AI analysis of gene expression with use of single blood samples can predict circadian phase and may be used for precision timing of chemotherapy to reduce toxicity and improve clinical outcomes [17]. Deep learning has been used to more efficiently diagnose REM behavior disorder (RBD), which holds great relevance given the high risk of developing alpha synucleinopathy neurodegenerative disease among RBD patients [18].

Widespread use of consumer wearable devices affords the opportunity to collect physiological data in free-living conditions. AI algorithms quantify sleep stages and breathing during sleep from signal collected by consumer wearable devices [19] and can further analyze the derived sleep parameters to understand sleep at scale. One of the most exciting applications of AI in sleep medicine is the potential to influence large-scale public health initiatives by informing screening programs and recommending tailored interventions [20].

In addition to informing the development of precision treatments for sleep disorders, AI powered patient facing mobile applications can assist with intervention delivery and enhance patient adherence and engagement [21,22].

Current challenges

While AI-assisted algorithms hold promise, there are plenty of obstacles to overcome before its widespread adoption in clinical sleep medicine. As deep learning continues to gain popularity, it is important to create heterogenous databases for algorithm training to increase generalizability. There is an acute need for improved regulation of AI-assisted algorithms that will include the utilization of external datasets for validation. While central agencies like the US Food and Drug Administration provide clear guidelines for medical device approval, such regulations are not well suited for software as a medical device, particularly when unlocked AI algorithms are utilized. Additionally, for AI to truly improve health, large-scale research trials on heterogenous populations are required to corroborate AI assisted algorithm data with clinical outcomes. Greater integration of other data sources, such as "omics," metadata and electronic medical record information will likely augment the health insights provided by AI. At minimum, issues surrounding security, technology infrastructure, staff workflows, and patient and provider education may interfere with implementation of clinical AI tools. Careful attention is required to ensure that the bias and inequity already evident in sleep medicine is reduced, not increased by algorithms that learn from available data.

Future directions

As there is increased awareness of AI-assisted scoring amongst clinicians, future directions should include promoting education on how AI learns. This will help clinicians determine the reliability of AI-generated data. Alongside improved transparency and regulation of AI assisted algorithms, larger, heterogenous datasets of individuals with sleep disorders are needed for more generalizable AI-assisted algorithms. Ambulatory tracking devices are a rich source for longitudinal data collection and integration of these devices with electronic health records on a HIPAA protected platform will be very useful. If used correctly, AI has the potential to streamline our analysis of PSG and other data streams but will also produce powerful clinical tools to further precision sleep medicine and deepen our understanding of the role of sleep in health and disease.

Major takeaway points

- AI assisted algorithms have been used successfully to score physiological data in PSGs.
- AI assisted algorithms can diagnose various sleep disorders from usual and nontraditional data sources.
- AI will better phenotype and endotype sleep disorders, which will improve outcome prediction and patient-specific treatment selection.
- AI can deliver behavioral interventions and improve patient engagement during treatment for sleep disorders.
- As in all healthcare AI applications, increased availability of large heterogenous databases is required to improve generalizability of AI-assisted algorithms.
- Sleep medicine faces challenges to AI implementation common to all specialties that include but are not limited to issues related to validity, safety, regulation, security, privacy, ethics, and equity.

References

[1] Goldstein CA, Berry RB, Kent DT, Kristo DA, Seixas AA, Redline S, et al. Artificial intelligence in sleep medicine: an American Academy of Sleep Medicine position statement. J Clin Sleep Med 2020;16:605–7.

[2] Goldstein CA, Berry RB, Kent DT, Kristo DA, Seixas AA, Redline S, et al. Artificial intelligence in sleep medicine: background and implications for clinicians. J Clin Sleep Med 2020;16:609–18.

[3] Redline S, Purcell SM. Sleep and Big Data: harnessing data, technology, and analytics for monitoring sleep and improving diagnostics, prediction, and interventions—an era for Sleep-Omics? US: Oxford University Press; 2021. p. zsab107.

[4] Sun H, Jia J, Goparaju B, Huang GB, Sourina O, Bianchi MT, et al. Large-scale automated sleep staging. Sleep. 2017;40.

[5] Lee PL, Huang YH, Lin PC, Chiao YA, Hou JW, Liu HW, et al. Automatic sleep staging in patients with obstructive sleep apnea using single-channel frontal EEG. J Clin Sleep Med 2019;15:1411–20.

[6] Peter-Derex L, Berthomier C, Taillard J, Berthomier P, Bouet R, Mattout J, et al. Automatic analysis of single-channel sleep EEG in a large spectrum of sleep disorders. J Clin Sleep Med 2021;17:393–402.

[7] Korkalainen H, Aakko J, Duce B, Kainulainen S, Leino A, Nikkonen S, et al. Deep learning enables sleep staging from photoplethysmogram for patients with suspected sleep apnea. Sleep. 2020;43.

[8] Fiorillo L, Puiatti A, Papandrea M, Ratti PL, Favaro P, Roth C, et al. Automated sleep scoring: a review of the latest approaches. Sleep Med Rev 2019;48:101204.

[9] Liu D, Pang Z, Lloyd SR. A neural network method for detection of obstructive sleep apnea and narcolepsy based on pupil size and EEG. IEEE Trans Neural Netw 2008;19:308–18.

[10] Tsuiki S, Nagaoka T, Fukuda T, Sakamoto Y, Almeida FR, Nakayama H, et al. Machine learning for image-based detection of patients with obstructive sleep apnea: an exploratory study. Sleep Breath 2021;25:2297–305.

[11] Huang WC, Lee PL, Liu YT, Chiang AA, Lai F. Support vector machine prediction of obstructive sleep apnea in a large-scale Chinese clinical sample. Sleep. 2020;43.

[12] Zinchuk A, Yaggi HK. Phenotypic subtypes of OSA: a challenge and opportunity for precision medicine. Chest. 2020;157:403–20.

[13] Edwards BA, Redline S, Sands SA, Owens RL. More than the sum of the respiratory events: personalized medicine approaches for obstructive sleep apnea. Am J Respir Crit Care Med 2019;200:691–703.

[14] Dutta R, Delaney G, Toson B, Jordan AS, White DP, Wellman A, et al. A novel model to estimate key obstructive sleep apnea endotypes from standard polysomnography and clinical data and their contribution to obstructive sleep apnea severity. Ann Am Thorac Soc 2021;18:656–67.

[15] Stephansen JB, Olesen AN, Olsen M, Ambati A, Leary EB, Moore HE, et al. Neural network analysis of sleep stages enables efficient diagnosis of narcolepsy. Nat Commun 2018;9:5229.

[16] Aydın S, Saraoğlu HM, Kara S. Singular spectrum analysis of sleep EEG in insomnia. J Med Syst 2011;35:457–61.
[17] Hesse J, Malhan D, Yalçin M, Aboumanify O, Basti A, Relógio A. An optimal time for treatment—predicting circadian time by machine learning and mathematical modelling. Cancers. 2020;12:3103.
[18] Ruffini G, Ibañez D, Castellano M, Dubreuil-Vall L, Soria-Frisch A, Postuma R, et al. Deep learning with EEG spectrograms in rapid eye movement behavior disorder. Front Neurol 2019;10:806.
[19] Schutte-Rodin S, Deak MC, Khosla S, Goldstein CA, Yurcheshen M, Chiang A, et al. Evaluating consumer and clinical sleep technologies: an American Academy of Sleep Medicine update. J Clin Sleep Med 2021;17:2275–82.
[20] Wallace ML, Stone K, Smagula SF, Hall MH, Simsek B, Kado DM, et al. Which sleep health characteristics predict all-cause mortality in older men? An application of flexible multivariable approaches. Sleep. 2018;41.
[21] Turino C, Benítez ID, Rafael-Palou X, Mayoral A, Lopera A, Pascual L, et al. Management and treatment of patients with obstructive sleep apnea using an intelligent monitoring system based on machine learning aiming to improve continuous positive airway pressure treatment compliance: randomized controlled trial. J Med Internet Res 2021;23:e24072.
[22] Philip P, Dupuy L, Morin CM, de Sevin E, Bioulac S, Taillard J, et al. Smartphone-based virtual agents to help individuals with sleep concerns during COVID-19 confinement: feasibility study. J Med Internet Res 2020;22:e24268.

CHAPTER

23

Artificial intelligence in nephrology

Shankara Anand[1] *and Ashish Verma*[2]

[1]Boston University Chobanian & Avedisian School of Medicine, Boston, MA, United States
[2]Section of Nephrology, Department of Medicine, Boston University Avedisian & Chobanian School of Medicine, Boston, MD, United States

Introduction

Artificial intelligence (AI) is a rapidly growing branch of applied statistics with an emerging set of clinical applications. AI is a general term that employs computational models to mimic intelligent behavior. Machine learning (ML), a subfield of AI, are algorithms that identify patterns in datasets to generate inferences. A joint growth in large biomedical datasets, statistical methodology [1], and hardware advances [2] have ushered in a myriad of opportunities for AI to impact patient care. Commonly used ML/AI algorithms in clinical contexts include clinical decision support, early detection of disease progression, patient subtyping, image analysis in pathology/radiology, and genotype-to-phenotype analyses [3]. Medical fields including neurology, cardiology, and pulmonology have out-paced nephrology in the use of ML techniques [4]. Here, we briefly discuss the role of AI/ML in the care of nephrology patients, current applications, and future directions/challenges.

Overview of machine learning

The main subtypes of ML include supervised, unsupervised, and reinforcement learning. Supervised learning algorithms "train" on labeled input data to assign these labels to unseen data. Supervised learning includes classification, where the prediction is a class or category, and regression, where the prediction is a continuous variable. For example, predicting whether a patient will experience an acute kidney injury (AKI) given a training dataset of prior patient hospitalizations is a supervised classification problem. Predicting a patient's eGFR given their baseline vitals and labs is a supervised regression problem. Within the realm of supervised learning are well-known algorithms including linear regression, logistic regression, generalized linear models, SVMs, boosted gradient trees, and artificial neural networks (ANNs) (Fig. 23.1).

Artificial Intelligence in Clinical Practice
DOI: https://doi.org/10.1016/B978-0-443-15688-5.00035-8

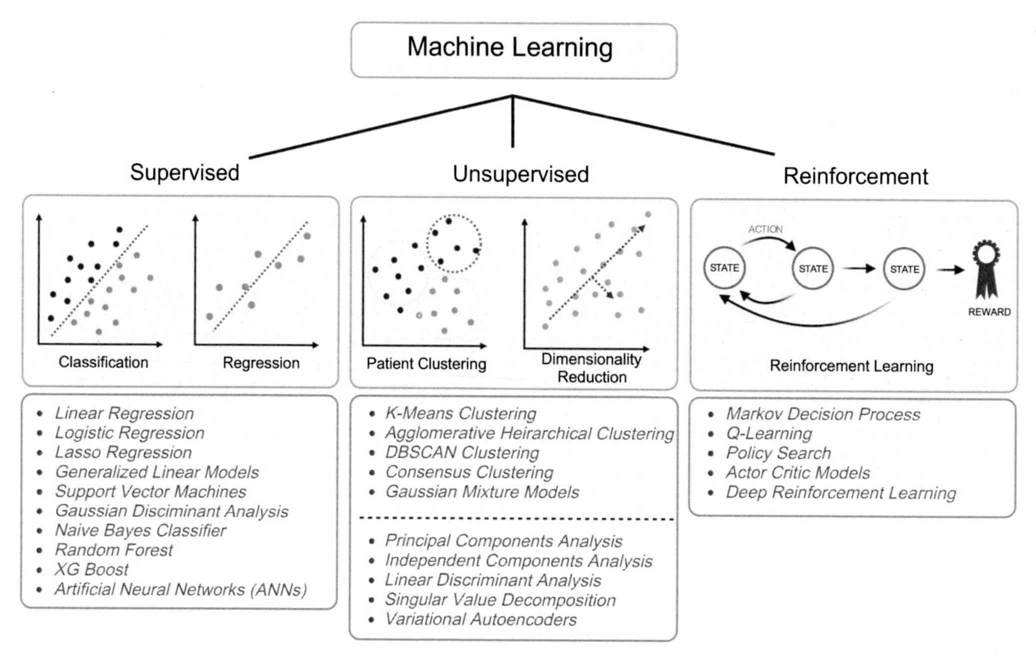

FIGURE 23.1 Types of machine learning.

Unsupervised learning algorithms are applied to unlabeled datasets with the goal of identifying underlying structure or patterns. Two notable unsupervised ML applications in medicine include (1) clustering patient groups with similar pathophysiology or clinical outcomes and (2) transforming large sets of clinical data to interpretable, smaller feature spaces known as dimensionality reduction (Fig. 23.1). For example, clustering hospitalized patients that experience AKI using clinical data may identify high-risk patients that benefit from early intervention. Dimensionality reduction may enable clinicians to distill large datasets into tractable features, such as identifying biomarkers to identify early end-stage renal disease (ESRD) development.

Reinforcement learning (RL) algorithms are trained using a reward function after a series of decisions rather than correctness of given inference (supervised learning) or metric of similarity (unsupervised learning). These are formalized in a framework known as a Markov decision process, where "states," such as a patient's current set of interventions, may be traversed with possible "actions," such as clinical decision. RL is gaining traction in medicine where decision support systems rely on complex sequences of decisions to reach a "reward," such as improved mortality, reduced hospital cost, or efficient use of resources.

Finally, "deep learning" is a computational technique applied to each of these ML disciplines with great promise in tackling complex problems in medicine. Deep learning algorithms, simply, are a stacked collection of individual regression models, or "neurons," connected by nonlinear functions. Both these neurons and their connections have optimizable parameters and together are known as an ANN. Optimization methods for ANNs leverage principles in calculus and linear algebra to efficiently compute cost functions on these large, highly parameterized systems. This has led to an explosion in prediction capability for complex problems, including supervised approaches, unsupervised approaches (ex. variational autoencoders), and RL (Fig. 23.1). Throughout this chapter, we highlight how these broad ML/AI subfields are impacting the practice of nephrology.

Examples of artificial intelligence implementation in nephrology

Supervised learning

Supervised learning is appealing in clinical nephrology given the increase in available, labeled electronic health records. Training supervised algorithms on prior clinical scenarios may predict outcomes for future patients, such as mortality risk in patients newly started on dialysis [5] or allograft loss in kidney transplant patients [6]. A notable use-case of these approaches is predicting inpatient AKI, which impacts one in five hospital admissions [7]. Approaches, such as gradient boosting and random forest, have shown success in predicting future AKI with admission lab data [8,9] and prior clinical health record data [10]. Using longitudinal data, a team at DeepMind (Google) successfully trained a recurrent neural network, a deep learning method for modeling temporal dynamic behavior, to predict in-patient AKI events up to 48 hours in advance, demonstrating impressive prediction power and clinical utility with opportunity for early intervention [11].

Another application of supervised learning is prediction of chronic kidney disease (CKD) [12]. Early clinical intervention in patients with CKD may decrease microvascular consequences responsible for increased mortality. Additionally, developing accurate, long-term progression models may be used in-place of long-lasting clinical trials for rapid innovation of therapeutic intervention. Challenges contributing to early detection of CKD include insensitive and nonspecific biomarkers, such as eGFR, and a lack of incorporation of longitudinal data spanning beyond a given hospital stay [13]. Multiple studies have used widely available lab values and ML approaches, including logistic regression and deep learning, to predict CKD progression [14–17].

AI may additionally provide utility in clinical decision support systems for nephrology patients. *Chen* et al. have demonstrated this with the prediction and risk stratification of patients with IgA nephropathy [18]. These researchers created a decision support system with an online risk calculator to better inform treatment and identify patients for clinical trials. Decision-support algorithms have also been used to suggest erythropoietin dosing in chronic hemodialysis patients and with external clinical verification of improved hemoglobin levels and decrease fluctuations [19,20]. Another example is prediction of postoperative nephrolithotomy to tailor optimal surgical treatment for newly diagnosed patients with kidney stones [21]. Supervised learning is a powerful tool that can be applied readily as an extension of a clinician's toolkit and will likely become a mainstay of parallel processing in healthcare.

Imaging processing

Convolutional neural network (CNN) architectures are a type of deep learning effective for image analysis, with novel use-cases in renal disease. CNNs use convolution filters within its ANN that permit spatial invariance, meaning they learn information about an image unspecific to its exact location in the image. For example, CNNs are invariant to a region of sclerosis in the top right vs. bottom left of a glomerular pathology slide. CNNs have been successful in classifying images, segmenting images (outlining within-image boundaries), and describing characteristics of images [22,23], and as a result, have become increasingly popular in radiology [24] and pathology [25].

In the field of nephropathology, CNNs have been used to aid pathologists in outlining and identifying, or segmenting, regions of interest in whole-slide images [26,27]. Ginley et al. use deep learning to segment renal biopsies of patients with diabetic glomerulosclerosis into glomerular components. Rather than rely on the effort and expertise of a pathologist to find glomeruli, their algorithm was capable of highlighting areas for further characterization to increase efficiency. Next, features extracted from these outlined areas were used to predict diabetic nephropathy progression and classify the extent of disease process [28]. These tools are not only highlighting areas of interest, but providing assistance in analyzing as well.

The development of nephropathology findings into clinically useful, data-driven risk scores is an emerging field evidenced by transplant kidney survival with Banff Lesion scores [29] and IgA glomerulonephritis progression risk [30]. These scores demonstrate a trend toward quantitative assessment of high information density whole-slide images in the hands of experts. AI may broaden the accessibility of these actionable metrics to clinics around the world [31]. CNNs have been used to evaluate intraoperative donor kidney biopsies for determining kidneys eligible for transplantation [32]. This task bears great importance as the criteria for accepting or rejecting donor kidneys relies on the expertise of a pathologist to determine the percent of sclerotic glomeruli. CNNs have also been used to score fibrosis and tubular atrophy in trichrome stained renal biopsies and shown comparable results to pathologist-estimated fibrosis scores [33]. This was shown to aid in risk stratification of CKD stage and renal survival.

These same tools apply to radiology as well. For example, the use of segmentation algorithms in CT/MRI scans of patients with autosomal dominant polycystic kidney disease can estimate total kidney volume, a potential proxy for renal function [34,35]. These algorithms could be deployed for early detection of kidney neoplasms [36], rapid differentiation of neoplasms, such as clear cell renal carcinoma and oncocytomas [37], and identification of kidney stones [38]. Imaging is a growing modality in the work-up and management for a host of nephrology pathologies. AI has the capability to aid pathologists and radiologists, assist in rapid detection, suggest clinical decisions, and yield quantitative features for risk scoring.

Genotypes and phenotypes

The rapid progress of nucleic acid sequencing has created an explosion of genomic data with the potential to impact clinical decision making [39]. Genomics, transcriptomics, proteomics, etc., highlight biological processes occurring in patients, and when paired with clinical data and AI, may aid in diagnosis or therapeutic development for kidney disease [40]. Gene sequencing has aided diagnosis in rare kidney disease [41]. This was first demonstrated when researchers linked the *ADPKD* locus to adult polycystic kidney disease [42]. Since then, up to 160 rare kidney diseases with genetic bases enable diagnosis and a biological understanding of their disease processes. These approaches may shed light on epidemiological phenomena in kidney disease as well. For example, in 2010, risk variants in the *APOL1* gene, encoding a high density lipoprotein, had increased odds of focal segmental glomerulosclerosis (FSGS) and hypertension-associated ESRD in patients with African ancestry [43].

Large, genome-wide association studies (GWAS) have further expanded the number of genomic variants explored in nephrology patients. Large consortia, such as the CKDGen consortia, have used GWAS to identify single-nucleotide polymorphisms (SNPs) associated with kidney disease [44,45]. The gene *UMOD*, which encodes a common protein in human urine, was shown to be associated with CKD. Alterations in *SHROOM3*, reported to have a role in epithelial cell shape regulation, suggested to have an association with GFR. *GATM*, which encodes a glycine amidinotransferase involved in creatinine biosynthesis, similarly associated with GFR. Identifying correlations between SNPs and kidney disease phenotypes is a growing field where ML/AI techniques may highlight shared, targetable biology in nephrology patients.

Other notable consortia generating multimodal data for kidney disease include the Nephrotic Syndrome Study Network (NEPTUNE), curating longitudinal, observational data paired with clinical, histopathological, and molecular phenotyping to better understand nephrotic syndromes [46,47]. The Cure Glomerulonephropathy (CureGN) consortium focuses on a variety of glomerular diseases, including minimal change disease, FSGS, membranous nephropathy, and IgA nephropathy in both children and adults over time [48]. The Chronic Renal Insufficiency Cohort (CRIC) seeks to elucidate the mechanisms tying cardiovascular disease to chronic renal insufficiency over time. The Transformative Research in Diabetic Nephropathy (TRIDENT) Study seeks to better understand disease pathogenesis of diabetic kidney disease through similar means [49]. Mentioned here are a few on-going efforts that demonstrate the interest, excitement, and future of combining genomic data with nephrology practice.

Unsupervised learning

In the context of nephrology, unsupervised learning can identify kidney disease patient groups with similar outcomes to highlight clinical features important for progression. Within in-patient settings, AKIs are a common, yet debilitating occurrence. Researchers have used unsupervised learning methods to group patients experiencing AKI in hospital and ICU settings with similar phenotypic and pathologic outcomes [50,51]. In-patient AKI has been stratified into (1) hypovolemia dominant, (2) contrast associated and sustained hypotension, (3) shock, and (4) hospital-acquired with distinct hospital courses [52]. Current implications of these subtypes is prognostic enrichment, such as identifying high-risk patients who require kidney replacement therapy. Future wide scale validation of these subtypes and subsequent clinical trials are needed to explore this.

Clustering approaches have also been applied to CKD, a heterogenous disease previously defined by eGFR, to identify relevant clinical features to disease progression [53].

One study applied unsupervised consensus clustering with 72 baseline characteristics in the CRIC study to identify three unique CKD subgroups: (1) early stage CKD with low prevalence of diabetes/obesity, (2) high prevalence of diabetes/obesity, and (3) low bone mineral density and poor cardiac/kidney function, and inflammation. Cluster membership had distinct risk of disease progression, cardiovascular events, and mortality. Yu et al. approach this problem by first dividing CKD status by clinical data, and training a random forest classifier, a supervised learning approach, to identify relevant clinical variables to CKD progression. Then, they use this subset of variables to perform dendrogram-based clustering, an unsupervised learning approach, to understand what drives CKD progression between patient groups [54]. Both of these approaches are limited by their retrospective nature, and future validation with longitudinal CKD progression data may improve clustering approaches for this disease's phenotypes.

Dimensionality reduction, another subset of unsupervised learning, has been used to identify biological features associated with ESRD development in patients. ESRD may be considered a low-grade, chronic inflammatory state. Lioulios et al. show that dimensionality reduction techniques like principal component analysis and uniform manifold approximation and projection on T- and B-cell markers may identify patterns of lymphocyte activation associated with ESRD [55]. Unsupervised methods may refine prior clinical group definitions, scale to larger feature spaces, and identify novel biomarkers associated with certain patient groups [56]. Identifying these disease progression markers may inspire new mechanistically derived therapies. This is under investigation in nephrotic syndromes, where using multiomic, integrated data with clustering approaches have identified tissue necrosis factor activation as a targetable biological process enriched in patients with poorer outcomes [55].

Challenges and future directions

We discuss numerous ML methods, most of which require massive amounts of training data to perform inference and cluster patient groups. Unfortunately, a challenge of applying ML globally is over-training algorithms in resource-rich areas with requisite funding that does not generalize to other settings [57]. This biases the utility of these methods in diverse, underserved populations and requires the utmost scrutiny as their use widens in the clinic. Additionally, the performance of ML methods depends entirely on available input data and prediction parameters. Social determinants of health may be unavailable or difficult to represent in training data or prediction labels. For example, one of the most widely used health care delivery algorithms predicts health care costs rather than illness, failing to reconcile unequal access between Black and White populations results in less money spent [58]. Thus the use of health care cost as a proxy for health in this context perpetuates large racial biases. Both healthcare and ML are fields subject to this.

Other important challenges preventing AI use in the clinic are explainability and uncertainty. Many algorithms, especially deep learning, work as "black boxes" when generating predictions. Despite incredible prediction accuracy, trained algorithms return numerical "weights" with little interpretability for clinicians to identify information crucial for generating a given result. A growing field known as "explainable AI" seeks to upend this problem and is worthy of its own separate review [59]. Finally, given the wide range of uncertainty inherent to clinical medicine, AI algorithms that predict distributions rather than point estimates is an important distinction that may broaden the use of these methods in the clinic. Bayesian statistics is a powerful tool that bases estimates on prior assumptions about a dataset to generate such distributions. Its use in AI models is a growing field that may broaden clinical application [60].

Major takeaway points

- Advances in hardware, innovations in applied statistics, and growing availability of biomedical data for patients with renal disease have set the stage for improved clinical decision support, early intervention, and targeted therapeutics.
- Acute kidney injuries (AKI), CKD, and ESRD, each hold their own unique burden on patient lives and the healthcare system; ML techniques can identify high risk patients, suggest early interventions, and predict the course of their hospital stays.
- Nephropathology and radiology are important for characterizing disease processes (e.g., sclerosis), determining viability of donor kidneys, and identifying renal masses, and deep learning may support clinicians in these tasks and broaden availability of their expertise.
- Advances in nucleic acid sequencing paired with high-throughput statistics enhance our understanding of renal disease pathology and may generate therapeutic targets in the coming years.
- AI in nephrology is subject to the source of its patient data and parameters set by the developer, rendering its application and interpretation limited and worthy of scrutiny in an inequitable, global system of healthcare.

References

[1] LeCun Y, Bengio Y, Hinton G. Deep learning. Nature 2015;521:436–44.
[2] Blythe D. Rise of the graphics processor. Proc IEEE 2008;96:761–78. Available from: https://doi.org/10.1109/jproc.2008.917718.
[3] Esteva A, et al. A guide to deep learning in healthcare. Nat Med 2019;25:24–9. Available from: https://doi.org/10.1038/s41591-018-0316-z.
[4] Verma A, Chitalia VC, Waikar SS, Kolachalama VB. Machine learning applications in nephrology: a bibliometric analysis comparing kidney studies to other medicine subspecialities. Kidney Med 2021;3:762–7.
[5] Akbilgic O, et al. Machine learning to identify dialysis patients at high death risk. Kidney Int Rep 2019;4:1219–29.
[6] Loupy A, et al. Prediction system for risk of allograft loss in patients receiving kidney transplants: international derivation and validation study. BMJ 2019;366:l4923.
[7] Wang HE, Muntner P, Chertow GM, Warnock DG. Acute kidney injury and mortality in hospitalized patients. Am J Nephrol 2012;35:349–55.
[8] Koyner JL, Carey KA, Edelson DP, Churpek MM. The development of a machine learning inpatient acute kidney injury prediction model. Crit Care Med 2018;46:1070–7.
[9] Lin K, Hu Y, Kong G. Predicting in-hospital mortality of patients with acute kidney injury in the ICU using random forest model. Int J Med Inf 2019;125:55–61.
[10] Shawwa K, et al. Predicting acute kidney injury in critically ill patients using comorbid conditions utilizing machine learning. Clin Kidney J 2021;14:1428–35.

[11] Tomašev N, et al. A clinically applicable approach to continuous prediction of future acute kidney injury. Nature 2019;572:116–19.
[12] Schena FP, Anelli VW, Abbrescia DI, Di Noia T. Prediction of chronic kidney disease and its progression by artificial intelligence algorithms. J Nephrol 2022;. Available from: https://doi.org/10.1007/s40620-022-01302-3.
[13] Chauhan K, et al. Initial validation of a machine learning-derived prognostic test (KidneyIntelX) integrating biomarkers and electronic health record data to predict longitudinal kidney outcomes. Kidney 2020;360 (1):731–9.
[14] Ravizza S, et al. Predicting the early risk of chronic kidney disease in patients with diabetes using real-world data. Nat Med 2019;25:57–9.
[15] Makino M, et al. Artificial intelligence predicts the progression of diabetic kidney disease using big data machine learning. Sci Rep 2019;9:1–9.
[16] Lei N, et al. Machine learning algorithms' accuracy in predicting kidney disease progression: a systematic review and meta-analysis. BMC Med Inf Decis Mak 2022;22:1–16.
[17] Norouzi J, Yadollahpour A, Mirbagheri SA, Mazdeh MM, Hosseini SA. Predicting renal failure progression in chronic kidney disease using integrated intelligent fuzzy expert system. Comput Math Methods Med 2016;2016:6080814.
[18] Chen T, Li X, Li Y, Xia E, Qin Y, Liang S, et al. Prediction and risk stratification of kidney outcomes in IgA nephropathy. Am J Kidney Dis 2019;74:300–9.
[19] Brier ME, Gaweda AE. Artificial intelligence for optimal anemia management in end-stage renal disease. Kidney Int 2016;90.
[20] Barbieri C, et al. An international observational study suggests that artificial intelligence for clinical decision support optimizes anemia management in hemodialysis patients. Kidney Int 2016;90:422–9.
[21] Shabaniyan T, et al. An artificial intelligence-based clinical decision support system for large kidney stone treatment. Australas Phys Eng Sci Med 2019;42:771–9.
[22] Voulodimos A, Doulamis N, Doulamis A, Protopapadakis E. Deep learning for computer vision: a brief review. Comput Intell Neurosci 2018;2018.
[23] Alex Krizhevsky Google Inc, Ilya Sutskever Google Inc, Hinton OpenAI GE. ImageNet classification with deep convolutional neural networks. Commun ACM 2017;. Available from: https://doi.org/10.1145/3065386.
[24] Yamashita R, Nishio M, Do RKG, Togashi K. Convolutional neural networks: an overview and application in radiology. Insights Imaging 2018;9:611–29.
[25] Wang S, Yang DM, Rong R, Zhan X, Xiao G. Pathology image analysis using segmentation deep learning algorithms. Am J Pathol 2019;189:1686–98.
[26] Hermsen M, et al. Deep learning-based histopathologic assessment of kidney tissue. J Am Soc Nephrol 2019;30:1968–79.
[27] Kannan S, et al. Segmentation of glomeruli within trichrome images using deep learning. Kidney Int Rep 2019;4:955–62.
[28] Ginley B, et al. Computational segmentation and classification of diabetic glomerulosclerosis. J Am Soc Nephrol 2019;30:1953–67.
[29] Roufosse C, et al. A 2018 reference guide to the Banff classification of renal allograft pathology. Transplantation 2018;102:1795–814.
[30] Barbour SJ, et al. Evaluating a new international risk-prediction tool in IgA nephropathy. JAMA Intern Med 2019;179:942–52.
[31] Becker JU, et al. Artificial intelligence and machine learning in nephropathology. Kidney Int 2020;98:65–75.
[32] Marsh JN, et al. Deep learning global glomerulosclerosis in transplant kidney frozen sections. IEEE Trans Med Imaging 2018;37:2718–28.
[33] Kolachalama VB, et al. Association of pathological fibrosis with renal survival using deep neural networks. Kidney Int Rep 2018;3:464–75.
[34] Sharma K, et al. Automatic segmentation of kidneys using deep learning for total kidney volume quantification in autosomal dominant polycystic kidney disease. Sci Rep 2017;7:2049.
[35] Goel A, et al. Deployed deep learning kidney segmentation for polycystic kidney disease MRI. Radiol Artif Intell 2022;4:e210205.

[36] Gharaibeh M, et al. Radiology imaging scans for early diagnosis of kidney tumors: a review of data analytics-based machine learning and deep learning approaches. Big Data Cognit Comput 2022;6:29.
[37] Nikpanah M, et al. A deep-learning based artificial intelligence (AI) approach for differentiation of clear cell renal cell carcinoma from oncocytoma on multi-phasic MRI. Clin Imaging 2021;77:291–8.
[38] Yildirim K, et al. Deep learning model for automated kidney stone detection using coronal CT images. Comput Biol Med 2021;135:104569.
[39] Goodwin S, McPherson JD, McCombie WR. Coming of age: ten years of next-generation sequencing technologies. Nat Rev Genet 2016;17:333–51.
[40] Groopman EE, et al. Diagnostic utility of exome sequencing for kidney disease. N Engl J Med 2019;380:142–51.
[41] Devuyst O, Knoers NVAM, Remuzzi G, Schaefer FBoard of the Working Group for Inherited Kidney Diseases of the European Renal Association and European Dialysis and Transplant Association. Rare inherited kidney diseases: challenges, opportunities, and perspectives. Lancet 2014;383:1844–59.
[42] Reeders ST, et al. A highly polymorphic DNA marker linked to adult polycystic kidney disease on chromosome 16. Nature 1985;317:542–4.
[43] Genovese G, et al. Association of trypanolytic ApoL1 variants with kidney disease in African Americans. Science 2010;329:841–5.
[44] Köttgen A, et al. Multiple loci associated with indices of renal function and chronic kidney disease. Nat Genet 2009;41:712–17.
[45] Köttgen A, et al. New loci associated with kidney function and chronic kidney disease. Nat Genet 2010;42:376–84.
[46] Gadegbeku CA, et al. Design of the Nephrotic Syndrome Study Network (NEPTUNE) to evaluate primary glomerular nephropathy by a multidisciplinary approach. Kidney Int 2013;83:749–56.
[47] Gillies CE, et al. An eQTL landscape of kidney tissue in human nephrotic syndrome. Am J Hum Genet 2018;103:232–44.
[48] Mariani LH, et al. CureGN study rationale, design, and methods: establishing a large prospective observational study of glomerular disease. Am J Kidney Dis 2019;73:218–29.
[49] Townsend RR, et al. Rationale and design of the Transformative Research in Diabetic Nephropathy (TRIDENT) study. Kidney Int 2020;97:10–13.
[50] Thongprayoon C, et al. Clinically distinct subtypes of acute kidney injury on hospital admission identified by machine learning consensus clustering. Med Sci (Basel) 2021;9.
[51] Castela Forte J, et al. Identifying and characterizing high-risk clusters in a heterogeneous ICU population with deep embedded clustering. Sci Rep 2021;11:12109.
[52] Vaara ST, et al. Subphenotypes in acute kidney injury: a narrative review. Crit Care 2022;26:251.
[53] Zheng Z, et al. Subtyping CKD patients by consensus clustering: the Chronic Renal Insufficiency Cohort (CRIC) study. J Am Soc Nephrol 2021;32:639–53.
[54] Yu C-S, et al. Clustering heatmap for visualizing and exploring complex and high-dimensional data related to chronic kidney disease. J Clin Med Res 2020;9.
[55] Lioulios G, et al. Clustering of end stage renal disease patients by dimensionality reduction algorithms according to lymphocyte senescence markers. Front Immunol 2022;13:841031.
[56] Mariani LH, et al. Multidimensional data integration identifies tumor necrosis factor activation in nephrotic syndrome: a model for precision nephrology. bioRxiv 2021;. Available from: https://doi.org/10.1101/2021.09.09.21262925.
[57] Celi LA, et al. Sources of bias in artificial intelligence that perpetuate healthcare disparities—a global review. PLoS Digital Health 2022;1:e0000022.
[58] Obermeyer Z, Powers B, Vogeli C, Mullainathan S. Dissecting racial bias in an algorithm used to manage the health of populations. Science 2019;366:447–53.
[59] Holzinger A, et al. Towards the augmented pathologist: challenges of explainable-AI in digital pathology. arXiv 2017;. Available from: https://doi.org/10.48550/arXiv.1712.06657.
[60] van de Schoot R, et al. Bayesian statistics and modelling. Nat Rev Methods Prim 2021;1:1–26.

CHAPTER

24

Artificial intelligence in surgery

Simon Laplante[1,2] *and Amin Madani*[1,2,3]

[1]Surgical Artificial Intelligence Research Academy, University Health Network, Toronto, ON, Canada [2]Department of Surgery, University of Toronto, Toronto, ON, Canada [3]Division of General Surgery, University Health Network, Toronto, ON, Canada

Introduction

Artificial intelligence (AI) is a field that uses computer science and datasets to train machines to reason and perform like humans. AI is a broad term and often used to describe technological innovations that do not necessarily use methodologies in AI. In this chapter, we will focus primarily on two subsets of the parent term AI: machine learning (ML) and deep learning (DL). Although there are numerous examples of the potential benefits of ML for clinical applications in surgery, very few have made it to patients' bedside as the process of procuring and annotating datasets, as well as training, testing, validating, and implementing algorithms is an extremely complex and resource-intensive process. Understanding how to transition from research to real clinical use of various AI tools will require more consideration and time investment if we are to justify the use of AI for surgical patients. At its current stage, ML applications in surgery are still far from replacing human judgment, or surgeons' technical and nontechnical skills and should rather be seen as a supplement.

Two of the most common applications of ML and DL in surgery are in the field of computer vision (CV) and natural language processing (NLP). CV is relatively in its infancy compared to other subfields of computer science, with major achievements, such as accurate image recognition of common objects only possible since 2012 [1]. CV in surgery is used primarily to understand images and videos based on pixels. As of now, most advances in CV have been in image interpretation in both the fields of radiology and pathology, given the readily available images able to generate large datasets. CV can be used to detect objects within an image and allow segmentation of these objects as surgical tools or different anatomical structures. Segmentation of surgical images combined with our understanding of the safe steps of an operation can lead to the development of tools able to validate critical steps of an operation [2]. Additionally, CV with or without kinematic data can be used for technical skill assessment of surgeons with accuracy rates of up to 80% [3].

Artificial Intelligence in Clinical Practice
DOI: https://doi.org/10.1016/B978-0-443-15688-5.00019-X

In comparison, NLP attempts to understand syntax, semantics, and the meaning of phrases [4]. The main use of NLP is for the analysis of unstructured free text in electronic medical records (EMR). In surgery, examples of these would include interpretation of consult notes, progress notes, operative reports, and discharge summaries. The records can be analyzed, and various features extracted and automatically reformatted in a structured manner. An application of this is to better capture complication rates at a large scale for system level analysis, without relying exclusively on administrative data which comes with its own bias.

The goal of ML in surgery should be to help optimize the preoperative and postoperative phases of care, and most importantly, help make the operating room (OR) safer by merging innovations in AI with the high-tech environment of the OR.

Preoperative planning

Preoperative risk prediction

All operations come with some degree of avoidable and unavoidable risk. Up to 20% of surgeries have complications [5,6]. Therefore developing a ML model for risk prediction to assess a patient's candidacy for surgery and to anticipate possible postoperative complications is important. A well-known classification system like the American Society of Anesthesiologists can help communicate a patient's preanesthesia medical comorbidities to the surgical team. This system alone is unable to predict perioperative risks, and relies on combination with other factors (e.g., type of surgery). Such a system has significant built-in subjectiveness which can lead to variability amongst anesthesiologists, and sometimes up to more than a third of patients [7,8]. National level data can be used to create models, such as the American College of Surgeons – National Surgical Quality Improvement Program (ACS-NSQIP) risk calculator to estimate the chances of an unfavorable outcome 30 days after surgery. This calculator uses a generalized linear mixed model to predict the risk of mortality and various other complications [9].

Preoperative risk calculation is evolving and can now combine large databases using EMR, such as the ACS-NSQIP, with ML algorithms to increase the objectiveness and accuracy of predictions of patient outcomes. Examples of automated predictive models of complications in surgery include the MySurgeryRisk score developed at the University of Florida and the Predictive Optimal Trees in Emergency Surgery Risk calculator developed in Boston for emergency operations [10,11].

Intraoperative use

In the preoperative phase, the utility of ML is clear and mainly used for risk prediction or to tailor the care to patients. However, given the complexities and workflows of the operating theater, ML in the intraoperative phase is in its infancy. ML has significant potential, but current clinical applications are scarce. Given the gradual shift from open to laparoscopic and robotic surgery over the last decades, the rise of available endoscopic videos has helped researchers in CV. Using ML and DL techniques for intraoperative video analysis is a hot topic for research in various disciplines given its potential for intraoperative guidance and coaching. Intraoperative near misses and complications are not infrequent and can often be attributed to errors in human visual perception leading to errors in judgment. Consequently, AI has promises when it comes to making the OR a safer environment. This can be achieved by helping surgeons and surgical trainees with identification of landmark anatomy and by augmentation of a surgeon's mental model [2,12].

Examples include a DL algorithm able to detect landmark anatomy during laparoscopic cholecystectomy (e.g., critical view of safety) [13]. Additionally, DL can be used to replicate the mental model of expert surgeons and help identify areas of safe and unsafe dissection during laparoscopic cholecystectomies [12] (Fig. 24.1). Both models have yet to be used intraoperatively to guide surgeons but in theory could help reduce bile duct injuries, a rare but serious complication during cholecystectomy. To develop such models for various operations requires sufficient training data, which can be alleviated by sharing and pooling of surgical videos across institutions. Groups, such as the Global Surgical AI Collaborative (https://www.surgicalai.org/), are providing the means to support and manage a global data-sharing platform of surgical videos and annotated datasets and provides the data pipeline for the surgical community to train and deploy algorithms in a transparent and collaborative process. While training DL models in CV, there are several factors that can affect the quality of predictions. Those include bias, and over or under fitting. Hence, ML models should be trained on heterogeneous datasets (from various sources) and with some variation in anatomy (e.g., various degrees of inflammation or tissue distortion). This will ensure the model can be used on a new external dataset and that the results are reproducible. Also, efforts should be made early to ensure interpretability by various surgeons so that the model can be externally validated. Lastly, the ML model should be validated by a panel of expert surgeons with the aim of establishing a ground truth "gold standard," such as for target anatomy or zone of safe or unsafe dissection whose boundaries are often not clearly defined and controversial amongst experts.

Another use of ML is analysis of surgical phases during an operation to anticipate events, length of an operation and for overall better workflow and logistics. Groups have worked on identification of surgical phase with good accuracy across laparoscopic

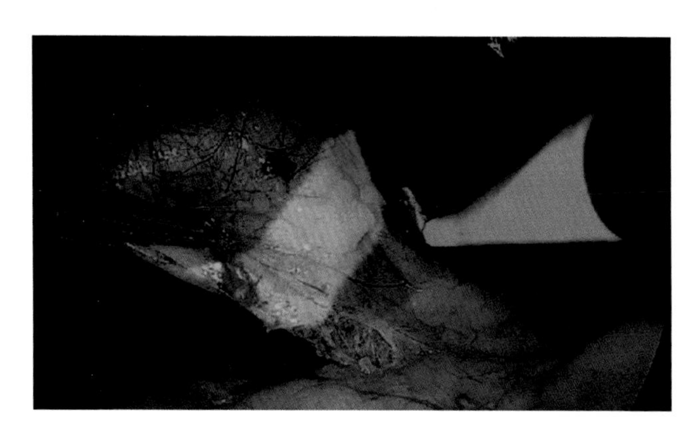

FIGURE 24.1 The GoNoGoNet for laparoscopic cholecystectomy by Madani et al. Here, the safe (green) and unsafe (red) zones of dissection are displayed in real time during dissection of the hepatocystic triangle.

cholecystectomy (92%), sleeve gastrectomy (85.6%), sigmoidectomy (91.9%), and inguinal hernia (88.8%) [14–17]. A group was able to use ML for prediction of remaining operative time for laparoscopic cholecystectomy and gastric bypass operations using intraoperative videos alone [18].

Collecting videos for ML applications comes with its own regulatory and legal considerations. These regulations will affect how surgeons use and share images and videos for research and for clinical applications. In surgery, these are likely to play a major role given the large amount of video data required to make scalable AI models. Given the rise in robotic surgery and streamlined surgical video capture as a result, considerations will be on who owns the video, and will various robotic platform suppliers agree to share videos for research purposes. Therefore policies governing data acquisition, storage, sharing and utilization will have to be made [19].

Postoperative care

NLP is a promising and innovative approach to collect EMR data that conventional labor-intensive data collection methods do not capture. NLP is quicker but also less expensive, less time-consuming, and more objective [20,21]. NLP for extracting information from patient free-text notes often outperforms other conventional methods. A recent systematic review and meta-analysis showed that both NLP and traditional non-NLP models can effectively rule in outcomes (i.e., high specificity), but only NLP models can effectively rule out outcomes (i.e., high sensitivity), and particularly in the case of postoperative complications [20]. Other examples of the use of NLP is in the detection of surgical site infection from provider notes [22]. Lastly other uses of NLP includes; during dictation of operative notes to automatically detect potential errors, computer-assisted coding, automated registry reporting, and population surveillance.

During morbidity and mortality (M&M) rounds, surgeons are expected to present a case and recall from memory and the patients' chart what led to a certain unexpected outcome. However, errors can also occur from factors outside a surgeon's direct control. The operating room (OR) Black Box developed in Toronto, Canada, simultaneously captures audiovisual data from the operation, the surgical tools and from the entire OR. It also captures physiologic data from the patient and members of the surgical team [23]. This information can be used to enhance the M&Ms by providing objective and quantifiable data from almost all intraoperative factors. This can then help various stakeholders have all the elements needed to really understand the root-cause of surgical but also trauma room near misses or complications [24].

Conclusion

ML applications in surgery remain in early development and most projects are still in the preclinical phase. However, research ideas and possible use of CV and ML in surgery are increasing rapidly. This is true given the rise of minimally invasive approaches, such as laparoscopy, endoscopy, and robotic surgery, which all provide high resolution recordable videos. There are now greater and cheaper storage options making it easier to save operative videos for education and research. Eventually, we should expect real time intraoperative decision support, and other applications, such as advanced autonomous agents for robotic surgery, performance assessment of surgical skills, and improved operating room workflow. All of which will help transform the operating room of the future.

Major takeaway points

- At its current stage, ML applications in surgery are still far from replacing human judgment, or surgeons' technical and nontechnical skills and should rather be seen as a supplement.
- Preoperative risk calculation is evolving and can now combine large databases with ML algorithms to increase the objectiveness and accuracy of predictions of patient outcomes.
- In surgery, CV can lead to the development of several applications, such as segmentation of landmark anatomy and tools, identification of safe and unsafe zones of dissections, intraoperative guidance, identification of phases of an operation, enhanced coaching, and technical skills assessment.
- Regulatory and legal considerations of sharing operative videos between institutions for ML applications will require new policies governing data acquisition, storage, sharing and utilization.

References

[1] Krizhevsky A, Sutskever I, Hinton GE. ImageNet classification with deep convolutional neural networks. Commun ACM 2017;60(6):84–90. Available from: https://doi.org/10.1145/3065386.

[2] Hashimoto DA, Ward TM, Meireles OR. The role of artificial intelligence in surgery. Adv Surg 2020;54:89–101. Available from: https://doi.org/10.1016/j.yasu.2020.05.010.

[3] Lam K, et al. Machine learning for technical skill assessment in surgery: a systematic review. npj Digital Med 2022;5(1). Available from: https://doi.org/10.1038/s41746-022-00566-0.

[4] Eisenstein J. Introduction to natural language processing. MIT Press; 2019.

[5] Ludbrook GL. The hidden pandemic: the cost of postoperative complications. Curr Anesthesiol Rep 2022; 12(1):1–9.

[6] Ghaferi AA, Birkmeyer JD, Dimick JB. Variation in hospital mortality associated with inpatient surgery. N Engl J Med 2009;361(14):1368–75. Available from: https://doi.org/10.1056/nejmsa0903048.

[7] Haynes SR, Lawler PG. An assessment of the consistency of ASA physical status classification allocation. Anaesthesia 1995;50(3):195–9.

[8] Owens WD, Felts JA, Spitznagel Jr EL. ASA physical status classifications: a study of consistency of ratings. Anesthesiology 1978;49(4):239–43.

[9] New ACS NSQIP Surgical Risk Calculator offers personalized estimates of surgical complications. Bull Am Coll Surg 2013;98(10):72–73.

[10] Bihorac A, et al. MySurgeryRisk: development and validation of a machine-learning risk algorithm for major complications and death after surgery. Ann Surg 2019;269(4):652–62.

[11] El Hechi MW, et al. Validation of the artificial intelligence-based Predictive Optimal Trees in Emergency Surgery Risk (POTTER) calculator in emergency general surgery and emergency laparotomy patients. J Am Coll Surg 2021;232(6):912–19.e1.

[12] Laplante S, et al. Validation of an artificial intelligence platform for the guidance of safe laparoscopic cholecystectomy. Surg Endosc 2022. Available from: https://doi.org/10.1007/s00464-022-09439-9.

[13] Mascagni P, et al. Artificial intelligence for surgical safety: automatic assessment of the critical view of safety in laparoscopic cholecystectomy using deep learning. Ann Surg 2022;275(5):955–61.
[14] Golany T, et al. Artificial intelligence for phase recognition in complex laparoscopic cholecystectomy. Surg Endosc 2022. Available from: https://doi.org/10.1007/s00464-022-09405-5.
[15] Hashimoto DA, et al. Computer vision analysis of intraoperative video: automated recognition of operative steps in laparoscopic sleeve gastrectomy. Ann Surg 2019;270(3):414–21.
[16] Kitaguchi D, et al. Real-time automatic surgical phase recognition in laparoscopic sigmoidectomy using the convolutional neural network-based deep learning approach. Surg Endosc 2020;34(11):4924–31. Available from: https://doi.org/10.1007/s00464-019-07281-0.
[17] Takeuchi M, et al. Automatic surgical phase recognition in laparoscopic inguinal hernia repair with artificial intelligence. Hernia 2022. Available from: https://doi.org/10.1007/s10029-022-02621-x.
[18] Twinanda AP, Yengera G, Mutter D, Marescaux J, Padoy N. RSDNet: learning to predict remaining surgery duration from laparoscopic videos without manual annotations. IEEE Trans Med Imaging 2019;38(4):1069–78.
[19] Hashimoto DA, Rosman G, Rus D, Meireles OR. Artificial intelligence in surgery: promises and perils. Ann Surg 2018;268(1):70–6. Available from: https://doi.org/10.1097/sla.0000000000002693.
[20] Mellia JA, et al. Natural language processing in surgery. Ann Surg 2021;273(5):900–8. Available from: https://doi.org/10.1097/sla.0000000000004419.
[21] Morris MP, et al. Feasibility of natural language processing in surgery: sensitivity and specificity compared to manual extraction. J Am Coll Surg 2021;233(5):S93. Available from: https://doi.org/10.1016/j.jamcollsurg.2021.07.173.
[22] Shen F, Larson DW, Naessens JM, Habermann EB, Liu H, Sohn S. Detection of surgical site infection utilizing automated feature generation in clinical notes. Int J Healthc Inf Syst Inf 2019;3(3):267–82.
[23] Jung JJ, Jüni P, Lebovic G, Grantcharov T. First-year analysis of the operating room black box study. Ann Surg 2020;271(1):122–7.
[24] Nolan B, Hicks CM, Petrosoniak A, Jung J, Grantcharov T. Pushing boundaries of video review in trauma: using comprehensive data to improve the safety of trauma care. Trauma Surg Acute Care Open 2020;5(1):e000510.

CHAPTER

25

Artificial intelligence in cardiothoracic surgery: current applications and future perspectives

Mahdi Ebnali[1,2], *Marco A. Zenati*[3,4] *and Roger D. Dias*[1,2]

[1]Department of Emergency Medicine, Harvard Medical School, Boston, MA, United States [2]STRATUS Center for Medical Simulation, Mass General Brigham, Boston, MA, United States [3]Department of Surgery, Harvard Medical School, Boston, MA, United States [4]Division of Cardiac Surgery, VA Boston Healthcare System, West Roxbury, MA, United States

Introduction

Cardiothoracic surgery (CTS) takes place in a complex operative environment, where multiple specialized professionals work together to deliver effective care to patients. As a high-risk intervention, CTS is undergoing rapid evolution, and many of the changes involve the application of artificial intelligence (AI) technologies to tackle critical challenges in risk management, operation planning, and delivering personalized care [1–4].

AI encompasses a number of components ranging from supervised, unsupervised, and reinforcement machine learning (ML) technologies, as well as its subcomponents, such as deep learning (DL) [5]. ML involves the process of analyzing data and training algorithms to recognize specific patterns or predict certain outcomes [5]. In supervised ML, inputs and outcomes are clearly labeled by humans before they are presented to an algorithm. Unlike traditional statistical methods, where predefined rules are necessary for processing data, supervised ML methods identify patterns and trends based on labeled inputs and outputs. In unsupervised ML, the data are processed without labels, with the goal of finding similarities between datasets by clustering the data to identify patterns [5]. Reinforcement learning consists of adjusting algorithms based on rewards and punishments, whereby rewards are given for actions that increase the likelihood of achieving the desired outcome, and punishments are given for actions that hinder the achievement of the objective, leading the algorithm to subsequently learn the optimal strategy for a given task [5]. DL algorithms, one of the most

Artificial Intelligence in Clinical Practice
DOI: https://doi.org/10.1016/B978-0-443-15688-5.00030-9

recent advances in AI, are capable of learning very complex nonlinear mathematical functions using a sequential chain of features derived from input data. These algorithms usually are based on deep neural networks that can take input data and learn them sequentially [5].

Applications of artificial intelligence in cardiothoracic surgery

AI has been investigated in previous studies for improving the accuracy of risk assessment, diagnosis, surgical intervention planning, and predicting interventional outcomes in CTS operations, with the ultimate goal of improving and personalizing care as well as reducing complications associated with CTS procedures. In this chapter, we provide an overview of the implementation of AI in *preoperative*, *intraoperative*, and *postoperative* phases of CTS (Fig. 25.1). The final section also summarizes the major challenges associated with

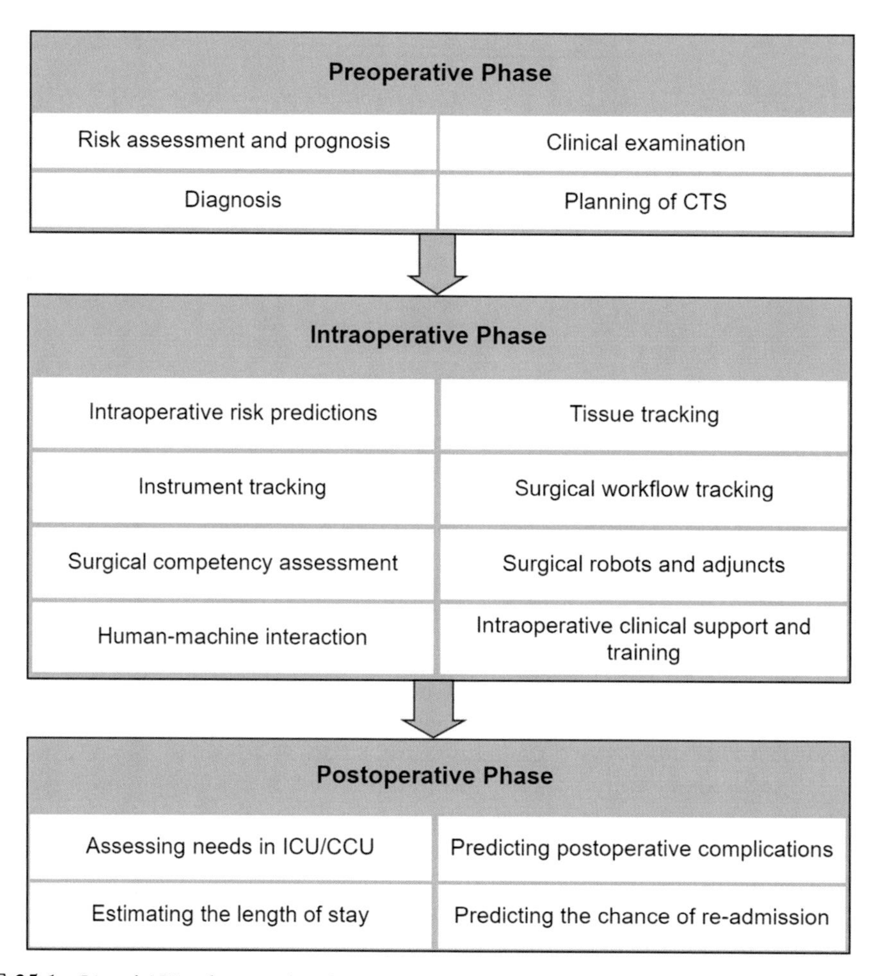

FIGURE 25.1 List of AI implementations in preoperative, intraoperative, and postoperative phases of CTS.

the application and operationalization of AI in CTS and suggests potential future research directions.

Preoperative phase

AI models can process large amounts of data collected before CTS, and because of their comprehensive and nonlinear properties, they can be useful for detecting underlying patterns without explicit instructions for several critical preoperative tasks including (1) risk assessment and prognosis, (2) clinical examination and diagnosis, and (3) planning for CTS interventions.

Risk assessment and prognosis

The perioperative mortality rate in the United States continues to be the most prevalent cause of death of patients with CTS complications [6], and identifying and predicting risks associated with cardiothoracic diseases and surgical interventions remains a major challenge. Traditional methods, such as The Society of Thoracic Surgeons risk score and the European System for Cardiac Operative Risk Evaluation [7], require modeler input to outline complex interactions between variables, resulting in less accurate risk assessment [8]. AI approaches, such as artificial neural network (ANN), have been increasingly investigated to determine mortality rates, make decisions on the course of action for CTS procedures, and provide more personalized care following surgery [9]. The findings from the current literature suggest that AI models built based on preoperative data can achieve better discrimination in the prediction of mortality after CTS when compared with traditional methods, such as logistic regression (LR) [8,10].

Clinical examination and diagnosis

AI methods have been also extensively investigated as a tool for improving diagnosis as they can make inferences from complex, high-dimensional, and often multimodal data; and clinical studies have shown their excellent performance in a number of diagnosis applications of CT cardiothoracic before surgical intervention, including measurement of aortic diameter [11,12] and volumetric segmentation of the left ventricle to determine the cardiac function [13]. Particularly, models based on DL, such as convolutional neural networks (CNNs) and recurrent neural networks, are capable of detecting subtle patterns in medical images in order to detect anomalies quickly and accurately. For example, these algorithms were used to detect distinct pathologies from chest X-rays [9,14], segmentizing cardiac MRI data [15], and enhancing image quality of coronary angiography data [16]. A CNN algorithm also was able to detect wall motion abnormalities more accurately in echocardiographic images, achieving an area under the receiver curve (AUC) of 0.99, outperforming physicians on the same task [17]. The AUC is one of the most widely used metrics to evaluate the performance of AI classifiers.

Furthermore, early diagnosis of thoracic and abdominal aneurysms would be particularly beneficial due to the fact that the majority of patients have no symptoms prior to the development of life-threatening complications [18]. AI techniques integrating genomic and electronic health record data were recently demonstrated to be effective in diagnosing abdominal aortic aneurysms [19]. In addition, a random forest, an ML-based classifier, trained on a large data set of patients was able to accurately identify ruptures of the ascending aorta within the hospital setting for patients with thoracic ascending aortic aneurysms with an AUC value of 0.752 and a sensitivity of 0.99 [20]. Using only the ECG data as input for the DL model, a previous study achieved an AUC of 0.756 for heart failure classification using multiple preoperative predictors, such as patients' age, gender, race, body mass index, smoking status, prevalent coronary heart disease, diabetes mellitus, systolic blood pressure, and heart rate, as predictors [21].

Planning of cardiothoracic surgery

Currently CTS are planned and executed primarily according to treatment protocols and guidelines. The lack of tailored interventions could negatively impact outcomes and prognosis after intervention [22]. Recently AI models have been evaluated and validated for personalizing surgical intervention plans and managing potential postoperative care based on patients' phenotype and preoperative data [23,24]. Compared to traditional methods, AI models are capable of capturing nonlinearity and interactions among features without requiring the modeler to manually specify all interactions [9]; they are also capable of handling missing data more efficiently since they do not require assumptions about data distribution and are capable of more complex computations [25]. In addition, several forms of patient data are available in unstructured formats, including images, electronic health record data, and clinical notes.

It is extremely time-consuming and error-prone to process these data resources manually. However, AI methods have been found to be useful in analyzing this unstructured and imbalance data without or with very little human input to improve risk estimation preoperatively and to enable data-driven and objective decision-making tools for CTS [23,24].

Intraoperative phase

Previous studies have shown promising applications of AI in intraoperative CTS due to their ability to dynamically process and produce predictions based on data across different phases of the operation, all of which differ in data type and density. The intraoperative phase, in particular, produces large quantities of continuous data including recorded video and audio, time series of hemodynamic, temperature, and cardiopulmonary bypass perfusion measurements, and machine/robots/medical devices/wearable devices. As discussed in this section, researchers have used AI models intraoperatively to predict interventional risks, identify surgical workflow and assessment of surgical competency, track tissue and surgical instruments, optimize surgical robots/adjuncts, and provide intraoperative clinical support and training.

Intraoperative risk predictions

AI has been shown as a potential approach to continuously monitor various components of the OR, such as patient's hemodynamic status, provide intuitive and real-time overview, and warn the surgical team earlier regarding critical episodes. In a previous study, fuzzy logic was used to provide surgeons with early warnings of critical changes in patients' hemodynamic status with an accuracy of 99.5% over traditional methods [26]. As opposed to a simple threshold alarm, this system may allow the physician to interpret changes more quickly. Hypotension prediction index is another AI-based monitoring tool using a logistic regression model to predict an impending hypotensive episode 15 minutes in advance with high accuracy in CTS [27,28]. In high-risk procedures, such as CTS, early warning systems can be particularly beneficial, reducing the chances of adverse events occurring [29].

The condition of hypoxemia, or low arterial blood oxygen tension, is an undesirable physiological condition that is associated with cardiac arrest, cardiac arrhythmias, postoperative infections, wound healing impairments, delirium, decreased cognitive function, and cerebral ischemia [30]. ML methods have been shown as a potential approach to predict the future risk of intraoperative hypoxemia and augment the current anesthesiologist's prediction of hypoxemia during surgery [31]. Massive bleeding is another major complication that may occur during CTS, leading to increased chances of mortality and prolonged hospital stays [29]. Some recent efforts have focused on the application of CV analysis to evaluate the likelihood of blood stains during CTS [32] using high throughput CV algorithms that could process many CTS videos simultaneously.

Tissue tracking

Dynamic and accurate tracking of tissue is essential in optimizing intraoperative surgical guidance and navigation during robot-assisted and minimally invasive surgeries [33], and AI algorithms have been utilized to estimate and track soft tissue in CTS. Lu et al. [33] developed a DL-based online learning framework for updating soft tissue features to optimize minimally invasive cardiac surgery and their findings showed significantly lower feature engineering efforts to fully perceive a surgical scene. Based on the independent component analysis method, Mountney and Yang [34] also proposed and validated an algorithm to deal with drift and occlusion during decoupling cardiac and respiratory motion in robotic-assisted surgery, helping in more accurate retargeting of regions on soft tissues. These efforts demonstrate that learning strategies can enhance the robustness of tissue tracking in various situations, including deformations, soft tissue differentiation from the background, and variations in illumination [35].

Instrument tracking

Tracking and segmenting instruments also have been investigated to optimize CTS workflow to facilitate surgical team communication, assess surgical skills, and improve surgeon-robot interaction. In recent years, DL has made significant progress in the area of instrument tracking, allowing to distinguish instruments from their backgrounds, segment tools, and identify their components (e.g., shaft, gripper, wrist). Most of the previous works relied on labeled features, such as color and texture, and were built on the basis of detection algorithms to better segmentize instruments. For example, fully convolutional neural networks, an ML-based classifier, and Fuse-Net architecture have significantly improved the accuracy of CTS instruments segmentation and tracking using pixel-level binary classification analysis [36].

Surgical workflow tracking

An interruption in the procedural flow during CTS can result in surgical errors and increase the risk of complications [37]. Errors associated with teamwork and communication, technological failures, and training-related distractions are among the most common causes of surgical workflow interruptions [37]. To minimize flow interruptions and maximize coordination between the OR team AI models can be used in identifying surgical phases and steps, and their order [9]. In another effort to manage CTS workflow, Avrunin et al. [38] developed smart checklists to help the CTS team, including surgical, anesthesiology, perfusion, and nursing staff [38] by reminding the team about the next step in the procedure and by alerting the other team members not to disturb specific members of the team during some high cognitive load episodes.

Surgical competency assessment

Assessment of surgical competency at individual and team levels has become a priority for both surgical educators and licensing boards [39]. Currently methods for competency assessment are based on observation and rating by experts, which is extremely time-consuming, costly, and sometimes biased [40]. Due to the nonlinear, multidimensional, and dynamic nature of the OR as well as the complexity of various surgical procedures, AI techniques can be useful to assess trainees' and surgeons' CTS technical and nontechnical skills objectively.

Technical skills assessment

AI methods, particularly CV-based techniques, has been investigated in technical assessment of cardiac surgical performance and providing surgeons with quantitative and actionable feedback on some tasks, such as suturing, needle passing, and knot tying [41–43]. In one study, DL was applied in laparoscopic surgery to track surgical tools as a proxy for surgical skill and quality [44]. Integration of these methods in warning systems has the potential to alert surgeons if their performance deviates from other surgeons in a database, or to provide real-time feedback on technique during surgery [9]. Nontechnical competency: In addition to technical skills, the practice of CTS requires nontechnical skills, such as teamwork, communication, and an awareness of the surrounding environment. Another new area of application of CV is to assess nontechnical surgical skills by monitoring team dynamics in the OR, particularly for complex surgeries [3,45,46]. AI has been utilized to enhance machines' capabilities [29] by acquiring transportable teamwork competencies that are critical to teams, creating better coordination with the human counterparts in a team, and allowing other team members to determine when and how best to communicate with teammates, further enhancing ability and trust in human-machines [47].

Surgical robots and adjuncts

Surgical robots

Robotic surgery has advanced substantially in recent decades, with examples including the Da Vinci Surgical System (Intuitive Surgical, Sunnyvale, CA, USA) [48] and the Sensei X robotic catheter system (Hansen Medical Inc., Mountain View, CA, USA) for cardiac catheter insertion [49]. Surgical robotics is already capable of performing simple surgical subtasks, such as simple suturing and precise surgical cutting; however, due to the complexity and dynamic nature of the tasks cardiothoracic surgeons perform, surgical robots will not be completely independent of human control in the near future, and they will require continuous or near-continuous human intervention [50]. By integrating AI, surgical robotics will be able to perceive and understand complex surroundings, beyond simple tasks, such as suturing, make real-time decisions, and perform surgical tasks more accurately, safely, and efficiently [9].

Adjuncts

AI may also benefit image guidance semiautomated adjunct systems, providing aid in planning operative trajectory by utilizing either preoperative or intraoperative imaging data [51]. In addition, a robot scrub nurse that understands surgeons' needs using Kinect sensors were designed to facilitate the delivery of surgical instruments [52]. Telesurgery also has already taken place in surgical interventions in various types of operations [53]; and AI methods can be integrated into cardiothoracic telesurgical technologies to augment images, improve diagnosis accuracy, and optimize robot motion [54].

Human–machine interaction

The field of human–machine interaction combines knowledge and techniques from a variety of disciplines to facilitate effective human interaction with advanced machines and robots. More specifically, human–robot interaction requires robots to understand the intentions of humans in order to perform appropriate actions. To this end, AI has been used to track surgeons' eye-gaze data in real time and provide useful information for more accurate navigation of robots [55]. In addition to human gaze, head movements [56], body gestures, and voice commands [57] have been processed using AI methods, such as supervised and unsupervised linear models, classification, and DL algorithms, to interpret surgeon intentions for more accurate teaming with robots. Incorporating DL into speech recognition has significantly improved the precision and accuracy of speech recognition, thereby resulting in more reliable robotic surgery [58]. It still remains challenging to use robots during surgery due to the noisy environment in the operating room.

Intraoperative clinical support and training

Intraoperative clinical support has always been an integral component of complex surgical procedures, including CTS. Various AI learning methods have been extensively applied to the development of intraoperative guidance localization in surgery in order to improve visualization and provide enhanced guidance. Rapid technological advancements in the field of augmented reality (AR) and virtual reality (VR) have also provided new solutions for a wide variety of industries [59–63] including surgical application, such as intraoperative clinical guidance and surgical training. Although integration of AR/VR and AI methods not yet broadly used in CTS, it has already shown great potential to improve preoperative planning, intraoperative navigation, patient education, and surgical training. For example, AI and VR were found as effective solutions to give a 3D view of the whole pulmonary anatomy with its intricate parts to the cardiothoracic surgeon to preoperatively get a better insight into the individual patient's anatomy [64]. Further research is required to develop the application of AI-driven AR and VR technologies in CTS.

Postoperative phase

Clinical medicine often involves the use of patient data to predict future outcomes of surgical interventions and manage follow-up care accordingly. The decision-making process for postoperative care has traditionally been based on medical literature and clinical experience; however, AI has the potential to facilitate various aspects of this process, including assessing needs in intensive care unit (ICU)/coronary care unit (CCU), predicting postoperative complications, such as kidney injury and delirium, estimating the length of stay in the hospital, and predicting the chance of re-admission.

Assessing needs in ICU/CCU

To assess patients' needs immediately after CTS for high-concentration oxygenation, ICU care, and ventilator, AI methods, such as Naïve Bayes algorithm, were found to predict risks adequately and reported as a useful assisted prediction system [65]. Additionally, both ANN and advanced LR models accurately predicted the need for early continuous venovenous hemofiltration after cardiac surgery [66], the need for prolonged mechanical ventilation, and the chance of re-intubation in postcoronary artery bypass graft surgery [67,68]. They found that the ANN method yielded a higher prediction accuracy than the LR method.

Predicting postoperative complications

Early prediction of the occurrence of postoperative complications is critical to reduce the further deterioration of the patient's condition. Several studies have shown that compared to traditional statistical models, AI-based models show a higher advantage in predictive power [69]. For example, a recent study showed the accuracy of XGboost (a supervised ML algorithm) in the prediction of septic shock, thrombocytopenia, and liver dysfunction after open-heart surgery [70].

Kidney injury

There is a high incidence of acute kidney injury after cardiac surgery [71], and it is important to identify the patients who are at risk so that postoperative care can be tailored to their specific needs [72]. As compared to traditional risk score systems, AI-based predictive models, such as ANN, are superior in predicting acute kidney injury [73]. An analysis of ML was successful in predicting cardiac surgery-related acute kidney injury. The study also showed that intraoperative time series and other features are important to acute kidney injury prediction (Tseng et al., 2020).

Delirium

Prevention and early recognition of delirium are essential components of CTS intervention [74], and recently AI researchers and clinicians have applied AI methods in predicting delirium in postcoronary artery bypass graft surgery [75,76]. Mufti et al., for example, reported that ML methods can enable the investigation of hidden patterns in delirium causation as well as predict its occurrence following cardiac surgery [75].

Estimating the length of stay

Having an accurate estimate of the length of stay in an ICU is beneficial both for providing better postoperative care for patients and hospitals particularly considering that cardiovascular intensive care resources are limited and waiting lists for cardiac surgery exist [77]. ML, such as ANN and logistic regression models outperformed traditional statistical and score-based models in predicting the length of stay of patients and stratifying CTS patients at risk of extended stay [77,78].

Predicting the chance of readmission

Predicting the chance of readmission is one of the most essential elements of preventive measures for patients who have undergone CTS after being discharged from the hospital. Various ML-based classification approaches, such as XGBoost and advanced LR, have been used to predict patients with a higher chance of re-admission [79,80]. These algorithms over-performed compared to the conventional LR method. In particular, AI-based methods can perform better when patient monitoring data (time series) is incorporated into preoperative and intraoperative data sets. The development of an artificial intelligence algorithm based on ECG signals was recently found to be capable of predicting the long-term mortality of cardiac surgery and ventricular dysfunction postsurgery [81].

Current challenges and future direction of artificial intelligence applications in cardiothoracic surgery

As discussed in previous sections, AI methods have significantly improved preoperative diagnosis, prognosis and planning, intraoperative intervention and guidance, and postoperative care. Despite these advancements and benefits, there are several critical challenges that impede the translation of AI to practice in CTS; and extensive research remains needed in order to develop interpretable, robust, and generalizable models for clinical application. Here we summarize the major challenges and suggest an outline of achievable future research directions.

Overfitting and external validation

The lack of generalizability of AL algorithms is considered to be one of the major barriers to their adoption in healthcare [82]. The performance of AI algorithms is likely to differ across centers, settings and times due to model overfitting resulting from small training and testing samples, heterogeneity of testing populations, and operational heterogeneity [83]. The clinical utility of predictive algorithms for decision-making may vary greatly between hospitals, for example, due to different protocols, imaging machines, and instruments. To mitigate this limitation, algorithms must be rigorously validated using external data [84]. A recent review study on the application of ML and DL in cardiothoracic imaging reported that only a small portion (15%) of studies that utilized AI methods in cardiothoracic imaging performed external validation to test the generalizability of their models [82]. External validation can be performed via publicly available datasets when an algorithm has been trained on institutional data, allowing for a fair comparison of the algorithms. It is important that external validation be extensive, and it should take place at various sites in cohorts of patients from the targeted population [83]. Moreover, it is imperative that external validation should be evaluated over time [85] and carried out with independent investigators [86] to reduce biases.

Algorithmic black boxes and artificial intelligence explainability

AI often involves the computation of multiple layers of interconnected networks in datasets, making it difficult to determine how the input is transformed into the output, a phenomenon known as "algorithm black box" [87]. The AI algorithms used in clinical applications are also often characterized by black box problems, making it difficult to trust their performance and outcomes, particularly in highly critical scenarios, such as cardiothoracic surgeries. Nevertheless, the accuracy and interpretability of these methods are tradeoffs, resulting in a continuing debate regarding the best course of action for clinical use [88]. In addition, when big data are poorly organized and managed in healthcare, inaccurate models are produced which include erroneous data [89]. It is imperative that multidisciplinary teams collaborate closely to overcome these issues, particularly between surgeons and AI researchers, to generate large-scale annotated data that will be used as training data for AI algorithms [90].

Ethical considerations

Beyond the promising application of AI in CTS, ethical and legal issues must be carefully considered in all aspects of obtaining patients' data, creating data repositories, training algorithms, interpreting results, and implementing AI-driven decisions [87]. Privacy concerns regarding patient data are becoming increasingly important as more clinical data is being made available for algorithm training. Although excessive privacy protection can halt or impede technological progress, neglecting privacy can erode trust in technological advances [87]. For clinical algorithms to be efficient and versatile, it is necessary to overcome these barriers and plan for challenges in data sharing across institutions and countries [91]. It may be possible to mitigate these problems by employing different transfer learning techniques and developing more explainable AI algorithms to improve its decision-making performance.

Data imbalance

Several studies have observed an output class imbalance due to a lack of data from complex CTS and rare complications, which may challenge the performance and generalizability of AI algorithms due to their bias toward predicting the more prominent outcome state [9]. Future AI research must adequately address this issue to overcome the limitations of applied predictive methods, particularly when dealing with highly specialized surgical procedures, such as CTS. Oversampling, sample weighting, synthesis of new data, and prediction postprocessing can counteract this bias [92]. Researchers have also used some particular methods, such as decision ensembles [93] and the synthetic minority class oversampling technique [94], to optimize class imbalance.

Localization and mapping performance

Enhancing the localization and mapping performance of computer-assisted guidance from visual observations is crucial when dealing with a texture-less surface, variations in illumination, and limited field of view [95]. The deformation of organs and tissues is another major challenge in a dynamic and uncertain surgical environment. Despite the success of AI technologies in detection, segmentation, tracking, and classification, further research is needed to extend these processes to more sophisticated 3D applications [96]. Additionally, to achieve a more precise perception of the complex environment, future AI technologies must fuse multimodal data from various sensors. As micro- and nanorobotics become increasingly popular in surgery, new issues will arise in terms of guidance [97]. The ability to provide real-time assistance to surgeons is one of the most important requirements for an AI algorithm during surgery. Such requirements have been also highlighted in AI-integrated robotic-assisted and tele-surgeries where interaction between surgeons and autonomous guidance is crucial, particularly during remote surgery involving multidisciplinary teams located in different geographical locations.

Overtrust and deskilling

While interventional robotic capability is still in its infancy, there are already several adjunct systems and semiautonomous machines that are likely to see a growing number of applications in the future of CTS [98]. Since many of these solutions, such as automated microscopes, endoscopes, or adjunct robots, enable clinicians to perform their duties more effectively, frequent usage may result in de-skilling, and possibly increase the proportion of errors associated with the deskilling [99]. Moreover, if trainees become overly dependent on these AI-informed technologies, their ability to develop technical skills may be adversely affected, leading to technical errors during unexpected scenarios and equipment malfunctions [99,100]. Several lessons can be drawn from fields where AI has been transformative, such as aviation [101] and automated driving [61,62,102,103]. In recent years, pilots have been exposed to a number of incidents due to their overreliance on computer-based flight systems and assistive driving technologies. While AI has increased safety within the transportation industry, over-trust and over-reliance on AI-based technologies have resulted in an alarming de-skilling of operators, which is only apparent when unexpected, novel or emergency events occur and AI fails to respond. AI-based technologies may only be used as adjuncts by CTS teams in order to prevent the loss of their surgical skills. In order to ensure high levels of competency among surgical staff, frequent surgical competency assessments can also be planned at the organizational level.

Human factors considerations in integrating artificial intelligence into cardiothoracic surgery

Most current AI models developed for CT operations and other healthcare applications overly focus on engineering technology (technical aspects) without sufficiently incorporating human factors and usability best practices and methods. In a recent review study, Asan and Choudhury (2021) highlighted the importance of including human factors researchers in AI design and implementation, as well as in dynamic assessments of AI systems' impacts on interaction, workflow, and patient outcomes [104]. Although following a standardized protocol for developing user-centered healthcare technologies is challenging due to the complexity of systems and the diversity of users, a recent work proposed a framework for incorporating usability assessment in developing digital health technologies [60,63]. The authors used this framework to evaluate an electronic cognitive aid (Smart Checklist) that was developed to guide cardiac surgery teams during common cardiac procedures in the OR. The application of AI in surgical workflows must be designed using a user-centered approach, involving clinicians throughout the entire design cycle as users, moving AI research beyond pure technical development into a sociotechnical system model, and effectively integrating human factors principles into its design–development–research life cycle.

Conclusion

Several previous studies have examined how artificial intelligence can improve risk assessment, diagnosis, surgical intervention planning, and interventional outcome prediction in CTS operations, ultimately aiming to reduce complications associated with CTS procedures as well as improve and personalize care. The studies reviewed in this chapter suggest that there is substantial potential for further development of research outputs in this area. AI algorithms appeared to outperform other traditional methods in pre-, intra-, and postoperative phases of CTS, accurately predicting a wide range of operative outcomes. A continued deployment of artificial intelligence via human-centered clinical decision support and adjunct robots will likely contribute to a further decrease in complication rates and an improvement in quality and safety in CTS. We also summarized the major challenges associated with the application and operationalization of AI in CTS. In the future, research should explore these hurdles and attempt to address them in the development, validation, and deployment of ML models in CTS.

Major takeaway points

- AI applications have shown promising results in pre-, intra-, and postoperative phases of CTS.
- AI is likely to be a big part of the future of CTS operations.
- The development of interpretable, robust, and generalizable AI models for CTS applications still requires extensive research.

Acknowledgments

This work was supported by the National Heart, Lung, and Blood Institute of the National Institutes of Health (R01HL126896, R01HL157457). The content is solely the responsibility of the authors and does not necessarily represent the official views of the National Institutes of Health.

References

[1] Dearani Joseph A, Todd KRosengart, Blair Marshall M, Mack Michael J, Jones David R, Prager Richard L, et al. Incorporating innovation and new technology into cardiothoracic surgery. Ann Thorac Surg 2019;107 (4):1267–74.

[2] Molina JA, Heng Bee Hoon. Global trends in cardiology and cardiothoracic surgery—an opportunity or a threat. Ann Acad Med Singap 2009;38(6):541–5.

[3] Dias Roger D, Julie AShah, Marco AZenati. Artificial intelligence in cardiothoracic surgery. Miner Cardioangiol 2020;68(5):532–8.

[4] Dias RD, Zenati MA, Rance G, Srey Rithy, Arney D, Chen L, et al. Using machine learning to predict perfusionists' critical decision-making during cardiac surgery. Comput Methods Biomech Biomed Eng Imaging Vis 2022;10(3):308–12.

[5] Hallinan Dara, Leenes Ronald, Hert Paul De. Data protection and privacy, volume 13: data protection and artificial intelligence. New York: Bloomsbury Publishing; 2021.

[6] Bhatnagar Prachi, Wickramasinghe Kremlin, Williams Julianne, Rayner Mike, Townsend Nick. The epidemiology of cardiovascular disease in the UK 2014. Heart 2015;101(15):1182–9.

[7] Sullivan Patrick G, Joshua DWallach, Ioannidis John PA. Meta-analysis comparing established risk prediction models (EuroSCORE II, STS score, and ACEF score) for perioperative mortality during cardiac surgery. Am J Cardiol 2016;118(10):1574–82.

[8] Kilic Arman, Goyal Anshul, Miller James K, Gjekmarkaj Eva, Lam Tam Weng, Gleason Thomas G, et al. Predictive Utility of a machine learning algorithm in estimating mortality risk in cardiac surgery. Ann Thorac Surg 2020;109(6):1811–19.

[9] Raghu Vineet K, Moonsamy Philicia, Sundt Thoralf M, Siang Ong Chin, Singh Sanjana, Cheng Alexander, et al. Deep learning to predict mortality after cardiothoracic surgery using preoperative chest radiographs. Ann Thorac Surg 2022;2023(115):257–65.

[10] Nilsson Johan, Ohlsson Mattias, Thulin Lars, Höglund Peter, Nashef Samer AM, Brandt Johan. Risk factor identification and mortality prediction in cardiac surgery using artificial neural networks. J Thorac Cardiovasc Surg 2006;132(1):12–19.

[11] Chen Chen Chen, Qin Chen, Qiu Huaqi, Tarroni Giacomo, Duan Jinming, Bai Wenjia, et al. Deep learning for cardiac image segmentation: a review. Front Cardiovasc Med 2020;. Available from: https://doi.org/10.3389/fcvm.2020.00025.

[12] Bai Wenjia, Suzuki Hideaki, Qin Chen, Tarroni Giacomo, Oktay Ozan, Matthews Paul M, et al. Recurrent neural networks for aortic image sequence segmentation with sparse annotations. Med image comput computer assist intervention – MICCAI 2018. Springer International Publishing; 2018. p. 586–94.

[13] Ouyang David, He Bryan, Ghorbani Amirata, Yuan Neal, Ebinger Joseph, Langlotz Curtis P, et al. Video-based AI for beat-to-beat assessment of cardiac function. Nature. 2020;. Available from: https://doi.org/10.1038/s41586-020-2145-8.

[14] Wang Xiaosong, Peng Yifan, Lu Le, Lu Zhiyong, Bagheri Mohammadhadi, Summers Ronald M. ChestX-Ray8: hospital-scale chest X-ray database and benchmarks on weakly-supervised classification and localization of common thorax diseases. 2017 IEEE Conf Computer Vis Pattern Recognit (CVPR). IEEE; 2017. p. 2097–106.

[15] Ankenbrand Markus J, Shainberg Liliia, Hock Michael, Lohr David, Schreiber Laura M. Sensitivity analysis for interpretation of machine learning based segmentation models in cardiac MRI. BMC Med Imaging 2021;21(1):27.

[16] Tatsugami Fuminari, Higaki Toru, Nakamura Yuko, Yu Zhou, Zhou Jian, Lu Yujie, et al. Deep learning-based image restoration algorithm for coronary CT angiography. Eur Radiol 2019;29(10):5322–9.

[17] Kusunose Kenya, Abe Takashi, Haga Akihiro, Fukuda Daiju, Yamada Hirotsugu, Harada Masafumi, et al. A deep learning approach for assessment of regional wall motion abnormality from echocardiographic images. JACC Cardiovasc Imaging 2020;. Available from: https://doi.org/10.1016/j.jcmg.2019.02.024.

[18] Elefteriades John A, Sang Adam, Kuzmik Gregory, Hornick Matthew. Guilt by association: paradigm for detecting a silent killer (thoracic aortic aneurysm). Open Heart 2015;2(1):e000169.

[19] Li Jingjing, Pan Cuiping, Zhang Sai, Spin Joshua M, Deng Alicia, Leung Lawrence LK, et al. Decoding the genomics of abdominal aortic aneurysm. Cell 2018;174(6):1361–72 e10.

[20] Wu Jinlin, Qiu Juntao, Xie Enzehua, Jiang Wenxiang, Zhao Rui, Qiu Jiawei, et al. Predicting in-hospital rupture of type A aortic dissection using random forest. J Thorac Dis 2019;11(11):4634–46.
[21] Akbilgic Oguz, Butler Liam, Karabayir Ibrahim, Chang Patricia P, Kitzman Dalane W, Alonso Alvaro, et al. ECG-AI: electrocardiographic artificial intelligence model for prediction of heart failure. Eur Heart J Digital Health 2021;2(4):626–34.
[22] Van den Eynde Jef, Manlhiot Cedric, Van De Bruaene Alexander, Diller Gerhard-Paul, Frangi Alejandro F, Budts Werner, et al. Medicine-based evidence in congenital heart disease: how artificial intelligence can guide treatment decisions for individual patients. Front Cardiovasc Med 2021;. Available from: https://doi.org/10.3389/fcvm.2021.798215.
[23] Golas Sara, Bersche Takuma, Shibahara Stephen, Agboola Hiroko, Otaki Jumpei, Sato Tatsuya, et al. A machine learning model to predict the risk of 30-day readmissions in patients with heart failure: a retrospective analysis of electronic medical records data. BMC Med Inform Decis Mak 2018;18(1):44.
[24] Thongprayoon Charat, Pattharanitima Pattharawin, Kattah Andrea G, Mao Michael A, Keddis Mira T, Dillon John J, et al. Explainable preoperative automated machine learning prediction model for cardiac surgery-associated acute kidney injury. J Clin Med Res 2022;11(21). Available from: https://doi.org/10.3390/jcm11216264.
[25] Benedetto U, et al. Machine learning improves mortality risk prediction after cardiac surgery: systematic review and meta-analysis. The Journal of Thoracic and Cardiovascular Surgery 2022;163(6):2075–87.
[26] Becker K, Thull B, Käsmacher-Leidinger H, Stemmer J, Rau G, Kalff G, et al. Design and validation of an intelligent patient monitoring and alarm system based on a fuzzy logic process model. Artif Intell Med 1997;11(1):33–53.
[27] Hatib Feras, Jian Zhongping, Buddi Sai, Lee Christine, Settels Jos, Sibert Karen, et al. Machine-learning algorithm to predict hypotension based on high-fidelity arterial pressure waveform analysis. Anesthesiology 2018;129(4):663–74.
[28] Shin Brian, Maler Steven A, Reddy Keerthi, Fleming Neal W. Use of the hypotension prediction index during cardiac surgery. J Cardiothorac Vasc Anesth 2021;35(6):1769–75.
[29] Mumtaz Hassan, Saqib Muhammad, Ansar Farrukh, Zargar Durafshan, Hameed Madiha, Hasan Mohammad, et al. The future of cardiothoracic surgery in artificial intelligence. Ann Med Surg 2022;80(August):104251.
[30] Huffmyer Julie L, Groves Danja S. Pulmonary complications of cardiopulmonary bypass. Best Pract Res Clin Anaesthesiol 2015;29(2):163–75.
[31] Lundberg Scott M, Nair Bala, Vavilala Monica S, Horibe Mayumi, Eisses Michael J, Adams Trevor, et al. Explainable machine-learning predictions for the prevention of hypoxaemia during surgery. Nat Biomed Eng 2018;2(10):749–60.
[32] Xu Hao, Han Tingxuan, Wang Haifeng, Liu Shanggui, Hou Guanghao, Sun Lina, et al. Detection of blood stains using computer vision-based algorithms and their association with postoperative outcomes in thoracoscopic lobectomies. Eur J Cardiothorac Surg 2022;62(5). Available from: https://doi.org/10.1093/ejcts/ezac154.
[33] Lu Jingpei, Jayakumari Ambareesh, Richter Florian, Li Yang, Yip Michael C. SuPer deep: a surgical perception framework for robotic tissue manipulation using deep learning for feature extraction. 2021 IEEE int conf robot autom (ICRA). 2021. p. 4783–9.
[34] Mountney Peter, Yang Guang-Zhong. Soft tissue tracking for minimally invasive surgery: learning local deformation online. Medical image computing and computer-assisted intervention – MICCAI 2008. Springer Berlin Heidelberg; 2008. p. 364–72.
[35] Stoyanov D, Guang-Zhong Y. Soft tissue deformation tracking for robotic assisted minimally invasive surgery. In: annual international conference of the IEEE engineering in medicine and biology society, IEEE; 2009.
[36] Yang Hongxu, Shan Caifeng, Bouwman Arthur, Kolen Alexander F, de With Peter HN. Efficient and robust instrument segmentation in 3D ultrasound using patch-of-interest-FuseNet with hybrid loss. Med Image Anal 2021;67(January):101842.
[37] Cohen TN, Cabrera JS, Sisk OD, Welsh KL, Abernathy JH, Reeves ST, et al. Identifying workflow disruptions in the cardiovascular operating room. Anaesthesia 2016;71(8):948–54.
[38] Avrunin George S, Clarke Lori A, Conboy Heather M, Osterweil Leon J, Dias Roger D, Yule Steven J, et al. Toward improving surgical outcomes by incorporating cognitive load measurement into process-driven guidance. Softw eng healthc syst (SEHS), IEEE/ACM int workshop 2018. 2018. p. 2–9.
[39] Liu Daochang, Li Qiyue, Jiang Tingting, Wang Yizhou, Miao Rulin, Shan Fei, et al. Towards unified surgical skill assessment. 2021 IEEE/CVF conf computer vis pattern recognit (CVPR). IEEE; 2021. p. 9522–31.

[40] Mitchell Erica L, Arora Sonal, Moneta Gregory L, Kret Marcus R, Dargon Phong T, Landry Gregory J, et al. A systematic review of assessment of skill acquisition and operative competency in vascular surgical training. J Vasc Surg 2014;59(5):1440–55.
[41] Jin Amy, Yeung Serena, Jopling Jeffrey, Krause Jonathan, Azagury Dan, Milstein Arnold, et al. Tool detection and operative skill assessment in surgical videos using region-based convolutional neural networks.". arXiv [csCV] 2018;. Available from: http://arxiv.org/abs/1802.08774.
[42] Louis Nathan, Zhou Luowei, Yule Steven J, Dias Roger D, Manojlovich Milisa, Pagani Francis D, et al. Temporally guided articulated hand pose tracking in surgical videos. Int J Comput Assist Radiol Surg 2022;. Available from: https://doi.org/10.1007/s11548-022-02761-6.
[43] Wang Ziheng, Fey Ann Majewicz. Deep learning with convolutional neural network for objective skill evaluation in robot-assisted surgery. Int J Comput Assist Radiol Surg 2018;13(12):1959–70.
[44] Lodge Daniel, Grantcharov Teodor. Training and assessment of technical skills and competency in cardiac surgery. Eur J Cardiothorac Surg 2011;39(3):287–93.
[45] Kennedy-Metz Lauren R, Mascagni Pietro, Torralba Antonio, Dias Roger D, Perona Pietro, Shah Julie A, et al. Computer Vision in the Operating Room: Opportunities and Caveats. IEEE Trans Med Robot Bionics 2021;3(1):2–10.
[46] Dias Roger D, Kennedy-Metz Lauren R, Yule Steven J, Gombolay Matthew, Zenati Marco A. Assessing team situational awareness in the operating room via computer vision. IEEE Conf Cognit Comput Asp Situat Manag 2022;2022(June):94–6.
[47] Carroll Micah, Shah Rohin, Ho Mark K, Griffiths Tom, Seshia Sanjit, Abbeel Pieter, et al. On the utility of learning about humans for human-Ai coordination. Adv Neural Inf Process Syst 2019;32. Available from: https://proceedings.neurips.cc/paper/2019/hash/f5b1b89d98b7286673128a5fb112cb9a-Abstract.html.
[48] Morgan Jeffrey A, Barbara AThornton, Peacock Joy C, Hollingsworth Karen W, Smith Craig R, Oz Mehmet C, et al. Does robotic technology make minimally invasive cardiac surgery too expensive? A hospital cost analysis of robotic and conventional techniques. J Card Surg 2005;. Available from: https://doi.org/10.1111/j.1540-8191.2005.200385.x.
[49] Park Jun, Woo Jaesoon, Choi Yongdoo Park, Sun Kyung. Haptic virtual fixture for robotic cardiac catheter navigation. Artif Organs 2011;35(11):1127–31.
[50] O'Sullivan S, et al. Legal, regulatory, and ethical frameworks for development of standards in artificial intelligence (AI) and autonomous robotic surgery. The international journal of medical robotics and computer assisted surgery 2019;15(1)e1968.
[51] Nakamura Y., Kishi K., Kawakami H.. Heartbeat synchronization for robotic cardiac surgery. In: Proceedings 2001 ICRA. IEEE international conference on robotics and automation (cat. no. 01CH37164), vol. 2; 2001. p. 2014–19.
[52] Jacob Mithun, George Yu-Ting, Li George A, Akingba, Juan PWachs. Collaboration with a robotic scrub nurse. Commun ACM 2013;. Available from: https://doi.org/10.1145/2447976.2447993.
[53] Çavuşoğlu M, Cenk MCenk Çavuşoğlu, Williams Winthrop, Tendick Frank, Sastry SShankar. Robotics for telesurgery: second generation Berkeley/UCSF Laparoscopic Telesurgical Workstation and looking towards the future applications. Ind Robot: An Int J 2003;. Available from: https://doi.org/10.1108/01439910310457670.
[54] Xu Xiaowei, Qiu Hailong, Jia Qianjun, Dong Yuhao, Yao Zeyang, Xie Wen, et al. AI-CHD: an AI-based framework for cost-effective surgical telementoring of congenital heart disease. Commun ACM 2021; 64(12):66–74.
[55] Yang Guang-Zhong, Dempere-Marco Laura, Hu Xiao-Peng, Rowe Anthony. Visual search: psychophysical models and practical applications. Image Vis Comput 2002;20(4):291–305.
[56] Zuo Siyang, Chen Teng, Chen Xin, Chen Baojun. A wearable hands-free human-robot interface for robotized flexible endoscope. IEEE Robot Autom Lett 2022;7(2):3953–60.
[57] Abdelaal Alaa, Eldin Prateek, Mathur, Septimiu ESalcudean. Robotics in vivo: a perspective on human–robot interaction in surgical robotics. Annu Rev Control, Robotics, Autonomous Syst 2020;3(1):221–42.
[58] Graves Alex, Mohamed Abdel-Rahman, Hinton Geoffrey. Speech recognition with deep recurrent neural networks. 2013 IEEE int conf acoustics, speech signal process. 2013. p. 6645–9.
[59] Burian BK, et al. Using extended reality (XR) for medical training and real-time clinical support during deep space missions. Appl Ergonomics 2023;106:103902.
[60] Ebnali M, et al. AR-Coach: using augmented reality (AR) for real-time clinical guidance during medical emergencies on deep space exploration missions. Healthc Med Devices 2022;51:67.

[61] Ebnali M, Kian C, Ebnali-Heidari M, Mazloumi A. User experience in immersive VR-based serious game: an application in highly automated driving training. Int conf appl hum factors ergonomics. Cham: Springer; 2019. p. 133–44.
[62] Ebnali M, Hulme K, E-Heidari A, Mazloumi A. How does training effect users' attitudes and skills needed for highly automated driving? Transp Res F: Traffic Psychol Behav 2019;66:184–95.
[63] Ebnali M, Kennedy-Metz LR, Conboy HM, Clarke LA, Osterweil LJ, Avrunin G, et al. A coding framework for usability evaluation of digital health technologies. Int conf human-computer interact. Cham: Springer; 2022. p. 185–96.
[64] Sadeghi Amir H, Maat Alexander PW, Taverne Yannick JH, Cornelissen Robin, Dingemans Anne-Marie C, Bogers Ad JJ, et al. Virtual reality and artificial intelligence for 3-dimensional planning of lung segmentectomies. JTCVS Tech 2021;. Available from: https://doi.org/10.1016/j.xjtc.2021.03.016.
[65] Chang Ying-Jen, Hung Kuo-Chuan, Wang Li-Kai, Yu Chia-Hung, Chen Chao-Kun, Tay Hung-Tze, et al. A real-time artificial intelligence-assisted system to predict weaning from ventilator immediately after lung resection surgery. Int J Environ Res Public Health 2021;18(5). Available from: https://doi.org/10.3390/ijerph18052713.
[66] Bapat Vinayak, Sabetai Michael, Roxburgh Jamers, Young Christopher, Venn Graham. Early and intensive continuous veno-venous hemofiltration for acute renal failure after cardiac surgery. Interact Cardiovasc Thorac Surg 2004;3(3):426–30.
[67] Mendes Renata G, de Souza César R, Machado Maurício N, Correa Paulo R, Di Thommazo-Luporini Luciana, Arena Ross, et al. Clinical research predicting reintubation, prolonged mechanical ventilation and death in post-coronary artery bypass graft surgery: a comparison between artificial neural networks and logistic regression models. Arch Med Sci 2015;. Available from: https://doi.org/10.5114/aoms.2015.48145.
[68] Wise Eric Stephen, Stonko David P, Glaser Zachary A, Garcia Kelly L, Huang Jennifer J, Kim Justine S, et al. Prediction of prolonged ventilation after coronary artery bypass grafting: data from an artificial neural network. Heart Surg Forum 2017;20(1):E007–14.
[69] Peng S-Y, Peng S-K. Predicting adverse outcomes of cardiac surgery with the application of artificial neural networks. Anaesthesia 2008;63(7):705–13.
[70] Zhong Zhihua, Yuan Xin, Liu Shizhen, Yang Yuer, Liu Fanna. Machine learning prediction models for prognosis of critically ill patients after open-heart surgery. Sci Rep 2021;11(1):3384.
[71] Rosner Mitchell H, Okusa Mark D. Acute kidney injury associated with cardiac surgery. Clin J Am Soc Nephrol 2006;1(1):19–32.
[72] Huen Sarah C, Parikh Chirag R. Predicting acute kidney injury after cardiac surgery: a systematic review. Ann Thorac Surg 2012;93(1):337–47.
[73] Penny-Dimri Jahan C, Bergmeir Christoph, Reid Christopher M, Williams-Spence Jenni, Cochrane Andrew D, Smith Julian A. Machine learning algorithms for predicting and risk profiling of cardiac surgery-associated acute kidney injury. SemThorac Cardiovasc Surg 2021;33(3):735–45.
[74] Kazmierski Jakub, Kowman Maciej, Banach Maciej, Fendler Wojciech, Okonski Piotr, Banys Andrzej, et al. Incidence and predictors of delirium after cardiac surgery: results from the IPDACS study. J Psychosom Res 2010;69(2):179–85.
[75] Mufti Hani Nabeel, Hirsch Gregory Marshal, Abidi Samina Raza, Abidi Syed Sibte Raza. Exploiting machine learning algorithms and methods for the prediction of agitated delirium after cardiac surgery: models development and validation study. JMIR Med Inform 2019;7(4):e14993.
[76] Xue Xin, Chen Wen, Chen Xin. A novel radiomics-based machine learning framework for prediction of acute kidney injury-related delirium in patients who underwent cardiovascular surgery. Comput Math Methods Med 2022;2022(March):4242069.
[77] Tu JV, Guerriere MR. Use of a neural network as a predictive instrument for length of stay in the intensive care unit following cardiac surgery. Comput Biomed Res 1993;26(3):220–9.
[78] Triana Austin J, Vyas Rushikesh, Shah Ashish S, Tiwari Vikram. Predicting length of stay of coronary artery bypass grafting patients using machine learning. J Surg Res 2021;264(August):68–75.
[79] Benuzillo Jose, Caine William, Scott Evans R, Roberts Colleen, Lappe Donald, Doty John. Predicting readmission risk shortly after admission for CABG surgery. J Card Surg 2018;33(4):163–70.
[80] Manyam Rameshbabu, Zhang Yanqing, Carter Seth, Binongo Jose N, Rosenblum Joshua M, Keeling William B. Unraveling the impact of time-dependent perioperative variables on 30-day readmission after coronary artery bypass surgery. J Thorac Cardiovasc Surg 2022;164(3):943–55 e7.

[81] Mahayni AA, et al. Electrocardiography-based artificial intelligence algorithm aids in prediction of long-term mortality after cardiac surgery. Mayo Clinic Proceedings 2021;96:12.
[82] Khosravi Bardia, Rouzrokh Pouria, Faghani Shahriar, Moassefi Mana, Vahdati Sanaz, Mahmoudi Elham, et al. Machine learning and deep learning in cardiothoracic imaging: a scoping review. Diagnostics (Basel, Switzerland) 2022;12(10). Available from: https://doi.org/10.3390/diagnostics12102512.
[83] Van Calster Ben, Wynants Laure, Timmerman Dirk, Steyerberg Ewout W, Collins Gary S. Predictive analytics in health care: how can we know it works? J Am Med Inform Assoc 2019;26(12):1651–4.
[84] Cabitza Federico, Campagner Andrea, Soares Felipe, Guadiana-Romualdo Luis García de, Challa Feyissa, Sulejmani Adela, et al. The importance of being external. methodological insights for the external validation of machine learning models in medicine. Comput Methods Prog Biomed 2021;208(September):106288.
[85] Davis Sharon E, Lasko Thomas A, Chen Guanhua, Siew Edward D, Matheny Michael E. Calibration drift in regression and machine learning models for acute kidney injury. J Am Med Inform Assoc 2017; 24(6):1052–61.
[86] Steyerberg Ewout W, Moons Karel GM, van der Windt Danielle A, Hayden Jill A, Perel Pablo, et al. PROGRESS Group. Prognosis Research Strategy (PROGRESS) 3: prognostic model research. PLoS Med 2013;10(2):e1001381.
[87] Mirnezami R, Ahmed A. Surgery 3.0, artificial intelligence and the next-generation surgeon. Br J Surg 2018;105(5):463–5.
[88] Van den Eynde, Mark Lachmann Jef, Laugwitz Karl-Ludwig, Manlhiot Cedric, Kutty Shelby. Successfully implemented artificial intelligence and machine learning applications in cardiology: state-of-the-art review. Trends Cardiovasc Med 2022;January. Available from: https://doi.org/10.1016/j.tcm.2022.01.010.
[89] Muthalaly Rahul G, Robert MEvans. Applications of machine learning in cardiac electrophysiology. Arrhythm Electrophysiol Rev 2020;9(2):71–7.
[90] Zhou Xiao-Yun, Guo Yao, Shen Mali, Yang Guang-Zhong. Application of artificial intelligence in surgery. Front Med 2020;14(4):417–30.
[91] Shokri Reza, Vitaly Shmatikov. Privacy-preserving deep learning. Proc 22nd ACM SIGSAC conf computer commun security, CCS '15. New York, NY, USA: Association for Computing Machinery; 2015. p. 1310–21.
[92] Abouzari Mehdi, Rashidi Armin, Zandi-Toghani Mehdi, Behzadi Mehrdad, Asadollahi Marjan. Chronic subdural hematoma outcome prediction using logistic regression and an artificial neural network. Neurosurg Rev 2009;32(4):479–84.
[93] Alotaibi Nawal N, Sasi Sreela. Stroke in-patients' transfer to the ICU using ensemble based model. 2016 int conf electrical, ElectrOptim Tech (ICEEOT). 2016. p. 2004–10.
[94] Awad Aya, Bader-El-Den Mohamed, McNicholas James, Briggs Jim. Early hospital mortality prediction of intensive care unit patients using an ensemble learning approach. Int J Med Inform 2017;108(December):185–95.
[95] Firouznia Marjan, Feeny Albert K, LaBarbera Michael A, McHale Meghan, Cantlay Catherine, Kalfas Natalie, et al. Machine learning-derived fractal features of shape and texture of the left atrium and pulmonary veins from cardiac computed tomography scans are associated with risk of recurrence of atrial fibrillation postablation. Circ Arrhythm Electrophysiol 2021;14(3):e009265.
[96] Martel Anne L, Abolmaesumi Purang, Stoyanov Danail, Mateus Diana, Zuluaga Maria A, Kevin Zhou S, et al. Medical Image Computing and Computer Assisted Intervention – MICCAI 2020: 23rd International Conference, Lima, Peru, October 4–8, 2020, Proceedings, Part II. Springer Nature; 2020.
[97] Legeza Peter, Britz Gavin W, Loh Thomas, Lumsden Alan. Current utilization and future directions of robotic-assisted endovascular surgery. Expert Rev Med Devices 2020;17(9):919–27.
[98] Doulamis Ilias P, Spartalis Eleftherios, Machairas Nikolaos, Schizas Dimitrios, Patsouras Dimitrios, Spartalis Michael, et al. The role of robotics in cardiac surgery: a systematic review. J Robot Surg 2019;13(1):41–52.
[99] Darbari Anshuman, Kumar Krishan, Darbari Shubhankar, Patil Prashant L. Requirement of artificial intelligence technology awareness for thoracic surgeons. Cardiothorac Surg 2021;29(1):1–10.
[100] Williams Simon, Hugo Layard Horsfall Jonathan P, Funnell John G, Hanrahan Danyal Z, Khan William, Muirhead Danail, et al. Artificial intelligence in brain tumour surgery-an emerging paradigm. Cancers 2021;13(19):510–35. Available from: https://doi.org/10.3390/cancers13195010.
[101] Salas Eduardo, Maurino Dan, Curtis Michael. Chapter 1—human factors in aviation: an overview. In: Eduardo Salas, Dan Maurino, editors. Human factors in aviation. 2nded. San Diego: Academic Press; 2010. p. 3–19.

[102] Ebnali M, Lamb R, Fathi R, Hulme K. Virtual reality tour for first-time users of highly automated cars: comparing the effects of virtual environments with different levels of interaction fidelity. Appl Ergon 2021;90 (January):103226.
[103] Ebnali M, Fathi R, Lamb R, Pourfalatoun S, Motamedi S. Using augmented holographic uis to communicate automation reliability in partially automated driving. AutomationXP@ CHI. 2020.
[104] Asan O, Avishek C. Research trends in artificial intelligence applications in human factors health care: mapping review. JMIR human factors 2021;8(2):e28236.

CHAPTER

26

Artificial intelligence in orthopedics

Hashim J.F. Shaikh[1], *Evan Polce*[2], *Jorge Chahla*[2], *Kyle Kunze*[3] *and Thomas G. Myers*[1]

[1]Department of Orthopaedics, University of Rochester Medical Center, Rochester, NY, United States [2]Rush University Medical Center, Chicago, IL, United States [3]Hospital for Special Surgery, New York, NY, United States

Introduction

Within the last 5-year artificial intelligence (AI) has become a fast-growing area of research within orthopedics [1]. Fig. 26.1 shows a steady increase in the number of PubMed citations included in the search term "artificial intelligence AND orthopedics." Nearly every orthopedic subspecialty is exploring ways to incorporate the potential benefits of this technology in order to improve patient care [2–9]. Makhni et al. and others have performed reviews exemplifying the wide spectrum of benefits that AI can provide [10,11]. Cited benefits include but are not limited to wearable technology for remote patient monitoring, predicting risk of prolonged opioid use after a total hip arthroplasty, and identifying osteoarthritis on radiographic imaging [10,11].

Artificial intelligence in clinical orthopedic practice

Imaging

In a 2020 review, the field of radiology was second only to pathology citations in PubMed regarding machine learning (ML) studies across all medical specialties [12]. This trend appears to be true among potential AI applications within orthopedics [9]. One explanation for radiographic and pathologic predominance lies in the fact that radiographic and pathologic images have a much higher degree of clinical fidelity or "ground truth" present within the imaging data (e.g., fracture or tumor present "yes" or "no"). Image based applications created from image datasets lack the degree of clinical variability and data integrity issues present in other clinical datasets.

Artificial Intelligence in Clinical Practice
DOI: https://doi.org/10.1016/B978-0-443-15688-5.00023-1

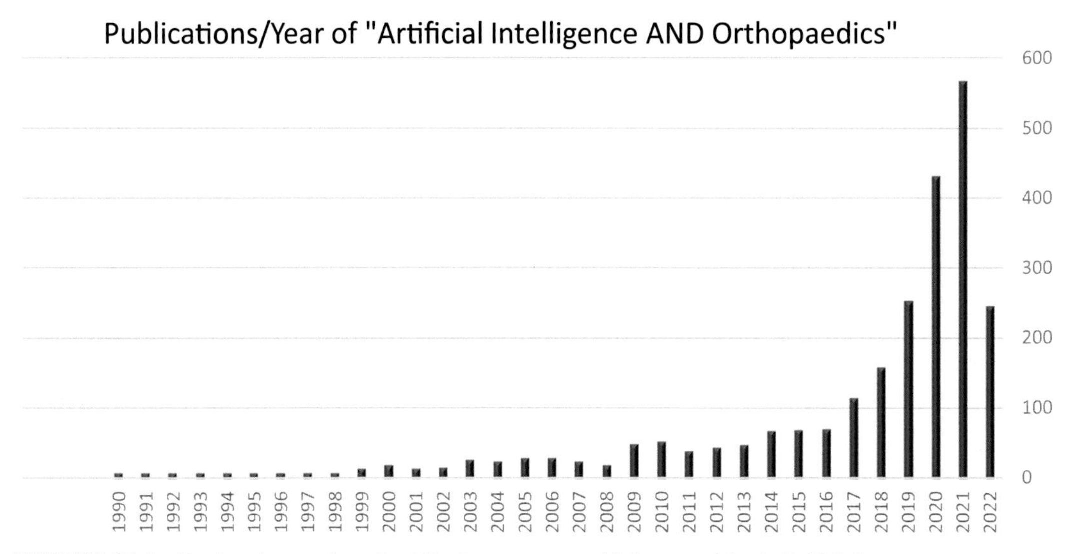

FIGURE 26.1 Depicts the number of publications per year of full text articles in PubMed.

Shim et al. used a 3D conventional neural network (3D CNN) to train an ML model on MRI data from 2412 patients with a diagnosed rotator cuff tear [13]. When tested on a random set of data, the 3D CNN was able to outperform a clinical shoulder specialist in its ability to identify a rotator cuff tear on MRI. The model was able to perform at an accuracy level of 92.5% (versus 76.5%), a sensitivity of 0.92 (vs 0.89), and a specificity of 0.86 (vs 0.61). This performance of a 3D CNN over a clinical specialist provides support for the suggestion that AI has a role in identify imaging pathology within orthopedics.

Karnuta et al. further demonstrated AI use in imaging by assessing if total knee arthroplasty (TKA) implants can be classified from radiographs [14]. The need to identify discontinued or unfamiliar implants when planning for revision surgery is critical as medical records detailing the manufacturer may not be available. The inability to identify the correct implant may substantially alter the surgical treatment plan. When trained by ML Karnuta's group demonstrated that the model was able to determine implant model and manufacturer with an accuracy level of 99%.

Predict in-patient cost and length of stay in total joint arthroplasty

AI has demonstrated significant importance in terms of identifying factors that increase a patient length of hospital stay (LOS) and higher inpatient cost following TKA surgery [15]. Navarro et al. trained a Bayesian ML model on 141,446 patients who underwent TKA. The model showed that race, age, gender, and comorbidity scores were strongly associated for an increase in LOS and inpatient cost as demonstrated by an area under the curve (AUC) of 0.78 and 0.73, respectively.

Predict risk for revisions in sports medicine

AI has been used to identify factors for an increased risk for an anterior cruciate ligament (ACL) revision. A nationwide database of individuals who suffered an ACL retear up to 5 years from their initial surgery was used to train Cox Lasso, survival random force, general additive mode, and gradient boosted regression ML models [16]. Overall, the use of the Cox Lasso model was able to predict ACL retear with the highest level of accuracy (AUC = 0.68) while using the least amount of variables (5) when compared to the other models. With the use of this model, a risk calculator for ACL revisions was developed to identify patients who would be at low risk (0%) or high risk (20%) for an ACL revision at 5 years their index surgery.

Artificial intelligence and postoperative rehabilitation

AI has significant potential to provide patient level data for rehabilitation and treatment after orthopedic procedures. Ramkumar et al. studied a combination knee sensor and smartphone application platform [17]. This platform can track compliance with a rehabilitation program, knee range of motion, narcotic medication usage, and collect patient reported outcome measures. The provider is given access to real-time data as the patient progresses through their recovery. The group used AI technology (motion algorithms in sensors) within this platform on 25 patients who underwent a TKA to validate knee range of motion measurements taken by a provider during postoperative clinic appointments. The use of AI for remote patient monitoring has tremendous potential to develop and validate rehabilitation protocols and provide telehealth following orthopedic surgery. Furthermore, the ability to aggregate accurate data on thousands of patients allows the creation of clinically accurate large-scale databases necessary for predictive modeling.

Natural language processing in electronic medical record

AI analysis of the electronic medical record (EMR) may assist with the identification of clinical conditions. Thirukumaran et al. used natural language processing (NLP) to determine if the model can detect orthopedic surgical site infections (SSI) from retrospective EMR data analysis. [18]. The group incorporated keywords and grammatical variants indicative for a SSI into the model. The model was able to analyze free text EMR notes to provide a binary decision regarding the presence or absence of an SSI. Overall, the model was able to achieve SSI with a diagnostic accuracy of 97% equaling that of manual abstraction. This study serves as a proof of concept of the ability for NLP to automate the identification of SSI which can be a labor intensive and time-consuming process. Automating EMR analysis with NLP has the potential to aid in the prevention and early identification of surgical complications, thereby reducing their adverse clinical and economic impact.

Commercially available artificial intelligence applications within healthcare

The most immediately useful AI tools are the ones that have current commercial availability. These AI based applications tools do not assist with clinical decision making but assist with the mundane, repetitive, and time-consuming yet critical tasks involved with practice management. AI allows providers to maximize their efficiency by automating processes throughout the spectrum of a clinical visit including the previsit, registration, exam documentation, orders, checkout, billing, and patient satisfaction. Table 26.1 lists commercially available AI applications categorized by their utility in a particular phase of care [19].

Limitations

Many current clinical AI studies are limited proof of concepts that lack *external* validation with a new unseen dataset. Investigators frequently take a dataset and perform an 80/20 split of the entire dataset. The 80% of the split goes into training the ML algorithm (training set) while the 20% is used to *internally* validate the trained algorithm (testing set). The performance of the models on the testing dataset is typically very good—because the training and testing datasets are identical in every way including noise within the data that does not represent clinical reality. However, when *external* validation analysis is performed on new data (which contains its own detail and noise) the accuracy of the model decreases. This phenomenon is known as *overfitting*. *Overfitting* typically occurs because the algorithms learn the detail and noise of the training data so well that when these associations are not present in a new dataset the model's performance is negatively impacted. Recently more studies have begun to externally validate their trained algorithms which is critical if AI applications are ever going to assist in clinical decision-making.

Another limitation is the lack of standardized reporting for ML research. As a result, many studies fail to properly discuss their methodology. Without a proper understanding of a study's methods the study's findings that cannot be reproduced, which ultimately lead to invalid findings regardless of model's performance. Polce et al. presented a review of 55 studies that focused on the use of ML in orthopedic total joint arthroplasty [20]. The group investigated the methodology regarding algorithm development and testing. The authors found 50% of the studies failed to disclose how issues regarding missing data were managed. Furthermore, over 75% of studies failed to describe how their models were calibrated. Without proper calibration model performance can suffer and lead to inaccurate conclusions regarding clinical utility.

Advancing AI in orthopedics will require journals to adopt guidelines, standards, and quality reporting criteria regarding both the development of the ML algorithms and methodology for clinical integration. Guidelines which currently exist include CONSORT-AI and SPIRIT-AI for the clinical testing and regulatory phase of trials, respectively. The CONSORT-AT and SPIRIT-AI guidelines are analogous to CONSORT reporting for clinical trials and PRISMA reporting for meta-analyses [21–23]. Collins et al. have reported on the protocol for guidelines to address the development and validation phase of ML based diagnostic and prognostic algorithms which have yet to be formalized [24]. Standardization of reporting can also help to determine if more basic ML algorithms (e.g., logistic regression) can help classify clinical scenarios as effectively as more advanced and time consuming (e.g., neural networks) algorithms as demonstrated by Oosterhoff et al. following orthopedic trauma surgery [25].

TABLE 26.1 Descriptive summary of commercial AI products within healthcare.

Product	Phase of clinical care	Function
Klara	Previsit	Providing multichannel communication in one place
Patient Pop	Previsit	Combines telehealth, communications, and patient intake in a streamlined platform
Lifelink	Previsit	Platform for chatbot solutions that use conversational AI to help patients navigate the healthcare
Ensofia	Previsit	AI-based patient journey engagement, communication and automation solutions
Patient Point	Previsit	Create more effective doctor–patient interactions from patient acquisition to in-office and hospital engagement to telehealth and remote care.
"Dash" by Relatient	Previsit	Appointment reminders, secure chat, 2-way messaging, rules-based scheduling.
Buey Health	Previsit	Chatbot listens to a patient's symptoms and health: 1. What's going on? 2. How to fix it 3. How much is it
"Nuance Virtual Assistance"	Visit and exam documentation	AI assistant named kara to listen in real time and create clinical notes, fill orders or queue up a patient referral.
DeepScribe	Visit and exam documentation	Extract medically relevant information from a natural patient conversation, use it to write a complete note, and then sync that note directly into the discrete fields in her
Suki Assistant	Visit and exam documentation	Intelligent and responsive speech recognition program is trained on specialty specific data
"Health OST" Nanox.AI	Imaging	Detection of vertebral compression fracture for detection of osteoporosis
Zebra Medical Vision	Imaging	Chest X-rays and CT brain-automated analysis
Caption Health	Imaging	Perform ultrasound examinations accurately and quickly, provides automatic quality assessment and smart interpretation
"Subtle Mr"-Subtle Medical	Imaging	Enhancement software improves the quality of noisy medical images
Aidoc	Imaging	Analyzes medical imaging for flagging acute abnormalities across the body
"OsteoDetect"-Imagen	Imaging	Detection of wrist fractures-FDA cleared
Olive AI	Prior authorization	Automates preauthorizations, eligibility checks to unadjudicated claims and data migrations
Cohere	Prior authorization	Automate preauthorizations
CorroHealth	Checkout and billing	Extracts metadata, generates charges and codes
Infinx	Checkout and billing	Optimize patient access and revenue cycle workflows
Olive AI	Checkout and billing	Minimize touchpoints, clean claims, standardize submission, and return remittance

Future directions

Creating advancements in clinical decision applications featuring AI in orthopedics requires standardization of inputs, collection, and reporting of data in studies [26,27]. Creation of devices which are based upon ML models is overseen by United States Food and Drug Administration. Adopting standardized guidelines for orthopedic AI research will provide the required process fidelity necessary to withstand scrutiny from the FDA and produce a product that will benefit patient care.

Considerable advancements in data management alongside the application of AI have reached the field of orthopedic surgery. The landmark studies have solidified the framework for the potential capabilities and clinical applications of AI in this setting. For example, deep learning (DL), which uses artificial neural networks designed to emulate the neuronal circuitry of the human brain, is a subdomain of AI that utilizes hierarchal algorithms and nonlinear functions to model increasingly complex relationships among data. DL has demonstrated great potential and robust accuracies in imaging-based tasks, such as hip joint measurements for preoperative planning [28] and implant identification [29]. Ultimately, this technology may help enhance the diagnostic capabilities of orthopedic surgeons and improve workflow.

Another subdomain of AI, called NLP, is used to translate both structured and unstructured written medical data (e.g., medical notes in the electronic health record) to identify an outcome or event of interest. Researchers have already applied NLP to solve a variety of clinically relevant tasks in orthopedics; for example, the surveillance of free-text notes for better detection of complications after spine surgery [30]. Given the increasing cost and administrative burden within orthopedics and health care more generally, NLP represents a valuable tool that may mitigate redundant activities by automating documentation and review of the electronic health record [31].

Given that proof-of-concept for several clinically important applications has been established, the integration of AI into EMRs and clinical workflow represents the next frontier [32]. The future of AI in orthopedics surgery will likely leverage such applications to automate data extraction and analysis, where a patient's likelihood of experiencing a clinically meaningful outcome or complication will be automatically calculated and presented to the clinician. In cases of referrals, a radiograph of the hip or knee of a new patient presenting to the clinic may also auto-populate the dimensions and manufacturer of the existing implant, decreasing time and effort associated with attempting to obtain information about implant design. This may allow the surgeon to be better prepared entering a revision surgery.

Current challenges

A particularly arduous challenge faced by current AI research is that of model explainability. Clinical decision making by physicians often relies on decades of education and countless patient encounters. Often, diagnoses and treatment decisions are guided by tangible, objective data, such as physical examination and laboratory findings. In contrast, AI algorithms represent "black boxes," meaning that the exact process by which recommendations are made by the model is not easily discernable. Considering treatment decisions in medicine can be a matter of life and death, insight into the process underlying AI recommendations is crucial for developing physician trust and buy-in. An AI model that is accurate but poorly understood is unlikely to be meaningfully deployed in clinical practice.

Given the complexity of modern AI and electronic data, one popular approach to human-comprehensible AI involves analysis of local model behavior to gain insight into the global decision making process, known as post hoc explainability [33]. Methods used for this purpose include local interpretable model-agnostic explanations [34], Shapley additive explanations [35], and saliency maps for imaging tasks [36]. However, research has demonstrated that these explanation methods are imperfect and even can mislead human users in certain situations [33,37]. Specifically, these methods provide some elucidation on the reasoning behind AI decision making, but fail to answer the question of whether the logic used to arrive at a recommendation was reasonable. This so called "interpretability gap" [8] forces the human user to then decide the validity of a given explanation, which may introduce harmful cognitive biases and a false sense of confidence in AI recommendations. Ultimately, the overarching goal of achieving reliable means for assessing AI decision making should primarily rely on rigorous internal and external validation of model performance during the development phase.

Major takeaway points

- The application of AI research in the field of orthopedic surgery can be broadly categorized into four domains: (1) clinical outcomes and resource utilization, (2) medical imaging, (3) activity and kinematics, and (4) natural language interpretation.
- The future integration of AI into the electronic health record will automate redundant tasks and serve as a data-driven tool to supplement clinical decision making.
- Currently available methods for explaining AI decision making are imperfect and should be used judiciously.
- Engendering trust in the validity of AI recommendations should primarily rely on robust internal and external validation of model performance during the development phase.

Conclusions

- AI has shown significant promise in orthopedics.
- AI can currently provide support for repetitive administrative tasks with commercially available applications.
- Clinical AI research is limited by the uncertainty of clinical data and external validation which hamper the development of clinical decision support applications.
- Rigorous standardized guidelines are needed to develop AI applications safe for clinical use.

References

[1] Federer SJ, Jones GG. Artificial intelligence in orthopaedics: a scoping review. PLoS One 2021;16(11):e0260471.
[2] Katsuura Y, et al. A primer on the use of artificial intelligence in spine. Surg Clin Spine Surg 2021;34 (9):316–21.

[3] Ramkumar PN, et al. Sports medicine and artificial intelligence: a primer. Am J Sports Med 2022;50 (4):1166–74.
[4] Ajmera P, et al. Real-world analysis of artificial intelligence in musculoskeletal trauma. J Clin Orthop Trauma 2021;22:101573.
[5] Haeberle HS, et al. Artificial intelligence and machine learning in lower extremity arthroplasty: a review. J Arthroplasty 2019;34(10):2201–3.
[6] Li MD, et al. Artificial intelligence applied to musculoskeletal oncology: a systematic review. Skelet Radiol 2022;51(2):245–56.
[7] Day J, et al. Evaluation of a weightbearing CT artificial intelligence-based automatic measurement for the M1-M2 intermetatarsal angle in hallux valgus. Foot Ankle Int 2021;42(11):1502–9.
[8] Zhang SC, et al. Clinical application of artificial intelligence-assisted diagnosis using anteroposterior pelvic radiographs in children with developmental dysplasia of the hip. Bone Joint J 2020;102-b(11):1574–81.
[9] Mutasa S, Yi PH. Clinical artificial intelligence applications: musculoskeletal. Radiol Clin North Am 2021; 59(6):1013–26.
[10] Makhni EC, Makhni S, Ramkumar PN. Artificial intelligence for the orthopaedic surgeon: an overview of potential benefits, limitations, and clinical applications. J Am Acad Orthop Surg 2021;29(6):235–43.
[11] Myers TG, et al. Artificial intelligence and orthopaedics: an introduction for clinicians. J Bone Jt Surg Am 2020;102(9):830–40.
[12] Meskó B, Görög M. A short guide for medical professionals in the era of artificial intelligence. npj Digit Med 2020;3:126.
[13] Shim E, et al. Automated rotator cuff tear classification using 3D convolutional neural network. Sci Rep 2020;10(1):15632.
[14] Karnuta JM, et al. Artificial intelligence to identify arthroplasty implants from radiographs of the knee. J Arthroplasty 2021;36(3):935–40.
[15] Navarro SM, et al. Machine learning and primary total knee arthroplasty: patient forecasting for a patient-specific payment model. J Arthroplasty 2018;33(12):3617–23.
[16] Martin RK, et al. Predicting anterior cruciate ligament reconstruction revision: a machine learning analysis utilizing the norwegian knee ligament register. J Bone Jt Surg Am 2022;104(2):145–53.
[17] Ramkumar PN, et al. Remote patient monitoring using mobile health for total knee arthroplasty: validation of a wearable and machine learning-based surveillance platform. J Arthroplasty 2019;34(10):2253–9.
[18] Thirukumaran CP, et al. Natural language processing for the identification of surgical site infections in orthopaedics. J Bone Jt Surg Am 2019;101(24):2167–74.
[19] Sachdev, R. Clincal applications of AI in orthopaedic workflow. In: Am assoc orthopaedic surg annu meet; 2022. Chicago.
[20] Polce EM, et al. Efficacy and applications of artificial intelligence and machine learning analyses in total joint arthroplasty: a call for improved reporting. J Bone Jt Surg Am 2022;.
[21] de Hond AAH, et al. Guidelines and quality criteria for artificial intelligence-based prediction models in healthcare: a scoping review. npj Digital Med 2022;5(1):2.
[22] Liu X, et al. Reporting guidelines for clinical trial reports for interventions involving artificial intelligence: the CONSORT-AI extension. Nat Med 2020;26(9):1364–74.
[23] Cruz Rivera S, et al. Guidelines for clinical trial protocols for interventions involving artificial intelligence: the SPIRIT-AI extension. Nat Med 2020;26(9):1351–63.
[24] Collins GS, et al. Protocol for development of a reporting guideline (TRIPOD-AI) and risk of bias tool (PROBAST-AI) for diagnostic and prognostic prediction model studies based on artificial intelligence. BMJ Open 2021;11(7):e048008.
[25] Oosterhoff JHF, et al. Feasibility of machine learning and logistic regression algorithms to predict outcome in orthopaedic trauma surgery. J Bone Jt Surg Am 2022;104(6):544–51.
[26] Kunze KN, et al. Potential benefits, unintended consequences, and future roles of artificial intelligence in orthopaedic surgery research: a call to emphasize data quality and indications. Bone Jt Open 2022;3(1):93–7.
[27] Kurmis AP, Ianunzio JR. Artificial intelligence in orthopedic surgery: evolution, current state and future directions. Arthroplasty 2022;4(1):9.
[28] Jang SJ, Kunze KN, Vigdorchik JM, Jerabek SA, Mayman DJ, Sculco PK. John Charnley Award: deep learning prediction of hip joint center on standard pelvis radiographs. J Arthroplasty 2022;.

[29] Karnuta JM, Murphy MP, Luu BC, et al. Artificial intelligence for automated implant identification in total hip arthroplasty: a multicenter external validation study exceeding two million plain radiographs. J Arthroplasty 2022;.

[30] Karhade AV, Oosterhoff JHF, Groot OQ, et al. Can we geographically validate a natural language processing algorithm for automated detection of incidental durotomy across three independent cohorts from two continents? Clin Orthop Relat Res 2022;.

[31] Ramkumar PN, Kunze KN, Haeberle HS, et al. Clinical and research medical applications of artificial intelligence. Arthroscopy. 2020.

[32] Greenstein AS, Teitel J, Mitten DJ, Ricciardi BF, Myers TG. An electronic medical record-based discharge disposition tool gets bundle busted: decaying relevance of clinical data accuracy in machine learning. Arthroplast Today 2020;6:850–5.

[33] Ghassemi M, Oakden-Rayner L, Beam AL. The false hope of current approaches to explainable artificial intelligence in health care. Lancet Digit Health 2021;3:e745–50.

[34] Ribeiro M.T., Singh S., Guestrin C. Why should I trust you?": Explaining the predictions of any classifier. In: Proc 22nd SIGKDD Int Conf Knowl Discovery Data Min; 2016:1135–44.

[35] Lundberg SM, Lee S-I. A unified approach to interpreting model predictions. Adv Neural Inf Process Syst 2017;4765–74.

[36] Pierson E, Cutler DM, Leskovec J, Mullainathan S, Obermeyer Z. An algorithmic approach to reducing unexplained pain disparities in underserved populations. Nat Med 2021;27:136–40.

[37] Slack D, Hilgard S, Jia E, Singh S, Lakkaraju H. Fooling LIME and SHAP: adversarial attacks on post hoc explanation methods New York, NY: Association for Computing Machinery Proc AAAI/ACM Conf AI, ethics, Soc. 2020. p. 180–6.

CHAPTER

27

Artificial intelligence in plastic surgery

Chad M. Teven and Michael A. Howard

Division of Plastic Surgery, Northwestern Medicine, Chicago, IL, United States

Introduction

The potential to improve patient care across medical disciplines with technology based on artificial intelligence (AI) is becoming an increasingly exciting opportunity. Although AI has been an important topic of investigation for over 50 years, its use in medicine is a relatively recent phenomenon. For example, it may offer improvements in disease diagnosis and prognosis, ultimately facilitating improved medical decision-making [1]. Moreover, AI-based technology has demonstrated effectiveness in surgery, both in the form of preoperative planning and intraoperative guidance and precision [2].

Plastic and reconstructive surgery is a field known for and dependent upon creativity and innovation. Plastic surgeons manage wide-ranging indications and conditions, routinely operating across the entire body (i.e., head to toe) in the treatment of defects related to cancer, burn injury, congenital malformations, and trauma. Plastic surgery also requires a high degree of precision, particularly in the case of aesthetic procedures. These characteristics of plastic surgery make it particularly well-suited for and receptive to the integration of innovative technologies like AI into its armamentarium. Although the use of AI in plastic surgery is limited to this point, several recent studies point to its potential to positively impact outcomes in multiple areas within plastic surgery [3]. For example, big data analytics may be used for quality improvement [4]; machine learning may facilitate early diagnosis [5], outcome prediction [6], and surgical planning in orthognathic surgery [7]; and facial recognition technology may aid in optimizing cosmetic outcomes [8].

Increasingly, researchers and clinicians within plastic surgery are developing and testing AI-based methods that may prove useful for improving patient care. As advancements continue and new technologies prove safe and effective, it is important that the plastic surgery community stays abreast of forthcoming innovations. Indeed, pioneering AI technology is likely to be well-received by both plastic surgeons and patients. Key to the implementation of AI is an awareness by surgeons of the benefits and indications of AI in various clinical contexts. Similarly, an understanding of the challenges, limitations, and

Artificial Intelligence in Clinical Practice
DOI: https://doi.org/10.1016/B978-0-443-15688-5.00012-7

ethical implications related to AI is important. This chapter provides an overview of AI in plastic surgery, specifically highlighting its clinical utility, potential barriers to implementation, and future implications for surgeons.

Artificial intelligence implementation in clinical practice

While AI has not yet been adopted on a widescale basis within plastic surgery, several studies have demonstrated effectiveness and proof of concept with respect to AI. An early example of AI in plastic surgery consisted of using machine learning to develop a method by which healing time after a burn injury could be predicted with a high degree of certainty [6]. The authors used reflectance spectrometry and an artificial neural network to generate a model that would predict whether a burn would require less than or greater than 14 days to heal. This information was used to aid in assessing the level of burn depth, which ultimately informed surgical planning [6]. Machine learning methods have also been employed to develop algorithms that extract and analyze large sets of radiographic data in order to provide a more precise diagnosis. Mendoza et al. found that such algorithms could diagnose distinct types of craniosynostosis based on computed tomography data with a high degree of accuracy (sensitivity 92.7%, specificity 98.9%) [9]. Similarly, AI has been used to guide surgical decision-making in cases of head and neck oncologic extirpation and reconstruction. For example, surgeons wearing a headset that facilitates the projection of three-dimensional radiographic images onto a patient are able to better visualize whether their planned surgical approach is optimal or requires adjustment [10].

Recently commercially available facial recognition technology has been used in the context of cosmetic surgery. Specifically, pattern recognition models, image analysis, and deep neural networks have provided an objective means of interpreting facial characteristics using biometric measurements [11]. In addition, Chen et al. trained four convolutional neural networks (from Amazon, IBM, Microsoft, and Face++) to identify gender based on a patient's facial features in the context of gender-affirming surgery [8]. Preoperative assessment of images of patients undergoing facial feminization demonstrated correct classification in only 53% of patients. In contrast, postoperative classification of patient images after facial feminization was significantly more accurate (98%) [8]. These results highlight a potential role for AI in measuring clinical effectiveness and outcomes after surgery.

In addition to clinical utility, additional applications of AI-based technology have been observed in the context of patient education [12], resident/student evaluation and teaching [13], and efforts to effectively market as well as influence public perception of plastic surgery [14]. Similar to clinical applications, however, the overall use of AI in these contexts is still in its infancy. Ongoing research and innovation will be necessary to drive the use of AI on a more regular basis.

Current challenges

Similar to other technologies before it, the implementation of AI within plastic surgery is subject to key hurdles and challenges. First, the actualization of clinical benefits from the application of novel technology is an arduous process. At present, few applications of AI are common in plastic surgical practice. Many opportunities exist, however, that are currently under investigation. Because of substantial costs and resource requirements associated with performing AI-based research in surgical subpopulations, the timeframe from bench to bedside is not insignificant. A second hurdle constraining AI application in the surgical setting relates to inherent limitations with respect to the technology itself. Data input must be accurate, unbiased, and comprehensive to ensure that various outputs provided by AI-based methods are reliable. Systemic biases regarding minorities, for example, may be contained within the algorithms and data underpinning diagnostic and prognostic assessments, resulting in predictions that are nonrepresentative [15].

An additional present-date challenge associated with AI is ethical concerns that arise with its use. Several potential ethical issues exist, including informed consent, patient anonymity, shared-decision making, the patient-physician relationship, and equity [3]. In particular, the use of big data puts patient-level data at risk for loss of confidentiality and other quality issues. In the realm of aesthetic surgery, machine learning has been used to ostensibly predict facial attractiveness [13]. Concerns leveled at this approach include inhibition of shared decision-making, biases in the model's predictions, and a reduction in cultural values and diversity in judging beauty. Efforts to mitigate these ethical challenges are crucial moving forward. These include careful attention by providers to preserve the patient-physician relationship, a focus on shared decision-making, and consistent re-evaluation of methods, algorithms, and data sets by clinicians and researchers to recognize and remove potential biases.

Future directions

The future of AI-based technologies within the field of plastic surgery is bright. Efforts are underway for its implementation in the preoperative, intraoperative, and postoperative settings. One area of tremendous potential is the use of AI-enabled decision-making tools to augment diagnosis, management, and planning of various plastic surgical conditions. Specifically, processes that employ data acquisition, predictive analytics, and integration with human (i.e., surgeon) decision-making are useful [3,16]. One example that may be of particular importance to hand surgeons is the use of AI to facilitate the interpretation of radiographic images. Deep neural network-based algorithms have been successfully used to identify wrist fractures [17]. This could similarly be applied to detecting closed fracture injuries of the hand and fingers [10]. Similarly, image analysis using AI-based technology could provide a richer information set to guide surgical technique and decision-making than current methods (e.g., high-resolution computed tomography) [10,18].

In conjunction with preoperative surgical planning and anatomical localization, intraoperative AI-assisted navigation could improve surgical accuracy and precision, thereby reducing complications and suboptimal outcomes. Preliminary data have demonstrated that AI models can be effectively trained to identify safe and dangerous zones of dissection in general surgery procedures; however, ongoing clinical studies are required prior to the widespread adoption of this technology [19]. In addition to potentially improving operative and postoperative efficiency and efficacy, this technology could be particularly helpful in resource-poor environments (e.g., developing countries), where the number and training of surgeons may not meet the demand for services [10].

Another exciting opportunity regarding AI implementation in plastic surgery relates to the ability of big data to facilitate evidence-based practices on the individual level. Big data, which refers to information collected and analyzed on a large scale, has already been instituted in plastic surgery through several databases, including Tracking Operations and Outcomes in plastic surgery, General Registry of Autologous Fat Transfer, and CosmetAssure [20]. These and forthcoming databases will offer the opportunity to identify relevant patterns that aid in disease diagnosis, predictions related to both the risk of disease as well as the likelihood of successful versus suboptimal interventions, and the risk of postoperative complications. Big data therefore could improve clinical decision-making by plastic surgeons, thus facilitating a more individualized approach and precision medicine [21].

Major takeaway points

- AI has the potential to fundamentally improve clinical outcomes associated with the provision of plastic and reconstructive surgery.
- Though the implementation of AI within plastic surgery is in its relative infancy, ongoing research is paving the way toward the availability of AI-based technologies to plastic surgeons as a mechanism for improving patient care.
- Numerous AI-based technologies have demonstrated preliminary utility in plastic surgery, including the use of artificial neural networks (i.e., big data), machine learning, deep learning, and natural language processing.
- Despite the significant potential of AI technology to facilitate improved patient care and outcomes in plastic surgery, clinicians and researchers must be vigilant in the recognition and mitigation of unintended consequences that could arise with the use of new technologies (e.g., ethical issues).

References

[1] Gambhir S, Malik S, Kumar Y. Role of soft computing approaches in healthcare domain: a mini review. J Med Syst 2016;40(12):1–20.

[2] Hashimoto DA, Rosman G, Rus D, Meireles OR. Artificial intelligence in surgery: promises and perils. Ann Surg 2018;268(1):70–6.

[3] Jarvis T, Thornburg D, Rebecca AM, Teven CM. Artificial intelligence in plastic surgery: current applications, future directions, and ethical implications. Plast Reconstr Surg Glob Open 2020;8(10):e3200.

[4] Zhu VZ, Tuggle CT, Au AF. Promise and limitations of big data research in plastic surgery. Ann Plast Surg 2016;76(4):453–8.

[5] Kiranantawat K, Sitpahul N, Taeprasartsit P, et al. The first Smartphone application for microsurgery monitoring: SilpaRamanitor. Plast Reconstr Surg 2014;134(1):130–9.

[6] Yeong E, Hsiao T, Chiang H, Lin CW. Prediction of burn healing time using artificial neural networks and reflectance spectrometer. J Burn 2005;31(4):415–20.

[7] Knoops PGM, Papaioannou A, Borghi A, et al. A machine learning framework for automated diagnosis and computer-assisted planning in plastic and reconstructive surgery. Sci Rep 2019;9:13597.

[8] Chen K, Lu SM, Cheng R, et al. Facial recognition neural networks confirm success in facial feminization surgery. Plast Reconstr Surg 2020;145(1):203–9.
[9] Mendoza CS, Safdar N, Okada K, et al. Personalized assessment of craniosynostosis via statistical shape modeling. Med Image Anal 2014;18(4):635–46.
[10] Murphy DC, Saleh DB. Artificial intelligence in plastic surgery: what is it? Where are we now? What is on the horizon? Ann R Coll Surg Engl 2020;102(8):577–80.
[11] Zuo KJ, Saun TJ, Forrest CR. Facial recognition technology: a primer for plastic surgeons. Plast Reconstr Surg 2019;143(6):1298e–306e.
[12] Boczar D, Sisti A, Oliver JD, et al. Artificial intelligent virtual assistant for plastic surgery patient's frequently asked questions: a pilot study. Ann Plast Surg 2020;84(4):e16–21.
[13] Kanevsky J, Corban J, Gaster R, Kanevsky A, Lin S, Gilardino M. Big data and machine learning in plastic surgery: a new frontier in surgical innovation. Plast Reconstr Surg 2016;137(5):890e–7e.
[14] Levites HA, Thomas AB, Levites JB, et al. The use of emotional artificial intelligence in plastic surgery. Plast Reconstr Surg 2019;144(2):499–504.
[15] Koimizu J, Numajiri T, Kato K. Machine learning and ethics in plastic surgery. Plast Reconstr Surg Glob Open 2019;7(3):e2162.
[16] U.S. Food and Drug Administration. Artificial intelligence and machine learning in software. <http://www.fda.gov/medical-devices/software-medical-device-samd/artificial-intelligence-and-machine-learning-software-medical-device>; 2020 [accessed 25.04.22].
[17] Lindsey R, Daluiski A, Chopra S, et al. Deep neural network improves fracture detection by clinicians. Proc Natl Acad Sci U S A 2018;115(45):11591–6.
[18] Grenda TR, Pradarelli JC, Dimick JB. Using surgical video to improve technique and skill. Ann Surg 2016;64 (1):32–3.
[19] Madani A, Namazi B, Altieri M, et al. Artificial intelligence for intraoperative guidance: Using semantic segmentation to identify surgical anatomy during laparoscopic cholecystectomy. Ann Surg 2022;276(2):263–9.
[20] Chandawarkar A, Chartier C, Kanevsky J, Cress PE. A practical approach to artificial intelligence in plastic surgery. Aesthet Surg J Open Forum 2020;2(1). Available from: https://doi.org/10.1093/asjof/ojaa001.
[21] Kim YJ, Kelley BP, Nasser JS, Chung KC. Implementing precision medicine and artificial intelligence in plastic surgery. Plast Reconstr Surg Glob Open 2019;7(3):e2113.

CHAPTER

28

Artificial intelligence in obstetrics and gynecology

Elias Kassir[1], Veronica C. Kuhn[2], Melissa S. Wong[3,4] and Christina S. Han[1]

[1]Division of Maternal-Fetal Medicine, Department of Obstetrics and Gynecology, David Geffen School of Medicine, University of California at Los Angeles (UCLA), Los Angeles, CA, United States [2]University of Virginia, Charlottesville, VA, United States [3]Division of Maternal-Fetal Medicine, Department of Obstetrics and Gynecology, Cedars-Sinai Medical Center Department of Obstetrics and Gynecology, Los Angeles, CA, United States [4]Division of Informatics, Department of Biomedical Sciences, Cedars-Sinai Medical Center, Los Angeles, CA, United States

While artificial intelligence (AI) has made significant inroads into clinical care in radiology, pathology, and genomics, applications in obstetrics and gynecology have remained minimally explored despite the potential for tremendous impact. AI in obstetrics and gynecology has the potential to affect every aspect of the field: improve diagnostic capabilities, develop more accurate risk prediction models, personalize treatments, and accelerate much-needed research. The "Great Obstetrical Syndromes" of preterm labor, preterm prelabor rupture of membranes, stillbirth, preeclampsia, fetal growth restriction, abruption and placenta accreta remain enigmas despite decades of research, with current knowledge limited to only snapshot understanding of the complex pathophysiology and management recommendations based heavily on expert opinions [1]. Women's healthcare throughout the reproductive continuum, including puberty, contraception, menopause, cancer screening, and management, remains understudied and poorly understood.

Further, even at this nascent stage in AI research, output in obstetrics and gynecology lags behind other specialties. Focusing on image analysis, for example, a PubMed Medical Subject Heading terminology search in February 2022 revealed 5367 papers on "CT" and "artificial intelligence" and 4952 papers on "MRI" and "artificial intelligence." In contrast,

Artificial Intelligence in Clinical Practice
DOI: https://doi.org/10.1016/B978-0-443-15688-5.00004-8

a search for papers on "ultrasound" and "pregnancy" and "artificial intelligence" yielded only 137 results; a search for "ultrasound" and "ovarian" and "artificial intelligence" yielded only 37.

The goal of this chapter is to review current applications of AI in obstetrics and gynecology, to discuss challenges to AI research and provider adoption, and to suggest future applications and avenues for research.

Current applications of artificial intelligence in obstetrics

Uses of AI in obstetrics can be considered in three different domains: (1) ultrasound and imaging applications, (2) cardiotocography or monitoring of the fetal heart rate and contractions, and (3) electronic health record analysis to develop classification or prediction models. Each of these has several examples of where AI has benefited clinical care and where there is much work to be done.

Ultrasound applications of artificial intelligence

Ultrasound in obstetrics and gynecology has benefited from the progress made in imaging and computer vision, particularly by industry. Focusing on operations, AI can help improve workflow efficiency when conducting and interpreting obstetric ultrasounds. AI-driven automated biometrics, if perfected, may ultimately reduce scanning times, increase throughput and reduce repetitive physical strain in the workforce [2,3]. Multiple studies have evaluated the ability of AI to place calipers efficiently and effectively, though up to 1/3 require adjustment [3–5]. In addition, some groups have demonstrated the ability to simply acquire a volume or sweep and post hoc have the machine/model identify the intended scan planes [6].

Current and future ultrasound applications of artificial intelligence

Operations support

- Automatic caliper placement,
- Automatic image labeling,
- Reminder/trigger for completion of missed anatomic views, and
- Biometry plane identification.

Education

- Probe guidance/plane identification and
- Novel skills (e.g., providing guidance on less commonly used measurements, such as frontomaxillary facial angle for micrognathia).

Anomaly detection and augmentation

- Detection of anomalies and
- Recommendations for complementary views based on syndromes, for example, probing for other views when part of VACTERL is identified.

AI has the potential to augment the knowledge of the sonographer or clinician even while acquiring images. For the novice, AI's ability to identify intended scan planes could easily be modified to teach the technician what adjustments need to be made until a perfect image is achieved. In addition, because certain fetal anomalies are rare, it can be difficult to recall the exact way in which certain measurements or images should be obtained. It might be possible, for example, when evaluating a cleft for the machine to recommend a "reverse face" or "flipped face" view for better evaluation of the lip and palate [7,8].

Finally, AI has been applied to ultrasound images for anomaly detection. Using a database of 12,780 cases, Xie et al. were able to correctly identify abnormal neurosonographic images with 96.3% accuracy [9]. The deep learning algorithm was able further to highlight "hot spots" where the anomaly was suspected, suggesting deep learning could serve as an alert system for the clinician interpreting the images.

Cardiotocography applications of artificial intelligence

Fetal cardiotocography (CTG) utilizes an ultrasound transducer applied on the patient's abdomen to enable a continuous recording of the fetal heart rate in the antenatal period and during labor and delivery. Significant abnormalities in the fetal CTG output may indicate the presence of conditions, such as fetal metabolic acidemia or anemia, thereby informing clinical decision-making regarding delivery. However, CTG is subject to human error and subjectivity, leading to high inter- and intra-observer variability [10].

AI applications to CTG can be divided into (1) alert systems and (2) novel pattern identification. A 2017 study evaluated several machine learning algorithms (a deep learning classifier, a Fisher's linear discriminant analysis classifier, and random forest classifier) to determine whether these algorithms could help practitioners determine when fetal heart tracings warrant further intervention. After training, their results showed that a deep learning classifier could significantly improve the positive predictive value achieved by practitioners. This study provides a proof-of-concept that AI could be a vital diagnostic support tool when used in conjunction with CTG [10] (Fig. 28.1).

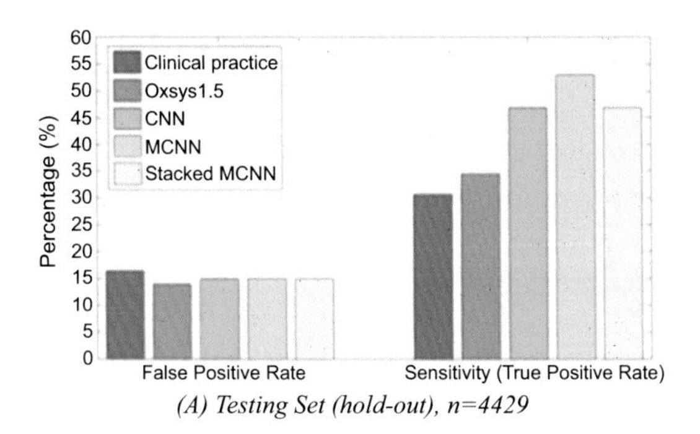

(A) Testing Set (hold-out), n=4429

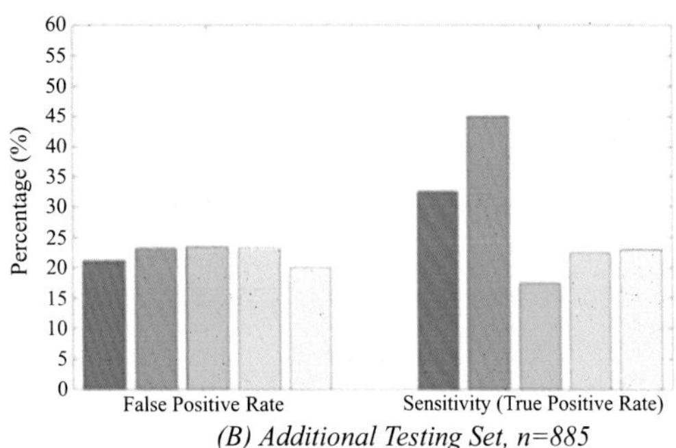

(B) Additional Testing Set, n=885

FIGURE 28.1 Performance on last 60 minutes of CTG: Clinical Practice, Oxsys 1.5, CNN, MCNN, Stacked MCNN (median of 5 runs). The FPR was fixed at 15% for the CNN and MCNN in order to be comparable to the FPR of Clinical Practice.

AI systems to identify novel findings in CTG are fewer. Based on their Oxford archive of 60,985 continuous labor CTG tracings, Georgieva and colleagues have developed an OxSys system to identify fetal compromise. Beyond clinical indicators that have been previously known to be associated with fetal acidemia (e.g., bradycardia, short term heart rate variability), OxSys was able to identify new features, such as decelerative capacity, which were strongly associated with adverse outcome [11].

Electronic health record analysis for applications of artificial intelligence

Electronic health record (EHR) analysis can be considered as tasks of either classification or prediction. In recent years a diversity of methods have been applied to each of these. One of the strongest uses of EHR data for classification has been in preeclampsia, where clinical and -omics data were combined to suggest different phenotypes of preeclampsia manifestation [12].

While predictive models are no stranger to obstetrics [13], in recent years AI methods have been applied to a number of common Obstetric conditions: cesarean delivery, vaginal birth after cesarean, postpartum hemorrhage, shoulder dystocia, and severe maternal morbidity [14–18]. Each system has shown promising results, but model effectiveness studies (in which a model is incorporated into clinical care) remain lacking.

Current applications of artificial intelligence in gynecology

Advances in the use of AI in gynecology have been less domain-specific, often benefiting from developments outside of the field. For example, robotics and the use of AI for haptics was developed initially for cardiac surgery but soon found applications in radical hysterectomy and other gynecologic surgeries [19,20]. Similarly, much of the advance in application of automated processing of Pap smears for cervical cancer screening has been a crossover from other applications in pathology [21].

Novel developments have come through varying applications of computer vision in infertility treatment and gynecologic oncology. In reproductive endocrinology and infertility, a persistent clinical conundrum is the significant interembryologist and intra-embryologist variability in scoring of the quality of embryos, driven primarily by the subjective nature morphologic grading. Use of AI in IVF can increase the consistency and objectivity of embryo assessments, which can improve implantation rates and decrease the risk of discarding a viable embryo. Furthermore, use of AI in interpretation of 3-D ultrasounds can provide automated volume counts to identify and measure follicles [22,23].

In gynecologic oncology, AI has been beneficial in radiomics, which quantifies imaging features to support clinical decision-making. Textural analysis can be used to potentially tailor treatment and inform prognosis for ovarian cancer. In this clinical setting, textural analysis uses AI to quantify tumor heterogeneity in CT imaging of ovarian cancer, which could potentially predict response to treatment [24]. More broadly, AI can analyze large datasets of demographics, tumor-related parameters and clinical data to predict prognosis.

In a 2017 study, of six intelligence methods tested, a probabilistic neural network performed the best, with an accuracy of 0.892 and a sensitivity of 0.975 in prediction of 5-year survival in cervical cancer patients treated with radical hysterectomy [25].

Future challenges and opportunities for artificial intelligence in obstetrics and gynecology

While AI has begun to make inroads in obstetrics and gynecology, the potential for its uses still far exceeds the applications. Childbirth remains the most common reason for hospitalization in the United States, exceeding the next two most common combined [26]. Additionally, gynecologic issues are pervasive, with the potential to affect 51% of the US population. Research and funding opportunities have not reflected this reality. Furthermore, data science teams may be more likely to identify applications of AI in areas which are pertinent to their interests. Unfortunately, there is, like in many areas of computer science, a significant gender gap in AI professionals, with a 72% gender gap favoring males [27] (Fig. 28.2).

Additionally, obstetrics and gynecology is marked by known disparities in care, including threefold increased rates of maternal mortality in Black women, disparities in access to infertility technologies, and disparities in outcome for treatment for gynecologic malignancy [28]. As AI is known to have the potential to further exacerbate these

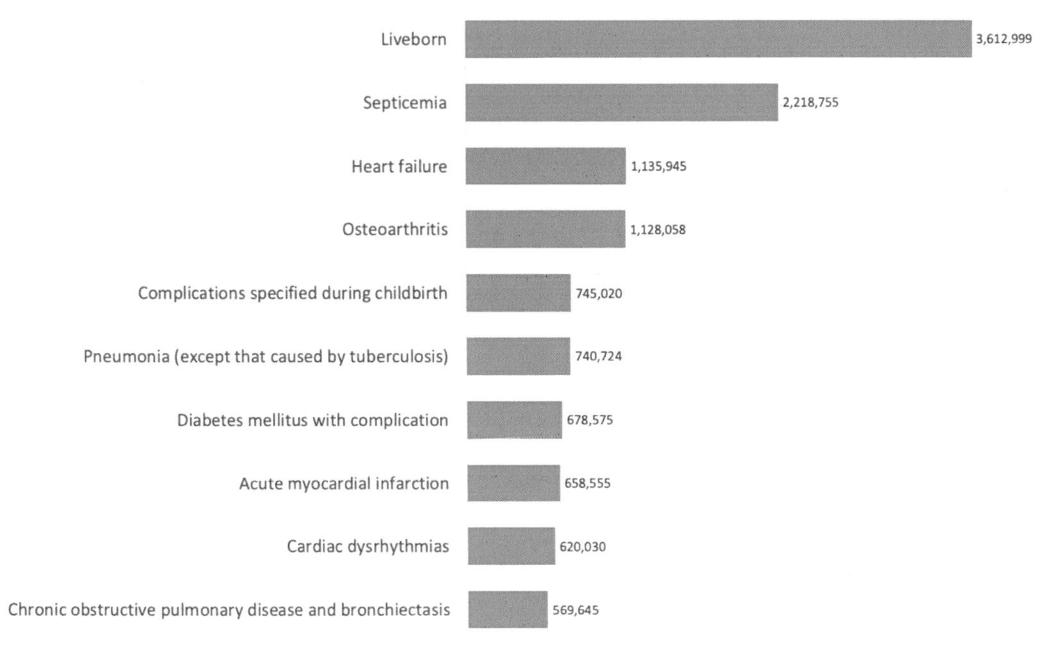

FIGURE 28.2 Most common diagnoses for inpatients stays, AHRQ 2018.

disparities, in some ways we are gifted from being in the early stages of our field's foray into AI [29]. By grounding our work in reduction of health disparities and actively mitigating bias in our algorithms, we have the potential to correct these problems before they become germane in our care [13].

References

[1] Brosens I, Pijnenborg R, Vercruysse L, Romero R. The "Great Obstetrical Syndromes" are associated with disorders of deep placentation. Am J Obstet Gynecol 2011;204(3):193–201.
[2] Volkov M., Hashimoto D.A., Rosman G., Meireles O.R., Rus D. Machine learning and coresets for automated real-time video segmentation of laparoscopic and robot-assisted surgery. In: 2017 IEEE International Conference on Robotics and Automation (ICRA). IEEE; 2017 May. p. 754–759.
[3] Salim I, Cavallaro A, Ciofolo-Veit C, Rouet L, Raynaud C, Mory B, et al. Evaluation of automated tool for two-dimensional fetal biometry. Ultrasound Obstet Gynecol 2019;54(5):650–4.
[4] Espinoza J, Good S, Russell E, Lee W. Does the use of automated fetal biometry improve clinical work flow efficiency? J Ultrasound Med 2013;32(5):847–50.
[5] Pluym ID, Afshar Y, Holliman K, Kwan L, Bolagani A, Mok T, et al. Accuracy of automated three-dimensional ultrasound imaging technique for fetal head biometry. Ultrasound Obstet Gynecol 2021; 57(5):798–803.
[6] Baumgartner CF, Kamnitsas K, Matthew J, Fletcher TP, Smith S, Koch LM, et al. SonoNet: real-time detection and localisation of fetal standard scan planes in freehand ultrasound. IEEE Trans Med Imaging 2017; 36(11):2204–15.
[7] Platt LD, Devore GR, Pretorius DH. Improving cleft palate/cleft lip antenatal diagnosis by 3-dimensional sonography: the "flipped face" view. J Ultrasound Med 2006;25(11):1423–30.
[8] Ten PM, Pedregosa JP, Santacruz B, Adiego B, Barron E, Sepulveda W. Three-dimensional ultrasound diagnosis of cleft palate: "reverse face," "flipped face" or "oblique face"- which method is best? Ultrasound Obstet Gynecol 2009;33(4):399–406.
[9] Xie HN, Wang N, He M, Zhang LH, Cai HM, Xian JB, et al. Using deep-learning algorithms to classify fetal brain ultrasound images as normal or abnormal. Ultrasound Obstet Gynecol 2020;56(4):579–87.
[10] Fergus P, Hussain A, Al-Jumeily D, Huang DS, Bouguila N. Classification of caesarean section and normal vaginal deliveries using foetal heart rate signals and advanced machine learning algorithms. Biomed Eng OnLine 2017;16(89). Available from: https://doi.org/10.1186/s12938-017-0378-z.
[11] Georgieva A, Redman CWG, Papageorghiou AT. Computerized data-driven interpretation of the intrapartum cardiotocogram: a cohort study. Acta Obstet Gynecol Scand 2017;96(7):883–91.
[12] Roberts JM, Rich-Edwards JW, McElrath TF, Garmire L, Myatt L. Subtypes of pre-eclampsia: recognition and determining clinical usefulness. Hypertension. 2021;77:1430–41.
[13] Grobman WA, Sandoval G, Rice MM, Bailit JL, Chauhan SP, Constantine MM, et al. Prediction of vaginal birth after cesarean in term gestations: a calculator without race and ethnicity. Am J Obstet Gynecol 2021; 225(6):664E.1–7.
[14] Wong MS, Wells M, Zamanzadeh D, Akre S, Pevnick JM, Bui AA, et al. Applying automated machine learning to predict mode of delivery using ongoing intrapartum data in laboring patients. Am J Perinatol 2022;. Available from: https://doi.org/10.1055/a-1885-1697.
[15] Lipschuetz M, Guedalia J, Rottenstreich A, Persky MN, Cohen SM, Kabiri D, et al. Prediction of vaginal birth after cesarean deliveries using machine learning. Am J Obstet Gynecol 2020;222(6) 613.e1–613.e12.
[16] Venkatesh KK, Strauss RA, Grotegut CA, Heine RP, Chescheir NC, Stringer JSA, et al. Machine learning and statistical models to predict postpartum hemorrhage. Obstet Gynecol 2020;135(4):935–44.
[17] Tsur A, Batsry L, Toussia-Cohen S, Rosenstein MG, Barak O, Brezinov Y, et al. Development and validation of a machine-learning model for prediction of shoulder dystocia. Ultrasound Obstet Gynecol 2020; 56(4):588–96.
[18] Clap MA, Kim E, James KE, Perlis RH, Kaimal AJ, McCoy Jr. TH. Natural language processing of admission notes to predict severe maternal morbidity during the delivery encounter. Am J Obstet Gynecol 2022;227(3) 511.e1–511.e8.

[19] Svoboda E. Your robot surgeon will see you now. Nature. 2019;573(7775):S110–11.
[20] George EI, Brand TC, LaPorta A, Marescaux J, Satava RM. Origins of robotic surgery: from skepticism to standard of care. JSLS. 2018;22(4) e2018.00039.
[21] Wentzensen N, Lahrmann B, Clarke MA, Kinnney W, Tokugawa D, Poitras N, et al. Accuracy and efficiency of deep-learning-based automation of dual stain cytology in cervical cancer screening. JNCI. 2021; 113(1):72–9.
[22] Bormann CL, Thirumalaraju P, Kanakasabapathy MK, Kanakasabapathy MK, Kandula H, Souter I, et al. Consistency and objectivity of automated embryo assessments using deep neural networks. Fertil Steril 2020;113(4) 781–787.e1.
[23] Zaninovic N, Rosenwaks Z. Artificial intelligence in human in vitro fertilization and embryology. Fertil Steril 2020;114(5):914–20.
[24] Nougaret S, Tardieu M, Vargas HA, Reinhold C, Vande Perre S, Bonanno N, et al. Ovarian cancer: an update on imaging in the era of radiomics. Diagn Interv Imaging 2019;100(10):647–55.
[25] Obrzut B, Kusy M, Semczuk A, Obrzut M, Kluska J. Prediction of 5-year overall survival in cervical cancer patients treated with radical hysterectomy using computational intelligence methods. BMC Cancer 2017; 17(1):840.
[26] HCUP Fast Stats. Healthcare Cost and Utilization Project (HCUP). April 2021. Agency for Healthcare Research and Quality, Rockville, MD. <http://www.hcup-us.ahrq.gov/faststats/national/inpatientcommondiagnoses.jsp?year1 = 2017&characteristic1 = 0&included1 = 1&year2 = &characteristic2 = 0&included2 = 1&expansionInfoState = hide&dataTablesState = hide&definitionsState = hide&exportState = hide> [accessed 16.09.22].
[27] World Economic Forum.The Global Gender Gap Report 2018. Assessing Gender Gaps in Artificial Intelligence; 2018.
[28] Arvizo C, Garrison E. Diversity and inclusion: the role of unconscious bias on patient care, health outcomes and the workforce in obstetrics and gynaecology. Curr Opin Obstet Gynecol 2019;31(5):356–62.
[29] Char DS, Shah NH, Magnus D. Implementing machine learning in healthcare—addressing ethical challenges. N Enl J Med 2018;378(11):981–3.

CHAPTER

29

Artificial intelligence in urology

Raghav Gupta, Adriana Marcela Pedraza Bermeo, Krunal Pandav and Ashutosh Kumar Tewari

Department of Urology, Icahn School of Medicine at Mount Sinai, New York, NY, United States

Introduction (role of artificial intelligence, landmark trials in urology)

The field of artificial intelligence (AI) emerged following advances in modern computing and has made a notable impact within the sphere of healthcare already. AI broadly refers to the ability of a machine to make decisions in response to various inputs by mimicking the cognitive processes traditionally ascribed to human beings [1]. There are multiple components to AI including language processing, machine learning (ML), artificial neural networks (ANNs) including deep learning networks, and computer vision [2–5]. The advantage of AI lies in its ability to learn from large datasets, and to be able to make independent predictions grounded in statistics without being explicitly programmed to provide certain outputs. The implementation of AI within urology thus far has largely concentrated on (1) diagnosing urological disease, (2) predicting outcomes following urologic intervention, and (3) assessing surgical performance/skill [1,2,6]. The use of AI has been documented in the literature across multiple urologic subspecialities including urologic oncology (i.e., prostate cancer, renal cancer, and urothelial cancer), endourology, and andrology. In this chapter, we aim to provide a brief but comprehensive summary of the ways in which AI has been used in urology to date, and what we anticipate AI will be used for in the years ahead.

Current examples of artificial intelligence implementation in urology

Diagnosis of disease

As noted above, AI has been used to aid clinicians in the diagnosis of urologic disease by extracting information from various imaging modalities, pathologic specimen, and

Artificial Intelligence in Clinical Practice
DOI: https://doi.org/10.1016/B978-0-443-15688-5.00051-6

genomics data. This is perhaps most robustly demonstrated within the field of urologic oncology, and particularly so for prostate cancer. Fehr et al. have previously reported on the use of ML-based classification of prostate cancer using multiparametric magnetic resonance imaging-based texture features from T2-weighted and apparent diffusion coefficient sequences [7]. The authors observed a 93% accuracy rate in differentiating between cancers with a Gleason score of 6 or higher on the basis of imaging alone [7]. Multiple other studies have similarly reported on using AI and radiomic features from MR imaging to detect prostate cancer [8–10]. Pathologic specimen from patients have even been subjected to analyses via AI-based algorithms to either identify prostate cancer or to predict Gleason scores based on tissue characteristics [11,12]. Other authors report on the use of ANNs. Tewari et al., for example, have previously used an ANN to predict the risk of extracapsular extension in the staging of clinically localized prostate cancer [3]. Similarly, ANNs have been used to predict prostate cancer biopsy outcomes [13–15], to assess the risk of lymph node involvement in clinically localized prostate cancer [16], and to evaluate the likelihood of biochemical failure/recurrence following robotic prostatectomy [17].

It is important to note that within the field of urologic oncology, the uses of AI are not limited to prostate cancer. AI-based algorithms have been applied to help diagnose renal cell carcinoma (RCC) on the basis of either computed tomography (CT) or MR imaging [18–28] as well as urothelial carcinoma [29–34] on the basis of either cystoscopy, CT, or MR imaging. Similarly, AI has been used to improve the accuracy with which hydronephrosis secondary to ureteropelvic junction obstruction can be diagnosed [35].

Predicting outcomes following urologic intervention

AI can be utilized to analyze large volumes of data (also known as "big data") to develop mathematical models which can predict clinical outcomes. For example, within the field of urologic oncology AI has been used to predict biochemical recurrence rates following robotic-assisted prostatectomy [36–40]. Wong et al. used ML to combine patient demographics, operative, clinical, and imaging data to build a model which performed better than conventional statistical regression techniques to predict prostate cancer recurrence [36]. AI has also been used to predict length of hospital stay [41] as well as recovery of urinary continence [42] following robotic radical prostatectomy. It has been applied to predict mortality rates after radical cystectomy surgery [43], to predict recurrence of urothelial carcinoma on the basis of microRNA data [44] and genomic data [45]. It has also been used to predict survival in patients with clear-cell RCC by evaluating patients' genetic profiles [46].

In the field of endourology, AI has been used to assess outcomes for the management of kidney stones. Aminsharifi et al. detailed their experience using an ANN to predict results following percutaneous nephrolithotomy [47]. Mannil et al., similarly reported on using CT data and 3D-texture analyses to develop an ML model which could predict success following shock wave lithotripsy [48].

In the field of andrology, there has been greater attention devoted toward predicting male infertility and identifying factors that can contribute to infertility. Girela et al. developed an ANN which could predict semen characteristics (such as sperm concentration and motility) using self-reported life habits, environmental factors, and health status [49]. A different group designed an AI-based image recognition system that assessed sperm motility and concentration. This system conveniently utilized a smartphone-based interface and a compatible at-home testing kit [50]. Using data on 310 patients with azoospermia, another study reported on the development of an ANN capable of predicting chromosomal abnormalities with greater than 95% accuracy using patient demographic and laboratory data [51].

Surgical training and performance assessment

Given the widespread adoption of the use of robotic-assisted laparoscopic surgery within urology, there is growing interest in finding ways to assess and evaluate surgical proficiency and performance. Hung et al. have previously reported on developing three ML-based algorithms which were trained using automated performance metrics (derived from instrument motion tracking and surgical footage) during robot-assisted radical prostatectomy surgery. The "label" was designed as hospital length of stay for training these algorithms. These algorithms were then able to successfully predict select postoperative outcomes, including differentiating between different length of stays. The authors proposed that intraoperative camera manipulation was an importance indicator of surgical expertise [41]. Though in its infancy, such data can have profound implications in the future with regards to evaluating trainee proficiency during residency and fellowship training.

Challenges preventing artificial intelligence application

The potential for AI to revolutionize the delivery of healthcare in the field of urology, is unquestionably evident. It leverages the vast processing power of computers to analyze large sums of clinical and imaging data. Nonetheless, there are several challenges to the widespread implementation of AI, that one should be aware of. There is significant heterogeneity with respect to the algorithms that have been created thus far, making it difficult for clinicians to be able to compare different studies. In addition, most studies report on algorithms that have been trained using patient data from individual institutions. This limits the generalizability of the reported results and can lead to "overfitting" of individual models. Larger multiinstitutional studies in which these algorithms have been externally validated, are needed. However, the cost associated with the tools needed to process these large volumes of patient data, can be prohibitively high.

In addition, there are natural variations in decision-making amongst practitioners, leading some physicians to question the utility, practicality, and/or applicability of AI-based algorithms in the clinical setting. We posit that clinicians need to be better educated on the science behind AI to be able to properly evaluate AI-related literature critically and judiciously. This will require that they work together with industry representatives and computer/software engineers to decipher the technological "jargon" conventionally associated with AI. Ultimately, the future of AI rests on the willingness of physicians to actively engage in cross-collaboration and to seek opportunities for self-education.

Future directions

As described above, AI has been applied in urology to help diagnose disease more accurately and efficiently, to predict outcomes following surgical intervention using clinical and imaging data, and to guide the assessment of surgical performance. We believe the future of AI will continue to build on these core pillars. More specifically, as the interplay of epigenetics and disease becomes better elucidated, we hypothesize that more AI-based algorithms will transition toward incorporating patients' genetic profiles in their predictive models. Additionally, as the stream of data that we can receive from surgical instruments expands, we predict that consolidation of this data with AI-based technologies will provide valuable guidance for both novice and expert surgeons with regard to surgical technique improvement. Ultimately, our hope is that the incorporation of AI into the clinical setting will provide clinicians with the knowledge and the tools needed to provide better care for our patients.

Major takeaway points

- AI has been used in the field of urology to help diagnose disease, to predict outcomes following surgical intervention, and to assess surgical performance and skill.
- There are multiple reports in the literature of AI being used to aid clinicians across various urologic subspecialities including urologic oncology, endourology, and andrology.
- Although AI has the power to be able to revolutionize the delivery of healthcare, challenges including (1) the paucity of multiinstitutional studies, (2) a lack of understanding of the "science behind AI" amongst clinicians, and (3) a significant cost associated with AI-related technologies, will each need to be addressed in the future to enable its widespread use and adoption.

Abbreviations

AI Artificial intelligence
ANNs Artificial neural networks
ML Machine learning
CT Computed tomography
RCC Renal cell carcinoma

References

[1] Goldenberg SL, Nir G, Salcudean SE. A new era: artificial intelligence and machine learning in prostate cancer. Nat Rev Urol 2019;16(7):391–403. Available from: https://doi.org/10.1038/s41585-019-0193-3.

[2] Chen J, Remulla D, Nguyen JH, Aastha D, Liu Y, Dasgupta P, et al. Current status of artificial intelligence applications in urology and their potential to influence clinical practice. BJU Int 2019;. Available from: https://doi.org/10.1111/bju.14852.

[3] Tewari A, Narayan P. Novel staging tool for localized prostate cancer: a pilot study using genetic adaptive neural networks. J Urol 1998;160(2):430–6. Available from: https://doi.org/10.1016/s0022-5347(01)62916-1.

[4] Wei JT, Tewari A. Artificial neural networks in urology: pro. Urology 1999;54(6):945–8. Available from: https://doi.org/10.1016/s0090-4295(99)00341-6.

[5] Errejon A, Crawford ED, Dayhoff J, O'Donnell C, Tewari A, Finkelstein J, et al. Use of artificial neural networks in prostate cancer. Mol Urol 2001;5(4):153–8. Available from: https://doi.org/10.1089/10915360152745821.

[6] Brodie A, Dai N, Teoh JY, Decaestecker K, Dasgupta P, Vasdev N. Artificial intelligence in urological oncology: an update and future applications. Urol Oncol 2021;39(7):379–99. Available from: https://doi.org/10.1016/j.urolonc.2021.03.012.

[7] Fehr D, Veeraraghavan H, Wibmer A, Gondo T, Matsumoto K, Vargas HA, et al. Automatic classification of prostate cancer Gleason scores from multiparametric magnetic resonance images. Proc Natl Acad Sci U S A 2015;112(46). Available from: https://doi.org/10.1073/pnas.1505935112 E6265-73.

[8] Algohary A, Viswanath S, Shiradkar R, Ghose S, Pahwa S, Moses D, et al. Radiomic features on MRI enable risk categorization of prostate cancer patients on active surveillance: Preliminary findings. J Magn Reson Imaging 2018;. Available from: https://doi.org/10.1002/jmri.25983.

[9] Ginsburg SB, Algohary A, Pahwa S, Gulani V, Ponsky L, Aronen HJ, et al. Radiomic features for prostate cancer detection on MRI differ between the transition and peripheral zones: preliminary findings from a multi-institutional study. J Magn Reson Imaging 2017;46(1):184–93. Available from: https://doi.org/10.1002/jmri.25562.

[10] Merisaari H, Movahedi P, Perez IM, Toivonen J, Pesola M, Taimen P, et al. Fitting methods for intravoxel incoherent motion imaging of prostate cancer on region of interest level: repeatability and gleason score prediction. Magn Reson Med 2017;77(3):1249–64. Available from: https://doi.org/10.1002/mrm.26169.

[11] Kwak JT, Hewitt SM. Multiview boosting digital pathology analysis of prostate cancer. Comput Methods Prog Biomed 2017;142:91–9. Available from: https://doi.org/10.1016/j.cmpb.2017.02.023.

[12] Nguyen TH, Sridharan S, Macias V, Kajdacsy-Balla A, Melamed J, Do MN, et al. Automatic Gleason grading of prostate cancer using quantitative phase imaging and machine learning. J Biomed Opt 2017;22(3):36015. Available from: https://doi.org/10.1117/1.JBO.22.3.036015.

[13] Porter CR, Gamito EJ, Crawford ED, Bartsch G, Presti Jr. JC, Tewari A, et al. Model to predict prostate biopsy outcome in large screening population with independent validation in referral setting. Urology 2005;65 (5):937–41. Available from: https://doi.org/10.1016/j.urology.2004.11.049.

[14] Chakraborty S, Ghosh M, Maiti T, Tewari A. Bayesian neural networks for bivariate binary data: an application to prostate cancer study. Stat Med 2005;24(23):3645–62. Available from: https://doi.org/10.1002/sim.2214.

[15] Porter CR, O'Donnell C, Crawford ED, Gamito EJ, Sentizimary B, De Rosalia A, et al. Predicting the outcome of prostate biopsy in a racially diverse population: a prospective study. Urology 2002;60(5):831–5. Available from: https://doi.org/10.1016/s0090-4295(02)01882-4.

[16] Batuello JT, Gamito EJ, Crawford ED, Han M, Partin AW, McLeod DG, et al. Artificial neural network model for the assessment of lymph node spread in patients with clinically localized prostate cancer. Urology 2001;57(3):481–5. Available from: https://doi.org/10.1016/s0090-4295(00)01039-6.

[17] Porter C, O'Donnell C, Crawford ED, Gamito EJ, Errejon A, Genega E, et al. Artificial neural network model to predict biochemical failure after radical prostatectomy. Mol Urol 2001;5(4):159–62. Available from: https://doi.org/10.1089/10915360152745830.

[18] Yan L, Liu Z, Wang G, Huang Y, Liu Y, Yu Y, et al. Angiomyolipoma with minimal fat: differentiation from clear cell renal cell carcinoma and papillary renal cell carcinoma by texture analysis on CT images. Acad Radiol 2015;22(9):1115–21. Available from: https://doi.org/10.1016/j.acra.2015.04.004.

[19] Yin Q, Hung SC, Wang L, Lin W, Fielding JR, Rathmell WK, et al. Associations between tumor vascularity, vascular endothelial growth factor expression and PET/MRI radiomic signatures in primary clear-cell-renal-cell-carcinoma: proof-of-concept study. Sci Rep 2017;7:43356. Available from: https://doi.org/10.1038/srep43356.

[20] Feng Z, Rong P, Cao P, Zhou Q, Zhu W, Yan Z, et al. Machine learning-based quantitative texture analysis of CT images of small renal masses: differentiation of angiomyolipoma without visible fat from renal cell carcinoma. Eur Radiol 2018;28(4):1625–33. Available from: https://doi.org/10.1007/s00330-017-5118-z.

[21] Cui EM, Lin F, Li Q, Li RG, Chen XM, Liu ZS, et al. Differentiation of renal angiomyolipoma without visible fat from renal cell carcinoma by machine learning based on whole-tumor computed tomography texture features. Acta Radiol 2019;60(11):1543–52. Available from: https://doi.org/10.1177/0284185119830282.
[22] Coy H, Hsieh K, Wu W, Nagarajan MB, Young JR, Douek ML, et al. Deep learning and radiomics: the utility of Google TensorFlow Inception in classifying clear cell renal cell carcinoma and oncocytoma on multiphasic CT. Abdom Radiol (NY) 2019;44(6):2009–20. Available from: https://doi.org/10.1007/s00261-019-01929-0.
[23] Bektas CT, Kocak B, Yardimci AH, Turkcanoglu MH, Yucetas U, Koca SB, et al. Clear cell renal cell carcinoma: machine learning-based quantitative computed tomography texture analysis for prediction of Fuhrman nuclear grade. Eur Radiol 2019;29(3):1153–63. Available from: https://doi.org/10.1007/s00330-018-5698-2.
[24] Kocak B, Durmaz ES, Ates E, Kaya OK, Kilickesmez O, Unenhanced CT. Texture analysis of clear cell renal cell carcinomas: a machine learning-based study for predicting histopathologic nuclear grade. AJR Am J Roentgenol 2019;W1–8. Available from: https://doi.org/10.2214/AJR.18.20742.
[25] Lin F, Cui EM, Lei Y, Luo LP. CT-based machine learning model to predict the Fuhrman nuclear grade of clear cell renal cell carcinoma. Abdom Radiol (NY) 2019;44(7):2528–34. Available from: https://doi.org/10.1007/s00261-019-01992-7.
[26] Shu J, Wen D, Xi Y, Xia Y, Cai Z, Xu W, et al. Clear cell renal cell carcinoma: machine learning-based computed tomography radiomics analysis for the prediction of WHO/ISUP grade. Eur J Radiol 2019;121:108738. Available from: https://doi.org/10.1016/j.ejrad.2019.108738.
[27] Sun X, Liu L, Xu K, Li W, Huo Z, Liu H, et al. Prediction of ISUP grading of clear cell renal cell carcinoma using support vector machine model based on CT images. Med (Baltim) 2019;98(14):e15022. Available from: https://doi.org/10.1097/MD.0000000000015022.
[28] Cui E, Li Z, Ma C, Li Q, Lei Y, Lan Y, et al. Predicting the ISUP grade of clear cell renal cell carcinoma with multiparametric MR and multiphase CT radiomics. Eur Radiol 2020;30(5):2912–21. Available from: https://doi.org/10.1007/s00330-019-06601-1.
[29] Lorencin I, Andelic N, Spanjol J, Car Z. Using multi-layer perceptron with Laplacian edge detector for bladder cancer diagnosis. Artif Intell Med 2020;102:101746. Available from: https://doi.org/10.1016/j.artmed.2019.101746.
[30] Ikeda A, Nosato H, Kochi Y, Kojima T, Kawai K, Sakanashi H, et al. Support system of cystoscopic diagnosis for bladder cancer based on artificial intelligence. J Endourol 2020;34(3):352–8. Available from: https://doi.org/10.1089/end.2019.0509.
[31] Shkolyar E, Jia X, Chang TC, Trivedi D, Mach KE, Meng MQ, et al. Augmented bladder tumor detection using deep learning. Eur Urol 2019;76(6):714–18. Available from: https://doi.org/10.1016/j.eururo.2019.08.032.
[32] Garapati SS, Hadjiiski L, Cha KH, Chan HP, Caoili EM, Cohan RH, et al. Urinary bladder cancer staging in CT urography using machine learning. Med Phys 2017;44(11):5814–23. Available from: https://doi.org/10.1002/mp.12510.
[33] Zheng J, Kong J, Wu S, Li Y, Cai J, Yu H, et al. Development of a noninvasive tool to preoperatively evaluate the muscular invasiveness of bladder cancer using a radiomics approach. Cancer 2019;125(24):4388–98. Available from: https://doi.org/10.1002/cncr.32490.
[34] Wang H, Hu D, Yao H, Chen M, Li S, Chen H, et al. Radiomics analysis of multiparametric MRI for the preoperative evaluation of pathological grade in bladder cancer tumors. Eur Radiol 2019;29(11):6182–90. Available from: https://doi.org/10.1007/s00330-019-06222-8.
[35] Blum ES, Porras AR, Biggs E, Tabrizi PR, Sussman RD, Sprague BM, et al. Early detection of ureteropelvic junction obstruction using signal analysis and machine learning: a dynamic solution to a dynamic problem. J Urol 2018;199(3):847–52. Available from: https://doi.org/10.1016/j.juro.2017.09.147.
[36] Wong NC, Lam C, Patterson L, Shayegan B. Use of machine learning to predict early biochemical recurrence after robot-assisted prostatectomy. BJU Int 2019;123(1):51–7. Available from: https://doi.org/10.1111/bju.14477.
[37] Harder N, Athelogou M, Hessel H, Brieu N, Yigitsoy M, Zimmermann J, et al. Tissue phenomics for prognostic biomarker discovery in low- and intermediate-risk prostate cancer. Sci Rep 2018;8(1):4470. Available from: https://doi.org/10.1038/s41598-018-22564-7.
[38] Zhang YD, Wang J, Wu CJ, Bao ML, Li H, Wang XN, et al. An imaging-based approach predicts clinical outcomes in prostate cancer through a novel support vector machine classification. Oncotarget 2016;7(47):78140–51. Available from: https://doi.org/10.18632/oncotarget.11293.

[39] Shiradkar R, Ghose S, Jambor I, Taimen P, Ettala O, Purysko AS, et al. Radiomic features from pretreatment biparametric MRI predict prostate cancer biochemical recurrence: Preliminary findings. J Magn Reson Imaging 2018;48(6):1626–36. Available from: https://doi.org/10.1002/jmri.26178.

[40] Lalonde E, Ishkanian AS, Sykes J, Fraser M, Ross-Adams H, Erho N, et al. Tumour genomic and microenvironmental heterogeneity for integrated prediction of 5-year biochemical recurrence of prostate cancer: a retrospective cohort study. Lancet Oncol 2014;15(13):1521–32. Available from: https://doi.org/10.1016/S1470-2045(14)71021-6.

[41] Hung AJ, Chen J, Che Z, Nilanon T, Jarc A, Titus M, et al. Utilizing machine learning and automated performance metrics to evaluate robot-assisted radical prostatectomy performance and predict outcomes. J Endourol 2018;32(5):438–44. Available from: https://doi.org/10.1089/end.2018.0035.

[42] Hung AJ, Chen J, Ghodoussipour S, Oh PJ, Liu Z, Nguyen J, et al. A deep-learning model using automated performance metrics and clinical features to predict urinary continence recovery after robot-assisted radical prostatectomy. BJU Int 2019;124(3):487–95. Available from: https://doi.org/10.1111/bju.14735.

[43] Wang G, Lam KM, Deng Z, Choi KS. Prediction of mortality after radical cystectomy for bladder cancer by machine learning techniques. Comput Biol Med 2015;63:124–32. Available from: https://doi.org/10.1016/j.compbiomed.2015.05.015.

[44] Sapre N, Macintyre G, Clarkson M, Naeem H, Cmero M, Kowalczyk A, et al. A urinary microRNA signature can predict the presence of bladder urothelial carcinoma in patients undergoing surveillance. Br J Cancer 2016;114(4):454–62. Available from: https://doi.org/10.1038/bjc.2015.472.

[45] Bartsch G Jr, Mitra AP, Mitra SA, Almal AA, Steven KE, Skinner DG, et al. Use of artificial intelligence and machine learning algorithms with gene expression profiling to predict recurrent nonmuscle invasive urothelial carcinoma of the bladder. J Urol 2016;195(2):493–8. Available from: https://doi.org/10.1016/j.juro.2015.09.090.

[46] Li P, Ren H, Zhang Y, Zhou Z. Fifteen-gene expression based model predicts the survival of clear cell renal cell carcinoma. Med (Baltim) 2018;97(33):e11839. Available from: https://doi.org/10.1097/MD.0000000000011839.

[47] Aminsharifi A, Irani D, Pooyesh S, Parvin H, Dehghani S, Yousofi K, et al. Artificial neural network system to predict the postoperative outcome of percutaneous nephrolithotomy. J Endourol 2017;31(5):461–7. Available from: https://doi.org/10.1089/end.2016.0791.

[48] Mannil M, von Spiczak J, Hermanns T, Poyet C, Alkadhi H, Fankhauser CD. Three-dimensional texture analysis with machine learning provides incremental predictive information for successful shock wave lithotripsy in patients with kidney stones. J Urol 2018;200(4):829–36. Available from: https://doi.org/10.1016/j.juro.2018.04.059.

[49] Girela JL, Gil D, Johnsson M, Gomez-Torres MJ, De Juan J. Semen parameters can be predicted from environmental factors and lifestyle using artificial intelligence methods. Biol Reprod 2013;88(4):99. Available from: https://doi.org/10.1095/biolreprod.112.104653.

[50] Tsai VF, Zhuang B, Pong YH, Hsieh JT, Chang HC. Web- and artificial intelligence-based image recognition for sperm motility analysis: verification study. JMIR Med Inf 2020;8(11):e20031. Available from: https://doi.org/10.2196/20031.

[51] Akinsal EC, Haznedar B, Baydilli N, Kalinli A, Ozturk A, Ekmekcioglu O. Artificial neural network for the prediction of chromosomal abnormalities in azoospermic males. Urol J 2018;15(3):122–5. Available from: https://doi.org/10.22037/uj.v0i0.4029.

CHAPTER

30

Artificial intelligence in neurosurgery—a focus on neuro-oncology

A. Boaro[1] *and O. Arnaout*[2]

[1]Section of Neurosurgery, Department of Neurosciences, Biomedicine and Movement Sciences, University of Verona, Verona, Italy [2]Department of Neurosurgery, Brigham and Women's Hospital, Harvard Medical School, Boston, MA, United States

Introduction

With a survival rate for patients with a malignant central nervous system tumor of around 36% beyond 5 years, neuro-oncology remains one of the most challenging and unresolved fields in medicine [1,2]. The journey of the neurooncological patient with surgical indication goes through different phases which include diagnosis, treatment, follow-up, and possibly secondary prevention. Artificial intelligence (AI) applications have the potential to significantly advance each of these phases, improving patient outcomes, surgical experience, and surgical safety [3–5].

Current examples

In the diagnostic phase, tumor detection and segmentation algorithms have extensively proven to be able to reach human-level performance, saving time, and reducing interexpert variability [6,7]. There is a particular interest in the realm of invasive lesions as low-grade and high-grade gliomas are the one that requires the highest resources expenditure in trying to provide the best and most durable care, while the focus on benign lesions as meningiomas and schwannomas has started to raise only recently [8,9]. While multiple navigation platforms have been providing semiautomated segmentation tools, the current AI-based algorithms demonstrated higher performances and can be implemented directly on the radiology suite, allowing more efficient and early surgical planning.

Artificial Intelligence in Clinical Practice
DOI: https://doi.org/10.1016/B978-0-443-15688-5.00018-8

Very good examples are the works stem from the BraTS challenge, in which the best performing glioma segmentations algorithms are presented each year at international conferences, and which allowed the development of some the most accurate models so far [7,10].

Along with the segmentation task, the ability to predict histological, molecular and genetic variances of any given tumor in the preoperative radiology data or intraoperative frozen sections is the other main challenge which is being tackled [9,11,12]. Again, great interest is evident with regards to gliomas, where the ability to correctly predict these lesion's features has the potential to modify the therapeutic plans, from avoiding surgery altogether up to push the surgeon to be more aggressive than anticipated, to tailor chemo and radiotherapy.

Beyond the opportunities provided by prediction and segmentation algorithms, there is a potential for more hardware-based, AI-embedded tools to improve surgical safety and efficiency in neurosurgical oncology; while robotic solutions are having a meaningful impact in the field of spine surgery and functional neurosurgery, the potentialities of augmented/virtual/mixed reality applications are being progressively translated into clinical tools in the field of neuro-oncology, providing unprecedented opportunities for planning, rehearsal and training as well as for allowing the surgeon to see "through" the lesion and "beyond" the corners for the first time [7,13].

In the follow-up phase, outcome prediction models are being developed and trained to consider multiple layers of patient's data, from demographics to histopathological features, to clinical picture and radiology data, to medication history and intraoperative monitoring data, with the final aim to improve accuracy in survival and quality of life predictions [5,14]. Such algorithms are already turned into user-friendly applications that can be easily used by surgeons in the clinic, directly providing additional prognostic information to the patients and their families.

Finally, the availability of great quantity of data generated by personal digital devices as smartphones and smartwatches are being used to provide the physician and the patient with continuous information about their physical and mental performances, allowing improved follow-up management [15,16].

Current challenges preventing artificial intelligence application

Some obstacles are in the way of allowing seamless implementation of AI applications in clinical practice. While the lack of multidisciplinary teamwork used to be one of the most important barriers to the development of clinically viable AI-based solutions, nowadays it seems that physicians and data scientists started to efficiently speak a common language. As a consequence, there has been an explosion of research papers as well as commercial deployments of more or less viable AI-based tools which more often than not lack a rigorous clinical readiness evaluation, making it more difficult for a health-care provider to feel comfortable in implementing these solutions [17–19].

With specific regards to neuro-oncology, the difficulty in having shared access to big repositories of data, in particular of radiology and pathology information, which would maximize heterogeneity and adequate representation of all possible disease variations, is one of the main obstacles in developing robust AI solutions.

From the conceptual and legal point of view, the fundamental difference existing between a completely explainable prediction model and a deep learning model where our ability of interpreting the reasoning behind the outcome is limited, is one of the most important matters of debate regarding ultimate liability and trustworthiness in using AI-based algorithms [17,18].

Future directions

As the field of AI in healthcare is exponentially growing, numerous are the potential applications in surgical neuro-oncology. The ability to correctly predict molecular or genetic variations from preoperative radiology exams would allow the tailoring of the whole therapeutic plan to each patient; the possibility to utilize AI-based clustering algorithms to discover new molecular features could unveil new therapeutic targets with the potential to improve survival or even cure specific types of neoplasms. In the follow-up phase, the exploitation of data generated by personal digital devices could allow such precise characterization of disease trajectory to be able to prevent the occurrence of complications through early intervention. Finally, one of the most interesting problems so far in neuro-oncology, consists in the ability to differentiate between treatment response and treatment effect in tumor recurrence in radiological data, as of today we have limited capabilities to differentiate between the two, with significant uncertainty in the determination of the best therapeutic and follow-up plan [20,21].

Major takeaway points

- AI-based solutions have the potential to improve all phases of neurooncological care, from diagnosis to follow-up.
- Diagnostic tools in the form of detection and segmentation algorithms represent the first AI-based applications ready to be implemented into clinical practice.
- Trustworthiness of the algorithms and shareability of data represents the cornerstones upon which develop safe and efficient AI-based clinical tools.

References

[1] Diamandis P, Aldape K. World Health Organization 2016 classification of central nervous system tumors. Neurol Clin 2018;36(3):439–47.

[2] Louis DN, Perry A, Wesseling P, Brat DJ, Cree IA, Figarella-Branger D, et al. The 2021 WHO classification of tumors of the central nervous system: a summary. Neuro-Oncol. 2021;23(8):1231–51.

[3] Galldiks N, Zadeh G, Lohmann P. Artificial intelligence, radiomics, and deep learning in neuro-oncology. Neurooncol Adv 2020;2(Supplement_4) iv1–2.

[4] Beig N, Bera K, Tiwari P. Introduction to radiomics and radiogenomics in neuro-oncology: implications and challenges. Neurooncol Adv 2020;2(Supplement_4) iv3–14.

[5] Senders JT, Staples PC, Karhade AV, Zaki MM, Gormley WB, Broekman MLD, et al. Machine learning and neurosurgical outcome prediction: a systematic review. World Neurosurg 2018;109 476–486.e1.
[6] van Kempen EJ, Post M, Mannil M, Witkam RL, ter Laan M, Patel A, et al. Performance of machine learning algorithms for glioma segmentation of brain MRI: a systematic literature review and meta-analysis. Eur Radiol 2021;31(12):9638–53.
[7] Kofler F, Berger C, Waldmannstetter D, Lipkova J, Ezhov I, Tetteh G, et al. BraTS Toolkit: translating BraTS brain tumor segmentation algorithms into clinical and scientific practice. Front Neurosci 2020;14:125.
[8] Boaro A, Kaczmarzyk JR, Kavouridis VK, Harary M, Mammi M., Dawood H, et al. Deep neural networks allow expert-level brain meningioma detection, segmentation and improvement of current clinical practice. Sci Rep 2022. <http://medrxiv.org/lookup/doi/10.1101/2021.05.11.21256429>.
[9] Chen C, Cheng Y, Xu J, Zhang T, Shu X, Huang W, et al. Automatic meningioma segmentation and grading prediction: a hybrid deep-learning method. J Pers Med 2021;11(8):786.
[10] Menze BH, Jakab A, Bauer S, Kalpathy-Cramer J, Farahani K, Kirby J, et al. The multimodal Brain Tumor Image Segmentation Benchmark (BRATS). IEEE Trans Med Imaging 2015;34(10):1993–2024.
[11] Chang K, Bai HX, Zhou H, Su C, Bi WL, Agbodza E, et al. Residual convolutional neural network for the determination of *idh* status in low- and high-grade gliomas from MR imaging. Clin Cancer Res 2018;24(5): 1073–81.
[12] Khalsa SSS, Hollon TC, Adapa A, Urias E, Srinivasan S, Jairath N, et al. Automated histologic diagnosis of CNS tumors with machine learning. CNS Oncol 2020;9(2):CNS56.
[13] Ivan ME, Eichberg DG, Di L, Shah AH, Luther EM, Lu VM, et al. Augmented reality head-mounted display-based incision planning in cranial neurosurgery: a prospective pilot study. Neurosurg Focus 2021;51(2):E3.
[14] Senders JT, Staples P, Mehrtash A, Cote DJ, Taphoorn MJB, Reardon DA, et al. An online calculator for the prediction of survival in glioblastoma patients using classical statistics and machine learning. Neurosurgery. 2020;86(2):E184–92.
[15] Richards R, Kinnersley P, Brain K, McCutchan G, Staffurth J, Wood F. Use of mobile devices to help cancer patients meet their information needs in non-inpatient settings: systematic review. JMIR MHealth UHealth 2018;6(12):e10026.
[16] Panda N, Solsky I, Hawrusik B, Liu G, Reeder H, Lipsitz S, et al. Smartphone global positioning system (GPS) data enhances recovery assessment after breast cancer surgery. Ann Surg Oncol 2021;28(2):985–94.
[17] Mathiesen T, Broekman M. Machine learning and ethics. In: Staartjes VE, Regli L, Serra C, editors. Machine learning in clinical neuroscience, 134. Cham: Springer International Publishing; 2022. p. 251–6. https://link.springer.com/10.1007/978-3-030-85292-4_28.
[18] Maher NA, Senders JT, Hulsbergen AFC, Lamba N, Parker M, Onnela JP, et al. Passive data collection and use in healthcare: a systematic review of ethical issues. Int J Med Inf 2019;129:242–7.
[19] Scott I, Carter S, Coiera E. Clinician checklist for assessing suitability of machine learning applications in healthcare. BMJ Health Care Inf 2021;28(1):e100251.
[20] Bacchi S, Zerner T, Dongas J, Asahina AT, Abou-Hamden A, Otto S, et al. Deep learning in the detection of high-grade glioma recurrence using multiple MRI sequences: a pilot study. J Clin Neurosci 2019;70:11–13.
[21] Peng L, Parekh V, Huang P, Lin DD, Sheikh K, Baker B, et al. Distinguishing true progression from radionecrosis after stereotactic radiation therapy for brain metastases with machine learning and radiomics. Int J Radiat Oncol 2018;102(4):1236–43.

CHAPTER

31

Artificial intelligence in vascular surgery

Uwe M. Fischer

Yale University School of Medicine, New Haven, CT, United States

Introduction

Technical advances in medicine have continuously evolved, mostly in an exponential rather than a linear fashion. However, we rarely have experienced such a rapid technical advance in medicine as that of artificial intelligence (AI) technologies and machine learning applications. Although the AI field is diverse, the basic foundations of AI include: (1) analyzing large amounts of data, (2) recognizing patterns, (3) predicting outcomes, and thus (4) aiding in drawing conclusions to improve workflows. Goal of every new technology should be to assist, improve and facilitate human life. Hence, AI applications in medicine and vascular surgery should be designed and implemented with the aim to assist, but not replace the healthcare provider. The goal is to efficiently utilize information beyond the processing capabilities of humans, thereby decreasing clinician cognitive load, expanding patient-centered treatment options, and improving patient outcomes.

Artificial intelligence applications in vascular surgery

Vascular surgery depends substantially on diagnostic imaging and large amounts of patient data. The ability of AI to analyze those data, detect patterns and draw conclusions surpasses human capacities and has already proven beneficial to patient treatment and outcomes. AI applications in vascular surgery can mainly be categorized into four key areas: vascular diagnostics, perioperative medicine, risk stratification, and outcome prediction.

Artificial Intelligence in Clinical Practice
DOI: https://doi.org/10.1016/B978-0-443-15688-5.00009-7

Artificial intelligence in vascular diagnostics

The prime application of AI in general and especially in vascular surgery is the ability to analyze large amounts of data. AI can accurately predict which patients are at risk and need of intervention using imaging platforms. Even before the need for an intervention, AI offers methods of diagnosing the severity of vascular pathology, such as peripheral artery disease (PAD), based on analysis of noninvasive diagnostics.

A study using deep learning analysis of arterial pulse waveforms was used to test proof-of-concept and potential challenges for this method. A deep convolutional neural network capable of detecting and assessing the severity of PAD based on analysis of brachial and ankle arterial pulse waveforms was constructed, evaluated for efficacy, and compared with the state-of-the art ankle-brachial index (ABI) using many virtual patients that were created to investigate the potential and challenges in DL-based pulse wave analysis for PAD diagnosis. This study demonstrated robust PAD detection performance superior to the ABI technique against a wide range of PAD severity threshold levels for labeling of healthy subjects and PAD patients [1]. This study showed that AI, in a theoretical setting with virtual patients, can diagnose patients earlier compared to traditional methods with higher sensitivity, specificity, and accuracy [1]. Consequently, with early PAD diagnosis treatment can be initiated even before symptoms appear both with medical treatment and lifestyle modifications.

AI also is a useful tool in the interpretation and analysis of abdominal aortic aneurysm (AAA) imaging. A comprehensive review of articles using AI in the diagnosis and treatment of patients with AAA showed that AI could be used to help surgeons in preoperative planning [2]. In addition, the potential of AI-driven data management was predicted to be crucial in the development of AAA evolution and risk of rupture evaluation as well as postoperative outcomes. AI can also aid in decision-making on types of surgical treatment. Especially, AI would allow investigators to detect the aneurysm more easily; to characterize its anatomic characteristics (including the presence of calcifications and intraluminal thrombus); and to automatically calculate the diameters, lengths, distances, and volumes of the aneurysm and vessels. This automated AAA assessment would be helpful in perioperative planning, such as sizing of the endograft. The authors also concluded that data derived from automatic analysis of AAA images could be combined with clinical and biologic characteristics of patients to develop multiple-variable scores, allowing identification of predictive patterns and better assessment of the prognosis of patients [2]. Thus *AI could aid in creation of an individualized surveillance program*. Lareyre et al. developed an automated software system to enable a fast and robust detection of the vascular system and detect AAA in computer tomography angiographies. The software automatically detects the aortic lumen and AAA characteristics including the presence of thrombus and calcifications [3]. These data can be used to improved surgical intervention planning including types of interventions. Recommendations for the best instrumentation and graft type are on the horizon.

Perioperative medicine, risk stratification, and outcome prediction

Although clinical applications are in their infancy the ability to analyze large amounts of patient data in an unbiased fashion enables AI to determine patient risk for the perioperative period providing tools to individualize and thus optimize patient care. Perioperative intelligence uses technologies, such as ML [4], AI [5], and big data [6], to provide appropriate and safe perioperative care [7]. Maheshwari et al. point out the need for high quality continuous data from multiple domains to make better predictions [7]. Hence, data acquisition and management are crucial for successfully applying the available AI tools which in turn can deliver vital information to assess patient disease and outcome risk.

Focusing on PAD risk stratification, AI can be used to efficiently and accurately mine data from electronic health records (EHRs) to create clinical databases from which models can be developed to predict PAD risk and the associated mortality risk. In this study, diverse clinical, demographic, imaging and genomic information of patients undergoing coronary angiography were utilized. The applied ML models outperformed the standard logistic regression models for the identification of patients with PAD and predicting future mortality [8]. Specifically, for limb-related outcomes Davis at al examined the ability of their model of algorithms to predict surgical site infections after lower extremity revascularization. Their model predicted with an AUC of 0.66 postbypass surgical site infections and could identify several patient and procedure risk factors [9].

Conclusions

AI and associated technologies have started their journey into daily diagnostics, treatment, and patient outcome strategies. Vascular medicine has enormous potential for using these technologies. It is possible to retrieve and assemble a disease picture that was previously incomplete and partially hidden in a large amount of patient data dispersed over different electronic systems. This data mining process will enable improved diagnostics enabling a tailored, personalized treatment algorithm that can predict and more importantly improve disease outcomes faster and more accurately. Despite the exponential increase in studies, publications and even review articles on the subject, the utilization of AI in healthcare has just begun. Keeping the focus on vascular medicine, vascular specialists are using the potential of AI less than neurologists, oncologists, and cardiologists. While a critical approach is certainly warranted, applications of AI in vascular surgery should be developed and implemented in patient care if rigorous testing in real-world settings proves to be beneficial.

Properly tested and carefully implemented AI has the potential to facilitate vascular medicine and add exponential value to our diagnostic and treatment arsenal. In addition, with the ability of effective data mining, imaging analysis across medical fields, and development of intelligent predictive models, AI can aid in improving team-based care with an improve patient-centered focus. Embracing this potential will be the first step in developing a leadership role in AI implementation and using this technology in our daily routine and quest to improve patient care.

Advancing AI in clinical vascular surgery practice will be an essential task in establishing leadership in vascular medicine. This will also include addressing criticism, caution, hesitancy and even bias openly. Limitations of AI in vascular medicine must be clearly recognized and the highest standard in patient safety must be met. Adhering to these principles will help establishing a leading role in healthcare AI.

References

[1] Kim S, Hahn JO, Youn BD. Detection and severity assessment of peripheral occlusive artery disease via deep learning analysis of arterial pulse waveforms: proof-of-concept and potential challenges. Front Bioeng Biotechnol 2020;8:720.

[2] Raffort J, Adam C, Carrier M, Ballaith A, Coscas R, Jean-Baptiste E, et al. Artificial intelligence in abdominal aortic aneurysm. J Vasc Surg 2020;72(1):321–33 e1.

[3] Lareyre F, Adam C, Carrier M, Dommerc C, Mialhe C, Raffort J. A fully automated pipeline for mining abdominal aortic aneurysm using image segmentation. Sci Rep 2019;9(1):13750.

[4] Obermeyer Z, Emanuel EJ. Predicting the future—big data, machine learning, and clinical medicine. N Engl J Med 2016;375(13):1216–19.

[5] Gambus P, Shafer SL. Artificial intelligence for everyone. Anesthesiology 2018;128(3):431–3.

[6] Sessler DI. Big Data—and its contributions to peri-operative medicine. Anaesthesia. 2014;69(2):100–5.

[7] Maheshwari K, Ruetzler K, Saugel B. Perioperative intelligence: applications of artificial intelligence in perioperative medicine. J Clin Monit Comput 2020;34(4):625–8.

[8] Ross EG, Shah NH, Dalman RL, Nead KT, Cooke JP, Leeper NJ. The use of machine learning for the identification of peripheral artery disease and future mortality risk. J Vasc Surg 2016;64(5):1515–22 e3.

[9] Davis FM, Sutzko DC, Grey SF, Mansour MA, Jain KM, Nypaver TJ, et al. Predictors of surgical site infection after open lower extremity revascularization. J Vasc Surg 2017;65(6):1769–78 e3.

CHAPTER

32

Artificial intelligence in neonatal and pediatric intensive care units

Avishek Choudhury[1] *and Estefania Urena*[2]

[1]Industrial and Management Systems Engineering, West Virginia University, Morgantown, WV, United States [2]Registered Nurse, Intensive Critical Unit, Lincoln Medical and Mental Health Centre, Bronx, NY, United States

Artificial intelligence in pediatrics

With increasing healthcare infrastructure and connected medical databases, clinicians have more data to inform clinical decision-making than ever before. However, when confronted with information that is (1) beyond the scope of the expertise of a clinician and/or (2) in such quantities that it becomes challenging for clinicians to comprehend on time, they are likely to resort to boundedly rational and in some cases incorrect diagnoses. One way to support complex clinical processes is to leverage artificial intelligence (AI) technologies, often known as AI-based clinical decision support systems. AI, as portrayed by the media, comes with surprising capabilities in healthcare. AI can be broadly defined as an intelligent system capable of performing human-like activities based on retrospective data. A typical AI system encompasses predefined rules, if-then statements, or is powered by dynamic statistical models that are proficient in capturing nonlinear relationships among several variables. More recently, wide arrays of unique AI technologies have been developed to augment the healthcare system and the US Food and Drug Administration approved several AI-based products, signaling the gradual integration of AI into healthcare [1,2].

Pediatric patients are typically at an increased risk of fatal decompensation and are sensitive to medications. That being said, any delay in treatment or minor errors in medication dosage can overcomplicate patient health. Under such an environment, clinicians are expected to quickly and effectively comprehend large volumes of medical information to diagnose and develop a treatment plan for a given baby. Being one of the most complex and sensitive healthcare domains, neonatal and pediatric intensive care units (NICUs and PICUs) are ideal environments for AI use where doctors and nurses can leverage AIs' computational capabilities to

Artificial Intelligence in Clinical Practice
DOI: https://doi.org/10.1016/B978-0-443-15688-5.00013-9

make well-informed and faster clinical decisions. The use of AI in pediatrics was first recorded in 1986 when Paycha developed SHELP a computer-assisted medical decision-making system, that diagnosed inborn errors of metabolism [3]. Soon after, Shortliffe developed an expert system named Mycin and identified bacteria causing severe blood infections among pediatric patients [4]. Since then, as AI developed, several randomized controlled trials have used the technology for various issues in pediatrics. For instance, a study implemented an automated AI-based decision support system to control glucose levels effectively and safely among pediatric patients [5]. Another study developed an AI-based wearable device known as Superpower Glass, to augment social outcomes of children with autism [6]. A study conducted in China successfully developed an AI-based disease risk prediction model for newborn babes with inherited metabolic diseases [7]. A study reported significant improvement in neurocognitive performance among children when an AI-based cognitive stimulation therapy was implemented [8]. Apart from clinical trials, several other AI technologies have been developed that play an active role in neonatal and pediatric ICUs. For example, AI-based models have been used in the NICU to predict birth asphyxia [9,10], and neonatal seizures [11] as well as diagnose neonatal sepsis [12,13] and respiratory distress syndrome [14]. Table 32.1 gives a snapshot of various applications of AI in NICU and PICU [15].

Overall, different studies have used AI either to directly improve patient health by allowing physicians "spend more time in direct patient care [while reducing provider burnout]" [24] or augment clinical processes to indirectly improve patient health. For instance, a study conducted in California reported AI's efficiency in identifying critically ill PICU patients with the underlying genetic disorder [25]. A study in Spain used AI-driven music to reduce stress levels among neonates [26]. Several studies used AI algorithms to develop an early warning system that provided timely detection of changes in health status and development of critical illness [12,16,18,20,27–29] and pathologic eye disease progression in preterm infants [30]. A recent review also reported several "indirect impacts" of AI on the pediatric patient [31] where AI was noted to augment clinical decision-making and diagnostic accuracy in the pediatric setting [22,29,32].

Current challenges preventing artificial intelligence application

Despite all the evidence supporting AI in pediatrics, its use and adoption have been limited. Even though no studies thus far have associated AI with worsened health outcomes or patient harm in a pediatric, why is it that doctors and healthcare management hesitate to integrate AI into their clinical workload? Of all possible reasons hindering the acceptance of AI in pediatrics, (1) lack of ecological validity and (2) low technology readiness level (TRL), two interrelated factors, seem to be prominent determinants that have not been sufficiently acknowledged in the literature.

Ecological validity

AI systems and technologies, as depicted across several studies, may facilitate a personalized approach to pediatric care through the augmentation of diagnostic processes.

TABLE 32.1 State of the art: artificial intelligence in PICU and NICU.

Author	Institution(s)	# Patients	Data source and type	Classification type	Model	Cross validation	Compared with clinicians	Accuracy	AUROC	Conclusion
He et al. [16]	Autism Brain Imaging Data Exchange Database	28	Research database: images	Binary	Support vector machine	10-fold	No	0.70	0.76	The study accurately predicted cognitive deficits/function in individual very preterm infants soon after birth. However, larger data size is required to achieve clinical gold standard.
Podda et al. [17]	Italian Neonatal Network	23,747	Research database: numerical	Binary	Artificial neural network	Fivefold	No	Not reported	0.91	The study shows that using only limited information available up to 5 min after birth, AI can have a significant advantage over current approaches in predicting survival of preterm infants.
Lamping et al. [18]	German Tertiary Care PICU	296	EHR: numerical	Binary	Random forest	threefold	No	Not reported	0.78	The study shows that AI can facilitate early detection of sepsis with an accuracy superior compared to traditional biomarkers. It can also potentially reduce antibiotic use by 30% in noninfectious cases.
Kayhanian et al. [19]	Cambridge University	94	EHR: numerical	Binary	Support vector machine	Fivefold	No	Not reported	Not reported	The study shows how AI algorithms can predict severe traumatic injury's outcomes at 6 months using just the three most informative parameters.

(*Continued*)

TABLE 32.1 (Continued)

Author	Institution(s)	# Patients	Data source and type	Classification type	Model	Cross validation	Compared with clinicians	Accuracy	AUROC	Conclusion
Kim et al. [20]	Severance Hospital and Samsung Medical Center	1,723	EHR: numerical	Binary	Convolutional neural network	Fivefold	No	0.84–0.77	0.89–0.97	The study demonstrated that machine learning-based model, the Pediatric Risk of Mortality Prediction Tool can outperform the conventional Pediatric Index of Mortality scoring system, in predictive ability.
Ruiz et al. [21]	University Hospital EHR	93	EHR: numerical	Multiclass	Naïve Bayesian models	Fivefold	Yes	Not reported	0.88	The study demonstrates the capability of AI models in augmenting clinicians' ability to identify infants with single-ventricle physiology at high risk of critical events. The study also reports that early prediction of critical events may improve the overall care quality and minimize health care expenses.
Fraiwan and Alkhodari [22]	University of Pittsburgh	37	Research database: EEG signals	Multiclass	Long short-term memory	10-fold	No	0.96	Not reported	The algorithm proposed in the study gave promising results in automatic sleep stage scoring in neonatal sleep signals.
Feng et al. [23]	St. Louis Children's Hospital	285	EHR: numerical	Binary	Novel deep learning model	External validation	No	0.88	0.89	The novel AI model developed in the study demonstrated efficacy in predicting the real-time mortality risk of preterm infants in initial NICU hospitalization. The proposed model also outperformed existing clinical risk index II scoring system for babies

The AI-based solution has the power to reinvigorate clinical practices. Although the advent of personalized patient treatment is provocative, and often crucial in a pediatric environment, there is a need to assess the true potential of AI when implemented in a real, uncontrolled, and chaotic healthcare scenario. In all the studies published around this topic, the experiments were either conducted retrospectively or by experts in a controlled setting, therefore lacking ecological validity. Recent systematic reviews [33,34] analyzing AIs' role and performance in healthcare acknowledged that AI systems or models were often evaluated under unrealistic conditions that had minimal relevance to routine clinical practice. The environment under which AI studies have been conducted does not represent an actual clinical environment. Therefore there is a lack of evidence exhibiting AIs' efficacy in a real clinical environment.

It is essential to understand that the working environment and cognitive workload are significant determinants of technology use. In a pediatric setting, clinicians are often assigned several patients, each with unique needs and health status. Given the global shortage of staff and the increasing burden on the healthcare industry, clinicians often experience burnout and fatigue. Individuals under such stress and discomfort might not be efficient in utilizing AI devices and comprehending its outcome in the same way as reported in several research articles.

Therefore studies must evaluate AI systems under a real scenario to ensure their effective use when integrated into a clinical workflow.

Technology readiness level

Recently, several innovations around medical AI have been associated with great performance in the literature. However, research breakthroughs do not necessarily translate into a technology that is ready to use in a high-risk environment, such as healthcare [34,35]. That said, most AI featuring prominent ability in research and literature, for the most part, would not be executable in a clinical environment. According to the TRL, most AI systems, at least in pediatric and neonatal intensive critical care (PICU and NICU) if not all, do not qualify for implementation. TRL is a gauging system developed to assess the maturity level of a particular technology [36]. TRL consists of nine categories (readiness levels), where a score of TRL 1 is the lowest and TRL 9 is the highest (see Box 32.1) [15]. Applying the TRL system to the articles involving AI in pediatrics, we can observe that most published articles are prototype testing in an operational environment with near-implementation readiness. Very few to none of the AI systems discussed in the literature have been deployed into a real ICU setting and evaluated longitudinally over a significant duration.

Recommendations

Concerns regarding ecological validity and TRL can be associated with AIs' usability. There is a lack of study evaluating the usability or user-centeredness of any AI technology

BOX 32.1 Technology readiness levels (1–9)

Technologies with TRL 1–4 are executable in laboratory setting, where the main object is to conduct research. This stage is the proof of concept.

- TRL 1: Basic principles of the technology observed.
- TRL 2: Technology concept formulated.
- TRL 3: Experimental proof of concept developed.
- TRL 4: Technology validated in a study laboratory.

Technologies with TRL 5–7 are in the development phase, where the functional prototype is ready.

- TRL 5: Technology validated in relevant environment (controlled setting in a real-life environment).
- TRL 6: Technology demonstrate in relevant environment.
- TRL 7: System prototype demonstrated in operational environment.

Lastly, technologies with TRL 8 and 9 are in the operational phase where the primary objective is implementation.

- TRL 8: System completed and certified for commercial use.
- TRL 9: System approved for and implemented in the actual operational environment.

in a pediatric setting. As acknowledged earlier in this chapter, clinicians are often overwhelmed with clinical responsibilities. Therefore to ensure the adoption of AI in pediatrics, it is essential to develop systems that are easy to use and fulfills pediatric nurses' and doctors' requirement. AI developers also need to consider the end-user of their products. Since most of the bedside tasks are performed by nurses, the AI system implemented at the bedside should be designed for nurses, as their digital literacy can be substantially different than other physicians or researchers (study participants) and may also vary across demographics.

Future work should also involve policymakers and address the concerns regarding accountability. How does the absence of AIs' accountability impact clinicians' intention to use the technology? This chapter explains "accountability" as a process in which healthcare practitioners have potential responsibilities to justify their "clinical actions" to patients (or families) and are held liable for any impending positive or negative impact on patient health. While using an AI-based decision support system, only clinicians are held accountable if they decide to follow AI, resulting in patient harm. Additionally, clinicians are also held responsible if they deviate from the standard protocols [37]. This may be worrisome because, under such circumstances, clinicians will only follow AI if it matches with their judgment and aligns with the standard protocol—making the AI underused. Furthermore, it might be difficult for clinicians, who are not necessarily trained in the subject, to effectively comprehend AIs' functioning under an existing burnout state and identify any technological flaw.

Lastly, future studies should include pediatric populations with multiple chronic complexities in randomized controlled trials. Current approaches to pediatric AI usually emphasize single diseases, which may have minimal relevance to a real complex scenario. Another consideration is to have an adaptive algorithm that can gauge patients' health status and evolve over time. Therefore future research efforts to integrate AI systems into pediatric settings need to match the measure and underlying disease trajectory to patients' situations.

Future steps

Until now all studies have been focused on the patient. What is missing in the literature is the use of AI to address clinicians' concerns. Addressing clinicians' problems can not only improve their clinical performance but also augment care quality.

The pediatric unit (PICU and NICU) is one of the most critical departments within any healthcare establishment. For example, while dealing with a pediatric patient, particularly in a NICU or PICU setting, the clinicians need to consider the body size differences that exist between every pediatric patient and consecutively be aware of all the continuous physical and cognitive development of their patients. That being said, the medication dosage (which largely depends on the body weight) might change over time for a pediatric patient (depending on their rate of physical growth). Additionally, clinicians need to have special consideration while intubating pediatric patients as they have larger tongues, along with uniquely positioned epiglottis and larynx. Pediatric patients also have subtle cardiovascular differences which make heart rate a critical clinical factor. They are also prone to pathogens and neurological disorders from poisoning. In other words, pediatric patients have a very low tolerance to any error and therefore clinicians are required to pay extra care and personalized treatment.

Apart from caring for patients, pediatric clinicians also have to dedicate a significant amount of their time and effort to educating patients' parents. Such work demand often takes a heavy toll on their cognitive workload and AI technologies can be developed to identify clinicians undergoing excessive cognitive load or burnout. Since clinicians in burnout state are prone to human errors, identifying and providing them with timely assistance can help ensure patient safety. Identifying cognitive workload will also help the floor manager to better schedule their staff and designate appropriate resources.

Night nurses, particularly those who are new in the profession, may feel exhausted during their shifts and in a setting where nurses have to keep a continuous watch on patient monitors (a critical aspect in NICU and PICU setting), performing efficiently often gets challenging. In such a scenario, AI in conjunction with eye trackers can be leveraged to measure nurses' attention span and identify the zone in the screen where they gaze. AI can then optimize the information being displayed on the clinical monitors to highlight the important data in real-time.

AI technology can be used to identify and record clinicians' behavior leading to near misses so that it can generate an alert in the future. It is important to acknowledge that in healthcare, outcomes are reasonable because clinicians make educated and just-in-time adjustments according to the fluctuating health condition. Future work should train AI on the critical adjustments made by clinicians, so that AI can adapt in real-time in the same manner as experienced clinicians do.

Major takeaway points

- Artificial Intelligence has great potential, but Human Factors Consideration is essential for its sustainability in pediatrics.
- The lack of AIs' ecological validity is hindering its adoption and usage in the clinical workflow.
- Artificial Intelligence if used appropriately, can improve clinical workflow and in turn augment the quality of care.
- All AI-based decision support systems should be exclusively designed for its end-users (doctors and nurses) to safeguard the technology as well as patient safety.

References

[1] Price 2nd WN, Gerke S, Cohen IG. Potential liability for physicians using artificial intelligence. JAMA 2019. Available from: https://doi.org/10.1001/jama.2019.15064; https://www.ncbi.nlm.nih.gov/pubmed/31584609.

[2] FDA. Patient engagement advisory committee meeting announcement. <https://www.fda.gov/advisory-committees/advisory-committee-calendar/october-22-2020-patient-engagement-advisory-committee-meeting-announcement-10222020-10222020>; 2020.

[3] Paycha F. Diagnosis with the aid of artificial intelligence: demonstration of the 1st diagnostic machine. Presse Therm Clim 1968;105(1):22–5.

[4] Shortliffe EH, Davis R, Axline SG, Buchanan BG, Green CC, Cohen SN. Computer-based consultations in clinical therapeutics: explanation and rule acquisition capabilities of the mycin system. Comput Biomed Res 1975;8(4):303–20.

[5] Nimri R, Battelino T, Laffel LM, Slover RH, Schatz D, Weinzimer SA, et al. Insulin dose optimization using an automated artificial intelligence-based decision support system in youths with type 1 diabetes. Nat Med 2020;26(9):1380–4. Available from: https://doi.org/10.1038/s41591-020-1045-7.

[6] Voss C, Schwartz J, Daniels J, Kline A, Haber N, Washington P, et al. Effect of wearable digital intervention for improving socialization in children with autism spectrum disorder: a randomized clinical trial. JAMA Pediatr 2019;173(5):446–54. Available from: https://doi.org/10.1001/jamapediatrics.2019.0285.

[7] Yang RL, Yang YL, Wang T, Xu WZ, Yu G, Yang JB, et al. Establishment of an auxiliary diagnosis system of newborn screening for inherited metabolic diseases based on artificial intelligence technology and a clinical trial. Zhonghua Er Ke Za Zhi 2021;59(4):286–93. Available from: https://doi.org/10.3760/cma.j.cn112140-20201209-01089.

[8] Medina R, Bouhaben J, de Ramón I, Cuesta P, Antón-Toro L, Pacios J, et al. Electrophysiological brain changes associated with cognitive improvement in a pediatric attention deficit hyperactivity disorder digital artificial intelligence-driven intervention: Randomized controlled trial. J Med Internet Res 2021;23(11):e25466. Available from: https://doi.org/10.2196/25466.

[9] Ubenwa: Cry-based diagnosis of birth asphyxia. In: Onu CC UI, Ndiomu E, Kengni U, Precup D, Sant'anna GM, Alikor EAD, et al., editors. Machine learning for development workshop, 31st conference on neural information processing systems. 2017.

[10] Neural transfer learning for cry-based diagnosis of perinatal asphyxia. In: Onu CC LJ, Hamilton WL, Precup D, editors. 20th annual conference of the International Speech Communication Association. INTERSPEECH; 2019.

[11] Si Y. Machine learning applications for electroencephalograph signals in epilepsy: a quick review. Acta Epileptol 2020;2(1):5. Available from: https://doi.org/10.1186/s42494-020-00014-0; https://doi.org/10.1186/s42494-020-00014-0.

[12] Mani S, Ozdas A, Aliferis C, Varol HA, Chen Q, Carnevale R, et al. Medical decision support using machine learning for early detection of late-onset neonatal sepsis. J Am Med Inf Assoc 2014;21(2):326–36. Available from: https://doi.org/10.1136/amiajnl-2013-001854; https://www.ncbi.nlm.nih.gov/pubmed/24043317.

[13] Masino AJ, Harris MC, Forsyth D, Ostapenko S, Srinivasan L, Bonafide CP, et al. Machine learning models for early sepsis recognition in the neonatal intensive care unit using readily available electronic health record

data. PLoS One 2019;14(2):e0212665. Available from: https://doi.org/10.1371/journal.pone.0212665; https://www.ncbi.nlm.nih.gov/pubmed/30794638.

[14] Verder H, Heiring C, Clark H, Sweet D, Jessen TE, Ebbesen F, et al. Rapid test for lung maturity, based on spectroscopy of gastric aspirate, predicted respiratory distress syndrome with high sensitivity. Acta Paediatr 2017;106(3):430–7. Available from: https://doi.org/10.1111/apa.13683; https://www.ncbi.nlm.nih.gov/pubmed/27886403.

[15] Choudhury A, Urena E. Artificial intelligence in NICU and PICU: a need for ecological validity, accountability, and human factors. Healthcare 2022;10(5):952–9. Available from: https://doi.org/10.3390/healthcare10050952.

[16] He L, Li H, Holland SK, Yuan W, Altaye M, Parikh NA. Early prediction of cognitive deficits in very preterm infants using functional connectome data in an artificial neural network framework. NeuroImage Clin 2018;18:290–7. Available from: https://doi.org/10.1016/j.nicl.2018.01.032; http://www.sciencedirect.com/science/article/pii/S2213158218300329.

[17] Podda M, Bacciu D, Micheli A, Bellu R, Placidi G, Gagliardi L. A machine learning approach to estimating preterm infants survival: development of the preterm infants survival assessment (pisa) predictor. Sci Rep 2018;8(1):13743. Available from: https://doi.org/10.1038/s41598-018-31920-6; https://www.ncbi.nlm.nih.gov/pubmed/30213963.

[18] Lamping F, Jack T, Rübsamen N, Sasse M, Beerbaum P, Mikolajczyk RT, et al. Development and validation of a diagnostic model for early differentiation of sepsis and non-infectious sirs in critically ill children—a data-driven approach using machine-learning algorithms. BMC Pediatr 2018;18(1):112. Available from: https://doi.org/10.1186/s12887-018-1082-2; https://www.ncbi.nlm.nih.gov/pubmed/29544449.

[19] Kayhanian S, Young AMH, Mangla C, Jalloh I, Fernandes HM, Garnett MR, et al. Modelling outcomes after paediatric brain injury with admission laboratory values: a machine-learning approach. Pediatr Res 2019;86(5):641–5. Available from: https://doi.org/10.1038/s41390-019-0510-9; https://www.ncbi.nlm.nih.gov/pubmed/31349360.

[20] Kim SY, Kim S, Cho J, Kim YS, Sol IS, Sung Y, et al. A deep learning model for real-time mortality prediction in critically ill children. Crit Care 2019;23(1):279. Available from: https://doi.org/10.1186/s13054-019-2561-z; https://doi.org/10.1186/s13054-019-2561-z.

[21] Ruiz VM, Saenz L, Lopez-Magallon A, Shields A, Ogoe HA, Suresh S, et al. Early prediction of critical events for infants with single-ventricle physiology in critical care using routinely collected data. J Thorac Cardiovasc Surg 2019;158(1):234–243.e3. Available from: https://doi.org/10.1016/j.jtcvs.2019.01.130; https://www.ncbi.nlm.nih.gov/pubmed/30948317.

[22] Fraiwan L, Alkhodari M. Neonatal sleep stage identification using long short-term memory learning system. Med Biol Eng Comput 2020;58(6):1383–91. Available from: https://doi.org/10.1007/s11517-020-02169-x. Retrieved from https://www.ncbi.nlm.nih.gov/pubmed/32281071.

[23] Feng J, Lee J, Vesoulis ZA, Li F. Predicting mortality risk for preterm infants using deep learning models with time-series vital sign data. npj Digital Med 2021;4(1):108. Available from: https://doi.org/10.1038/s41746-021-00479-4; https://doi.org/10.1038/s41746-021-00479-4.

[24] Spatharou A, Hieronimus S, Jenkins J. Transforming healthc AI: impact workforce organ 2020.

[25] Clark MM, Hildreth A, Batalov S, Ding Y, Chowdhury S, Watkins K, et al. Diagnosis of genetic diseases in seriously ill children by rapid whole-genome sequencing and automated phenotyping and interpretation. Sci Transl Med 2019;11(489):eaat6177. Available from: https://doi.org/10.1126/scitranslmed.aat6177; http://stm.sciencemag.org/content/11/489/eaat6177.abstract.

[26] Caparros-Gonzalez RA, de la Torre-Luque A, Diaz-Piedra C, Vico FJ, Buela-Casal G. Listening to relaxing music improves physiological responses in premature infants: a randomized controlled trial. Adv Neonatal Care 2018;18(1):58–69. Available from: https://doi.org/10.1097/ANC.0000000000000448; https://www.ncbi.nlm.nih.gov/pubmed/29045255.

[27] Ornek AH, Ceylan M, Ervural S. Health status detection of neonates using infrared thermography and deep convolutional neural networks. Infrared Phys Technol 2019;103:103044. Available from: https://doi.org/10.1016/j.infrared.2019.103044; http://www.sciencedirect.com/science/article/pii/S1350449519303123.

[28] Matam BR, Duncan H, Lowe D. Machine learning based framework to predict cardiac arrests in a paediatric intensive care unit: prediction of cardiac arrests. J Clin Monit Comput 2019;33(4):713–24. Available from: https://doi.org/10.1007/s10877-018-0198-0; https://www.ncbi.nlm.nih.gov/pubmed/30264218.

[29] Irles C, González-Pérez G, Carrera Muiños S, Michel Macias C, Sánchez Gómez C, Martínez-Zepeda A, et al. Estimation of neonatal intestinal perforation associated with necrotizing enterocolitis by machine learning reveals new key factors. Int J Env Res Public Health 2018;15(11). Available from: https://doi.org/10.3390/ijerph15112509; https://www.ncbi.nlm.nih.gov/pubmed/30423965.

[30] Campbell JP, Ataer-Cansizoglu E, Bolon-Canedo V, Bozkurt A, Erdogmus D, Kalpathy-Cramer J, et al. Expert diagnosis of plus disease in retinopathy of prematurity from computer-based image analysis. JAMA Ophthalmol 2016;134(6):651–7. Available from: https://doi.org/10.1001/jamaophthalmol.2016.0611; https://www.ncbi.nlm.nih.gov/pubmed/27077667.

[31] Adegboro CO, Choudhury A, Asan O, Kelly MM. Artificial intelligence to improve health outcomes in the nicu and picu: a systematic review. Hospital Pediatrics 2021;12(1):93–110. Available from: https://doi.org/10.1542/hpeds.2021-006094; https://doi.org/10.1542/hpeds.2021-006094.

[32] Clark MM, Hildreth A, Batalov S, Ding Y, Chowdhury S, Watkins K, et al. Diagnosis of genetic diseases in seriously ill children by rapid whole-genome sequencing and automated phenotyping and interpretation. Sci Transl Med 2019;11(489). Available from: https://doi.org/10.1126/scitranslmed.aat6177; https://www.ncbi.nlm.nih.gov/pubmed/31019026.

[33] Liu X, Faes L, Kale AU, Wagner SK, Fu DJ, Bruynseels A, et al. A comparison of deep learning performance against health-care professionals in detecting diseases from medical imaging: a systematic review and meta-analysis. Lancet Digital Health 2019;1(6):e271–97. Available from: https://doi.org/10.1016/S2589-7500(19)30123-2.

[34] Choudhury A, Asan O. Role of artificial intelligence in patient safety outcomes: systematic literature review. JMIR Med Inf 2020;8(7):e18599. Available from: https://doi.org/10.2196/18599; https://www.ncbi.nlm.nih.gov/pubmed/32706688.

[35] Choudhury A, Renjilian E, Asan O. Use of machine learning in geriatric clinical care for chronic diseases: a systematic literature review. JAMIA Open 2020;3(3):459–71. Available from: https://doi.org/10.1093/jamiaopen/ooaa034; https://doi.org/10.1093/jamiaopen/ooaa034.

[36] Straub J. In search of technology readiness level (trl) 10. Aerosp Sci Technol 2015;46:312–20. Available from: https://doi.org/10.1016/j.ast.2015.07.007; https://www.sciencedirect.com/science/article/pii/S127096381500214X.

[37] Price II WN, Gerke S, Cohen IG. Potential liability for physicians using artificial intelligence. JAMA. 2019;322(18):1765–6. Available from: https://doi.org/10.1001/jama.2019.15064; https://doi.org/10.1001/jama.2019.15064.

CHAPTER

33

Artificial intelligence in pediatrics

Lindsey A. Knake[1], Colin M. Rogerson[2], Meredith C. Winter[3,4] and Swaminathan Kandaswamy[5]

[1]Department of Pediatrics, Division of Neonatology, University of Iowa, IA, United States [2]Department of Pediatrics, Division of Critical Care, Indiana University, IN, United States [3]Department of Anesthesiology, Critical Care Medicine, Children's Hospital Los Angeles, CA, United States [4]Department of Pediatrics, Keck School of Medicine, University of Southern California, CA, United States [5]Department of Pediatrics, Emory University School of Medicine, Atlanta, GA, United States

Introduction

The specialty of Pediatrics is a rich area for implementing artificial intelligence (AI) because young patients and tech savvy guardians are already attuned to using technology in their daily lives. Wearable technology and internet access to health data are becoming an expectation of care for much of the older pediatric patient population [1]. However, as with other specialties within medicine, implementing AI or machine learning (ML) algorithms that clinicians are willing to trust and able to incorporate into their clinical decision making is a challenging endeavor.

A recent 2021 systematic review of ML models in pediatrics found 363 articles published in this area with the amount of research increasing exponentially over the past few years [2,3]. Neural networks and ensemble methods were the most commonly used ML algorithms with deep learning techniques most often used for radiology and cardiology models because of the large amount of data required for these algorithms [2]. The majority of these models focused on diagnostic modeling ($n = 232$, 78%) with a minority of models focusing on prediction and clinical management [2].

The clinical problems that are being approached with ML and AI in pediatric medicine are wide-ranging. Models have been trained to predict length of stay [4], hospital readmission [5], acceptable discharge physiology [6], and mortality [7]. Others have acted as clinical decision support tools to help providers anticipate acute clinical deterioration events, such as sepsis [8,9],

Artificial Intelligence in Clinical Practice
DOI: https://doi.org/10.1016/B978-0-443-15688-5.00017-6

transfer to the ICU [10], and in-hospital cardiac arrest [11]. ML has been used for prognostication of neurologic outcomes after traumatic brain injury [12,13] or cardiac arrest [14,15] and for identification of appropriate candidates to attempt organ donation after circulatory death [16].

Unsupervised ML has also been applied to several areas of pediatrics. The overall goal of most unsupervised ML studies has been the identification of novel disease phenotypes that can be recognized by AI, but not by pediatric clinicians. Pediatric clinicians have recognized that individual asthma patients respond differently to preventative medications, but this has been discovered through trial and error more than with precision medicine approaches. Researchers are now attempting to use ML clustering methods to identify unique phenotypes in pediatric asthma patients that may respond differently to therapeutic management [17,18]. A similar approach has been used in pediatric sepsis and multiorgan dysfunction syndrome in the pediatric intensive care unit. Researchers have attempted to use unsupervised ML to identify specific subgroups of sepsis that may respond differently to treatment with corticosteroids [19]. This is an area that holds promise for future research but has not yet had a strong impact on direct patient care.

The top pediatric specialty areas conducting ML research in a recent review article included neonatal medicine, psychiatry, neurology, and pulmonology [2]. Neonatology, specifically, has had ample opportunity for ML and AI research because of the large amounts of critical care data generated by the long inpatient stays (weeks to months). Additionally, the opportunity to positively affect neonatal outcomes and improve their quality of life for decades is particularly attractive to researchers and funding agencies. Many different groups are trying to use large electronic health record (EHR) databases to train ML algorithms to predict preterm birth, which is one of the leading causes of neonatal death [20–23]. Targeted interventions that prevent neonates from being born extremely preterm have significant potential to save lives and decrease morbidity of patients. However, algorithms based on administrative or EHR data are limited by the significant bias inherent in the increased collection of data for patients already recognized at high risk for the outcome of interest. This is one of the downsides of using EHR data for research. Thus many researchers are shifting to using multiomics to predict preterm birth earlier [24].

Imaging AI and ML models are frequently integrated and deployed in many adult specialties. However, AI heavy areas like ophthalmology and radiology have made less progress toward solving pediatric problems [25,26]. The most significant advances have been made in retinopathy of prematurity (ROP) which is a leading cause of blindness worldwide [27] with a shortage of trained providers to screen and diagnose ROP [25]. There are a number of different tools developed to help with screening and diagnosis of ROP but few are routinely incorporated clinically [28]. Ongoing research is evaluating the use of imaging data and AI technology to classify pediatric brain tumors [29], automate the calculation of bone age from hand radiographs [30], and predict autism spectrum disorder from brain MRI images before clinical symptoms are present [31].

Current implementation examples

Compared to the total number of publications on AI and ML in pediatrics, only a small percentage attempt to implement the published algorithms into clinical practice. Table 33.1 highlights articles published within the last five years that were implemented into clinical practice.

TABLE 33.1 Examples of publications of AI or ML algorithms implemented into clinical practice.

Paper title	Country of origin	AI task	Algorithm	Pediatric area of application	Model performance outcomes	Clinical outcome
Identifying child abuse through text mining and machine learning [32]	Netherlands	Identify and signal pediatricians on suspected child abuse.	Ensemble of random forest and SVM	General pediatrics	Retrospective performance Precision: 0.187 Accuracy: 0.822 Recall: 0.870 AUC: 0.914	None
Insulin dose optimization using an automated artificial intelligence-based decision support system in youths with type 1 diabetes [33]	Israel	Insulin dose adjustment through automated decision support	Unable to determine	Ambulatory care	None	Percentage of time spent within the target glucose range in AI-DSS arm were statistically noninferior to those in the physician arm.
Model-Informed Precision Dosing of Vancomycin in Hospitalized Children: Implementation and Adoption at an Academic Children's Hospital [34]	United States	Pharmacokinetic predictions and exposure metrics for vancomycin	Bayesian forecasting	NICU, PICU, and then all inpatient setting	None	853 patient courses ($n = 96$ neonates, $n = 757$ children) and 2148 therapeutic drug monitoring (TDM) levels Tool utilization: 54% (853/1587) vancomycin patient-courses and 62% (750/1217) of patient-courses Extreme trough concentrations as measured by a predicted steady-state trough <5 mg/L or >20 mg/L were low with 87.8% (64/82) of neonates and 79.0% (528/668) of children having achieved a predicted steady-state trough concentration between 5 and 20 mg/L by the second TDM. AUC24/MIC >400 at steady-state was achieved in 63.4% (52/82) of neonates and 46.7%

(*Continued*)

TABLE 33.1 (Continued)

Paper title	Country of origin	AI task	Algorithm	Pediatric area of application	Model performance outcomes	Clinical outcome
						(312/668) of children at the time of first TDM, and this improved to 78.0% (64/82) of neonates and 64.1%(428/668) of children by the second TDM. If a target predicted trough concentration of 10 to 20 mg/L was used, 53.7% (44/82) of neonates and 44.3% (296/668) of children achieved this target by the second TDM. Measured Ease of use (85% 22/26), satisfaction (81% 21/26), accessibility to tool (96% 25/26)
Artificial intelligence-assisted interpretation of bone age radiographs improves accuracy and decreases variability [35]	USA	Predict bone age	CNN—deep learning	General pediatrics or pediatric endocrinology	Exact accuracy and RMSE of bone age	Combined AI and radiologist interpretation resulted in higher accuracy than AI alone or radiologist alone. It also decreased variability in bone age No significant difference in model performance between Caucasian and non-Caucasian patients
Artificial intelligence-assisted reduction in patients' waiting time for outpatient process: a retrospective cohort study [36]	China	Recommend lab tests and examination to patient family based on medical history prior to seeing the physician	Deep learning	Outpatient pediatrics		Median waiting time with AI assisted group was 0.38 h compared to 1.97 h for traditional group The expenses of the AI-supported group were lower in terms of total cost

Artificial intelligence-assisted clinical decision support for childhood asthma management: A randomized clinical trial [37]	USA	A quarterly report to physicians including a ML based prediction of risk for asthma exacerbation	Bayesian classifier	Primary care		No difference in the proportion of duration when patients had well-controlled asthma during the study period between the intervention and the control groups. Significant reduction in time for reviewing EHRs for asthma management of each participant (median: 3.5 min, IQR: 2–5), compared to usual care without Asthma-Guidance (median: 11.3 min, IQR: 6.3–15); ($P < .001$) Mean health care costs with 95% CI of children during the trial (compared to before the trial) in the intervention group were lower than those in the control group (-$1036 [-$2177, $44] for the intervention group vs. + $80 [-$841, $1000] for the control group), though there was no significant difference ($P = .12$)

AUC, Area under the curve; *SVM*, support vector machine; *NICU*, neonatal intensive care unit; *PICU*, pediatric intensive care unit; *TDM*, therapeutic drug monitoring; *RMSE*, root mean squared error; *EHR*, electronic health records.

The studies highlighted in Table 33.1 display a variety of pediatric AI topics that are being studied and implemented into clinical practice including identifying risk for child abuse [32], drug dosing adjustments and pharmacokinetics [33,34], radiologist assisted interpretation of bone age [35], predicting imaging and laboratory tests to reduce patient wait time [36], and assistance with asthma management [37]. However, there are limited studies on these topics, and further research is needed in these areas to validate the initial findings by displaying that these results can be replicated at other sites.

Additionally, commercially available AI algorithms created by private companies and large EHR vendors that offer pediatric-specific clinical decision support tools are available for immediate implementation. These algorithms aim to predict critical deterioration [38], detect sepsis, and estimate oxygen delivery [39]. However, many of the algorithms that are available for immediate deployment do not have strong multicenter evidenced-based research reporting patient outcomes after algorithm implementation.

Current challenges

While many AI algorithms can accurately predict significant clinical outcomes or diagnoses, a major challenge is implementing them in a manner to facilitate a clinical intervention which then improves these outcomes. For example, a prediction model for 30-day readmissions of children and adolescents have been implemented into several major EHRs [40]. Patients with high-risk diagnoses and frequent admission were at highrisk of having another 30-day readmission [40]. However, no clinical intervention was tested after implementation of this algorithm. The authors suggest that the algorithm could help direct social workers or case management resources to focus on trying to prevent readmission for the most high-risk patients. However, the diagnoses and chronic problems of these patients may make readmission unavoidable, begging the question: *is this algorithm clinically useful*?

Another major challenge that affects all medical specialties is the lack of medical training in ML and AI leading to fear and uncertainty in trusting these models [41]. Even models that have been shown in randomized controlled trials (RCTs) to decrease mortality struggle to gain wide spread adoption. Third party software based on RCTs shown to decrease mortality in the NICU is available to display risk scores for early detection of neonatal sepsis using ML algorithms based on heart rate variability [42,43]. However, adoption of this software is slow on a national scale potentially because of clinicians' concern about how to handle the information provided. Should high risk for sepsis alerts prompt additional exams, laboratory work up, or antibiotics? What are the legal repercussions to clinicians who do not agree with the ML calculated risk of sepsis in patients that appear clinically stable?

Even with more simplistic prediction models, such as risk calculators based on logistic regression algorithms, there is inconsistent clinical use. If a clinician needs to use a third-party website or application to calculate a risk score, there is a high likelihood that clinicians will depend on clinical expertise instead. Most logistic regression algorithms are based on clinical experts choosing the parameters included in the model and many ML-based predictions may not provide any additional information about patient risk factors that clinicians were not already aware of.

Unsupervised ML approaches have similar hurdles to implementation. These methods require vast amounts of data to detect meaningful disease phenotypes. While some studies have been successful in identifying these phenotypes, they have been difficult to validate across centers due to the large amount of data involved and the technical and administrative barriers to data sharing that are currently in place.

Due to the highly regulated and protected nature of healthcare data, replicating predictions with an external dataset can be a significant challenge. A systematic review about ML models predicting neonatal mortality found that less than one-third of the articles performed external validation [44]. Most studies cite insufficient amounts of data as one of their major limitations to their models [45]. Multicenter studies with deidentified databases that possess significantly granular patient data (i.e., continuous physiologic monitoring data) are required. Instead of only a few local collaborations, large-scale collaborative networks and data marts should be created to facilitate algorithm development and testing on a national basis. Large EHR vendors and national funding sources are attempting to help facilitate these data collaborations.

Future directions

From a human factors perspective, there are many elements that should be considered for AI to be successful in healthcare. Sujan et al. described the 8 principles of Human factors for healthcare AI including situation awareness, workload, automation bias, training, relationship with people, ethical issues, and explanation and trust with AI [46]. While many have discussed importance of these issues separately, there is a dearth of consideration of these during all stages of the AI development cycle. Many AI systems fail in clinical implementation as the AI prediction task does not match the clinical requirements or address gaps where help is needed. For example, when addressing clinical deterioration, the actual challenge is identification of kids with bad outcomes and not in identification of kids at risk for deterioration, such broad predictions will have poor targeting of interventions and inappropriate overtreatment on misidentified patients [47]. It is crucial to spend more efforts during early stages of AI development and ensure appropriate understanding of work systems and requirements before developing the models. Much work has been done in this regard decades before and suggests we appropriately align AI tasks to one of four aspects (1) information acquisition; (2) information analysis; (3) decision and action selection; and (4) action implementation [48].

While many researchers want to use AI and ML to solve major pediatric problems, such as decreasing morbidities, mortality, and preterm birth, researchers should consider choosing a more focused task as suggested above with an opportunity for a clinical intervention based on the model [49]. For example, instead of predicting which patients are at high-risk of ventilator associated pneumonias, creating a predictive model to determine which patients are ready for extubation from the ventilator may be more beneficial. Prolonged mechanical ventilation is already known to be highly associated with ventilator associated pneumonias, thus helping clinicians decide when it is safe to remove the patient from mechanical ventilation would potentially be a more clinically useful prediction model.

Currently, one of the most promising techniques in AI utilize continuous vital sign or fetal monitoring data to create algorithms that can detect subtle physiologic changes and variations that may go unnoticed by expert clinicians [50–52]. Multiple studies have demonstrated that autonomic nervous system dysfunction is correlated with poor outcomes in children and adults, and AI algorithms are capable of identifying changes in heart rate variability or pulse oximetry variability [53–56]. Widespread use of these data is currently limited by complex extraction and processing [51], we anticipate that it will soon become more technically feasible to store and analyze continuous monitor data.

Another promising technique in pediatric research is using unsupervised ML approaches in the identification of distinct disease phenotypes. Much of current pediatric medicine involves practicing general treatment guidelines for all patients, but clinicians have learned over decades of experience that some patients respond differently to treatments than other patients with the same disease process. The next stage in pediatric medicine is to practice precision medicine, using a patient's genomic and clinical data to predict response to therapy earlier in the course of a disease. This approach holds promise to decrease the mortality of severe diseases, such as sepsis and acute respiratory distress syndrome, as well as improving efficiency in the treatment of common diseases, such as asthma. As large research networks are established across institutions with capable governance structures, the path to furthering this area of precision medicine will continue to open.

The future of AI in pediatrics is dependent upon the access to both multicenter data and multisource databases. A review of ML studies in neonatology found that most studies only used a single data source (defined as EHR data, continuous vital sign data, imagining data, environmental stimuli, genetic information, etc.) and those that attempted to use multiple data sources were limited by recurrent missingness which lowered the prediction accuracy [57]. Future databases should work on incorporating all available data sources to create a holistic clinical environment for predictive model development. This will likely require institutional investment in middleware infrastructure to facilitate data capture, implementation of a common data model, and expanding data storage capabilities [58]. Using these comprehensive databases will allow us to start developing truly personalized medicine and may allow for improved accuracy of predicting more complex outcomes [58].

Finally, to fully achieve the promise of AI, it is crucial to break down institutional silos and develop a multidisciplinary approach with considerations of clinical, ethical, human factors, and implementation science expertise in design, development, implementation and evaluation of AI. It is to be expected that users will not always do what AI designers expect them to do leading to many issues including automation bias [59] (propensity for humans to favor AI suggestions), complacency (satisfaction with current solution but may lack awareness of other safer or more efficient options), satisficing (behavior to accept most accessible solution that meets minimal level of performance), vigilance decrements, loss of situational awareness, skill degradation leading to misuse, and disuse and abuse of automated systems [60]. It is crucial to measure human performance measures during development as well as after implementation of AI to avoid unintended consequences [61] and to ensure the continued ethical use of these algorithms for all.

Major takeaway points

- Pediatric research and implementation of AI is generally behind adult literature and few algorithms have been implemented into clinical practice.

- More education for clinicians on this topic is needed to improve acceptance of ML or AI algorithms.
- Algorithms and models should be developed that are associated with a clinical intervention.
- Future directions include centralization of multi-institutional databases, increased use of continuous monitor and genomic data, and the implementation of precision medicine.

References

[1] Knake LA. Artificial intelligence in pediatrics: the future is now. Pediatr Res 2022.
[2] Hoodbhoy Z, Masroor Jeelani S, Aziz A, Habib MI, Iqbal B, Akmal W, et al. Machine learning for child and adolescent health: a systematic review. Pediatrics 2021;147(1).
[3] Kwok TC, Henry C, Saffaran S, Meeus M, Bates D, Van Laere D, et al. Application and potential of artificial intelligence in neonatal medicine. Semin Fetal Neonatal Med 2022;27(5):101346.
[4] Walczak S, Scorpio RJ. Predicting pediatric length of stay and acuity of care in the first ten minutes with artificial neural networks. Pediatr Crit Care Med 2000;1(1):42–7.
[5] Hogan AH, Brimacombe M, Mosha M, Flores G. Comparing artificial intelligence and traditional methods to identify factors associated with pediatric asthma readmission. Acad Pediatr 2022;22(1):55–61.
[6] Carlin CS, Ho LV, Ledbetter DR, Aczon MD, Wetzel RC. Predicting individual physiologically acceptable states at discharge from a pediatric intensive care unit. J Am Med Inf Assoc 2018;25(12):1600–7.
[7] Aczon MD, Ledbetter DR, Laksana E, Ho LV, Wetzel RC. Continuous prediction of mortality in the PICU: a recurrent neural network model in a single-center dataset. Pediatr Crit Care Med 2021;22(6):519–29.
[8] Lamping F, Jack T, Rübsamen N, Sasse M, Beerbaum P, Mikolajczyk RT, et al. Development and validation of a diagnostic model for early differentiation of sepsis and non-infectious SIRS in critically ill children—a data-driven approach using machine-learning algorithms. BMC Pediatr 2018;18(1):112.
[9] Le S, Hoffman J, Barton C, Fitzgerald JC, Allen A, Pellegrini E, et al. Pediatric severe sepsis prediction using machine learning. Front Pediatr 2019;7:413.
[10] Mayampurath A, Sanchez-Pinto LN, Hegermiller E, Erondu A, Carey K, Jani P, et al. Development and external validation of a machine learning model for prediction of potential transfer to the PICU. Pediatr Crit Care Med 2022;23(7):514–23.
[11] Pollack MM, Holubkov R, Berg RA, Newth CJL, Meert KL, Harrison RE, et al. Predicting cardiac arrests in pediatric intensive care units. Resuscitation 2018;133:25–32.
[12] Tunthanathip T, Oearsakul T. Application of machine learning to predict the outcome of pediatric traumatic brain injury. Chin J Traumatol 2021;24(6):350–5.
[13] McInnis C, Garcia MJS, Widjaja E, Frndova H, Huyse JV, Guerguerian AM, et al. Magnetic resonance imaging findings are associated with long-term global neurological function or death after traumatic brain injury in critically ill children. J Neurotrauma 2021;38(17):2407–18.
[14] Fung FW, Topjian AA, Xiao R, Abend NS. Early EEG features for outcome prediction after cardiac arrest in children. J Clin Neurophysiol 2019;36(5):349–57.
[15] Goto Y, Maeda T, Nakatsu-Goto Y. Decision tree model for predicting long-term outcomes in children with out-of-hospital cardiac arrest: a nationwide, population-based observational study. Crit Care 2014;18(3):R133.
[16] Winter MC, Day TE, Ledbetter DR, Aczon MD, Newth CJL, Wetzel RC, et al. Machine learning to predict cardiac death within 1 hour after terminal extubation. Pediatr Crit Care Med 2021;22(2):161–71.
[17] Brew BK, Chiesa F, Lundholm C, Örtqvist A, Almqvist C. A modern approach to identifying and characterizing child asthma and wheeze phenotypes based on clinical data. PLoS One 2019;14(12):e0227091.
[18] Oksel C, Haider S, Fontanella S, Frainay C, Custovic A. Classification of pediatric asthma: from phenotype discovery to clinical practice. Front Pediatr 2018;6:258.
[19] Sanchez-Pinto LN, Stroup EK, Pendergrast T, Pinto N, Luo Y. Derivation and validation of novel phenotypes of multiple organ dysfunction syndrome in critically ill children. JAMA Netw Open 2020;3(8):e209271.

[20] Sun Q, Zou X, Yan Y, Zhang H, Wang S, Gao Y, et al. Machine learning-based prediction model of preterm birth using electronic health record. J Healthc Eng 2022;2022:9635526.
[21] Abraham A, Le B, Kosti I, Straub P, Velez-Edwards DR, Davis LK, et al. Dense phenotyping from electronic health records enables machine learning-based prediction of preterm birth. BMC Med 2022;20(1):333.
[22] Weber A, Darmstadt GL, Gruber S, Foeller ME, Carmichael SL, Stevenson DK, et al. Application of machine-learning to predict early spontaneous preterm birth among nulliparous non-Hispanic black and white women. Ann Epidemiol 2018;28(11):783–9.e781.
[23] Gao C, Osmundson S, Velez Edwards DR, Jackson GP, Malin BA, Chen Y. Deep learning predicts extreme preterm birth from electronic health records. J Biomed Inf 2019;100:103334.
[24] Jehan F, Sazawal S, Baqui AH, Nisar MI, Dhingra U, Khanam R, et al. Multiomics characterization of preterm birth in low- and middle-income countries. JAMA Netw Open 2020;3(12):e2029655.
[25] Reid JE, Eaton E. Artificial intelligence for pediatric ophthalmology. Curr Opin Ophthalmol 2019;30(5):337–46.
[26] Davendralingam N, Sebire NJ, Arthurs OJ, Shelmerdine SC. Artificial intelligence in paediatric radiology: future opportunities. Br J Radiol 2021;94(1117):20200975.
[27] Quinn GE. Retinopathy of prematurity blindness worldwide: phenotypes in the third epidemic. Eye Brain 2016;8:31–6.
[28] Gensure RH, Chiang MF, Campbell JP. Artificial intelligence for retinopathy of prematurity. Curr Opin Ophthalmol 2020;31(5):312–17.
[29] Novak J, Zarinabad N, Rose H, Arvanitis T, MacPherson L, Pinkey B, et al. Classification of paediatric brain tumours by diffusion weighted imaging and machine learning. Sci Rep 2021;11(1):2987.
[30] Booz C, Yel I, Wichmann JL, Boettger S, Al Kamali A, Albrecht MH, et al. Artificial intelligence in bone age assessment: accuracy and efficiency of a novel fully automated algorithm compared to the Greulich-Pyle method. Eur Radiol Exp 2020;4(1):6.
[31] Chen T, Chen Y, Yuan M, Gerstein M, Li T, Liang H, et al. The development of a practical artificial intelligence tool for diagnosing and evaluating autism spectrum disorder: multicenter study. JMIR Med Inf 2020;8(5):e15767.
[32] Amrit C, Paauw T, Aly R, Lavric M. Identifying child abuse through text mining and machine learning. Expert Syst Appl 2017;88:402–18.
[33] Nimri R, Battelino T, Laffel LM, Slover RH, Schatz D, Weinzimer SA, et al. Insulin dose optimization using an automated artificial intelligence-based decision support system in youths with type 1 diabetes. Nat Med 2020;26(9):1380–4.
[34] Frymoyer A, Schwenk HT, Zorn Y, Bio L, Moss JD, Chasmawala B, et al. Model-informed precision dosing of vancomycin in hospitalized children: implementation and adoption at an Academic Children's Hospital. Front Pharmacol 2020;11:551.
[35] Tajmir SH, Lee H, Shailam R, Gale HI, Nguyen JC, Westra SJ, et al. Artificial intelligence-assisted interpretation of bone age radiographs improves accuracy and decreases variability. Skelet Radiol 2019;48(2):275–83.
[36] Li X, Tian D, Li W, Dong B, Wang H, Yuan J, et al. Artificial intelligence-assisted reduction in patients' waiting time for outpatient process: a retrospective cohort study. BMC Health Serv Res 2021;21(1):237.
[37] Seol HY, Shrestha P, Muth JF, Wi CI, Sohn S, Ryu E, et al. Artificial intelligence-assisted clinical decision support for childhood asthma management: a randomized clinical trial. PLoS One 2021;16(8):e0255261.
[38] Rothman MJ, Tepas 3rd JJ, Nowalk AJ, Levin JE, Rimar JM, Marchetti A, et al. Development and validation of a continuously age-adjusted measure of patient condition for hospitalized children using the electronic medical record. J Biomed Inf 2017;66:180–93.
[39] Futterman C, Salvin JW, McManus M, Lowry AW, Baronov D, Almodovar MC, et al. Inadequate oxygen delivery index dose is associated with cardiac arrest risk in neonates following cardiopulmonary bypass surgery. Resuscitation 2019;142:74–80.
[40] Goodman DM, Casale MT, Rychlik K, Carroll MS, Auger KA, Smith TL, et al. Development and validation of an integrated suite of prediction models for all-cause 30-day readmissions of children and adolescents aged 0 to 18 years. JAMA Netw Open 2022;5(11):e2241513.
[41] Wartman SA, Combs CD. Medical education must move from the information age to the age of artificial intelligence. Acad Med 2018;93(8):1107–9.
[42] Moorman JR, Carlo WA, Kattwinkel J, Schelonka RL, Porcelli PJ, Navarrete CT, et al. Mortality reduction by heart rate characteristic monitoring in very low birth weight neonates: a randomized trial. J Pediatr 2011;159(6):900–6.e901.

[43] Schelonka RL, Carlo WA, Bauer CR, Peralta-Carcelen M, Phillips V, Helderman J, et al. Mortality and neurodevelopmental outcomes in the heart rate characteristics monitoring randomized controlled trial. J Pediatr 2020;219:48–53.
[44] Mangold C, Zoretic S, Thallapureddy K, Moreira A, Chorath K, Moreira A. Machine learning models for predicting neonatal mortality: a systematic review. Neonatology 2021;118(4):394–405.
[45] Jeong H, Kamaleswaran R. Pivotal challenges in artificial intelligence and machine learning applications for neonatal care. Semin Fetal Neonatal Med 2022;27(5):101393.
[46] Sujan M, Pool R, Salmon P. Eight human factors and ergonomics principles for healthcare artificial intelligence. BMJ Health Care Inf 2022;29(1).
[47] Mehta SD, Muthu N, Yehya N, Galligan M, Porter E, McGowan N, et al. Leveraging EHR data to evaluate the association of late recognition of deterioration with outcomes. Hosp Pediatr 2022;12(5):447–60.
[48] Parasuraman R, Sheridan TB, Wickens CD. A model for types and levels of human interaction with automation. IEEE Trans Syst Man Cybern A Syst Hum 2000;30(3):286–97.
[49] Lindsell CJ, Stead WW, Johnson KB. Action-informed artificial intelligence-matching the algorithm to the problem. JAMA 2020;323(21):2141–2.
[50] Esteban-Escaño J, Castán B, Castán S, Chóliz-Ezquerro M, Asensio C, Laliena AR, et al. Machine learning algorithm to predict acidemia using electronic fetal monitoring recording parameters. Entropy (Basel) 2021;24(1).
[51] Monfredi O, Keim-Malpass J, Moorman JR. Continuous cardiorespiratory monitoring is a dominant source of predictive signal in machine learning for risk stratification and clinical decision support. Physiol Meas 2021;42(9).
[52] Kamaleswaran R, Akbilgic O, Hallman MA, West AN, Davis RL, Shah SH. Applying artificial intelligence to identify physiomarkers predicting severe sepsis in the PICU. Pediatr Crit Care Med 2018;19(10):e495–503.
[53] Goldstein B, Fiser DH, Kelly MM, Mickelsen D, Ruttimann U, Pollack MM. Decomplexification in critical illness and injury: relationship between heart rate variability, severity of illness, and outcome. Crit Care Med 1998;26(2):352–7.
[54] Ellenby MS, McNames J, Lai S, McDonald BA, Krieger D, Sclabassi RJ, et al. Uncoupling and recoupling of autonomic regulation of the heart beat in pediatric septic shock. Shock 2001;16(4):274–7.
[55] Badke CM, Marsillio LE, Weese-Mayer DE, Sanchez-Pinto LN. Autonomic nervous system dysfunction in pediatric sepsis. Front Pediatr 2018;6:280.
[56] Papaioannou VE, Maglaveras N, Houvarda I, Antoniadou E, Vretzakis G. Investigation of altered heart rate variability, nonlinear properties of heart rate signals, and organ dysfunction longitudinally over time in intensive care unit patients. J Crit Care 2006;21(1):95–103 discussion 103–104.
[57] McAdams RM, Kaur R, Sun Y, Bindra H, Cho SJ, Singh H. Predicting clinical outcomes using artificial intelligence and machine learning in neonatal intensive care units: a systematic review. J Perinatol 2022;42(12):1561–75.
[58] Wilson CG, Altamirano AE, Hillman T, Tan JB. Data analytics in a clinical setting: applications to understanding breathing patterns and their relevance to neonatal disease. Semin Fetal Neonatal Med 2022;27(5):101399.
[59] Goddard K, Roudsari A, Wyatt JC. Automation bias: a systematic review of frequency, effect mediators, and mitigators. J Am Med Inf Assoc 2012;19(1):121–7.
[60] Parasuraman R, Riley V. Humans and automation: use, misuse, disuse, abuse. Hum Factors 1997;39(2):230–53.
[61] Parasuraman R. Designing automation for human use: empirical studies and quantitative models. Ergonomics 2000;43(7):931–51.

CHAPTER

34

Artificial intelligence in pediatric congenital and acquired heart disease

Sowmith Rangu and Charitha D. Reddy

Division of Pediatric Cardiology, Department of Pediatrics, Stanford University School of Medicine, Lucile Packard Children's Hospital at Stanford, Stanford University, Palo Alto, CA, United States

Introduction

Artificial intelligence (AI) in healthcare has advanced considerably in the last few decades and has shown promise in diagnosis, imaging, risk stratification, and precision medicine [1]. The field of Pediatric Cardiology and congenital heart disease (CHD) already utilizes a broad array of technologies in daily practice and is poised to benefit greatly from AI-based implementation.

CHD affects 1 in every 100 infants, making it the most common congenital birth defect [2]. In comparison to other forms of congenital disease; CHD is also the leading cause of neonatal mortality [3]. A large proportion of these children will require significant interventions throughout their life, and early detection is key in accessing specialized cardiac programs and reducing morbidity and mortality [4]. Other acquired pediatric cardiac diseases include arrhythmias, cardiomyopathies, myocarditis, Kawasaki's disease, and other autoimmune disease processes.

Pediatric cardiologists face anatomically heterogenous and physiologically complex disease processes, making AI a particularly attractive domain in the field. In this chapter, we will review the currently available AI technologies as they apply to clinical assessment, diagnostic imaging, clinical decision-making, outcome prediction, as well as current barriers to evolution and future applications.

Clinical assessment

Undiagnosed and untreated pediatric heart disease results in avoidable morbidity and mortality, making accurate and timely diagnosis crucial [4]. The majority of pediatric

DOI: https://doi.org/10.1016/B978-0-443-15688-5.00038-3

cardiac screening begins with clinical assessment, including cardiac auscultation and electrocardiograms (ECGs).

Recent advances in computer-assisted cardiac auscultation permit recording, transmission, and analysis of heart sounds using AI, aiding accurate diagnosis and serving as a teaching tool [5]. Algorithms have been developed to identify murmurs associated with atrial septal defects, ventricular septal defects, patent ductus arteriosus, and other CHD [6–11]. Convolutional neural network (CNN)-based machine learning (ML) has also demonstrated the ability to identify murmurs as "rheumatic" with high accuracy, sensitivity and specificity [12], creating a potential for auscultation to be used in screening programs. Multiple "smart stethoscopes" are available on the market for this purpose [13], and they may be particularly useful for expanding access in resource-limited areas that may not have rapid access to subspecialty cardiac care.

ECGs are a mainstay of pediatric cardiac clinical care, used both for initial diagnosis and for ongoing surveillance. Deep learning algorithms have been used for interpretation of ECGs in adults for a few decades [14], but their utility is becoming increasingly recognized in pediatric cardiology. Identification of atrial septal defects [15] and hypertrophic cardiomyopathy [16] using ECG tracings, are a few examples of its utility. ECG data is ripe for the use of CNN to process an immense volume of data and potentially extract patterns that have not yet been identified [17]. The increasing availability of photoplethysmography on smart-phones and smart-watches has the potential to increase pediatric arrhythmia detection and improve prediction of cardiopulmonary arrests in high-risk populations [18]. However, most devices currently are not geared toward pediatric populations [19,20].

Imaging

Cardiac imaging, including echocardiography, magnetic resonance imaging (MRI), and computed tomography (CT), plays a key role in diagnosis, management, and screening of patients with pediatric cardiac disease. Image optimization, automated segmentation, disease detection, risk stratification, and surgical planning are areas in pediatric cardiology that are evolving with the aid of AI-based technologies.

Image acquisition and plane identification is often the first step in cardiac imaging. Due to the heterogenous nature of CHD, pediatric cardiologists are presented with a wide array of cardiac positioning and anatomy, thereby making identification and segmentation less straightforward than an anatomically normal heart often found in the adult population. Segmentation algorithms permit images to be subdivided into identifiable regions [21]. Integrating AI in the segmentation process could improve morphologic assessment and measurement of cardiac structures. Wang et al. designed a multichannel CNN to automatically interpret five-view echocardiograms and identify atrial and ventricular septal defects; their model was able to recognize abnormalities on a binary classification with a high degree of accuracy (AUC 0.922) [22]. More recently, Karim-Bidhendi et al. developed a model using synthetically augmented cardiac MRI images to produce a large synthetic training dataset that could be utilized to segment complex CHD [23].

Accurate antenatal diagnosis of cardiac disease allows for timely and informed decision-making during pregnancy. There is also increasing evidence that prenatal diagnosis improves neurodevelopmental outcomes, morbidity, and mortality for the fetus postnatally [24]. Despite the widespread use of fetal echocardiography, there remains a wide variation in diagnostic accuracy, sensitivity, and specificity of these scans, because of variable sonographer experience [25]. It is promising, however, that significant strides have been made in applying AI to fetal detection of heart disease. Arnout et al., used >100,000 images to train a model that successfully identifies standard fetal ECHO views, performs cardiac measurements (cardiothoracic ratio, cardiac axis, fractional area change for each chamber) and is able to differentiate between normal hearts and complex CHD; the model demonstrated high sensitivity and was successful even when applied to external scans [26]. Other models showed that previously acquired fetal images can be used to accurately segment cardiac chambers [27], and even identify ventricular septal defects [28]. Yeo et al. developed fetal intelligent navigation echocardiography (FINE) to automate the conversion of 3D fetal images to 2D images of interest, potentially improving the identification of CHD [29,30]. Integration of AI within fetal screening ultrasound has been demonstrated to improve detection of CHD [31], and ongoing work is being done to integrate these advances into clinical workflow.

ML models are being studied to decrease the need for invasive assessments by using cardiac imaging to predict physiologic findings and risk stratify patients. An MRI-based predictive model for risk of postcardiac surgical bleeding in neonates [32], a CT-based model to predict mean pulmonary artery pressures in patients post-Glenn palliation [33], and a chest radiography model that quantifies otherwise qualitative/subjective findings of pulmonary to systemic flow ratio [34] are a few examples.

Cardiac imaging plays a key role in surgical planning, but interventionalists are faced with a plethora of imaging data and highly heterogenous anatomy. Promising technologies are utilizing AI to simulate various interventions and predict outcomes prior to surgical intervention on the patient. For example, Pushparajah et al. demonstrated that AI can be integrated within virtual reality systems to permit 3D investigation of the anatomy prior to choosing the most appropriate surgical technique in repair of atrioventricular septal defects [35]. Similar applications have been used to optimize size, shape, and location of the transannular patch in repair of Tetralogy of Fallot [36], and to optimize aortic coarctation repair by applying ML to fluid dynamics [37].

Clinical decision-making

Pediatric cardiologists deal with highly heterogenous lesions and disease processes, hence formulation of prognosis and clinical outcomes becomes very challenging. The creation of large multicenter databases [38] has uncovered an opportunity for personalized risk stratification, adverse event prediction, and outcome prediction. The information extracted from these databases has the potential to predict long-term outcomes or even inform a clinician of a patient's clinical status in real time by integration with the electronic medical record. Physiologic data-driven models that notify clinicians with anticipatory warning of cardiorespiratory decompensation permit timely and targeted intervention. Rusin et al. have demonstrated one such model that identifies the risk of compromise 1–2 hours preceding overt extremis in children with single ventricle physiology during their interstage hospitalization [39]. Bose et al. applied a similar predictive model analyzing physiologic variables for all patients in their cardiac ICU and were able to predict an impending arrest an average of 17 hours prior to it occurring [40]. AI models, such as optimal classification trees, have been used in children postcardiac surgery to accurately predict mortality, duration of mechanical ventilation, and hospital length of stay [41]. Other such models that are being clinically assessed include the CORTEX "traffic light" system to stratify patients' based on vital signs into risk categories and determine if they are ready for transition into different phases of care. This initiative was successfully able to identify patients at risk of deterioration, while also reducing hospital length of stay by improving "flow" between phases of care [42].

Similarly, ML algorithms can be applied to evaluate longer term risks, such as malnutrition 1 year after surgery for CHD [43] or the risk of life-threatening events like a malignant arrhythmia [44,45]. Complex ML models have also been used to understand and predict more nuanced aspects of postoperative care, such as bleeding and coagulation [46].

Precision medicine is a developing strategy for individualized treatment of a wide range of diseases. A focus on omics and individualized environmental factors has been key to an evolved understanding the variation in morbidity and mortality in pediatric cardiology [47]. Methods, such as noninvasive screening for congenital heart defects, using serum metabolomics [48] or using DNA methylation to predict neonatal aortic coarctation [49] have demonstrated feasibility, thus creating a platform for further research into this domain. In the future, CHD patients may be triaged at an early stage in an evidence-based manner and treatment strategies can be streamlined through *in silico* predictions.

Kawasaki disease (KD) is an acquired pediatric vasculitis that leads to coronary artery vasculitis and complications, timely management with intravenous immunoglobulin (IVIG) is crucial in reducing morbidity and mortality [50]. ML-based algorithms that predict the risk of developing coronary aneurysms or likelihood of resistance to IVIG permit timely and targeted therapies and hence improve patient outcomes [51].

Future applications

The studies discussed in this chapter have demonstrated successful utilization of AI in the care of pediatric patients with cardiac disease, but relevant barriers remain that must be overcome prior to real-world implementation of these technologies.

Many of the studies investigated a singular diagnosis and used relatively small sample sizes, limiting the generalizability of the results. Broadening datasets with the use of multi-institutional databases may alleviate this particular challenge. There also remain significant ethical, legal, and proprietary implications of using AI in pediatric cardiac disease [52]. Appropriate safeguards should be in place during the implementation of these algorithms. Apprehension also exists around incorrect predictions made by AI algorithms, which could result in a lifetime of negative consequences for patients [53]. It is therefore crucial to use these algorithms to support clinician decision-making rather than supplant it [20], ensure appropriate measures have been taken to avoid bias, and improve "explainability" of models.

Conclusion

Pediatric cardiology is a subspecialty that is already rich in imaging and clinical data, and now with the advent of implantable monitors and biosensors, there is a high quantity of data that is available to develop AI algorithms. We anticipate broad utilization of AI in pediatric heart disease, ranging from clinical assessment, imaging optimization, intervention planning, prognosis stratification, adverse event prediction, and precision medicine. While there are expected barriers to the rapid evolution of AI in pediatric cardiology, impactful strides have already been made. Robust and ethical AI research in pediatric cardiology must continue to create a future wherein AI becomes a beneficial, safe, and routine aspect of clinical practice.

Summary table

Subspecialty area	AI application	Related papers	Description
Clinical cardiology	Computer-aided heart sound analysis	Xu et al. [5]	Screening for CHD with auscultation with a "smart stethoscope"
		Wang et al. [6]	Online screening for CHD using segmentation technology
		Lv et al. [7]	Remote auscultation in detecting abnormal heart sounds
		Xiao et al. [8]	Construction of a dataset of pediatric heart sounds
		Aziz et al. [9]	Distinguish between ASD, VSD, and normal patients
		Wang et al. [10]	Identify murmurs of VSD
		Gahrehbaghi et al. [11]	Distinguishing septal defects from valve regurgitation
		Asmare et al. [12]	Distinguishing between normal and rheumatic heart sounds
		Chowdhury et al. [13]	Detection of abnormal heart sounds in real time using smart stethoscopes

Subspecialty area	AI application	Related papers	Description
	Minimally invasive alternatives	Toba et al. [34]	Analysis of chest radiographs to predict pulmonary to systemic flow ratio in patients with CHD
	Patient monitoring	Rusin et al. [39]	Identify physiological precursors of cardiorespiratory deterioration in children with single-ventricle physiology during interstage hospitalization
		Bose et al. [40]	Identify impending cardiac arrest in neonates in the cardiac ICU based on multiple clinical parameters.
		Garcia-Canadilla et al. [42]	Risk stratification system based on patients' risk of deterioration, and identification of optimal timing to transition phases of care within the hospital
		Shi et al. [43]	Malnutrition risk prediction in children 1 year postsurgery for CHD
	Preventative cardiology	Wang et al. [51]	Intravenous immunoglobulin resistance prediction in patients with Kawasaki disease
Cardiac genetics	Screening for CHD	Troisi et al. [48]	Utilizing maternal serum metabolomics as a tool to screen for risk of CHD
		Bahado-Singh et al. [49]	Prediction of coarctation of aorta using a newborn blood spot
Fetal cardiology	Diagnostics	Arnaout et al. [26]	Identify recommended cardiac views, perform standard fetal cardiothoracic measurements, and distinguish between normal hearts and complex CHD
		Xu et al. [27]	Improve diagnostic accuracy and efficiency in prenatal CHD, by automatically segmenting the apical 4 chamber view using DW-NET
		Dozen et al. [28]	Improve accuracy of segmentation of the ventricular septum in fetal ECHO
		Yeo et al. [29]	Diagnosing complex fetal CHD using a novel method to generate and display nine standard fetal ECHO views
		Yeo et al. [30]	Applying FINE to distinguish between normal hearts and CHD
		Gong et al. [31]	Improve recognition of fetal heart disease
Electrophysiology	Identification of cardiac diseases	Mori et al. [15]	Diagnosing ASDs based on ECGs

Subspecialty area	AI application	Related papers	Description
		Siontis et al. [16]	Detect pediatric HCM using ECGs
		Aufiero et al. [17]	Diagnosis of long QT syndrome
		Vu et al. [18]	ST segment monitoring to detect early signs of ST instability hence identifying patients at increased risk of cardiopulmonary arrest
Transthoracic echocardiography	Diagnostics	Wang et al. [22]	Automatically interpret 5 view echocardiogram
Cardiac CT	Noninvasive alternative to cardiac catheterization	Huang et al. [33]	Predicting mean PA pressure in post-Glenn shunt patients, utilizing cardiac CT
Cardiac MRI	Image augmentation and diagnostics	Diller et al. [21]	Generate large quantity of MRI images, to create an image library to aid in supervised learning
		Karimi-Bidhendi et al. [23]	Synthetically segmented MRI images to produce a training dataset used for automated detection of complex congenital heart disease.
		Choi et al. [32]	Utilize T2Mr to detect risk for postcardiac surgery bleeding
Cardiac catheterization	Adverse outcome prediction	Sun et al. [44]	Predict risk of arrhythmias after interventional closure of ASD
Cardiac surgery	Optimizing surgical technique	Zhang et al. [36]	Optimize surgical patch in repair of Tetralogy of Fallot (TOF)
		Liu et al. [37]	Optimizing shape and geometry of patient specific tissue engineered vascular grafts in coarctation of aorta repair, resulting in decreased loss of blood flow energy
	Surgical decision-making and planning	Pushparajah et al. [35]	Improve clarity of anatomical structures, and hence clarifying surgical approach
	Adverse outcome prediction	Bertsimas et al. [41]	Predict mortality, postoperative mechanical ventilatory support time, and hospital length of stay for patients who underwent heart surgery
Cardiac critical care	Risk prediction	Guo et al. [46]	Predict postoperative blood coagulation function in children with CHD

References

[1] Briganti G, Le Moine O. Artificial intelligence in medicine: today and tomorrow. Front Med 2020;7:27. Available from: https://doi.org/10.3389/fmed.2020.00027.

[2] Sun R, Liu M, Lu L, Zheng Y, Zhang P. Congenital heart disease: causes, diagnosis, symptoms, and treatments. Cell Biochem Biophys 2015;72(3):857–60. Available from: https://doi.org/10.1007/s12013-015-0551-6.

[3] Lopes SAV, do A, Guimarães ICB, Costa SF, de O, Acosta AX, et al. Mortality for critical congenital heart diseases and associated risk factors in newborns. a cohort study. Arq Bras Cardiol 2018;111(5):666–73. Available from: https://doi.org/10.5935/abc.20180175.

[4] Eckersley L, Sadler L, Parry E, Finucane K, Gentles TL. Timing of diagnosis affects mortality in critical congenital heart disease. Arch Dis Child 2016;101(6):516–20. Available from: https://doi.org/10.1136/archdischild-2014-307691.

[5] Xu W, Yu K, Xu J, Ye J, Li H, Shu Q. [Artificial intelligence technology in cardiac auscultation screening for congenital heart disease: present and future]. Zhejiang Xue Xue Bao Yi Xue Ban J Zhejiang Univ Med Sci 2020;49(5):548–55. Available from: https://doi.org/10.3785/j.issn.1008-9292.2020.10.01.

[6] Wang J, You T, Yi K, et al. Intelligent diagnosis of heart murmurs in children with congenital heart disease. J Healthc Eng 2020;2020:1–9. Available from: https://doi.org/10.1155/2020/9640821.

[7] Lv J, Dong B, Lei H, et al. Artificial intelligence-assisted auscultation in detecting congenital heart disease. Eur Heart J Digit Health 2021;2(1):119–24. Available from: https://doi.org/10.1093/ehjdh/ztaa017.

[8] Xiao B, Xu Y, Bi X, et al. Follow the sound of children's heart: a deep-learning-based computer-aided pediatric CHDs diagnosis system. IEEE Internet Things J 2020;7(3):1994–2004. Available from: https://doi.org/10.1109/JIOT.2019.2961132.

[9] Aziz S, Khan MU, Alhaisoni M, Akram T, Altaf M. Phonocardiogram signal processing for automatic diagnosis of congenital heart disorders through fusion of temporal and cepstral features. Sensors 2020;20(13):3790. Available from: https://doi.org/10.3390/s20133790.

[10] Wang JK, Chang YF, Tsai KH, et al. Automatic recognition of murmurs of ventricular septal defect using convolutional recurrent neural networks with temporal attentive pooling. Sci Rep 2020;10(1):21797. Available from: https://doi.org/10.1038/s41598-020-77994-z.

[11] Gharehbaghi A, Sepehri AA, Babic A. Distinguishing septal heart defects from the valvular regurgitation using intelligent phonocardiography. Stud Health Technol Inf 2020;270:178–82. Available from: https://doi.org/10.3233/SHTI200146.

[12] Asmare MH, Woldehanna F, Janssens L, Vanrumste B. Rheumatic heart disease detection using deep learning from spectro-temporal representation of un-segmented heart sounds. Annu Int Conf IEEE Eng Med Biol Soc IEEE Eng Med Biol Soc Annu Int Conf 2020;2020:168–71. Available from: https://doi.org/10.1109/EMBC44109.2020.9176544.

[13] Chowdhury MEH, Khandakar A, Alzoubi K, et al. Real-time smart-digital stethoscope system for heart diseases monitoring. Sensors 2019;19(12):2781. Available from: https://doi.org/10.3390/s19122781.

[14] Zhou J, Du M, Chang S, Chen Z. Artificial intelligence in echocardiography: detection, functional evaluation, and disease diagnosis. Cardiovasc Ultrasound 2021;19(1):29. Available from: https://doi.org/10.1186/s12947-021-00261-2.

[15] Mori H, Inai K, Sugiyama H, Muragaki Y. Diagnosing atrial septal defect from electrocardiogram with deep learning. Pediatr Cardiol 2021;42(6):1379–87. Available from: https://doi.org/10.1007/s00246-021-02622-0.

[16] Siontis KC, Liu K, Bos JM, et al. Detection of hypertrophic cardiomyopathy by an artificial intelligence electrocardiogram in children and adolescents. Int J Cardiol 2021;340:42–7. Available from: https://doi.org/10.1016/j.ijcard.2021.08.026.

[17] Aufiero S, Bleijendaal H, Robyns T, et al. A deep learning approach identifies new ECG features in congenital long QT syndrome. BMC Med 2022;20(1):162. Available from: https://doi.org/10.1186/s12916-022-02350-z.

[18] Vu EL, Rusin CG, Penny DJ, et al. A novel electrocardiogram algorithm utilizing ST-segment instability for detection of cardiopulmonary arrest in single ventricle physiology: a retrospective study. Pediatr Crit Care Med 2017;18(1):44–53. Available from: https://doi.org/10.1097/PCC.0000000000000980.

[19] Tandon A, de Ferranti SD. Wearable biosensors in pediatric cardiovascular disease: promises and pitfalls toward generating actionable insights. Circulation 2019;140(5):350–2. Available from: https://doi.org/10.1161/CIRCULATIONAHA.119.038483.

[20] M T, K D, G M, F Ag. Orphan medical devices and pediatric cardiology – what interventionists in Europe need to know, and what needs to be done. Pediatr Cardiol 2022. Available from: https://doi.org/10.1007/s00246-022-03029-1.
[21] Diller GP, Vahle J, Radke R, et al. Utility of deep learning networks for the generation of artificial cardiac magnetic resonance images in congenital heart disease. BMC Med Imaging 2020;20(1):113. Available from: https://doi.org/10.1186/s12880-020-00511-1.
[22] Wang J, Liu X, Wang F, et al. Automated interpretation of congenital heart disease from multi-view echocardiograms. Med Image Anal 2021;69:101942. Available from: https://doi.org/10.1016/j.media.2020.101942.
[23] Karimi-Bidhendi S, Arafati A, Cheng AL, Wu Y, Kheradvar A, Jafarkhani H. Fully-automated deep-learning segmentation of pediatric cardiovascular magnetic resonance of patients with complex congenital heart diseases. J Cardiovasc Magn Reson 2020;22(1):80. Available from: https://doi.org/10.1186/s12968-020-00678-0.
[24] Holland BJ, Myers JA, Woods CR. Prenatal diagnosis of critical congenital heart disease reduces risk of death from cardiovascular compromise prior to planned neonatal cardiac surgery: a meta-analysis. Ultrasound Obstet Gynecol J Int Soc Ultrasound Obstet Gynecol 2015;45(6):631–8. Available from: https://doi.org/10.1002/uog.14882.
[25] Mozumdar N, Rowland J, Pan S, et al. Diagnostic accuracy of fetal echocardiography in congenital heart disease. J Am Soc Echocardiogr 2020;33(11):1384–90. Available from: https://doi.org/10.1016/j.echo.2020.06.017.
[26] Arnaout R, Curran L, Zhao Y, Levine JC, Chinn E, Moon-Grady AJ. An ensemble of neural networks provides expert-level prenatal detection of complex congenital heart disease. Nat Med 2021;27(5):882–91. Available from: https://doi.org/10.1038/s41591-021-01342-5.
[27] Xu L, Liu M, Shen Z, et al. DW-Net: a cascaded convolutional neural network for apical four-chamber view segmentation in fetal echocardiography. Comput Med Imaging Graph 2020;80:101690. Available from: https://doi.org/10.1016/j.compmedimag.2019.101690.
[28] Dozen A, Komatsu M, Sakai A, et al. Image segmentation of the ventricular septum in fetal cardiac ultrasound videos based on deep learning using time-series information. Biomolecules 2020;10(11):1526. Available from: https://doi.org/10.3390/biom10111526.
[29] Yeo L, Luewan S, Markush D, Gill N, Romero R. Prenatal diagnosis of dextrocardia with complex congenital heart disease using fetal intelligent navigation echocardiography (FINE) and a literature review. Fetal Diagn Ther 2018;43(4):304–16. Available from: https://doi.org/10.1159/000468929.
[30] Yeo L, Luewan S, Romero R. Fetal intelligent navigation echocardiography (FINE) detects 98% of congenital heart disease: FINE detection of congenital heart disease. J Ultrasound Med 2018;37(11):2577–93. Available from: https://doi.org/10.1002/jum.14616.
[31] Gong Y, Zhang Y, Zhu H, et al. Fetal congenital heart disease echocardiogram screening based on DGACNN: adversarial one-class classification combined with video transfer learning. IEEE Trans Med Imaging 2020;39(4):1206–22. Available from: https://doi.org/10.1109/TMI.2019.2946059.
[32] Choi PS, Emani S, Ibla JC, Marturano JE, Lowery TJ, Emani S. Magnetic resonance-based diagnostics for bleeding assessment in neonatal cardiac surgery. Ann Thorac Surg 2020;109(6):1931–6. Available from: https://doi.org/10.1016/j.athoracsur.2019.11.010.
[33] Huang L, Li J, Huang M, et al. Prediction of pulmonary pressure after Glenn shunts by computed tomography–based machine learning models. Eur Radiol 2020;30(3):1369–77. Available from: https://doi.org/10.1007/s00330-019-06502-3.
[34] Toba S, Mitani Y, Yodoya N, et al. Prediction of pulmonary to systemic flow ratio in patients with congenital heart disease using deep learning-based analysis of chest radiographs. JAMA Cardiol 2020;5(4):449. Available from: https://doi.org/10.1001/jamacardio.2019.5620.
[35] Pushparajah K, Chu KYK, Deng S, et al. Virtual reality three-dimensional echocardiographic imaging for planning surgical atrioventricular valve repair. JTCVS Tech 2021;7:269–77. Available from: https://doi.org/10.1016/j.xjtc.2021.02.044.
[36] Zhang G, Mao Y, Li M, Peng L, Ling Y, Zhou X. The optimal tetralogy of Fallot repair using generative adversarial networks. Front Physiol 2021;12:613330. Available from: https://doi.org/10.3389/fphys.2021.613330.
[37] Liu X, Aslan S, Hess R, et al. Automatic shape optimization of patient-specific tissue engineered vascular grafts for aortic coarctation. In: 2020 42nd Annual International Conference of the IEEE Engineering in

Medicine & Biology Society (EMBC). IEEE; 2020. p. 2319–23. Available from: https://doi.org/10.1109/EMBC44109.2020.9176371.

[38] Vener DF, Gaies M, Jacobs JP, Pasquali SK. Clinical databases and registries in congenital and pediatric cardiac surgery, cardiology, critical care, and anesthesiology worldwide. World J Pediatr Congenit Heart Surg 2017;8(1):77–87. Available from: https://doi.org/10.1177/2150135116681730.

[39] Rusin CG, Acosta SI, Vu EL, Ahmed M, Brady KM, Penny DJ. Automated prediction of cardiorespiratory deterioration in patients with single ventricle. J Am Coll Cardiol 2021;77(25):3184–92. Available from: https://doi.org/10.1016/j.jacc.2021.04.072.

[40] Bose SN, Verigan A, Hanson J, et al. Early identification of impending cardiac arrest in neonates and infants in the cardiovascular ICU: a statistical modelling approach using physiologic monitoring data. Cardiol Young 2019;29(11):1340–8. Available from: https://doi.org/10.1017/S1047951119002002.

[41] Bertsimas D, Zhuo D, Dunn J, et al. Adverse outcomes prediction for congenital heart surgery: a machine learning approach. World J Pediatr Congenit Heart Surg 2021;12(4):453–60. Available from: https://doi.org/10.1177/21501351211007106.

[42] Garcia-Canadilla P, Isabel-Roquero A, Aurensanz-Clemente E, et al. Machine learning-based systems for the anticipation of adverse events after pediatric cardiac surgery. Front Pediatr 2022;10:930913. Available from: https://doi.org/10.3389/fped.2022.930913.

[43] Shi H, Yang D, Tang K, et al. Explainable machine learning model for predicting the occurrence of postoperative malnutrition in children with congenital heart disease. Clin Nutr 2022;41(1):202–10. Available from: https://doi.org/10.1016/j.clnu.2021.11.006.

[44] Sun H, Liu Y, Song B, Cui X, Luo G, Pan S. Prediction of arrhythmia after intervention in children with atrial septal defect based on random forest. BMC Pediatr 2021;21(1):280. Available from: https://doi.org/10.1186/s12887-021-02744-7.

[45] Atallah J, Gonzalez Corcia MC, Walsh EP. Ventricular arrhythmia and life-threatening events in patients with repaired tetralogy of fallot. Am J Cardiol 2020;132:126–32. Available from: https://doi.org/10.1016/j.amjcard.2020.07.012.

[46] Guo K, Fu X, Zhang H, Wang M, Hong S, Ma S. Predicting the postoperative blood coagulation state of children with congenital heart disease by machine learning based on real-world data. Transl Pediatr 2021;10(1):33–43. Available from: https://doi.org/10.21037/tp-20-238.

[47] Kikano S, Kannankeril PJ. Precision medicine in pediatric cardiology. Pediatr Ann 2022;51(10). Available from: https://doi.org/10.3928/19382359-20220803-05.

[48] Troisi J, Cavallo P, Richards S, et al. Noninvasive screening for congenital heart defects using a serum metabolomics approach. Prenat Diagn 2021;41(6):743–53. Available from: https://doi.org/10.1002/pd.5893.

[49] Bahado-Singh RO, Vishweswaraiah S, Aydas B, et al. Precision cardiovascular medicine: artificial intelligence and epigenetics for the pathogenesis and prediction of coarctation in neonates. J Matern Fetal Neonatal Med 2022;35(3):457–64. Available from: https://doi.org/10.1080/14767058.2020.1722995.

[50] McCrindle BW, Rowley AH, Newburger JW, et al. Diagnosis, treatment, and long-term management of Kawasaki disease: a scientific statement for health professionals from the American Heart Association. Circulation 2017;135(17). Available from: https://doi.org/10.1161/CIR.0000000000000484.

[51] Wang T, Liu G, Lin H. A machine learning approach to predict intravenous immunoglobulin resistance in Kawasaki disease patients: a study based on a Southeast China population. PLoS One 2020;15(8):e0237321. Available from: https://doi.org/10.1371/journal.pone.0237321.

[52] Reddy CD, Van den Eynde J, Kutty S. Artificial intelligence in perinatal diagnosis and management of congenital heart disease. Semin Perinatol 2022;46(4):151588. Available from: https://doi.org/10.1016/j.semperi.2022.151588.

[53] Rajkomar A, Dean J, Kohane I. Machine learning in medicine. N Engl J Med 2019;380(14):1347–58. Available from: https://doi.org/10.1056/NEJMra1814259.

CHAPTER

35

Artificial intelligence in anesthesiology

Sean McManus, Reem Khatib and Piyush Mathur

Department of General Anesthesiology, Anesthesiology Institute, Cleveland Clinic, Cleveland, OH, United States

Introduction

The field of anesthesiology and perioperative medicine is uniquely positioned to leverage artificial intelligence to improve patient care and streamline the delivery of anesthesia. A tremendous amount of data is generated in the intraoperative and perioperative period. With the proper tools, such data can be used to create classification and prediction models that augment the delivery of perioperative care, identify high risk patients, and predict clinically significant events in real time. The potential benefits of utilizing artificial intelligence and machine learning techniques for the anesthesiologist include: decreasing cognitive load, reducing cost with increased access to care, providing large scale data-driven evidence for clinical decisions, and increasing standardization of care [1].

The phrase "perioperative intelligence" has been used to describe methods of artificial intelligence as they apply to perioperative medicine [2]. Perioperative intelligence consists of three foundational domains upon which current and future work aims to derive the most benefit [2]. The domains include identification of at-risk patients, early detection of complications, and timely and effective treatment [2].

At this point in the early life cycle of artificial intelligence applications for anesthesiology and perioperative medicine, there are many original publications both developing and validating machine-learning models and algorithms [3]. Generally speaking, the current state of research for artificial intelligence in anesthesiology can be thought of as falling into six distinct areas: (1) predictive analytics, (2) image analysis, (3) smart devices, (4) pain management, (5) education, and (6) quality and administration.

Artificial Intelligence in Clinical Practice
DOI: https://doi.org/10.1016/B978-0-443-15688-5.00026-7

Predictive analytics

Predictive analytics in anesthesiology broadly describes the implementation of artificial intelligence techniques to predict intraoperative adverse events accurately, risk stratify patients in the perioperative period, and predict depth of anesthesia.

Intraoperative hypotension results in worse postoperative outcomes, such as acute kidney injury, stroke, and myocardial infarction. Predicting intraoperative hypotension has become an increasingly popular area of interest for artificial intelligence models [4,5]. As such, predictive tools continue to be developed that alert the clinician to impending hypotensive events so that they may treat appropriately. Additionally, individualized intraoperative fluid responsiveness is another area in which models have demonstrated benefit to the anesthesiologist [6].

Measuring and predicting intraoperative depth of anesthesia has the potential to reduce intraoperative awareness and postoperative delirium [7]. Utilization of deep convolutional neural networks based on raw electroencephalogram (EEG) signal has been shown to differentiate between transition points in depth of anesthesia [8]. Similarly, a deep learning algorithm was leveraged with a data repurposing framework to predict anesthetic depth using sleep EEG [9]. As anesthetic states continue to be better characterized by artificial intelligence, pharmacokinetic drug modeling has also shown tremendous strides. The conventional compartment model of pharmacokinetics may be antiquated as shown by a neural network that was developed and accurately described high-resolution pharmacokinetic data for Propofol [10].

Perioperative risk prediction and stratification is a key area in which artificial intelligence can be deployed to improve morbidity, mortality, and patient satisfaction. For instance, a machine learning model was developed that identifies, with a high degree of accuracy, patients who are at high risk for opioid overdose [11]. Expanding on this idea, a machine learning model was developed to predict postoperative opioid requirements for ambulatory surgical patients [12]. Artificial intelligence models are particularly useful in the perioperative period for predicting patients at high risk for poor postoperative outcomes. A model predicting mortality within 48 hours of admission to the intensive care unit has also demonstrated a high degree of accuracy [13].

Image analysis

Artificial intelligence for image analysis as it applies to the anesthesiologist is an area that has shown rapid advances and innovation in recent years. As experts in airway management, there has been a great deal of effort to predict patients with a difficult airway diagnosis in order to properly prepare and allocate appropriate resources to manage these patients. One method of using artificial intelligence and deep learning to support this effort is by using facial recognition algorithms to predict an airway difficulty [14–16].

As ultrasound has become a staple for the practicing anesthesiologist, artificial intelligence applications have been developed to aid ultrasound-guided regional anesthesia procedures and echocardiography. One such application is using machine learning to determine ejection fraction on transthoracic echo [17]. Standard determination of ejection fraction by manually tracing endocardial borders is operator-dependent and can be time consuming. One group was able to measure ejection fraction with a high degree of accuracy in an average of 8 seconds using their artificial intelligence model [18]. The use of speckle-tracking echocardiographic data has also accelerated machine learning applications, as seen by the determination of physiologic versus pathologic hypertrophic remodeling patterns using a feature selection algorithm [17,19].

Regional anesthesia techniques with ultrasound guidance is an area in which machine learning algorithms can aid both experienced and less-experienced clinicians in the identification of anatomic landmarks [20]. However, accurate identification of nerve bundles remains a challenge as the size, shape, and echogenicity of nerve bundles is constantly changing while scanning with an ultrasound [21]. While still in its infancy, relying on robotic assistance for performance of ultrasound-guided nerve blocks may revolutionize the field of regional anesthesia as robotic laparoscopic procedures have done for surgery [22].

Smart devices

Broadly speaking, "smart" devices in the operating room have focused on closed loop systems for automated delivery of anesthesia, goal directed fluid therapy, vasopressor administration, mechanical ventilation, and titrating anesthetic delivery based on depth of anesthesia [23].

In early 2021 the first fully continuous deep reinforcement learning algorithm was successfully developed for automatic delivery and dosing of Propofol for general anesthesia [24]. The algorithm that was developed recommended Propofol bolus doses that were consistent with anesthesiologist-administered boluses, and the algorithm incentivized dose minimization. To date, there are two large randomized trials that have evaluated closed loop titration of vasopressors with spinal anesthesia [25,26]. A deep learning model was able to accurately predict bispectral index during target-controlled delivery of Propofol and Remifentanil, which helps ensure optimal depth of anesthesia on a continuous spectrum [27].

Pain management

Pain is an inherently subjective measure from patient-to-patient, but artificial intelligence techniques have been employed to systematically classify and predict those afflicted with certain types of pain. Additionally, prediction of response to certain pain treatments represents an area of ongoing interest in the pain management community. Big data analytics over the past half century for opioid treatment programs can predict which patients are most likely to succeed in an opioid treatment program and which patients may require further resource utilization [28].

Prediction and classification of pain intensity has rightfully received a great deal of attention from researchers in the artificial intelligence space. Machine learning algorithms have demonstrated benefit in predicting pain intensity across a variety of pain phenotypes [29]. However, standardization of reporting and external validation of results still remain a barrier to implementation into clinical workflows. One study identified type of surgery, anxiety, and pain with movement as three predictive factors for neuropathic pain after breast cancer surgery [30]. Facial image analysis has also shown potential for identification of patients experiencing pain [31,32].

Education

Education in the field of anesthesiology is ripe for harnessing the powers of artificial intelligence to improve feedback, identify gaps in knowledge, and predict performance on standard examinations [33]. Anesthesiology trainees receive performance feedback on a regular basis that helps identify strengths and weaknesses. High quality feedback for trainees is often times very beneficial and actionable, while low quality feedback has little utility in improving performance. Once such study used a machine learning model to rapidly screen for low quality feedback from attending anesthesiologists to anesthesiology residents, which improves the delivery of actionable feedback [34]. Prediction of standardized exam success in anesthesia is important in identifying trainees who may benefit from increased study time. One study used an artificial neural network to predict board exam scores based on in-training exam results [35].

Quality and administration

A tremendous amount of electronically captured data is generated in both the intraoperative and perioperative periods. Quality and billing measures are areas which can benefit from artificial intelligence to leverage the power and insight of large data sets. Understanding aspects of practice inefficiencies and identifying areas for quality improvement are vital toward improving value-driven care. A study using 1.1 million procedure records from 16 hospitals and implemented a supervised machine learning model to classify anesthesiology Current Procedural Terminology codes [36]. Understanding utilization of healthcare resources is also of great value when determining allocation of resources. Another study used machine learning techniques to classify "super-utilizers" of healthcare resources into five distinct cohorts among those patients undergoing total hip arthroplasty and total knee arthroplasty [37]. An important quality metric for hospitals is hospital readmission after discharge which is not as well-studied in the surgical patient population as it is in the medical patient population. One study implemented a machine learning model to predict postoperative readmission at various time points after discharge [38]. Amongst the ambulatory surgery patients who received anesthesia, analyzing patient survey comments using AI, also holds the potential for improving patient experience which is another important hospital quality metric [39].

Future directions

While artificial intelligence has demonstrated great value in the previously outlined arenas, an understanding of both the capabilities and limitations of artificial intelligence is critical toward directing future research. Recommendations focused on education of clinicians, development of platforms for collaboration and data sharing, validation and implementation of AI in perioperative medicine, provide us with a comprehensive approach toward successful implementation of AI in anesthesiology [40].

References

[1] Char DS, Burgart A. Machine-learning implementation in clinical anesthesia: opportunities and challenges. Anesth Analg 2020;130(6):1709–12. Available from: https://doi.org/10.1213/ANE.0000000000004656.

[2] Maheshwari K, Ruetzler K, Saugel B. Perioperative intelligence: applications of artificial intelligence in perioperative medicine. J Clin Monit Comput 2020;34:625–8. Available from: https://doi.org/10.1007/s10877-019-00379-9.

[3] Lonsdale H, Jalali A, Gálvez JA, Ahumada LM, Simpao AF. Artificial intelligence in anesthesiology: hype, hope, and hurdles. Anesth Analg 2020;130(5):1111–13.

[4] Maheshwari K, Shimada T, Yang D, Khanna S, Cywinski JB, Irefin SA, et al. Hypotension prediction index for prevention of hypotension during moderate- to high-risk noncardiac surgery. Anesthesiology 2020;133 (6):1214–22. Available from: https://doi.org/10.1097/ALN.0000000000003557.

[5] Maheshwari K, Buddi S, Jian Z, Settels J, Shimada T, Cohen B, et al. Performance of the Hypotension Prediction Index with non-invasive arterial pressure waveforms in non-cardiac surgical patients. J Clin Monit Comput 2021;35(1):71–8. Available from: https://doi.org/10.1007/s10877-020-00463-5.

[6] Maheshwari K, Malhotra G, Bao X, Lahsaei P, Hand WR, Fleming NW, et al. Assisted fluid management software guidance for intraoperative fluid administration. Anesthesiology 2021;135(2):273–83. Available from: https://doi.org/10.1097/ALN.0000000000003790.

[7] Evered LA, Chan MTV, Han R, Chu MHM, Cheng BP, Scott DA, et al. Anaesthetic depth and delirium after major surgery: a randomised clinical trial. Br J Anaesth 2021;127(5):704–12. Available from: https://doi.org/10.1016/j.bja.2021.07.021.

[8] Patlatzoglou K, Chennu S, Gosseries O, Bonhomme V, Wolff A, Laureys S. Generalized prediction of unconsciousness during propofol anesthesia using 3D convolutional neural networks. Annu Int Conf IEEE Eng Med Biol Soc 2020;2020:134–7.

[9] Belur Nagaraj S, Ramaswamy SM, Weerink MAS, Struys M. Predicting deep hypnotic state from sleep brain rhythms using deep learning: a data-repurposing approach. Anesth Analg 2020;130(5):1211–21.

[10] Ingrande J, Gabriel RA, McAuley J, Krasinska K, Chien A, Lemmens HJM. The performance of an artificial neural network model in predicting the early distribution kinetics of propofol in morbidly obese and lean subjects. Anesth Analg 2020;131(5):1500–9.

[11] Dong X, Rashidian S, Wang Y, et al. Machine learning based opioid overdose prediction using electronic health records. AMIA Annu Symp Proc 2019;2019:389–98.

[12] Nair AA, Velagapudi MA, Lang JA, et al. Machine learning approach to predict postoperative opioid requirements in ambulatory surgery patients. PLoS One 2020;15(7):e0236833.

[13] McManus Sean, Almuqati Reem, Khatib Reem, Khanna Ashish, Cywinski Jacek, Papay Francis, et al. 1214: Machine learning-based early mortality prediction at the time of ICU admission. Crit Care Med 2022;50 (1):607. Available from: https://doi.org/10.1097/01.ccm.0000811180.59073.fd.

[14] Cuendet GL, Schoettker P, Yücc A, Sorci M, Gao H, Perruchoud C, et al. Facial image analysis for fully automatic prediction of difficult endotracheal intubation. IEEE Trans Biomed Eng 2016;63(2):328–39. Available from: https://doi.org/10.1109/TBME.2015.2457032.

[15] Tavolara TE, Gurcan MN, Segal S, Niazi MKK. Identification of difficult to intubate patients from frontal face images using an ensemble of deep learning models. Comput Biol Med 2021;136:104737. Available from: https://doi.org/10.1016/j.compbiomed.2021.104737.

[16] Hayasaka T, Kawano K, Kurihara K, et al. Creation of an artificial intelligence model for intubation difficulty classification by deep learning (convolutional neural network) using face images: an observational study. J Intensive Care 2021;9:38. Available from: https://doi.org/10.1186/s40560-021-00551-x.

[17] Chen X, Owen CA, Huang EC, et al. Artificial intelligence in echocardiography for anesthesiologists. J Cardiothorac Vasc Anesth 2021;35(1):251–61.

[18] Knackstedt C, Bekkers SC, Schummers G, Schreckenberg M, Muraru D, Badano LP, et al. Fully automated versus standard tracking of left ventricular ejection fraction and longitudinal strain: the FAST-EFs multicenter study. J Am Coll Cardiol 2015;66(13):1456–66. Available from: https://doi.org/10.1016/j.jacc.2015.07.052.
[19] Narula S, Shameer K, Salem Omar AM, Dudley JT, Sengupta PP. Machine-learning algorithms to automate morphological and functional assessments in 2D echocardiography. J Am Coll Cardiol 2016;68(21):2287–95. Available from: https://doi.org/10.1016/j.jacc.2016.08.062.
[20] Bowness JS, El-Boghdadly K, Woodworth G, Noble JA, Higham H, Burckett-St Laurent D. Exploring the utility of assistive artificial intelligence for ultrasound scanning in regional anesthesia. Reg Anesth Pain Med 2022;47(6):375–9. Available from: https://doi.org/10.1136/rapm-2021-103368.
[21] McKendrick M, Yang S, McLeod GA. The use of artificial intelligence and robotics in regional anaesthesia. Anaesthesia 2021;76(Suppl 1):171–81.
[22] Morse J, Terrasini N, Wehbe M, Philippona C, Zaouter C, Cyr S, et al. Comparison of success rates, learning curves, and inter-subject performance variability of robot-assisted and manual ultrasound-guided nerve block needle guidance in simulation. Br J Anaesth 2014;112(6):1092–7. Available from: https://doi.org/10.1093/bja/aet440.
[23] Wingert T, Lee C, Cannesson M. Machine learning, deep learning, and closed loop devices-anesthesia delivery. Anesthesiol Clin 2021;39(3):565–81.
[24] Schamberg Gabriel, et al. Continuous action deep reinforcement learning for propofol dosing during general anesthesia. Artif Intell Med 2022;123:102227.
[25] Ngan Kee WD, Tam YH, Khaw KS, et al. Closed-loop feedback computer-controlled phenylephrine for maintenance of blood pressure during spinal anesthesia for cesarean delivery: a randomized trial comparing automated boluses versus infusion. Anesth Analg 2017;125:117–23.
[26] Sng BL, Tan HS, Sia ATH. Closed-loop double-vasopressor automated system vs manual bolus vasopressor to treat hypotension during spinal anaesthesia for caesarean section: a randomised controlled trial. Anaesthesia 2014;69:37–45.
[27] Lee HC, Ryu HG, Chung EJ, Jung CW. Prediction of bispectral index during target-controlled infusion of propofol and remifentanil: a deep learning approach. Anesthesiology 2018;128(3):492–501. Available from: https://doi.org/10.1097/ALN.0000000000001892.
[28] Cui W, Bachi K, Hurd Y, Finkelstein J. Using big data to predict outcomes of opioid treatment programs. Stud Health Technol Inform 2020;272:366–9.
[29] Mari T, Henderson J, Maden M, Nevitt S, Duarte R, Fallon N. Systematic review of the effectiveness of machine learning algorithms for classifying pain intensity, phenotype or treatment outcomes using electroencephalogram data. J Pain 2022;23(3):349–69. Available from: https://doi.org/10.1016/j.jpain.2021.07.011.
[30] Juwara L, Arora N, Gornitsky M, Saha-Chaudhuri P, Velly AM. Identifying predictive factors for neuropathic pain after breast cancer surgery using machine learning. Int J Med Inf 2020;141:104170.
[31] Thiam P, Kestler HA, Schwenker F. Two-stream attention network for pain recognition from video sequences. Sens (Basel, Switz) 2020;20(3).
[32] Xin X, Lin X, Yang S, Zheng X. Pain intensity estimation based on a spatial transformation and attention CNN. PLoS One 2020;15(8):e0232412.
[33] Arora VM. Harnessing the power of big data to improve graduate medical education: big idea or bust? Acad Med 2018;93:833–4.
[34] Neves SE, Chen MJ, Ku CM, Karan S, DiLorenzo AN, Schell RM, et al. Using machine learning to evaluate attending feedback on resident performance. Anesth Analg 2021;132(2):545–55.
[35] Amirhajlou L, Sohrabi Z, Alebouyeh MR, et al. Application of data mining techniques for predicting residents' performance on pre-board examinations: a case study. J Educ Health Promot 2019;8:108.
[36] Burns ML, Mathis MR, Vandervest J, et al. Classification of current procedural terminology codes from electronic health record data using machine learning. Anesthesiology 2020;132(4):738–49.
[37] Hyer JM, Paredes AZ, White S, Ejaz A, Pawlik TM. Assessment of utilization efficiency using machine learning techniques: a study of heterogeneity in preoperative healthcare utilization among super-utilizers. Am J Surg 2020;220(3):714–20.

[38] Mišić VV, Gabel E, Hofer I, Rajaram K, Mahajan A. Machine learning prediction of postoperative emergency department hospital readmission. Anesthesiology 2020;132(5):968–80.
[39] Mathur P, Cywinski JB, Maheshwari K, Niezgoda J, Mathew J, do Nascimento CC, et al. Automated analysis of ambulatory surgery patient experience comments using artificial intelligence for quality improvement: a patient centered approach. Intell Based Med 2021;5:100043.
[40] Maheshwari K, Cywinski JB, Papay F, Khanna AK, Mathur P. Artificial intelligence for perioperative medicine: perioperative intelligence. Anesth Analg 2022. Available from: https://doi.org/10.1213/ANE.0000000000005952.

CHAPTER

36

Artificial intelligence in emergency medicine

Sameer Masood

Division of Emergency Medicine, University Health Network, Department of Medicine, University of Toronto, Toronto, ON, Canada

Artificial intelligence (AI) use cases in medicine have largely focused on niche areas within medicine, such as genomics and oncology; however, more recently there has been a surge in the application of AI-based models in emergency medicine (EM) [1].

The emergency department (ED) has historically proven to be a challenging environment for traditional research, given that it inherently has highly variable and dynamic processes that intersect with diverse and complex patient presentations. ED providers are expected to make rapid decisions to risk stratify patients with very limited information available to them. As a result, there is a heavy reliance on using an algorithmic approach to decision-making in the ED (e.g., Canadian Triage and Acuity Score used by triage nurses to risk stratify and triage patients) [2]. As such, AI provides a unique value proposition for the ED environment as it has the potential to combine multiple variables that are constantly in flux and provide rapid information that can be leveraged to make real-time decisions.

The body of literature on the application on AI in EM is quite heterogenous, both in terms of its clinical application, but also the type of AI involved. 90% of published AI studies in EM have focused on acute care (ED and trauma bay), with the remaining 10% focusing on prehospital care [1]. Studies can be broadly categorized into five groups—(1) triage, (2) diagnosis, (3) prediction, (4) decision-making, and (5) operations (Fig. 36.1). Improving diagnosis in the ED is the largest category with 25% of published studies focusing on this aspect. About half of the studies that focus on diagnosis deal with the application of AI to diagnostic imaging in the acute setting.

In terms of type of AI used, the vast majority of AI applications in emergency medicine use a supervised machine learning framework (Fig. 36.2) and tackle well established clinical outcomes of interest. This generally entails using a large retrospective research database (often from a prior study) to produce more accurate models compared to current decision tools that been developed using traditional research methodology. An AI-based

Artificial Intelligence in Clinical Practice
DOI: https://doi.org/10.1016/B978-0-443-15688-5.00043-7

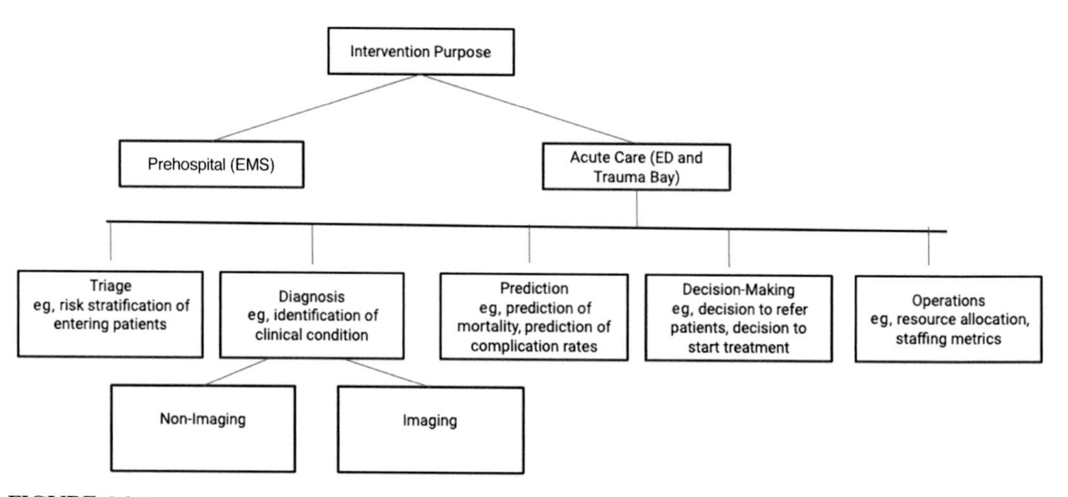

FIGURE 36.1 Purpose of intervention in emergency medicine. *ED*, emergency department; *EMS*, emergency medical services [1].

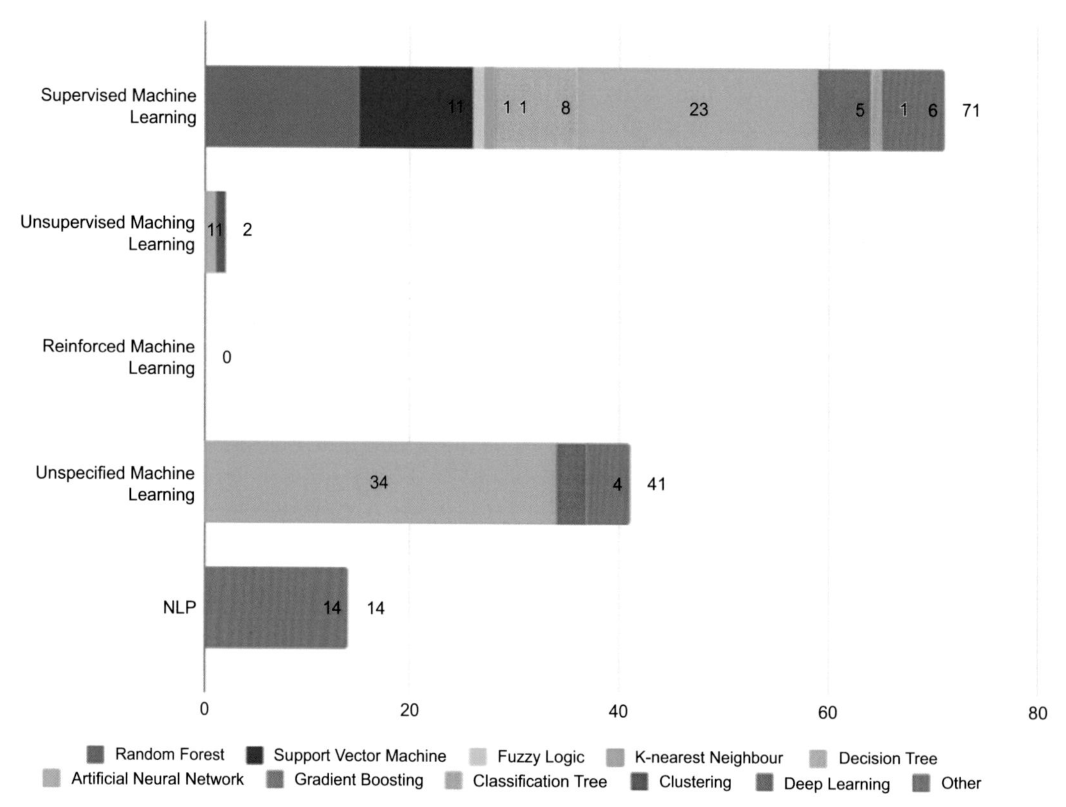

FIGURE 36.2 Types of artificial intelligence interventions in emergency medicine. *NLP*, Natural language processing [1].

model out-performed the quick sequential organ failure assessment score for sepsis detection in an ED cohort study [3]. Similarly, an AI model was superior to the well validated pneumonia severity index (PSI) in predicting mortality in ED patients with pneumonia [4]. In the pediatric trauma literature, Bertsimas et al. showed that a machine learning model had better specificity than the Pediatric Emergency Care Applied Research Network score in patients with minor head injuries [5].

A subset of AI-based research in EM has compared AI models to physicians directly, and in the majority of these studies, the AI model outperformed physicians [1]. For example, an AI model used to detect intracranial hemorrhage on computed tomography (T) scans was 82.2% sensitive, compared to 62.2% by emergency physicians [6]. As we deal with increasingly complex clinical scenarios, ability of AI to process and balance multiple clinical variables will likely enhance our decision-making ability, instead of relying on gestalt or existing decision tools, which can be quite heterogenous across clinicians. The study by Lindsay et al., which demonstrated that augmenting physician interpretation of X-rays for fracture detection was superior to physician interpretation alone is a good example of this [7]. These studies provide us with insights into the future of EM care, where AI will increasingly play a role in augmenting clinical care, without replacing the need for clinician expertise.

The adoption of AI in clinical EM is still in its infancy, as most of the work done till date is largely academic in nature with little impact on bedside clinical medicine. Future work needs to focus on moving from the retrospective domain to evaluating interventions in a prospective and pragmatic fashion with a focus on implementation. Careful consideration needs to be given to ensuring AI implementation uses a patient focused approach, and parallels work done in understanding the ethical and medico-legal ramifications of adopting AI-based care models.

References

[1] Kirubarajan A, Taher A, Khan Bhsc S, Masood S, General T. Artificial intelligence in emergency medicine: a scoping review. J Am Coll Emerg Phys Open 2020;1(6):1691–702. Available from: https://doi.org/10.1002/EMP2.12277.

[2] Murray MJ. The Canadian triage and acuity scale: a Canadian perspective on emergency department triage. Emerg Med 2003;15(1):6–10. Available from: https://doi.org/10.1046/J.1442-2026.2003.00400.X.

[3] Delahanty RJ, Alvarez JA, Flynn LM, Sherwin RL, Jones SS. Development and evaluation of a machine learning model for the early identification of patients at risk for sepsis. Ann Emerg Med 2019;73(4):334–44. Available from: https://doi.org/10.1016/j.annemergmed.2018.11.036.

[4] Bae Y, Moon H, Kim S. Predicting the mortality of pneumonia patients visiting the emergency demartment through machine learning. Eur. Respir. J. 2018;52:PA2635. Available from: https://doi.org/10.1183/13993003.congress-2018.pa2635.

[5] Bertsimas D, Dunn J, Steele DW, Trikalinos TA, Wang Y. Comparison of machine learning optimal classification trees with the pediatric emergency care applied research network head trauma decision rules. JAMA Pediatrics 2019;173(7):648–56. Available from: https://doi.org/10.1001/jamapediatrics.2019.1068.

[6] Sinha M, Kennedy C, Ramundo ML. Artificial neural network predicts CT scan abnormalities in pediatric patients with closed head injury. J Trauma 2001. <https://journals.lww.com/jtrauma/Fulltext/2001/02000/Artificial_Neural_Network_Predicts_CT_Scan.18.aspx>.

[7] Lindsey R, Daluiski A, Chopra S, Lachapelle A, Mozer M, Sicular S, et al. Deep neural network improves fracture detection by clinicians. Proc Natl Acad Sci U S A 2018;115(45):11591–6. Available from: https://doi.org/10.1073/PNAS.1806905115/SUPPL_FILE/PNAS.1806905115.SAPP.PDF.

CHAPTER

37

Artificial intelligence in allergy and immunology

Harold Shin and Nicholas L. Rider

Division of Clinical Informatics, Liberty University College of Osteopathic Medicine, Lynchburg, VA, United States

Introduction

The interest and role of artificial intelligence (AI) in the field of allergy–immunology has grown significantly over the past decade. With the advance of AI and machine learning (ML) in other fields of medicine, the American Academy of Allergy Asthma and Immunology (AAAAI) outlined a framework report to review what has been done and point toward future opportunities [1]. While the AAAAI has not stated any official or formal position regarding AI use and its implementation, it is acknowledged that AI will be a continued focus of emphasis in the following years to come.

A major application of AI across healthcare domains, including allergy–immunology, involves automated disease detection [1]. A 2019 study by Rider et al. clearly established the impact of AI in clinical immunology to improve disease diagnosis with the use of computational methods to identify risk of primary immunodeficiencies (PI) through analysis of claims data [2]. Availability of electronic health record (EHR) data has also proven to be an important catalyst for the advent, research, and possible application of AI in field of healthcare. Leveraging EHR data and merging this rich source of clinical information with multiomics datasets, such as genomics, via AI has propelled advances in clinical phenotyping [3]. Clinical phenotyping of PI specifically, has allowed for an exponential increase in the discovery of distinct, new PI disorders [4]. The use of either structured or unstructured EHR data continues to present itself as the largest target for AI in the field of allergy–immunology. Natural language processing (NLP) analysis of EHR text represents a tremendous opportunity for gleaning insight about PI and improving detection rates [1]. Use of NLP has also found impact into phenotyping a variety of conditions related to allergy and immunology, such as asthma [5,6].

Artificial Intelligence in Clinical Practice
DOI: https://doi.org/10.1016/B978-0-443-15688-5.00048-6

The application of AI in allergy–immunology also extends beyond automated disease detection and includes facilitating clinical decision support (CDS) [1]. Assisting clinicians with the most efficient treatment plans and optimal drug choices while minimizing risks and disease/symptom exacerbations is a key component of CDS. With the help of AI, clinicians can expect to forecast risk and trends and ultimately be able to treat patients earlier and better while saving costs for both the patient and the healthcare system [7]. Additional approaches include novel disease gene identification for clinical trial candidate selection, discovering and identifying disease biomarkers, and much more [8,9]. With the amalgamation of all that is AI, healthcare, and big data, it is important to develop a learning health system (LHS) that will organize and perpetuate the ideal system that both patients and clinicians long for [7].

Current examples of artificial intelligence implementation in allergy and immunology

Here we highlight a few current examples of AI implementation in five unique diseases: PI, asthma, drug allergies, contact dermatitis, and COVID-19. Some technical information concerning popular AI/ML models in this field will be explained in the first section, but are not all encompassing (Table 37.1).

Primary immunodeficiency

Epidemiologic assessments about the diagnostic rates for PI over the past few decades suggest a glaring gap in progress [4]. In the past few years the use of AI to improve screening, risk stratification, and disease phenotyping has the potential to improve time to diagnosis and outcomes [31]. A significant advance has been made in analyzing claims database information via a rule-based algorithm called the Software for Primary Immunodeficiency Recognition Intervention and Tracking (SPIRIT) to stratify an entire population for PI risk. Rider et al., [2] This would allow a significantly early timeframe to diagnosis and the algorithm also assigns a risk level (low, medium, high) to assist the physician's decision making. This rule-based classifier is generalizable and can enable CDS. SPIRIT is continuously updated to optimize diagnosis, reduce morbidity, mortality, and costs.

Other approaches for leveraging AI for PI include use of probabilistic models to help classify the various diseases and conditions in this field [32]. A Bayesian network (BN) is an example of a probabilistic model that is ideal to combine conditional probabilities alongside domain expertise to calculate probabilities of interest, for which in this case, to calculate the likelihood of disease [33]. BNs use a visual nodal structure between parent and child nodes to connect relevant features to predict a desired outcome. PI Prob is one such BN developed to use structured EHR data to classify PI versus controls with an impressive area under the curve (AUC) of 0.945 [11]. PI Prob was not only able to differentiate PI from the control, but also further classify these PI patients into the ten different International Union of Immunological Societies diagnostic categories of PI [4]. BNs have

TABLE 37.1 Artificial intelligence applications in allergy and immunology.

Area of allergy and immunology	AI application	Machine type	Study
Primary immunodeficiency	Disease and risk prediction using EHR	SPIRIT, LR, EN, RF, SVM, NLP	[2,10]
	Disease phenotyping	BN (PI Prob)	[11]
Asthma	Disease and risk prediction using EHR	NLP, TL	[12–16]
	Disease phenotyping	NLP (Predictor Pursuit)	[17]
	Disease and exacerbation control/management	LR, BN, SVM	[18,19]
	Quality of care	NLP	[20]
Atopic dermatitis	Disease and risk prediction using EHR	NLP, NN (ASCORAD)	[21,22]
	Disease and risk prediction using transcriptome and microbiota data	LR, SVM, RF	[23]
	Disease management and biomarker identification	RF, LR, EN, GBT	[8,24]
Drug allergy	Disease and risk prediction using EHR	LR, DT, NN, LR*	[25,26]
	Pharmacological development	DT, kNN, NN, SVM	[27]
	Epidemiological	NLP	[28]
COVID-19	Disease diagnosis	NN, RF, EN	[29]
	Epidemiological	LASSO	[30]

SPIRIT, Software for Primary Immunodeficiency Recognition Intervention and Tracking; LR, logistic regression; LR*, linear regression; EN, elastic nets; RF, random forest: GBT, gradient boosted trees; SVM, support vector machine; NLP, natural language processing; BN, Bayesian network; TL, transfer learning; NN, neural network; ASCORAD, automatic scoring atopic dermatitis; DT, decision tree; kNN, k-nearest neighbor classification; LASSO, least absolute shrinkage and selection operator. (This table is not all-encompassing but gives a brief summary of the literature reviewed for this chapter.)

the ability to continually learn with the inclusion of new and additional data into the algorithm to update weights of importance, optimize node connections and potentially improve the model's overall performance [34].

A recent study has taken a different step of diagnosing PI earlier with EHR data by creating an ML model specifically using EHR elements related to prior symptomatic treatments rather than ICD codes [10]. For this study, multiple learning models (e.g., logistic regression (LR), elastic nets (EN), Random forests (RF)) were utilized with a final classification made by averaging the predictions from these three models. It is important to note the utility of distinct ML algorithms and how they can be used as an ensemble to optimize performance. Mayampuranth et al. compared PI patients to a control group that consisted of asthma patients and noted that the frequency of radiological orders may be an important marker for the presence of PI in patients due to recurrent sinopulmonary infections.

Asthma

Given the high prevalence and global impact of asthma, use of AI to improve outcomes for patients with this condition is a natural fit. As such, NLP has proven effective for phenotyping asthma patients, ascertaining asthma severity, and asthma control via analysis of EHR text [12–14]. In one such study, investigators used NLP to automate assessment of the commonly used Asthma Predictive Index criteria to ascertain asthma status in children [15]. Apart from the use of EHR data, another study analyzed treatment response data to develop a machine named predictor pursuit to discover phenotypes of asthma and further predict asthma control [17].

In addition to predicting asthma severity and facilitating diagnosis, the development of systems for improving asthma outcomes has recently been showcased relating to the prediction of asthma exacerbations. One study uses an ML model that analyzes EHR data to specifically and accurately predict asthma exacerbations [18]. In contrast, instead of using EHR data, another machine using a BN and support vector machine (SVM) analyzed asthma patient telemonitoring data to aid in earlier prediction of asthma exacerbations [19]. Yet, another study examined use of several ML models for predicting changes in peak expiratory flow rate [16]. Both exacerbation prediction studies show great potential to aid in CDS. In the Finkelstein and Jeong study, investigators highlight the use case for telemonitoring other chronic diseases including asthma. Other asthma studies shift the perspective toward provider performance. For example, one study utilized an NLP model to target clinician's asthma guideline adherence by analyzing the clinical notes extracted from EHR free text [20]. Taken together, we note the emergence of AI/ML across many use cases aimed at improving asthma patient outcomes and quality improvements in care.

Atopic dermatitis

Atopic dermatitis (AD), also commonly known as eczema, is a complex chronic inflammatory skin disease that initially presents in children with much relation alongside allergies and asthma [35]. Broadly speaking, most research studies, especially those concerning AD, focus on individual-omics data alongside binary clinical outcomes. However, given the complex multifactorial nature of AD and potential for coupling to multiomic data, AI presents an ideal opportunity for gleaning insight to better understand AD. In one study, Ghosh et al. utilizes a multiomics approach for driving toward precision medicine in AD care. This study focuses on AD genotype-phenotype relationships through a big data analysis approach which would not have been possible without AI/ML. Like other disorders, use of unstructured EHR data to facilitated AD diagnosis has been explored. One study uses a NLP model to analyze unstructured data and compare its usefulness to using structure data alone; with results showing a 10-fold improvement in its sensitivity of identification of AD [21]. NLP has also been used to further understand patients' disease perception [36]. In yet another study, investigators used transcriptome and microbiota data for ML model building to predict AD risk successfully [23]. This method also allowed for the researchers to not only identify a set of related genes and microbiota, but also discover additional genes and microorganisms of relevance to AD pathogenesis.

One commonly used tool for diagnosing and monitoring AD is the SCOring Atopic Dermatitis (SCORAD) which has recently been reimagined in an automated version by using neural networks. Automatic SCORAD was able to measure AD severity by analyzing images and ultimately showing comparable results to the standard SCORAD conducted through manual human assessment [22]. ML is also being leveraged to improve AD disease management. Additionally, ML has been used to predict nonresponse indicators to dupilumab in patients with AD and for identification of gene expression patterns from skin biopsy materials [8,24]. This allows for a better understanding of the mechanisms behind these allergies through the identification of biomarkers.

Drug allergy and adverse drug events

Drug allergies and related adverse events have been another target by researchers seeking to utilize AI/ML in the field of allergy–immunology. ML is ideally suited and used to aggregate and analyze considerations about patient history, drug–drug interactions, and what other drugs are currently available. Linear regression and decision tree (DT) models have been able to accurately predict patients with future risk of beta-lactam allergic reactions in adults with similar performances [25]. Another study developed an artificial neural network (ANN) model to also predict beta-lactam hypersensitivity, that in which performed superiorly in comparison to their LR model [26]. In addition to superior accuracy in predicting beta-lactam hypersensitivity, the ANN was also able to classify reaction risk, thus contributing to AI's role in CDS which in this case leads to the suggestion of an initiation of an open challenge for low-risk patients. The study continues to also highlight the need for further research when it comes to finding the ideal AI model that best fits the needs of the unique data. Lastly, another study used ML to predict drug-induced liver injury and provide inference about drug mechanisms/interactions [27]. Here we see ML used to develop a screening tool aid in drug development and pharmacological practice. In addition to pharmacological applications, AI has the ability to further understand the epidemiology of an allergic drug reaction [28].

COVID-19

In light of recency, the COVID pandemic itself brough forth the use of AI in various immunological applications. At the patient level, a study used ANNs and random forest learning algorithms in order to analyze blood count data to help diagnose SARS-Cov-2 infections with an AUC of 95% [29]. Additionally, data science approaches may be used at the population level to help track the spread of COVID-19 and the impact such a spread can make on a vulnerable population [37]. The use of AI in immunology extends to other specialties including epidemiology and public health. This highlights the need of a concerted effort from the healthcare community (clinicians, researchers, public health workers) as a whole in order to fully utilize AI to its full potential [37,38]. ML has also shown the potential to predict how the virus spreads and identify which patients are more likely to die [30].

Challenges with artificial intelligence in allergy and immunology

While optimism about AI is warranted, challenges are evident and shared similarly by most medical specialties. These shared challenges include the complexities of EHR data, health inequities, ethical complications, and biases, and much more [1]. The wide variety of nonstandardized data sources and ethical/training biases require a constant, careful fine tuning and weighting in order for the AI to work properly as desired.

Asides from the general challenges AI faces, challenges specific to the field of allergy and immunology are not as well apparent due to the nature of its recency and being in its initial stages. However, one unique challenge in this field concerns the use of AI as a diagnostic tool to assist clinicians in decision making. It is a challenge to accurately diagnose PI due to the nature of the disease concerning its low prevalence and clinical heterogeneity [39]. This same challenge applies to many other diseases mentioned throughout this chapter. As much of a challenge this poses to healthcare providers, it will pose the same to AI. With the advance of AI applications in allergy and immunology, new challenges are sure to come.

Future directions

The future is bright for the field of allergy–immunology, and while still in the early stages, exponential growth in the utility of AI/ML and associated improvements is expected. Due to the flexible and learning nature of AI, aforementioned machines, such as SPIRIT and PI Prob, alongside many others are continuously being updated and learning new datasets. Advances in accompanying technologies, such as data systems and EHR designs, are sure to come, and with it, an avenue for better streamlining data for input into AI [1]. New technologies, such as the increased use of smartphones, and specific healthcare apps, such as "The Drug Allergy App," have the potential to facilitate and streamline CDS [40]. With the rapid increase of various forms of data constantly being added today, finding a way to integrate this multiomic data will pave the way as the norm for clinical/molecular phenotyping [41]. With healthcare being a multidisciplinary approach of multiple specialties, a broader implementation of AI even in other fields of pathology and radiology will be able to help improve care of patients with allergy and immunology [42,43]. Big data, the vast amount of multiomic data, only continues to grow larger and we must find a way to harness all this data using clinical informatics and AI to develop an efficient LHS [44].

Conclusion

The specialty of allergy–immunology has become a new player for advancing use of AI. Further research in developing new forms of AI and improving current forms of AI alongside new relevant technologies, such as genomic analyses, prove to be the immediate future of improving patient outcomes. Ultimately, it is imperative that current and future allergy–immunology specialists stay engaged with other stakeholders to transition to a future where AI in healthcare stays relevant and essential.

Major takeaways

- Artificial intelligence is driving improvements in patient diagnosis, CDS and integration of omic-data within the field of allergy–immunology.
- Subdomains of allergy–immunology where AI is having particular impact include—PI, asthma, drug allergy, and AD.
- Use of AI to create an LHS for patients with allergic and immunologic disease is an emerging opportunity.

References

[1] Khoury P, Srinivasan R, Kakumanu S, Ochoa S, Keswani A, Sparks R, et al. A framework for augmented intelligence in allergy and immunology practice and research—a work group report of the AAAAI health informatics, technology, and education committee. J Allergy Clin Immunol Pract 2022;10:1178–88. Available from: https://doi.org/10.1016/j.jaip.2022.01.047.

[2] Rider NL, Miao D, Dodds M, Modell V, Modell F, Quinn J, et al. Calculation of a primary immunodeficiency "risk vital sign" via population-wide analysis of claims data to aid in clinical decision support. Front Pediatr 2019;7:70. Available from: https://doi.org/10.3389/fped.2019.00070.

[3] Richesson RL, Sun J, Pathak J, Kho AN, Denny JC. Clinical phenotyping in selected national networks: demonstrating the need for high-throughput, portable, and computational methods. Artif Intell Med 2016;71:57–61. Available from: https://doi.org/10.1016/j.artmed.2016.05.005.

[4] Tangye SG, Al-Herz W, Bousfiha A, Chatila T, Cunningham-Rundles C, Etzioni A, et al. Human inborn errors of immunity: 2019 update on the classification from the International Union of Immunological Societies Expert Committee. J Clin Immunol 2020;40:24–64. Available from: https://doi.org/10.1007/s10875-019-00737-x.

[5] Juhn Y, Liu H. Artificial intelligence approaches using natural language processing to advance EHR-based clinical research. J Allergy Clin Immunol 2020;145:463–9. Available from: https://doi.org/10.1016/j.jaci.2019.12.897.

[6] Sharma H, Mao C, Zhang Y, Vatani H, Yao L, Zhong Y, et al. Developing a portable natural language processing based phenotyping system. BMC Med Inf Decis Mak 2019;19:78. Available from: https://doi.org/10.1186/s12911-019-0786-z.

[7] Chin A, Rider NL. Artificial intelligence in clinical immunology. In: Lidströmer N, Ashrafian H, editors. Artificial intelligence in medicine. Cham: Springer International Publishing; 2022. p. 1397–410. Available from: https://doi.org/10.1007/978-3-030–64573-1_83.

[8] Fortino V, Wisgrill L, Werner P, Suomela S, Linder N, Jalonen E, et al. Machine-learning–driven biomarker discovery for the discrimination between allergic and irritant contact dermatitis. Proc Natl Acad Sci U S A 2020;117:33474–85. Available from: https://doi.org/10.1073/pnas.2009192117.

[9] Sevim Bayrak C, Itan Y. Identifying disease-causing mutations in genomes of single patients by computational approaches. Hum Genet 2020;139:769–76. Available from: https://doi.org/10.1007/s00439-020-02179-7.

[10] Mayampurath A, Ajith A, Anderson-Smits C, Chang S-C, Brouwer E, Johnson J, et al. Early diagnosis of primary immunodeficiency disease using clinical data and machine learning. J Allergy Clin Immunol Pract 2022;10:3002–7. Available from: https://doi.org/10.1016/j.jaip.2022.08.041 e5.

[11] Rider NL, Cahill G, Motazedi T, Wei L, Kurian A, Noroski LM, et al. PI Prob: a risk prediction and clinical guidance system for evaluating patients with recurrent infections. PLoS One 2021;16:e0237285. Available from: https://doi.org/10.1371/journal.pone.0237285.

[12] Sauer BC, Jones BE, Globe G, Leng J, Lu C-C, He T, et al. Performance of an NLP tool to extract PFT reports from structured and semi-structured VA data. EGEMs Gener Evid Methods Improve Patient Outcomes 2016;4:10. Available from: https://doi.org/10.13063/2327-9214.1217.

[13] Seol HY, Rolfes MC, Chung W, Sohn S, Ryu E, Park MA, et al. Expert artificial intelligence-based natural language processing characterises childhood asthma. BMJ Open Respir Res 2020;7:e000524. Available from: https://doi.org/10.1136/bmjresp-2019-000524.

[14] Sohn S, Wi C-I, Wu ST, Liu H, Ryu E, Krusemark E, et al. Ascertainment of asthma prognosis using natural language processing from electronic medical records. J Allergy Clin Immunol 2018;141:2292–4. Available from: https://doi.org/10.1016/j.jaci.2017.12.1003 e3.

[15] Kaur H, Sohn S, Wi C-I, Ryu E, Park MA, Bachman K, et al. Automated chart review utilizing natural language processing algorithm for asthma predictive index. BMC Pulm Med 2018;18:34. Available from: https://doi.org/10.1186/s12890-018-0593-9.

[16] Bae WD, Kim S, Park C-S, Alkobaisi S, Lee J, Seo W, et al. Performance improvement of machine learning techniques predicting the association of exacerbation of peak expiratory flow ratio with short term exposure level to indoor air quality using adult asthmatics clustered data. PLoS One 2021;16:e0244233. Available from: https://doi.org/10.1371/journal.pone.0244233.

[17] Ross MK, Yoon J, van der Schaar A, van der Schaar M. Discovering pediatric asthma phenotypes on the basis of response to controller medication using machine learning. Ann Am Thorac Soc 2018;15:49–58. Available from: https://doi.org/10.1513/AnnalsATS.201702-101OC.

[18] Zein JG, Wu C-P, Attaway AH, Zhang P, Nazha A. Novel machine learning can predict acute asthma exacerbation. Chest 2021;159:1747–57. Available from: https://doi.org/10.1016/j.chest.2020.12.051.

[19] Finkelstein J, Jeong Icheol. Machine learning approaches to personalize early prediction of asthma exacerbations: personalized prediction of asthma exacerbation. Ann N Y Acad Sci 2017;1387:153–65. Available from: https://doi.org/10.1111/nyas.13218.

[20] Sagheb E, Wi C-I, Yoon J, Seol HY, Shrestha P, Ryu E, et al. Artificial intelligence assesses clinicians' adherence to asthma guidelines using electronic health records. J Allergy Clin Immunol Pract 2022;10:1047–56. Available from: https://doi.org/10.1016/j.jaip.2021.11.004 e1.

[21] Gustafson E, Pacheco J, Wehbe F, Silverberg J, Thompson W. A machine learning algorithm for identifying atopic dermatitis in adults from electronic health records. In: 2017 IEEE International Conference on Healthcare Informatics (ICHI). Presented at the 2017 IEEE International Conference on Healthcare Informatics (ICHI). Park City, UT: IEEE; 2017. p. 83–90. https://doi.org/10.1109/ICHI.2017.31.

[22] Medela A, Mac Carthy T, Aguilar Robles SA, Chiesa-Estomba CM, Grimalt R. Automatic SCOring of atopic dermatitis using deep learning: a pilot study. JID Innov 2022;2:100107. Available from: https://doi.org/10.1016/j.xjidi.2022.100107.

[23] Jiang Z, Li J, Kong N, Kim J-H, Kim B-S, Lee M-J, et al. Accurate diagnosis of atopic dermatitis by combining transcriptome and microbiota data with supervised machine learning. Sci Rep 2022;12:290. Available from: https://doi.org/10.1038/s41598-021-04373-7.

[24] Wu JJ, Hong C, Merola JF, Gruben D, Güler E, Feeney C, et al. Predictors of nonresponse to dupilumab in patients with atopic dermatitis. Ann Allergy Asthma Immunol 2022;129:354–9. Available from: https://doi.org/10.1016/j.anai.2022.05.025 e5.

[25] Chiriac AM, Wang Y, Schrijvers R, Bousquet PJ, Mura T, Molinari N, et al. Designing predictive models for beta-lactam allergy using the drug allergy and hypersensitivity database. J Allergy Clin Immunol Pract 2018;6:139–48. Available from: https://doi.org/10.1016/j.jaip.2017.04.045 e2.

[26] Moreno EM, Moreno V, Laffond E, Gracia-Bara MT, Muñoz-Bellido FJ, Macías EM, et al. Usefulness of an artificial neural network in the prediction of β-lactam allergy. J Allergy Clin Immunol Pract 2020;8:2974–82. Available from: https://doi.org/10.1016/j.jaip.2020.07.010 e1.

[27] Hammann F, Schöning V, Drewe J. Prediction of clinically relevant drug-induced liver injury from structure using machine learning: prediction of DILI. J Appl Toxicol 2019;39:412–19. Available from: https://doi.org/10.1002/jat.3741.

[28] Banerji A, Lai KH, Li Y, Saff RR, Camargo CA, Blumenthal KG, et al. Natural language processing combined with ICD-9-CM codes as a novel method to study the epidemiology of allergic drug reactions. J Allergy Clin Immunol Pract 2020;8:1032–8. Available from: https://doi.org/10.1016/j.jaip.2019.12.007 e1.

[29] Banerjee A, Ray S, Vorselaars B, Kitson J, Mamalakis M, Weeks S, et al. Use of machine learning and artificial intelligence to predict SARS-CoV-2 infection from full blood counts in a population. Int Immunopharmacol 2020;86:106705. Available from: https://doi.org/10.1016/j.intimp.2020.106705.

[30] Li M, Zhang Z, Cao W, Liu Y, Du B, Chen Canping, et al. Identifying novel factors associated with COVID-19 transmission and fatality using the machine learning approach. Sci Total Env 2021;764:142810. Available from: https://doi.org/10.1016/j.scitotenv.2020.142810.

[31] Rider NL, Srinivasan R, Khoury P. Artificial intelligence and the hunt for immunological disorders. Curr Opin Allergy Clin Immunol 2020;20:565–73. Available from: https://doi.org/10.1097/ACI.0000000000000691.

[32] Ferrante G, Licari A, Fasola S, Marseglia GL, La Grutta S. Artificial intelligence in the diagnosis of pediatric allergic diseases. Pediatr Allergy Immunol 2021;32:405–13. Available from: https://doi.org/10.1111/pai.13419.

[33] Korb KB, Nicholson AE. Bayesian artificial intelligence. Chapman & Hall/CRC computer science and data analysis series. 2nd ed. Boca Raton, FL: CRC Press; 2011.

[34] McLachlan S, Dube K, Hitman GA, Fenton NE, Kyrimi E. Bayesian networks in healthcare: distribution by medical condition. Artif Intell Med 2020;107:101912. Available from: https://doi.org/10.1016/j.artmed.2020.101912.

[35] Ghosh D, Bernstein JA, Khurana Hershey GK, Rothenberg ME, Mersha TB. Leveraging multilayered "omics" data for atopic dermatitis: a road map to precision medicine. Front Immunol 2018;9:2727. Available from: https://doi.org/10.3389/fimmu.2018.02727.

[36] Falissard B, Simpson EL, Guttman-Yassky E, Papp KA, Barbarot S, Gadkari A, et al. Qualitative assessment of adult patients' perception of atopic dermatitis using natural language processing analysis in a cross-sectional study. Dermatol Ther 2020;10:297–305. Available from: https://doi.org/10.1007/s13555-020-00356-0.

[37] Cahill G, Kutac C, Rider NL. Visualizing and assessing US county-level COVID19 vulnerability. Am J Infect Control 2021;49:678–84. Available from: https://doi.org/10.1016/j.ajic.2020.12.009.

[38] Malik YS, Sircar S, Bhat S, Ansari MI, Pande T, Kumar P, et al. How artificial intelligence may help the Covid-19 pandemic: Pitfalls and lessons for the future. Rev Med Virol 2021;31:1–11. Available from: https://doi.org/10.1002/rmv.2205.

[39] Yarmohammadi H, Estrella L, Doucette J, Cunningham-Rundles C. Recognizing primary immune deficiency in clinical practice. Clin Vacc Immunol 2006;13:329–32. Available from: https://doi.org/10.1128/CVI.13.3.329-332.2006.

[40] Elkhalifa S, Bhana R, Blaga A, Joshi S, Svejda M, Kasilingam V, et al. Development and validation of a mobile clinical decision support tool for the diagnosis of drug allergy in adults: the drug allergy app. J Allergy Clin Immunol Pract 2021;9:4410–18. Available from: https://doi.org/10.1016/j.jaip.2021.07.057 e4.

[41] GENYO, Centre for Genomics and Oncological Research: Pfizer, University of Granada, Andalusian Regional Government, Granada, Spain, Martorell-Marugán, J., Tabik, S., Department of Computer Science and Artificial Intelligence, University of Granada, Granada, Spain, Benhammou, Y., Department of Computer Science and Artificial Intelligence, University of Granada, Granada, Spain, del Val, C., Department of Computer Science and Artificial Intelligence, University of Granada, Granada, Spain, Zwir, I., Department of Computer Science and Artificial Intelligence, University of Granada, Granada, Spain, Herrera, F., Department of Computer Science and Artificial Intelligence, University of Granada, Granada, Spain, Carmona-Sáez, P., GENYO, Centre for Genomics and Oncological Research: Pfizer, University of Granada, Andalusian Regional Government, Granada, Spain. Deep learning in omics data analysis and precision medicine. In: Division of Biomedical Science, University of the Highlands and Islands, UK, Husi, H. (Eds.), Computational biology. Codon Publications; 2019. p. 37–53. Available from: https://doi.org/10.15586/computationalbiology.2019 ch3.

[42] Pantanowitz L, Wu U, Seigh L, LoPresti E, Yeh F-C, Salgia P, et al. Artificial intelligence–based screening for mycobacteria in whole-slide images of tissue samples. Am J Clin Pathol 2021;156:117–28. Available from: https://doi.org/10.1093/ajcp/aqaa215.

[43] van Leeuwen KG, de Rooij M, Schalekamp S, van Ginneken B, Rutten MJCM. How does artificial intelligence in radiology improve efficiency and health outcomes? Pediatr Radiol 2022;52:2087–93. Available from: https://doi.org/10.1007/s00247-021-05114-8.

[44] Schüssler-Fiorenza Rose SM, Contrepois K, Moneghetti KJ, Zhou W, Mishra T, Mataraso S, et al. A longitudinal big data approach for precision health. Nat Med 2019;25:792–804. Available from: https://doi.org/10.1038/s41591-019-0414-6.

CHAPTER

38

Artificial intelligence in medical genetics

Rebekah L. Waikel, Dat Duong and Benjamin D. Solomon

National Human Genome Research Institute, Bethesda, MD, United States

Artificial intelligence (AI) is rapidly advancing the field of genetics and genomics in general and holds the potential to revolutionize the practice of clinical genetics [1–3]. Broadly speaking, AI comprises of a variety of techniques that can be loosely grouped into three subbranches, ranging from intuitive deductions to complex deep learning (DL) models (Fig. 38.1). Among the various AI subbranches, DL in particular has produced highly impactful methods to study many genetic conditions under different settings, and thus has yielded the greatest growth for AI in the context of medical genetics [4]. This chapter briefly discusses four main AI applications in medical genetics (mostly under the DL subbranch) ranging from the relatively well-understood analyses of (1) medical images, (2) biology-related text datasets, (3) genomic including other "omic and biologic" data, and to the more recently explored (4) generation of synthetic data observations (e.g., images, text, 'omics data), which has applications in education and could further enhance the methods in the three previous applications. These use-cases are not meant to be comprehensive, but instead intended to provide a glimpse into some of the high-impact areas in which AI is used in medical genetics.

Before describing key uses of AI in this field, it may help to discuss aspects of medical genetics that are relevant to the AI. First, there are thousands of different conditions a genetics clinician may encounter. Though common in aggregate, these conditions are largely individually rare, and even the most experienced practitioner will manage patients with conditions they have never seen before [9]. Given the rarity of individual conditions, gathering sufficiently large datasets for AI training and testing can be difficult, and specific methodological considerations may be necessary [10,11]. Conversely, this challenge may mean that AI can be especially useful in medical genetics, since practitioners cannot be expected to be highly experienced with every condition and may appreciate support from AI-based tools [12]. Second, there is a dearth of most types of medical genetics

Artificial Intelligence in Clinical Practice
DOI: https://doi.org/10.1016/B978-0-443-15688-5.00007-3

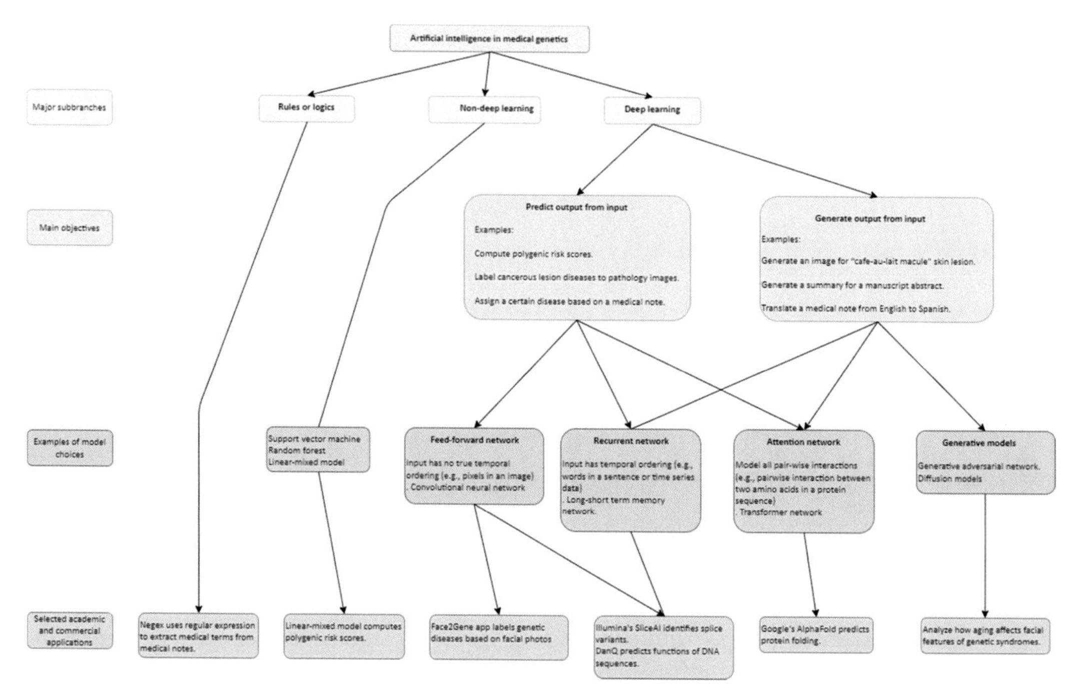

FIGURE 38.1 Overview of deep learning in medical genetics. AI methods can be broadly grouped into three subbranches: (1) those using rules or logics, (2) non-deep learning, and (3) deep learning methods. There are tasks in which approaches of the first subbranch excel; for example, string-matching algorithms are effective at extracting human phenotype ontology (HPO) terms or other phenotypic features from medical notes. The non-deep learning category consists of machine learning methods like support vector machine, and statistical approaches like linear-mixed model, which is widely used in genome-wide association studies. The deep learning subbranch contains four main groups: feed-forward, recurrent, attention, and generative network. Here, we highlight a few key examples mentioned in the figure. A feed-forward network built from convolutional neural network is effective at predicting a label for an input image; for example, the Face2Gene app identifies genetic conditions based on facial photos. Recurrent network (such as long-short term memory) has been designed for clinical time-series data and DNA sequences [5]. Google's Alpha Fold uses attention network to predict a protein folding for a given amino acid sequence [6]. We note that feed-forward networks also perform well for various tasks related to omic-data; a few examples include SliceAI and DeepVariant [7,8]. Generative network has been applied to different types of medical images; one example is the creation of hypothetical images for the same person over time to illustrate how aging affects the facial features of certain genetic syndromes.

practitioners, especially (but not only) physicians [13–15]. Supplementing the workforce with AI-based support may help meet the need for experts, including in remote geographic regions where specialists may be less available [16].

Third, and thanks to advances in genomic technologies and bioinformatic approaches, many genetic conditions have known molecular or cytogenomic causes [17,18]. This ability to precisely categorize patients based on the underlying etiologies is helpful for precisely labeling the data used to train AI models. Finally, the field of medical genetics is currently primarily a diagnostic specialty (see further discussion on this topic toward the conclusion of this section). Like radiology, ophthalmology, dermatology, pathology, and other "pattern recognition" disciplines, AI may be especially useful in medical genetics, where optimal practice rests on recognizing what condition, a person has [19].

As with many others of biomedicine—though perhaps more frequently than in most other areas - determining a clear dividing line between research exploration and clinical practice can be difficult in medical genetics. For example, a patient may be seen under a formally approved research protocol but may receive clinical genetic testing through that research study, which may have been otherwise unavailable to the patient due to financial or insurance constraints. Conversely, a patient seen at a clinical genetics practice may undergo traditional clinical genomic testing and may share the genomic data with researchers for academic research purposes, particularly if the result is initially uncertain or unrevealing. With those caveats in mind, this section will discuss research investigations of AI in medical genetics but will attempt to emphasize clinical implementation (Fig. 38.2).

The first area involves the analysis of images in which state-of-the-art methods are DL models built on the convolutional neural network. Many genetic conditions involve specific characteristics that can be recognizable [20]. Coupling this facet of medical genetics with DL models pretrained on very large (possibly closely related dataset) provides a powerful avenue for AI [21–23]. One example is the outstanding application of DL approaches to help assess the craniofacial characteristics of people suspected to have genetic conditions (Fig. 38.3) [24–27]. These tools can help generate a differential diagnosis based on images of a patient; moreover, clinicians can provide extra inputs, such as the presence or absence of specific clinical features apart from the facial appearance, to help the models refine the differential diagnosis [25]. Additionally, these differential diagnoses can help guide molecular testing, including to enable more thorough investigations of possible causative genes, and to help resolve uncertain genetic testing results. Much of the clinical implementation of this approach has been driven by a specific mobile app that enables clinicians to apply these approaches in real time [26]. The publication record emphasizes the impact of this app. In one recent review of the use of DL in medical genetics, almost half of all publications using DL in medical genetics involved this app for both diagnosis and phenotyping (Fig. 38.4). These publications include large studies involving many patients, as well as smaller studies of specific syndromes and case reports describing the use of this app as part of the routine clinical workflow. Though gathering and sharing standardized information is not new in medical genetics, the success of this app may hinge on its user-friendly interface that provides easy access to clinicians who are not experts in computer programming [28]. The widespread use of this app may help acclimate medical geneticists to other AI uses [29,30].

This use of computer vision highlights a key issue in medical genetics as well as across medicine: ensuring that methods work equitably across different groups and populations. For example, the accuracy of classifiers can vary in people of different ages or in people from different geographic regions and ancestries [24,27,31]. To address this issue, efforts must be made to gather diverse and representative datasets and to carefully examine results – this problem relates to the data collection caveat mentioned previously.

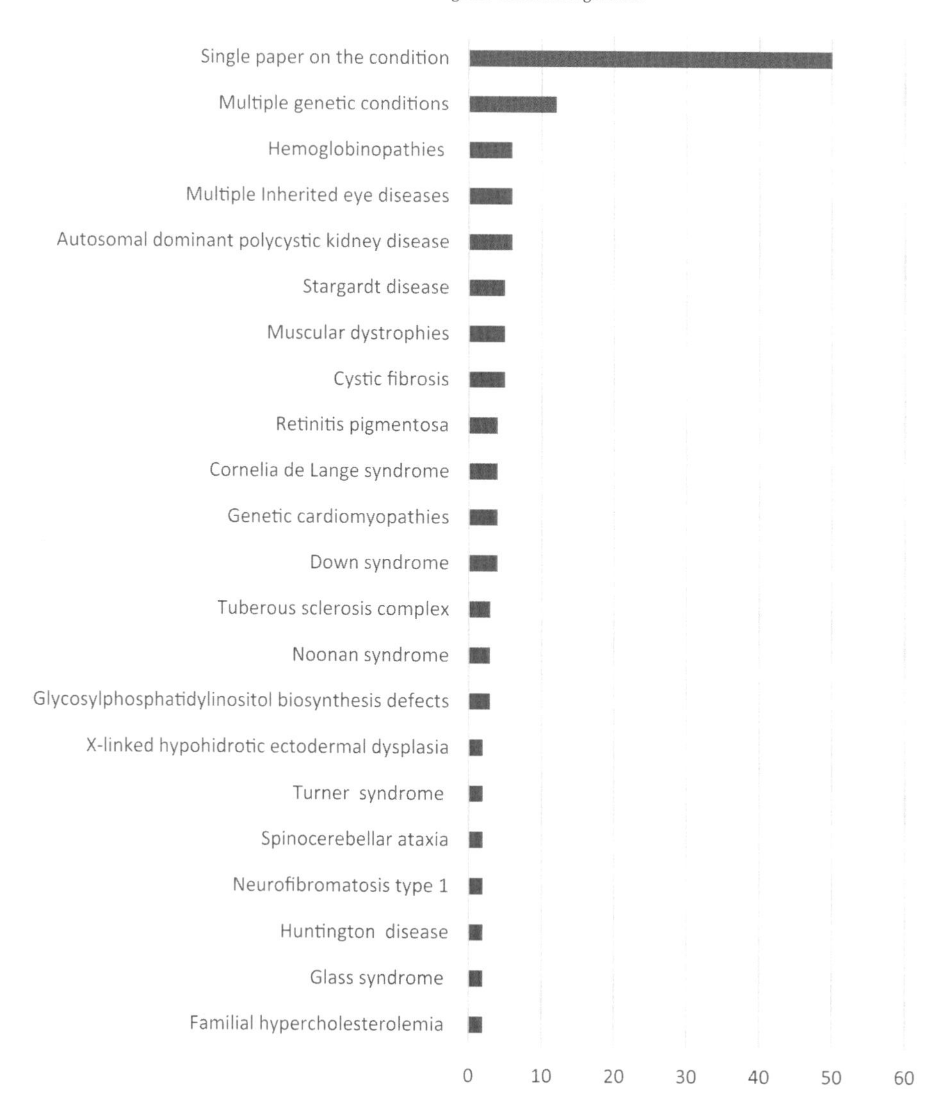

FIGURE 38.2 Diversity of genetic conditions studied via DL. A recent review showed that by 2021, publications on DL had been described for over 75 different genetic conditions. The most common conditions studied by DL included hemoglobinopathies, inherited eye conditions, and polycystic kidney disease, though some studies examined a wider array of conditions. Interestingly, and unlike studies in oncology, this analysis showed that, in medical genetics, DL tended to be used to analyze phenotypic or genomic data, but less commonly analyzed such data types together. Source: *Ledgister Hanchard SE, et al. Scoping review and classification of deep learning in medical genetics. Genet Med 2022;24:1593–1603. https://doi.org/10.1016/j.gim.2022.04.025 [4].*

In addition to assessment of facial features, there have been publications exploring other types of image datasets relevant to genetic conditions, such as radiologic images, microscopy, ophthalmologic images, skin images (Fig. 38.5), and others [30,32]. The example of facial image-based AI in medical genetics provides a roadmap for the use of these

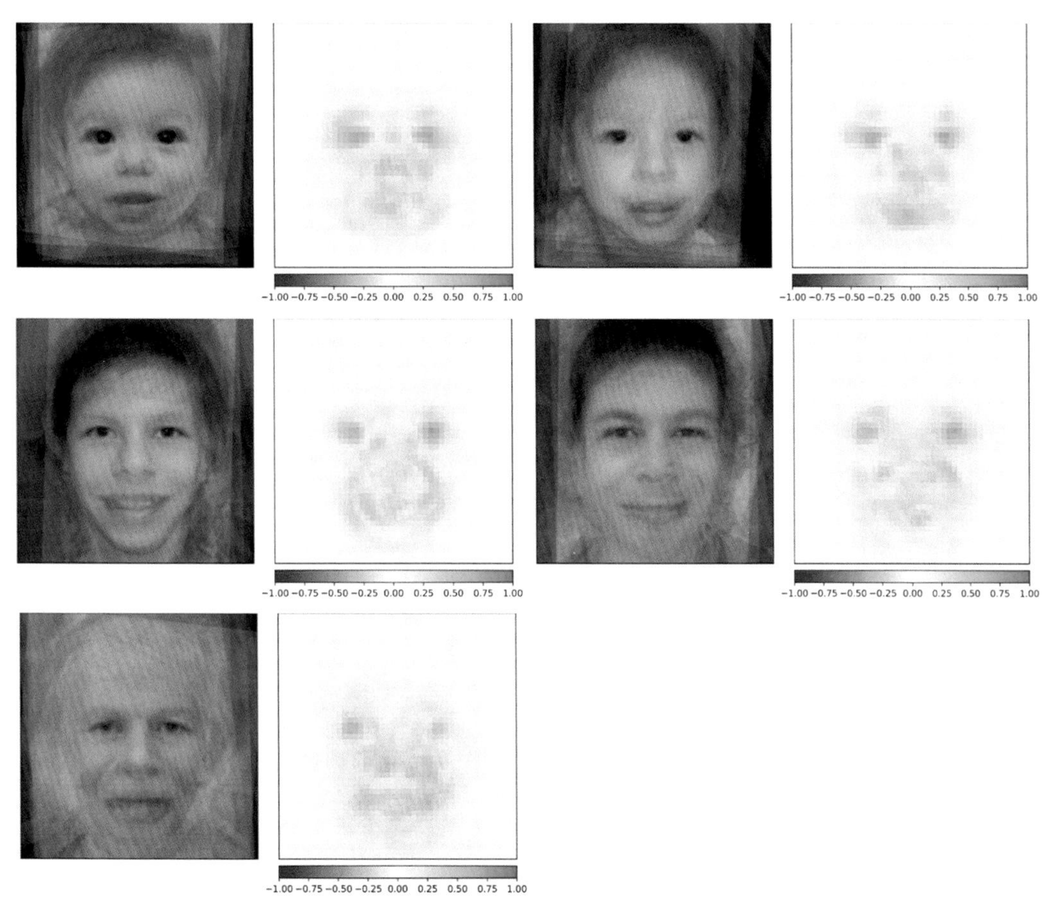

FIGURE 38.3 Occlusion analysis of facial images of people with Williams syndrome (WS) across the lifespan. Composite saliency maps were constructed from results from publicly available images in each of the represented age groups in the study: infant, child, adolescent, young adult, and older adult (reading left to right). In the figure, green indicates a positive contribution and red indicates a negative contribution to classification of the image. These types of approaches may be clinically helpful, as they can be used to identify important features for diagnosis as well as to fine-tune and check the output of DL models. Source: *Duong D, et al. Neural networks for classification and image generation of aging in genetic syndromes. Front Genet 2022;13:864092. https://doi.org/10.3389/fgene.2022.864092 [24].*

other image types. Further, AI with images can be used for more than diagnostic approaches. For example, generative adversarial networks (GANs) may help with classification accuracy but might eventually also be used to predict outcomes or therapeutic response (described further below).

The second area is the analysis of clinically oriented text datasets via AI; typically, these methods may be string-matching algorithms or DL models built from the Transformer network. For example, there have been attempts to extract information from electronic health records (EHRs) and other types of clinical records to efficiently determine which individuals might have a genetic condition or who [33–37]. Clinically oriented examples include

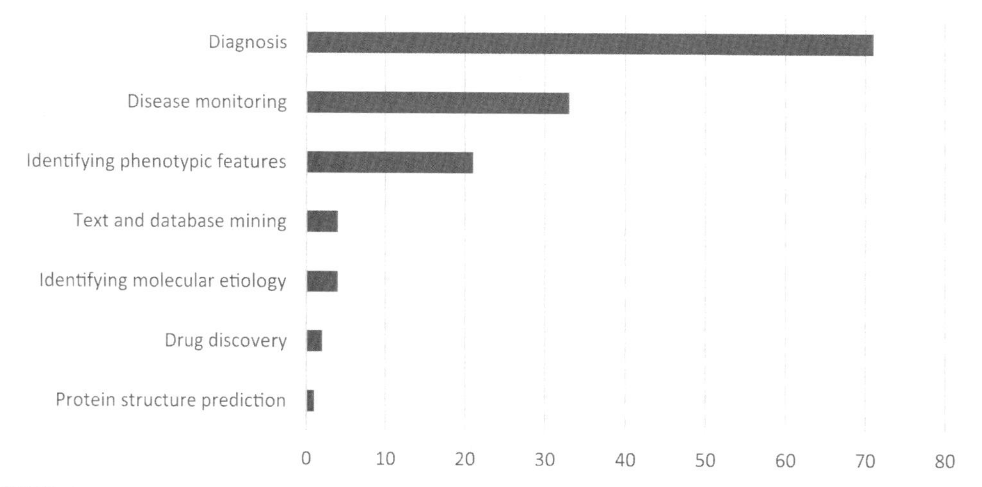

FIGURE 38.4 Categorization of the primary methods by which DL was used in each article on medical genetics, based on a review of DL in medical genetics from 2015 to 2021. Source: *Ledgister Hanchard SE, et al. Scoping review and classification of deep learning in medical genetics. Genet Med 2022;24:1593–1603. https://doi.org/10.1016/j.gim.2022.04.025 [4].*

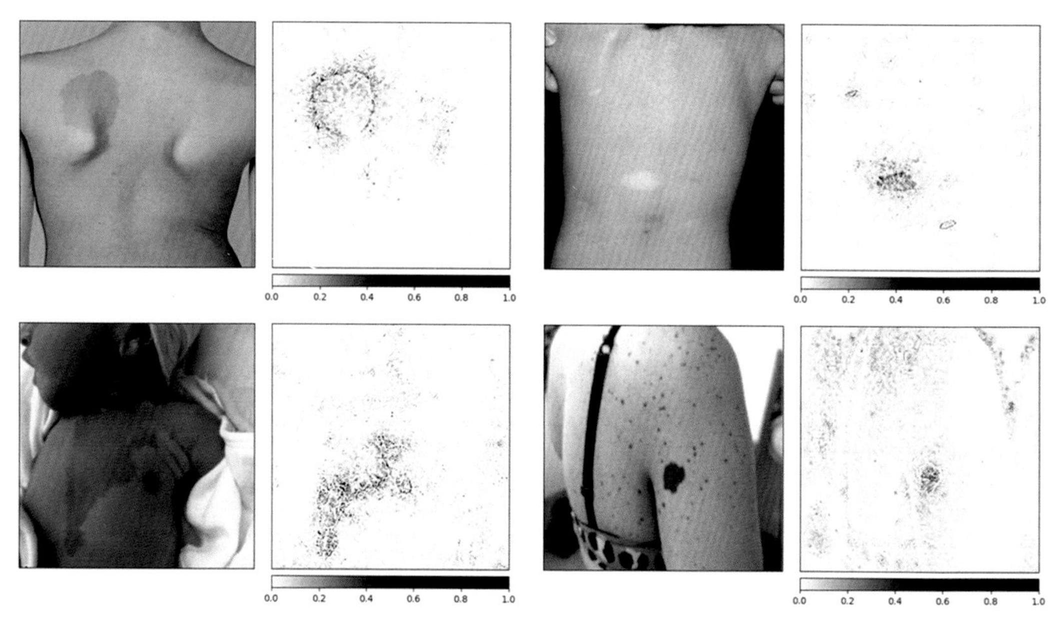

FIGURE 38.5 Attribution images showing which pixels the classifier weights when "deciding" which condition the image represents. Darker color indicates pixels with higher importance. These methods can be used during classifier training and testing to help examine output, as well as to potentially fine-tune results, such as related to adjusting the neural network hyperparameters. Clockwise, from top left: Neurofibromatosis type 1 (NF1); Tuberous Sclerosis Complex (TSC); Noonan syndrome with Multiple Lentigines (ML); McCune-Albright syndrome (MA). Source: *Duong D, Waikel RL, Hu P, Tekendo-Ngongang C, Solomon BD. Neural network classifiers for images of genetic conditions with cutaneous manifestations. Hum Genet Genomics Adv 2021:100053 [10].*

mining of EHR information to determine which patients in a healthcare system might benefit from microarray testing (and who may have been overlooked as being eligible for such testing), and analyzing a cohort of people [37,38] These efforts have been aided by ongoing efforts [39–41].

As with the computer vision example described earlier, having access to relevant text-based data is necessary to build and test AI models. To help address this issue, investigators have created and used deidentified, synthetic, or aggregated, publicly available datasets [35,38,42]. Again, large-scale collaborations involving international approaches may help further the use of AI in this area and address some of the related data collection challenges, especially to ensure that methods can be provided accurately in different populations.

Another separate but intriguing example of text-based AI in medical genetics involves the use of "chatbots" to interface with patients (as well as research subjects). These chatbots can perform a variety of activities, including answering patient questions about the inheritance of certain genetic conditions, triaging individuals for further conversations with a genetic counselors, and explaining issues like informed consent in genetic testing [43–46]. Some of these types of activities already occur in clinical practice via integration of AI-based tools into laboratory-based genetic testing workflows [43,44]. Recent developments related to large-language models is likely to supercharge this facet of genetics.

The third area involves the use of algorithms based on non-DL and DL AI methods to analyze genomic data. These may be used in a traditional clinical manner, such as to help diagnose a patient with a suspected genetic condition who undergoes genetic or genomic testing. AI can be used for overall variant assessment, to analyze specific types of genetic changes, and to help prioritize and filter causative variants [1,7,8,47].

Recent developments in rapid exome and genome sequencing rest upon evolving sequencing technologies (like next-generation sequencing and long-read sequencing) and other laboratory-based innovations, as well as the integration of AI approaches in the analytic pipelines. These provide a dramatic example of how much medical genetics has changed in the last several decades. Sequencing the first human genome took over a decade (the exact amount of time is an issue of debate!) [48]. Currently, critically ill patients have been sequenced and received diagnoses within a few hours [49,50]. These diagnoses can sometimes lead to life-saving interventions, but also yield more cost-effective and efficient care in general (including for people with negative findings) [47,51,52]. Scaling these technologies will require increased leveraging of automated, AI-based approaches, but hold great promise, especially as they are applied to many different clinical scenarios.

The fourth area involves the use of generative models, such as those built from GANs or diffusion models, to create synthetic data observations that are similar to the real samples in the training set. For example, the image discriminator in a GAN separates fake from real images. To fool the discriminator, the image generator in a GAN learns regularities or patterns from facial images of individuals with a particular genetic condition, such as William syndrome, thus allowing for the generation of realistic synthetic images of individuals with William syndromes. This approach can be taken further to demonstrate how facial feature change during the aging of a fake individual with a genetic condition, which can be clinically valuable [24]. For example, synthetic images could be used for educational purposes, to train genetics professionals, as well as for knowledge assessment purposes, as the GAN images can be made for any given age, gender, and other demographic features, including new images that a student has not previously viewed [11]. There are many advantages to artificially generated images including demonstrating clinically relevant characteristics without the concern of privacy issues with real patient images, supplementing datasets (Fig. 38.6), particularly for underrepresented groups, to improve accuracy of image classifiers, and to potentially predict disease prognosis and progression.

Interestingly, and in general in medical genetics, one recent review showed that AI analyses to date seem to be used to either investigate genomic or phenotypic data but are less often used to consider multiple data types together [4]. In oncology (in this case, referring to the study of somatic cancer, versus studies of heritable cancer predisposition syndrome), on the other hand, there is a much stronger track record of AI being used to analyze these types of datasets together. One reason may the availability of datasets in oncology versus in medical genetics [30]. That is, efforts to collate and share curated, clinically-annotated genomic datasets relevant to cancer have enabled the application and development of AI methods [53]. As mentioned earlier, broad data aggregation and sharing has yielded high-impact results in medical genetics [54,55]. However, and for a variety of reasons, the availability of such datasets in medical genetics lags behind oncology. The successes in oncology suggest

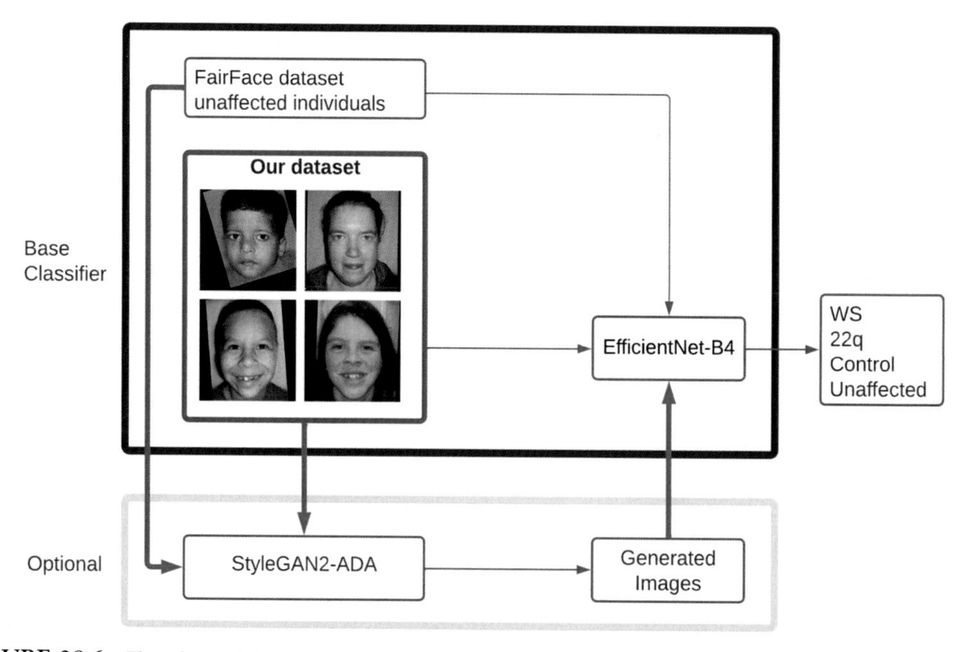

FIGURE 38.6 Flowchart of GAN image production and its application with real images to train a neural network (EfficientNet-B4). The training set of real images consisted of publicly available images of individuals with 22q11.2 deletion syndrome, Williams syndrome, other conditions (control) dataset and the publicly available FairFace dataset (unaffected). EfficientNet-B4 can be trained on the real images (top box) to develop a classifier. In order to test the effects of this approach on the performance of the classifier, GAN images were made via StyleGAN2-ADA and were used along with real (non-GAN) images to trainEfficientNet-B4 (bottom box). Source: *Duong D, et al. Neural networks for classification and image generation of aging in genetic syndromes. Front Genet 2022;13:864092. https://doi.org/10.3389/fgene.2022.864092 [24].*

the potential to use AI to help provide better insights for patients and families affected by genetic conditions, as well as the providers who care for them.

References

[1] Dias R, Torkamani A. Artificial intelligence in clinical and genomic diagnostics. Genome Med 2019;11:70. Available from: https://doi.org/10.1186/s13073-019-0689-8.

[2] Brasil S, et al. Artificial intelligence (AI) in rare diseases: is the future brighter? Genes (Basel) 2019;10. Available from: https://doi.org/10.3390/genes10120978.

[3] Schaefer J, Lehne M, Schepers J, Prasser F, Thun S. The use of machine learning in rare diseases: a scoping review. Orphanet J Rare Dis 2020;15:145. Available from: https://doi.org/10.1186/s13023-020-01424-6.

[4] Ledgister Hanchard SE, et al. Scoping review and classification of deep learning in medical genetics. Genet Med 2022;24:1593–603. Available from: https://doi.org/10.1016/j.gim.2022.04.025.

[5] Harutyunyan H, Khachatrian H, Kale DC, Ver Steeg G, Galstyan A. Multitask learning and benchmarking with clinical time series data. Sci Data 2019;6:1–18.

[6] Jumper J, et al. Highly accurate protein structure prediction with AlphaFold. Nature 2021;596:583–9. Available from: https://doi.org/10.1038/s41586-021-03819-2.

[7] Jaganathan K, et al. Predicting splicing from primary sequence with deep learning. Cell 2019;176:535–48. Available from: https://doi.org/10.1016/j.cell.2018.12.015 e524.

[8] Poplin R, et al. A universal SNP and small-indel variant caller using deep neural networks. Nat Biotechnol 2018;36:983–7. Available from: https://doi.org/10.1038/nbt.4235.

[9] Ferreira CR. The burden of rare diseases. Am J Med Genet A 2019;179:885–92. Available from: https://doi.org/10.1002/ajmg.a.61124.

[10] Duong D, Waikel RL, Hu P, Tekendo-Ngongang C, Solomon BD. Neural network classifiers for images of genetic conditions with cutaneous manifestations. Hum Genet Genomics Adv 2021;100053.

[11] Malechka VV, et al. Investigating determinants and evaluating deep learning training approaches for visual acuity in foveal hypoplasia.

[12] Solomon BD. Can artificial intelligence save medical genetics? Am J Med Genet A 2022;188:397–9. Available from: https://doi.org/10.1002/ajmg.a.62538.

[13] Jenkins BD, et al. The 2019 US medical genetics workforce: a focus on clinical genetics. Genet Med 2021;. Available from: https://doi.org/10.1038/s41436-021-01162-5.

[14] Maiese DR, et al. Current conditions in medical genetics practice. Genet Med 2019;21:1874–7. Available from: https://doi.org/10.1038/s41436-018-0417-6.

[15] Penon-Portmann M, Chang J, Cheng M, Shieh JT. Genetics workforce: distribution of genetics services and challenges to health care in California. Genet Med 2020;22:227–31. Available from: https://doi.org/10.1038/s41436-019-0628-5.

[16] Solomon BD. Can artificial intelligence save medical genetics? Am J Med Genet A 2021;. Available from: https://doi.org/10.1002/ajmg.a.62538.

[17] Bamshad MJ, Nickerson DA, Chong JX. Mendelian gene discovery: fast and furious with no end in sight. Am J Hum Genet 2019;105:448–55. Available from: https://doi.org/10.1016/j.ajhg.2019.07.011.

[18] Solomon BD, Lee T, Nguyen AD, Wolfsberg TG. A 2.5-year snapshot of Mendelian discovery. Mol Genet Genomic Med 2016;4:392–4. Available from: https://doi.org/10.1002/mgg3.221.

[19] Topol EJ. 1 online resource. New York: Basic Books; 2019.

[20] Solomon BD, Nguyen AD, Bear KA, Wolfsberg TG. Clinical genomic database. Proc Natl Acad Sci U S A 2013;110:9851–5. Available from: https://doi.org/10.1073/pnas.1302575110.

[21] Deng J, et al. ImageNet: A large-scale hierarchical image database. In: 2009 IEEE conference on computer vision and pattern recognition, IEEE; 2009. p. 248–55.

[22] Kermany DS, et al. Identifying medical diagnoses and treatable diseases by image-based deep learning. Cell 2018;172(1122–1131):e1129. Available from: https://doi.org/10.1016/j.cell.2018.02.010.

[23] Kärkkäinen K, Joo J. Fairface: Face attribute dataset for balanced race, gender, and age. arXiv:1908.04913, 2019.

[24] Duong D, et al. Neural networks for classification and image generation of aging in genetic syndromes. Front Genet 2022;13:864092. Available from: https://doi.org/10.3389/fgene.2022.864092.
[25] Hsieh TC, et al. GestaltMatcher facilitates rare disease matching using facial phenotype descriptors. Nat Genet 2022;54:349–57. Available from: https://doi.org/10.1038/s41588-021-01010-x.
[26] Gurovich Y, et al. Identifying facial phenotypes of genetic disorders using deep learning. Nat Med 2019;25:60–4. Available from: https://doi.org/10.1038/s41591-018-0279-0.
[27] Porras AR, Rosenbaum K, Tor-Diez C, Summar M, Linguraru MG. Development and evaluation of a machine learning-based point-of-care screening tool for genetic syndromes in children: a multinational retrospective study. Lancet Digit Health 2021;. Available from: https://doi.org/10.1016/S2589-7500(21)00137-0.
[28] Guest SS, Evans CD, Winter RM. The online London dysmorphology database. Genet Med 1999;1:207–12. Available from: https://doi.org/10.1097/00125817-199907000-00007.
[29] James CA, Wachter RM, Woolliscroft JO. Preparing clinicians for a clinical world influenced by artificial intelligence. JAMA 2022;. Available from: https://doi.org/10.1001/jama.2022.3580.
[30] Ledgister Hanchard S., Dwyer M.C., Liu S., Hu P., Tekendo-Ngongang C., Waikel R.L., et al. Scoping review and classification of deep learning in medical genetics; 2022.
[31] Tekendo-Ngongang C, et al. Rubinstein-Taybi syndrome in diverse populations. Am J Med Genet A 2020;182:2939–50. Available from: https://doi.org/10.1002/ajmg.a.61888.
[32] Verdu-Diaz J, et al. Accuracy of a machine learning muscle MRI-based tool for the diagnosis of muscular dystrophies. Neurology 2020;94:e1094–102. Available from: https://doi.org/10.1212/WNL.0000000000009068.
[33] Son JH, et al. Deep phenotyping on electronic health records facilitates genetic diagnosis by clinical exomes. Am J Hum Genet 2018;103:58–73. Available from: https://doi.org/10.1016/j.ajhg.2018.05.010.
[34] Williams MS, et al. Genomic information for clinicians in the electronic health record: lessons learned from the clinical genome resource project and the electronic medical records and genomics network. Front Genet 2019;10:1059. Available from: https://doi.org/10.3389/fgene.2019.01059.
[35] Bastarache L, et al. Phenotype risk scores identify patients with unrecognized Mendelian disease patterns. Science 2018;359:1233–9. Available from: https://doi.org/10.1126/science.aal4043.
[36] Luo L, et al. PhenoTagger: a hybrid method for phenotype concept recognition using human phenotype ontology. Bioinformatics 2021. Available from: https://doi.org/10.1093/bioinformatics/btab019.
[37] Dingemans AJM, et al. Phenotype based prediction of exome sequencing outcome using machine learning for neurodevelopmental disorders. Genet Med 2022;24:645–53. Available from: https://doi.org/10.1016/j.gim.2021.10.019.
[38] Morley TJ, et al. Phenotypic signatures in clinical data enable systematic identification of patients for genetic testing. Nat Med 2021;27:1097–104. Available from: https://doi.org/10.1038/s41591-021-01356-z.
[39] Kohler S, et al. Expansion of the Human Phenotype Ontology (HPO) knowledge base and resources. Nucleic Acids Res 2019;47:D1018–27. Available from: https://doi.org/10.1093/nar/gky1105.
[40] Kohler S, et al. The Human Phenotype Ontology in 2017. Nucleic Acids Res 2017;45:D865–76. Available from: https://doi.org/10.1093/nar/gkw1039.
[41] Dingemans AJM, et al. Human disease genes website series: an international, open and dynamic library for up-to-date clinical information. Am J Med Genet A 2021. Available from: https://doi.org/10.1002/ajmg.a.62057.
[42] Gottesman O, et al. The Electronic Medical Records and Genomics (eMERGE) network: past, present, and future. Genet Med 2013;15:761–71. Available from: https://doi.org/10.1038/gim.2013.72.
[43] Nazareth S, et al. Hereditary cancer risk using a genetic chatbot before routine care visits. Obstet Gynecol 2021;138:860–70. Available from: https://doi.org/10.1097/AOG.0000000000004596.
[44] Snir M, et al. Democratizing genomics: leveraging software to make genetics an integral part of routine care. Am J Med Genet C Semin Med Genet 2021;187:14–27. Available from: https://doi.org/10.1002/ajmg.c.31866.
[45] Nazareth S, Nussbaum RL, Siglen E, Wicklund CA. Chatbots & artificial intelligence to scale genetic information delivery. J Genet Couns 2021;30:7–10. Available from: https://doi.org/10.1002/jgc4.1359.
[46] Siglen E, et al. Ask Rosa—the making of a digital genetic conversation tool, a chatbot, about hereditary breast and ovarian cancer. Patient Educ Couns 2021. Available from: https://doi.org/10.1016/j.pec.2021.09.027.
[47] Clark MM, et al. Diagnosis of genetic diseases in seriously ill children by rapid whole-genome sequencing and automated phenotyping and interpretation. Sci Transl Med 2019;11. Available from: https://doi.org/10.1126/scitranslmed.aat6177.

[48] Lander ES, et al. Initial sequencing and analysis of the human genome. Nature 2001;409:860–921. Available from: https://doi.org/10.1038/35057062.
[49] Gorzynski JE, et al. Ultrarapid nanopore genome sequencing in a critical care setting. N Engl J Med 2022;386:700–2. Available from: https://doi.org/10.1056/NEJMc2112090.
[50] Goenka SD, et al. Accelerated identification of disease-causing variants with ultra-rapid nanopore genome sequencing. Nat Biotechnol 2022. Available from: https://doi.org/10.1038/s41587-022-01221-5.
[51] Kingsmore SF, et al. A randomized, controlled trial of the analytic and diagnostic performance of singleton and trio, rapid genome and exome sequencing in ill infants. Am J Hum Genet 2019;105:719–33. Available from: https://doi.org/10.1016/j.ajhg.2019.08.009.
[52] Farnaes L, et al. Rapid whole-genome sequencing decreases infant morbidity and cost of hospitalization. npj Genom Med 2018;3:10. Available from: https://doi.org/10.1038/s41525-018-0049-4.
[53] Cancer Genome Atlas Research N, et al. The Cancer Genome Atlas Pan-Cancer analysis project. Nat Genet 2013;45:1113–20. Available from: https://doi.org/10.1038/ng.2764.
[54] McWalter K, Torti E, Morrow M, Juusola J, Retterer K. Discovery of over 200 new and expanded genetic conditions using GeneMatcher. Hum Mutat 202243(6):760–4. Available from: https://doi.org/10.1002/humu.24351.
[55] Kaplanis J, et al. Evidence for 28 genetic disorders discovered by combining healthcare and research data. Nature 2020;586:757–62. Available from: https://doi.org/10.1038/s41586-020-2832-5.

CHAPTER

39

Artificial intelligence in healthcare: a perspective from Google

Lisa Soleymani Lehmann[1,2], *Vivek Natarajan*[1] *and Lily Peng*[1,3]

[1]Google, LLC, Mountain View, CA, United States [2]Harvard Medical School and Brigham and Women's Hospital, Boston, MA, United States [3]Verily Life Sciences, South San Francisco, CA, United States

Artificial intelligence (AI) holds the promise of transforming healthcare by improving patient outcomes, increasing accessibility and efficiency, and decreasing the cost of care. Realizing this vision of a healthier world for everyone everywhere requires partnerships and trust between healthcare systems, clinicians, payers, technology companies, pharmaceutical companies, and governments to help bring innovations in machine learning and AI to patients. Google is one example of a technology company that is partnering with healthcare systems, clinicians, and researchers to develop technology solutions to directly improve the lives of patients. In this chapter we share landmark studies of the use of AI in healthcare. We also describe the application of our novel system of organizing information to unify data in electronic health records (EHRs) and bring an integrated view of patient records to clinicians. We discuss our consumer-focused innovation in dermatology to help guide search journeys for personalized information about skin conditions. Finally, we share a perspective on how to embed ethics and a concern for all patients into the development of AI.

One of our first research efforts to illustrate the potential of AI in healthcare was an AI model that demonstrated high sensitivity and specificity in detecting diabetic retinopathy from retinal fundus images [1]. Similarly compelling results were achieved in other medical imaging domains like radiology [2–4], pathology [5], dermatology [6], and colonoscopy [7] with AI models demonstrating clinically applicable performance on-par with or exceeding experts in controlled research settings. Beyond predictive modeling for diagnosis in medical imaging, AI models have also demonstrated the ability to extract hidden signals and make novel discoveries like cardiovascular risk factor predictions from fundus images [8].

Despite the promise and potential demonstrated by AI in controlled research settings [6,9,10], responsible implementation of AI into clinical settings can be a challenge. Guided by our AI Principles, Google has focused on developing tools to comprehensively evaluate

Artificial Intelligence in Clinical Practice
DOI: https://doi.org/10.1016/B978-0-443-15688-5.00037-1

models in a prospective setting while striving to create AI models and products that are safe, effective, explainable, and generalizable to diverse populations [11,12]. For example, we conducted [13] a human-centered observational study of the AI model for diabetic retinopathy screening in clinics in Thailand and observed that several socio-economic factors impacted the model's performance in the real world. This finding led our partner, Rajavithi Hospital, to conduct a prospective interventional study of real-time diabetic retinopathy screening using AI in national screening programs in Thailand [14]. This prospective study demonstrated that accuracy of the AI system was on par with retina specialists, and it had the advantage of providing clinically actionable results in real time. A related study suggested that real-time screening results provided by AI systems at the point-of-care correlate with improved referral adherence [15]. These studies demonstrated the importance of designing medical AI systems centered around human experience, accounting for the wider clinical context and socio-economic factors. These studies are an important first step in tracking the overall health impact of AI-based screening. As AI models are developed, prospective studies in partnership with healthcare systems will be essential to develop the evidence base needed to safely scale world class healthcare to everyone.

While medical imaging is perhaps the most prominent application of AI in healthcare, Google is also advancing the application of AI to EHRs. Clinician stress and burnout is associated with poorly designed electronic health records [16]. Clinicians are forced to spend too much time sifting through an overwhelming amount of information in EHRs to find relevant information when they could be spending more time directly engaging with patients. EHRs that are often designed for billing purposes are difficult to quickly navigate and often result in clinicians worrying that they may be missing something important for their patients' care. Leveraging AI to organize information in an EHR, Google developed Care Studio to help clinicians spend more time directly caring for patients. Clinicians can use Care Studio to quickly find information anywhere in an EHR, whether it is from a scanned PDF from an outside hospital, an inpatient admission, a laboratory test, or a progress note. It provides an intuitive interface to visualize health data and trends in tables and graphs. Traditionally, clinicians are taught to recognize patterns in patients' clinical presentations and data to diagnose and treat patients. By using machine learning to help clinicians find the information they need from a sea of data, Google is working to augment clinicians' acumen. In addition, there is clear evidence that machine learning can enhance our ability to make predictions [17] and surface information in ways that were not previously possible [18].

In addition to evaluating AI algorithms in healthcare research settings and improving clinicians' ability to find and use data in electronic health records, Google is empowering people to learn more about dermatology conditions. Relying on computer vision AI and image search capabilities, Google is building dermatological tools to help individuals research their skin, hair, and nail conditions. While the tool is not yet available in the United States and it is undergoing further market testing, the goal is to allow individuals to obtain easy access to personalized information on more than 90% of the most commonly searched conditions [6,19,20]. Since skin conditions carry substantial health burden and are commonly searched for, we are exploring approaches to use this technology to empower individuals with relevant information and educate them on dermatological conditions.

As we continue to make progress in AI it is essential that we ethically evaluate and deploy it in ways that benefit all members of society. There are multiple efforts to develop ethical frameworks for the use of AI in society [21–24]. While these principles are a foundation for the ethical development of AI, there is a pressing need for more specific guidance on how to translate these high-level principles into practice, methods to embed ethics into the design of AI, tools to audit and evaluate AI for fairness and bias in different clinical settings, as well as governance and accountability mechanisms. Google has open sourced a growing number of tools to help AI developers ethically design and drive a human-centered approach to AI, such as tools to analyze fairness and bias, improve data quality, and technical interventions to address ethical concerns in machine learning models [25–29]. While these technical tools are an important dimension of our ability to ethically implement AI in healthcare, we should not assume that all ethical concerns can be addressed solely by a technical tool or a good design. Bioethics is a process of translating and implementing our principles into practice. It will require the courage to confront complex questions and engage in open debate, so that our values are reflected in the AI that we implement.

References

[1] Gulshan V, Peng L, Coram M, Stumpe MC, Wu D, Narayanaswamy A, et al. Development and validation of a deep learning algorithm for detection of diabetic retinopathy in retinal fundus photographs. JAMA 2016;316(22):2402–10.

[2] McKinney SM, Sieniek M, Godbole V, Godwin J, Antropova N, Ashrafian H, et al. International evaluation of an AI system for breast cancer screening. Nature 2020;577(7788):89–94.

[3] Ardila D, Kiraly AP, Bharadwaj S, Choi B, Reicher JJ, Peng L, et al. End-to-end lung cancer screening with three-dimensional deep learning on low-dose chest computed tomography. Nat Med 2019;25(6):954–61.

[4] Nabulsi Z, Sellergren A, Jamshy S, Lau C, Santos E, Kiraly AP, et al. Deep learning for distinguishing normal versus abnormal chest radiographs and generalization to two unseen diseases tuberculosis and COVID-19. Sci Rep 2021;11(1):1–15.

[5] Liu Y, Gadepalli K, Norouzi M, Dahl GE, Kohlberger T, Boyko A, et al. Detecting cancer metastases on gigapixel pathology images. arXiv:1703.02442, 2017.

[6] Liu Y, Jain A, Eng C, Way DH, Lee K, Bui P, et al. A deep learning system for differential diagnosis of skin diseases. Nat Med 2020;26(6):900–8.

[7] Livovsky DM, Veikherman D, Golany T, Aides A, Dashinsky V, Rabani N, et al. Detection of elusive polyps using a large-scale artificial intelligence system (with videos). Gastrointest Endosc 2021;94(6):1099–109.

[8] Poplin R, Varadarajan AV, Blumer K, Liu Y, McConnell MV, Corrado GS, et al. Prediction of cardiovascular risk factors from retinal fundus photographs via deep learning. Nat Biomed Eng 2018;2(3):158–64.

[9] Steiner DF, MacDonald R, Liu Y, Truszkowski P, Hipp JD, Gammage C, et al. Impact of deep learning assistance on the histopathologic review of lymph nodes for metastatic breast cancer. Am J Surg Pathol 2018;42(12):1636.

[10] Raumviboonsuk P, Krause J, Chotcomwongse P, Sayres R, Raman R, Widner K, et al. Deep learning vs. human graders for classifying severity levels of diabetic retinopathy in a real-world nationwide screening program. arXiv:1810.08290, 2018.

[11] D'Amour A, Heller K, Moldovan D, Adlam B, Alipanahi B, Beutel A, et al. Underspecification presents challenges for credibility in modern machine learning. arXiv:2011.03395, 2020.

[12] Roy AG, Ren J, Azizi S, Loh A, Natarajan V, Mustafa B, et al. Does your dermatology classifier know what it doesn't know? detecting the long-tail of unseen conditions. Med Image Anal 2022;75:102274.

[13] Beede E, Baylor E, Hersch F, Iurchenko A, Wilcox L, Ruamviboonsuk P, et al. A human-centered evaluation of a deep learning system deployed in clinics for the detection of diabetic retinopathy. Proceedings of the 2020 CHI conference on human factors in computing systems 2020;1–12.

[14] Ruamviboonsuk P, Tiwari R, Sayres R, Nganthavee V, Hemarat K, Kongprayoon A, et al. Real-time diabetic retinopathy screening by deep learning in a multisite national screening programme: a prospective interventional cohort study. Lancet Digital Health 2022;4(4):e235–44.
[15] Pedersen ER, Cuadros J, Khan M, Fleischmann S, Wolff G, Hammel N, et al. Redesigning clinical pathways for immediate diabetic retinopathy screening results. NEJM Catal Innov Care Deliv 2021;2(8).
[16] Kroth PJ, Morioka-Douglas N, Veres S, et al. Association of electronic health record design and use factors with clinician stress and burnout. JAMA Netw Open 2019;2(8):e199609. Available from: https://doi.org/10.1001/jamanetworkopen.2019.9609.
[17] Tomašev N, Glorot X, Rae JW, Zielinski M, Askham H, Saraiva A, et al. A clinically applicable approach to continuous prediction of future acute kidney injury. Nature 2019;572(7767):116–19.
[18] Rajkomar A, Oren E, Chen K, et al. Scalable and accurate deep learning with electronic health records. npj Digital Med 2018;1:18. Available from: https://doi.org/10.1038/s41746-018-0029-1.
[19] Jain A, Way D, Gupta V, Gao Y, de Oliveira Marinho G, Hartford J, et al. Development and assessment of an artificial intelligence–based tool for skin condition diagnosis by primary care physicians and nurse practitioners in teledermatology practices. JAMA Network Open 2021;4(4):e217249.
[20] https://health.google/consumers/dermassist/. [accessed 24.05.22].
[21] Artificial Intelligence at Google: Our Principles. http://www.ai.google/prinicples [accessed 02.05.22].
[22] Jobin A, Ienca M, Vayena E. The global landscape of AI ethics guidelines. Nat Mach Intell 2019;1:389–99.
[23] Floridi L, et al. AI4People—an ethical framework for a good AI society: opportunities, risks, principles, and recommendations. Minds Mach 2018;28:689–707.
[24] High Level Expert Group on Artificial Intelligence. Ethics guidelines for trustworthy AI. European Commission; 2019.
[25] What-If Tool. https://pair-code.github.io/what-if-tool/ [accessed 02.05.22].
[26] Language Interpretability Tool for Natural Language Processing models. https://pair-code.github.io/lit/ [accessed 02.05.22].
[27] Know Your Data tool. https://knowyourdata.withgoogle.com/ [accessed 02.05.22].
[28] Open-source library of common fairness metrics. https://www.tensorflow.org/tfx/guide/fairness_indicators [accessed 02.05.22].
[29] Model Remediation. https://www.tensorflow.org/responsible_ai/model_remediation [accessed 02.05.22].

CHAPTER

40

Artificial intelligence drives the digital transformation of pharma

Stefan Harrer[1], *Jeffrey Menard*[2], *Michael Rivers*[2,3], *Darren V.S. Green*[4], *Joel Karpiak*[5], *Jeliazko R. Jeliazkov*[5], *Maxim V. Shapovalov*[5], *Diego del Alamo*[5] and *Matt C. Sternke*[5]

[1]**Digital Health Cooperative Research Centre Ltd., Strategic Business Insights, Melbourne, VIC, Australia** [2]**Roche Diagnostics, Rotkreuz, Switzerland** [3]**Roche Diagnostics, Santa Clara, CA, United States** [4]**GSK R&D, Stevenage, United Kingdom** [5]**GSK R&D, Upper Providence, PA, United States**

From protein to prescription

For decades artificial intelligence (AI) has been heralded as a revolutionary technology poised to transform countless industry sectors and fundamentally changing job roles and professions. But it took until the beginning of the 2010s, for these predictions to take shape in the real world and for AI-driven applications to gain traction in the pharmacological industry. The change was predominantly triggered by two key developments: digital data became available in abundance, and custom-developed computer hardware allowed the running of machine learning (ML) algorithms with unprecedented efficiency at scale. As a result, for the first time AI models could be trained to learn certain tasks in a specific domain exhibiting superhuman performance [1]. This development stage is called narrow AI. The first prominent examples of such narrow AI algorithms include deep learning (DL) models for pattern recognition in imagery data, and natural language processing (NLP) algorithms for analyzing spoken or written text. First deployment domains of such AI systems were the medical sector [2,3] and search engines.

The field has since advanced and begun to evolve into what is called broad AI. In this phase AI technology can be trained to master a series of tasks in a specific domain, and insights from doing so can be applied and transferred to other domains. For example, an

Artificial Intelligence in Clinical Practice
DOI: https://doi.org/10.1016/B978-0-443-15688-5.00049-8

object recognition algorithm can be trained to detect tumor samples in X-ray images and to find structural damage locations in imagery of airplane wings. While narrow and broad AI systems are extremely good at learning, they lack common sense and planning agency [4]. These features are subject of current research toward building so-called Artificial General Intelligence (AGI) systems which would be capable of reasoning, learning, and interacting with humans like humans interact with each other. Although several generative AI technologies have given rise to the suggestion that they might present a viable path to AGI, it does not exist at present, and the projected roadmap and timeline to achieving it as well as the question whether AGI is achievable at all are subject of debate and experimentation in the AI research community [5,6].

On this premise, the role of AI systems is to assist and complement the human decision maker, not to replace them. Applied AI technology can be seen as a fast-growing repository of automated digital decision support and knowledge management tools allowing humans to analyze data more efficiently and more accurately [7,8] Regulatory bodies govern the deployment and integration of AI systems into real-world workflows in certain industry sectors, such as medicine and healthcare [9]. In parallel, ethical frameworks for guiding the responsible design and operations of AI systems have been built and instated in several geographies [10,11,12]. Box 40.1 provides an overview of some of the most relevant AI techniques powering present-day and future data analytics applications.

BOX 40.1 Different methods used in AI. (Box published with permission from *Cell Press*. Original in Harrer S, et al. Artificial intelligence for clinical trial design. Trends Pharmacol Sci 2019;40(8):577–591 [13].) [14,15].

Different methods used in AI

AI: machine simulation of human intelligence processes including learning, reasoning, and self-correction [14]. The ultimate goal of AI is to build machines that can perceive the world and make decisions in the same way as humans do.

Association rule mining: ML algorithms for discovering interesting relations between variables in large databases to help a machine to mimic the extraction and abstract association capabilities of the human brain from new uncategorized data.

Brain–machine interface: a direct communication pathway between an enhanced or wired brain and an external device. Also, referred to as a brain–computer interface, a mind–machine interface, or a direct neural interface.

Deep learning (DL): a class of ML methods based on artificial neural networks, inspired by information processing and distributed communication nodes in biological systems, that use multiple layers to progressively extract higher level features from raw input [15]. The "deep" in "DL" refers to the number of layers through which the data is transformed.

Deep reinforcement learning (DRL): reinforcement learning (RL) is an area of ML that is concerned with building software agents that can take actions in an environment so as to maximize some notion of cumulative reward. DRL combines DL and RL principles to create efficient algorithms to achieve this task.

Human–machine interface (HMI): a direct communication pathway between a human

BOX 40.1 (*cont'd*)

and a device. For example, an artificial system capable of automatically understanding and responding to spoken or written human language constitutes an HMI.

Machine Learning (ML): the scientific study of algorithms that build a mathematical model of sample data to make predictions or decisions without being explicitly programmed to perform the task [15]. ML is often considered to be a branch of AI.

Natural language processing (NLP): a subfield of AI concerned with the interactions between computers and human (natural) languages, in particular how to program computers to process and analyze large amounts of natural language data. NLP draws from many disciplines including computer science and computational linguistics. Large language models (LLMs) also called foundation models are a key component of generative AI applications for creating new content including text, imagery, audio, code, and videos in response to textual instructions.

Optical character recognition: a field of research in AI, pattern recognition, and computational vision aimed at the electronic conversion of images of typed, handwritten, or printed text into machine-encoded text, whether from a scanned document, a photo of a document, a scene-photo, or from subtitle text superimposed on an image.

The biopharma and life sciences sectors are among the most profoundly affected by AI. The entire pharma value chain offers a plethora of opportunities for AI to not only boost the traditional pharma mission of bringing new drugs to market but to also shape new business models and identify new market segments along the digital patient data journey [16]. From discovering novel drug targets by studying proteins to designing clinical trials more efficiently to streamlining supply chains for manufacturing drugs to optimizing distribution and marketing channels for bringing new personalized therapies to patients: the pharmaceutical industry is poised like few others to transform itself by using AI to draw insights from data.

The drug development cycle

Bringing new drugs to market is the foundational mission of the pharma industry. The process of doing so spans several stages of the so-called drug development cycle depicted in Fig. 40.1. It takes on average 10–15 years and between 1.5 and 2.0 billion USD to bring one single new drug to market. Thereby, the success chances are low: only about 1 out of ten drug candidates entering the clinical trial stage receives regulatory approval [13]. Repurposing approved drugs for treatment of diseases other than the indications they had originally been approved for can accelerate the path to market to some extent but does not remove the need to go through clinical trial and regulatory approval phases. For decades the pharma industry has continuously increased its investment into research and development while at the same time the rate with which new drugs hit the market has declined. This economically unsustainable trend proved to be so notorious that it was termed "Eroom's Law," a play upon words on the

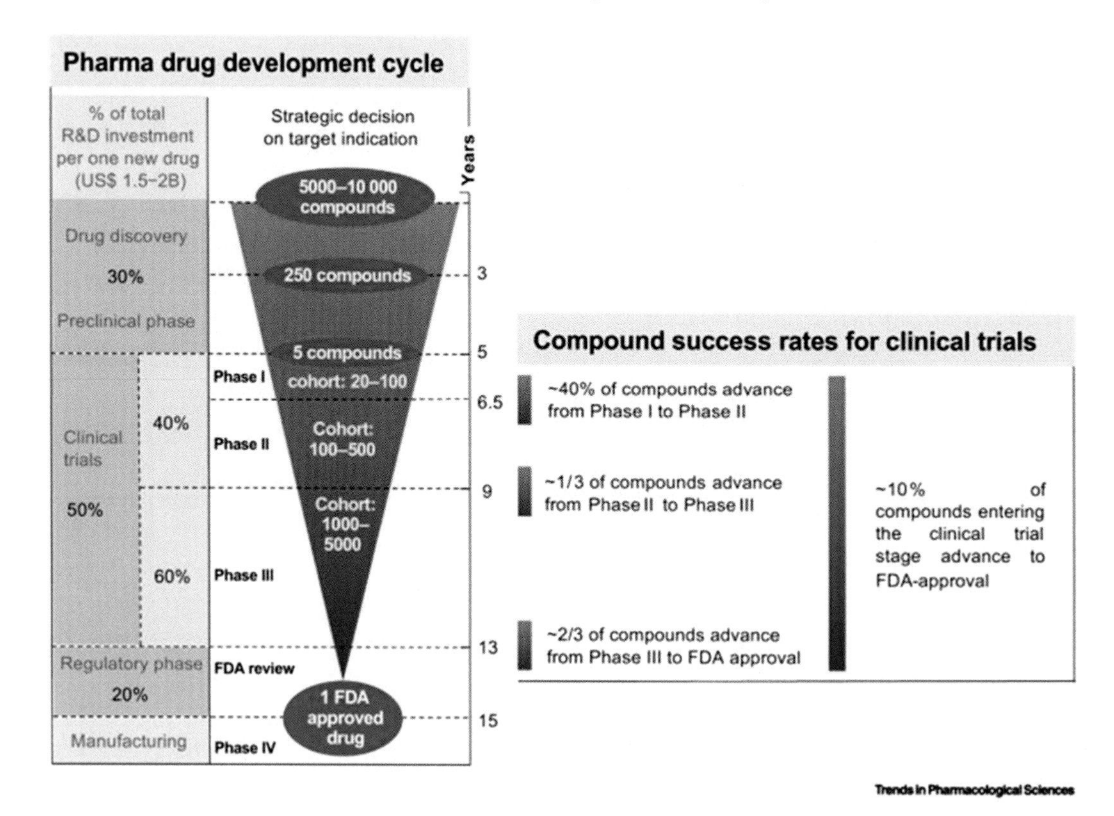

FIGURE 40.1 The drug development cycle. It takes up to 15 years and an average total R&D expenditure of 1.5–2 billion (B) USD to bring a single new drug to market. About half of this investment is spent on clinical trials, with phase III trials being the most complex and most expensive. Probabilities of success for compounds to proceed through the clinical trial stages vary from phase to phase and lead to a situation where only one of 10 compounds entering clinical trials advances to regulatory approval. High clinical trial failure rates are one major cause for the prevailing inefficiency of the drug development cycle. Source: *Figure and caption published with permission from* Cell Press. *Original in Harrer S, et al. Artificial intelligence for clinical trial design. Trends Pharmacol Sci 2019;40(8):577–591 [13].*

infamous "Moore's Law" from semiconductor industry which since 1965 has correctly postulated that the power of computing would increase while its relative cost would decrease at an exponential pace. It was not until 2014 that the pace of Eroom's Law started to slow down in some pharma sectors, indicating that the drug development cycle was at least heading for a point of break-even between R&D investments and the number of new drugs that reached the market [17]. However, the drug development cycle overall remains a highly inefficient process. Even incremental improvements of efficiency throughout its stages can yield savings in the order of billions of dollars and have profound impact on patients. AI technology is one of the best leads for the pharma sector to achieve such improvements [18]. Sections "Artificial intelligence for drug discovery and design", "Artificial intelligence for clinical trial design", and "Artificial intelligence for drug manufacturing and distribution" of this chapter will explain various ways in which AI technology could be used to increase the efficiency of the drug development cycle and provide examples of real-world use cases.

The digital health value generation cycle

As the digitization of healthcare is beginning to take shape and more patients embrace digital health technology, pharma finds itself in a prime position to play an important role on the digital health market which is expected to surpass the trillion-dollar mark in the next decade [19].

Digital health is a consumer-driven services business in which data circularly flows from patients to data managers on to data analyzers and from there via insights generators and carers back to patients in the form of services for treating and managing disease, health, and wellness (Fig. 40.2) [20]. As data moves along that pathway it is incrementally transformed into downstream value by each stakeholder group. For example, data managers can curate patient data and make it available to data analyzers. Those will in turn extract information and provide it to insight generators who will then incorporate the information in models of care and disease management which will be provided back to patients in the form of therapies and health management services. Such data transactions and transformations can only happen if users and providers find secure, trusted, and efficient ways to exchange health data and insights across technical, regulatory, and commercial interfaces between all stakeholder groups.

In this value chain the roles of user and provider are fluid concepts depending on the interface: for example, an electronic health record (EHR) manager acts as user when receiving data from patients and will then be the provider when transferring such data on to developers of analytical models. Data flow is enabled through integrated scalable cloud platforms. Commercial value is extracted by operating the digital health data ecosystem as a Healthcare as a Service platform with other business models also being in operation [21]. Clinical value in the form of health services is returned to patients by carers and providers closing the digital health value generation cycle. Digital transformation of healthcare is foremostly characterized by a transformation of the healthcare business model away from a provider-centric monopolized scheme to a patient-centric democratized one [22,23].

While this describes the direction in which the digital health market is headed and the underlying dynamics by which it functions, in present day reality many hurdles still need to be overcome before the full potential of digital health can be realized. Some of the most pressing tasks at hand are (1) the instantiation of commonly used standards for efficient data collection, linkage and sharing, (2) the adoption and enforcement of trusted ethical and governance structures for secure and privacy-preserving data management, (3) the introduction of attractive reimbursement models as part of an inclusive health economic model, (4) the digitization of health data in the first place, and (5) consumer-centric design of digital health services ensuring that real, quantifiable value is delivered to patients.

AI technology plays an integral role as part of the data analytics segment of the digital health value chain. Digital health data serves as a prime example for exhibiting unprecedented levels of the "five V's" of data: volume, variety, velocity, volatility, and veracity. AI is not the only piece of technology shaping the value proposition of digital health, but it is a crucial one: running in the analytical engine room of cloud-based digital health

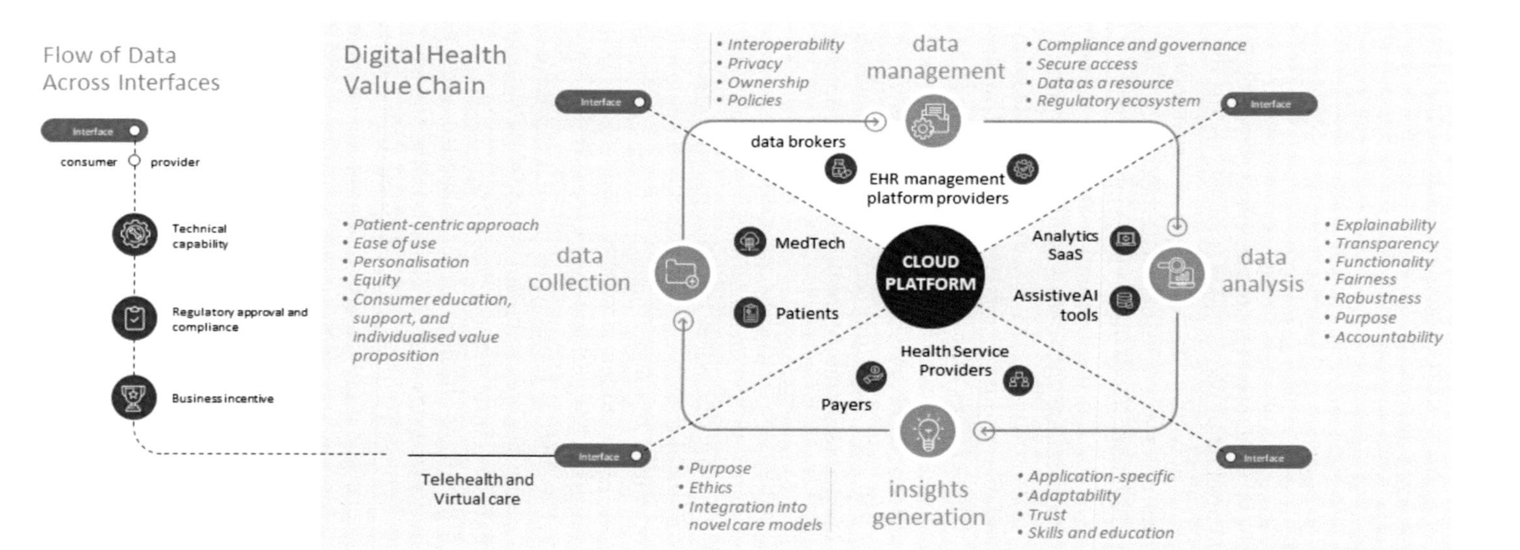

FIGURE 40.2 The digital health value generation cycle. Digital health data flows from patients back to patients and is transformed into value in the form of services by market stakeholders along the way. Each sector follows specific design features but all need to enable secure, efficient, and scalable flow of data through their segments to bring value to patients. Cloud and AI technology are key enablers in this scheme. By expanding its share of the data management and caregiver sectors into data collection and data analytics sectors the pharma industry undergoes a path of digital transformation that changes its business models and places it in the digital health market alongside big tech, MedTech, clinical, and healthcare agents.

services, AI technology offers a way to find patterns in and to extract meaningful information from large, unstructured datasets and to make it available to those who can translate it into actionable medical insights.

With traditionally already holding large stakes in the health data management segment and being closely involved with clinicians, caregivers and providers, pharma has made strategic strides in the form of collaborations and investments to fill the AI-centered data analytics and collection gaps in between (Fig. 40.3) [17]. Embracing new business models and targeting evolving data- and AI-driven market segments, such as telehealth, virtual care, and digital therapeutics (DTx), the pharma sector explores digital technologies for patient identification, stratification, and monitoring, for care management and precision medicine, and for pretreatment and diagnosis through to posttreatment and prognosis. In parallel ethical frameworks for the responsible use of AI are being developed and integrated in product and service development workflows [24]. The section "Artificial intelligence in digital health—digital pathology is leading the way" describe how pharma leverages AI technology to gain and grow market share in the digital health sector.

Artificial intelligence in digital health—digital pathology is leading the way

Healthcare has been built up over the years addressing episodic moments and treating those. Today a shift is underway from a provider-centric monopolized scheme to a patient-centric democratized one. People are moving from "something is wrong, let me ask my doctor" to wanting to manage their health and be proactive and preventive. Expectations are changing, driven by a growing global middle class and further democratization of health: patients expect proactive and personalized health management and best-in-class health services irrespective of location. These shifts in a post-COVID world strain healthcare systems and create new challenges and opportunities. How can healthcare move from a predominant reactive episodic system to a preventive health continuum? Can digital technology be part of the answer? How does digital health and specifically AI (or augmented intelligence) help safeguard and contribute to health for the future? How does that change the modus operandi of diagnostics and the underlying healthcare business model?

SARs-COV-2 has highlighted the importance of diagnostics in healthcare [25]. While lab results drive about 70% of clinical decision making, labs carry less than 5% of hospital costs [26]. Expanding capabilities of in-vitro and in-vivo diagnostics combined with digital tools, such as the digitization and linkage of longitudinal records and their analysis with AI systems, have triggered a profound evolution of the field of diagnostics.

A rapidly growing area of diagnostics enabled by digitalization and powered by AI is digital pathology. Tissue diagnostics has traditionally been a laborious practice driven by the need to process biopsies and surgical specimens into thinly sliced tissue sections mounted and stained on glass slides for pathologist review. The most common workflows for this process require many manual steps that culminate in the review of the sample by a pathologist. This subjective review, albeit by a highly trained practitioner, is a foundational element of most diagnoses today and is especially critical for cancer diagnosis.

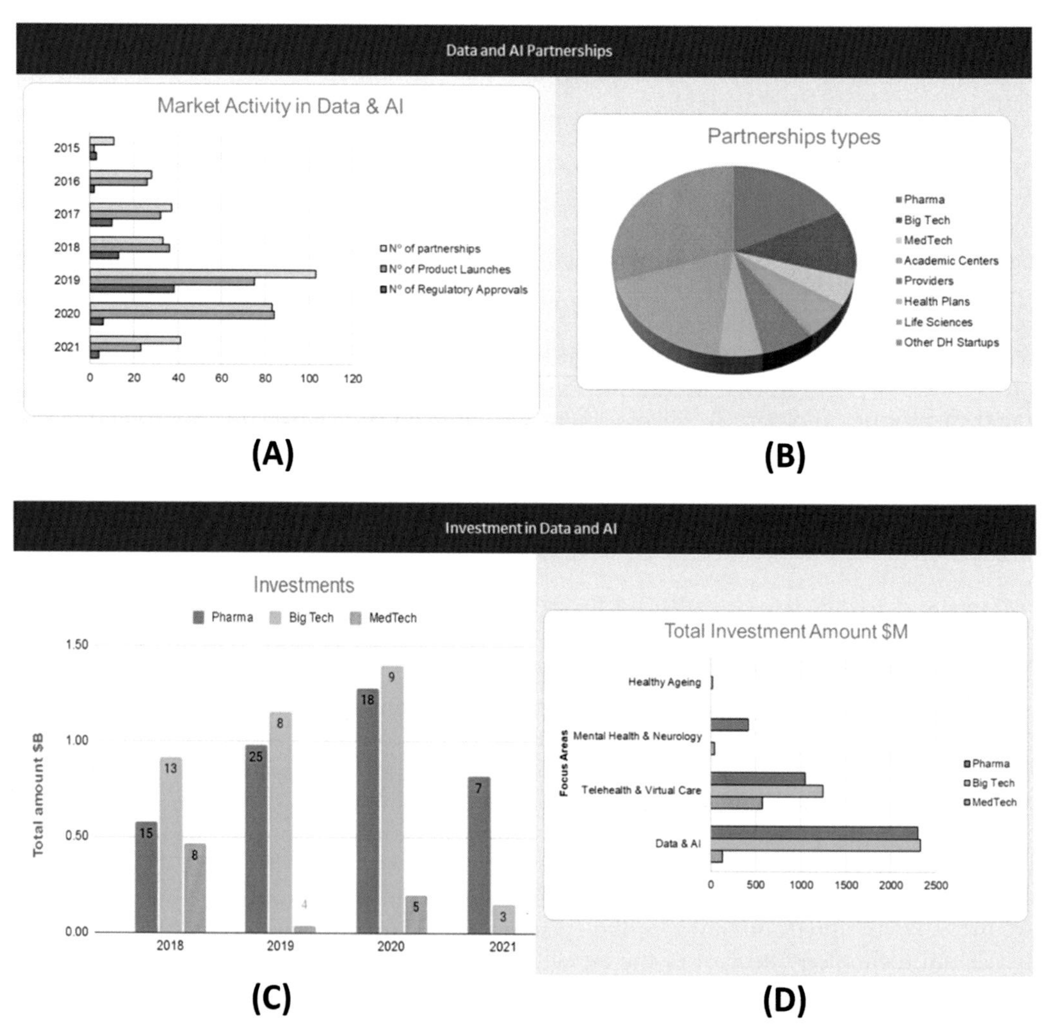

FIGURE 40.3 (A and B) Pharma is heavily involved in collaborations across big tech, healthcare, and life sciences sectors to tap into the potential of the digital health value chain. The data and AI segment is one of the most active areas within digital health, having the highest value of investments (D), the highest number of product launches, and the highest number of regulatory approvals (A). (C) Pharma and BigTech are investing heavily in digital health. Pharma has a larger number of investments in digital health, while BigTech shows a higher investment value. All numbers in this figure were derived from analyzing the investment activities between January 2018 and August 2021 of the following enterprise cohorts: Pharma: Novartis, Merck, Bayer, BMS, Pfizer, Roche, AstraZeneca, GSK, Amgen, Gilead, Sanofi; BigTech: Amazon, Google, Microsoft, Apple, IBM, Intel, GE, Samsung, Sony, Alphabet; MedTech: Medtronic, J&J, Smith & Nephew, Boston Scientific, Zimmer Biomet, Baxter, Stryker, Siemens Healthineers, BD, Philips. Source: *DHCRC Ltd./AUSTRADE 2021 Study on the Investment Value Proposition of Digital Health.*

The digital transformation of pathology is supported by more than two decades of investment and technology advancement from industry and is driven broadly by two categories of benefits: access to care and insights from the introduction of AI.

Digital pathology offers significant advantages in terms of both patient access to pathologists and in pathologist access to their samples. The SARS-COV-2 pandemic highlighted the need to enable remote diagnosis capabilities for both patients and pathologists. Digital pathology frees clinicians from their need to be connected physically to the glass slide and opens up the possibility of a digital review of the patient sample from anywhere and at any time. The potential of this advancement cannot be overstated for the many geographies that are underserved with trained pathology resources. Digital pathology offers a tremendous opportunity to level supply and demand and ensure that all patients get the best diagnostic information in the shortest amount of time from the right medical expert. It also facilitates the sharing of patient case information between pathologists and the rendering of consultations and second opinions.

The application of AI within digital pathology can improve patient treatment, generate new insights into a patient's health earlier, and create new commercial opportunities.

Artificial intelligence augments pathologists

A key application of AI in digital pathology arises from the growing complexity of tissue staining. Identification of cancer has steadily increased in complexity due to scientific advances in oncology. Cancer is no longer identified by a specific body part (e.g., breast, lung cancer) but through specific biomarker(s) (e.g., breast cancer—HER2, PD-L1).

The advancement of staining toward multiplex immunohistochemistry and immunofluorescence has created emerging opportunities for better outcomes and more efficient patient care but also introduced new challenges for a trained pathologist to interpret samples (Fig. 40.4).

The use of AI for quantitative image analysis can empower a pathologist to capture data from tissue slides that may not be accessible during manual routine microscopy inspection [27]. Digital pathology will become an essential part of any pathologist's multiplex stain reading workflow.

Pathologists aiming to leverage the expanding library of digital pathology algorithms are looking for feasible ways to integrate these analytical models into their clinical workflows. This creates opportunities for different algorithm providers to follow a consumer centric design process for their AI models and to support the overall growth of a services and API ecosystem for the benefit of pathologists and patients. The approach can also lead to commercial pathways and reimbursement frameworks for providers and users of such analytical services: employing quantitative image analysis tools for interpretation support will incur usage fees payable to AI service providers.

Digital pathology reduces the barriers of staining complexity and can improve the timing of biopsies. It thus plays a critical role for the expansion of precision medicine aiding the conception of personalized medication and care regimes. The use of pathology for precision medicine to advance cancer treatment is not novel and has been essential in finding the best treatment for an individual patient. The promise of AI in this context lies in enabling more complex pathological diagnostics and interpretation (Fig. 40.5). AI can support pathologists during clinical trials and later for specific treatment identification and optimization. Pharmaceutical companies continue to financially support the development

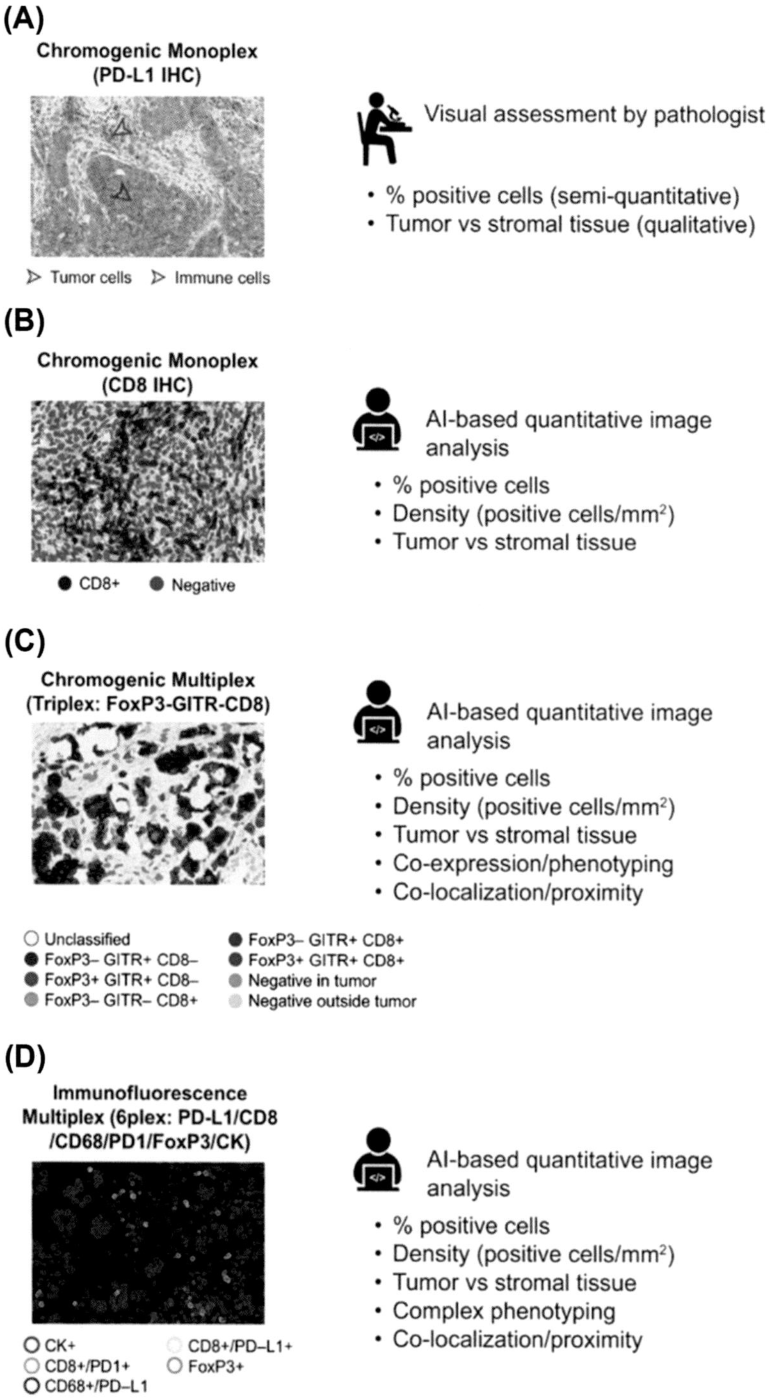

(A)

Chromogenic Monoplex
(PD-L1 IHC)

▷ Tumor cells ▷ Immune cells

Visual assessment by pathologist

- % positive cells (semi-quantitative)
- Tumor vs stromal tissue (qualitative)

(B)

Chromogenic Monoplex
(CD8 IHC)

● CD8+ ● Negative

AI-based quantitative image analysis

- % positive cells
- Density (positive cells/mm^2)
- Tumor vs stromal tissue

(C)

Chromogenic Multiplex
(Triplex: FoxP3-GITR-CD8)

○ Unclassified
● FoxP3– GITR+ CD8–
● FoxP3+ GITR+ CD8–
● FoxP3– GITR– CD8+
● FoxP3– GITR+ CD8+
● FoxP3+ GITR+ CD8+
● Negative in tumor
● Negative outside tumor

AI-based quantitative image analysis

- % positive cells
- Density (positive cells/mm^2)
- Tumor vs stromal tissue
- Co-expression/phenotyping
- Co-localization/proximity

(D)

Immunofluorescence
Multiplex (6plex: PD-L1/CD8
/CD68/PD1/FoxP3/CK)

○ CK+
○ CD8+/PD1+
○ CD68+/PD–L1
○ CD8+/PD–L1+
○ FoxP3+

AI-based quantitative image analysis

- % positive cells
- Density (positive cells/mm^2)
- Tumor vs stromal tissue
- Complex phenotyping
- Co-localization/proximity

FIGURE 40.4 Applications of digital pathology in immunohistochemistry. Source: *Figure published with permission from* Nature. *Original in Baxi V, Edwards R, Montalto M, et al. Digital pathology and artificial intelligence in translational medicine and clinical practice. Mod Pathol 2022;35:23–32. https://doi.org/10.1038/s41379-021-00919-2 [27].*

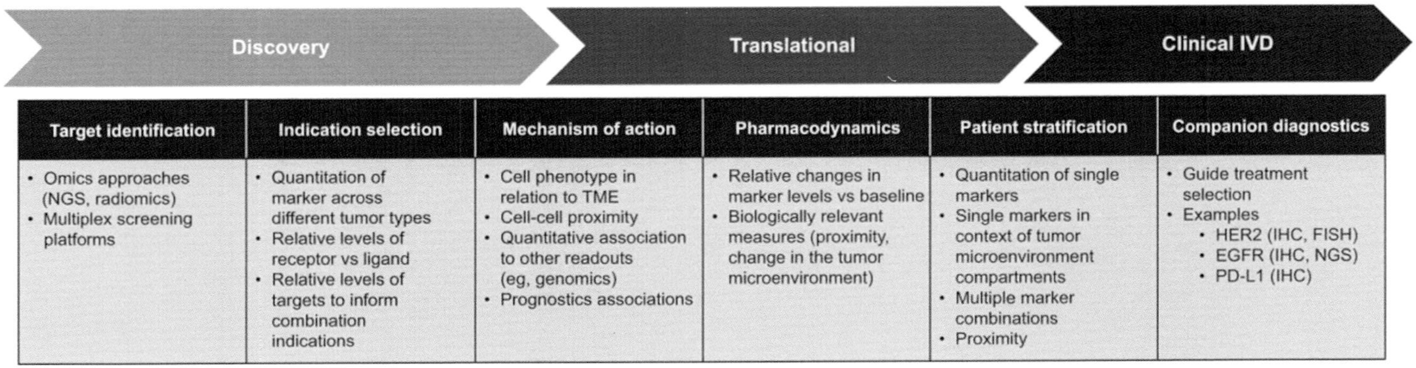

Discovery		Translational		Clinical IVD	
Target identification	Indication selection	Mechanism of action	Pharmacodynamics	Patient stratification	Companion diagnostics

• Omics approaches (NGS, radiomics)
• Multiplex screening platforms
	• Quantitation of marker across different tumor types
	• Relative levels of receptor vs ligand
	• Relative levels of targets to inform combination indications
		• Cell phenotype in relation to TME
		• Cell-cell proximity
		• Quantitative association to other readouts (eg, genomics)
		• Prognostics associations
			• Relative changes in marker levels vs baseline
			• Biologically relevant measures (proximity, change in the tumor microenvironment)
				• Quantitation of single markers
				• Single markers in context of tumor microenvironment compartments
				• Multiple marker combinations
				• Proximity
					• Guide treatment selection
					• Examples
					  • HER2 (IHC, FISH)
					  • EGFR (IHC, NGS)
					  • PD-L1 (IHC)

FIGURE 40.5 Digital pathology: from drug discovery to clinical diagnostics. Source: *Figure published with permission from* Nature. *Original in Baxi V, Edwards R, Montalto M, et al. Digital pathology and artificial intelligence in translational medicine and clinical practice. Mod Pathol 2022;35:23–32. https://doi.org/10.1038/s41379-021-00919-2 [27].*

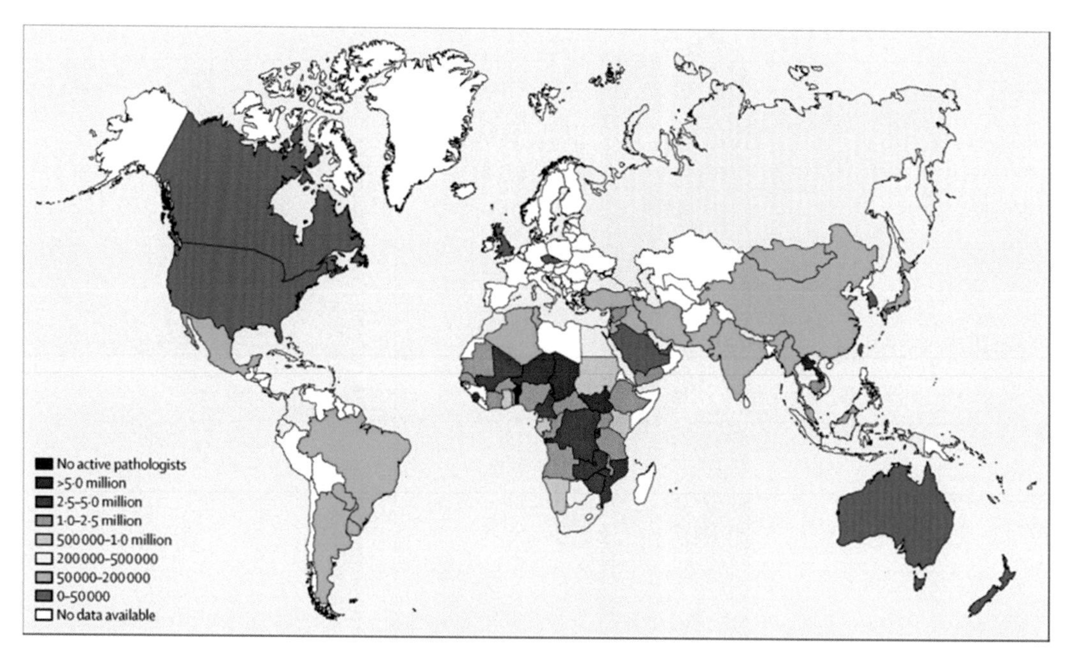

FIGURE 40.6 Global workforce capacity in pathology and laboratory medicine. Source: *Figure published with permission from* The Lancet. *Original in Wilson ML, Fleming KA, Kuti MA, Looi LM, Lago N, Ru K. Access to pathology and laboratory medicine services: a crucial gap. 2018;391:1933.*

of assays targeted at specific biomarkers and increasingly companion digital pathology algorithms to facilitate their analysis.

Digital pathology has the capability to open improved cancer diagnosis to a broader population around the world. Many geographies are underserved and do not have access to qualified pathologists or pathologists specialized in certain types of cancer. Digital pathology creates opportunities to advance patient care in rural areas as well as in low-income areas with poor levels of diagnostics support and often limited access to skilled pathologists (Fig. 40.6).

New business and services models can be utilized to expand pathology offerings to such underserved populations. Instead of needing to bring pathologists to patients, scans can be virtually made available to pathologists regardless of their location. This removes the need to mail tissue samples on glass slides and enables timely second and third opinions from peers as needed. As AI algorithms are trained and validated by leading pathologists from around the world their knowledge is embedded in the algorithms and workflows and can be leveraged and shared to peers around the world. This crowd-sourced spread of AI-inherent expertise in turn can lift the quality of pathological services across countries and institutional borders. Ethically designed assistive AI algorithms can be part of the answer to help reduce potential individual pathologist biases and elevate the overall quality of care.

Success of all these service models will require many different elements of the digital health value generation cycle to come together. Among those, reimbursement frameworks, data availability, privacy and connectivity, cyber security, and consumer centric workflow design are paramount. The global pandemic has elevated the need for remote, digital peer-to-peer collaboration. Lab budget constraints and staffing challenges, from hiring to retirement, will continue to drive the need for a more strategic laboratory workflow. In digital pathology, the continued building of trust across the industry will be essential. This means that AI algorithms do not only need to yield the desired technical performance—models also need to be built and integrated in responsible ways. This aspect of designing AI systems is referred to as the ethics of AI and spans the entire MLOps cycle: from observing data ownership and privacy to creating fair, robust, transparent, accountable and explainable AI algorithms to ensuring that AI is used for a clearly defined purpose: AI ethics is just as important as algorithmic technical development to gain the trust of those who are intended to adopt AI tools and to make sure that real value is delivered to users [28].

Today many use cases of AI within healthcare focus on streamlining operational and administrative inefficiencies and on optimizing resources. Involving AI in treatment and diagnosis requires overcoming steep regulatory hurdles and carries risks which investors and consumers are only gradually beginning to accept. Only approximately 30% of investments into AI and ML companies are backing ventures which support providers in direct patient care in areas\ such as treatment of disease, clinical decision support, and precision medicine [29]. The true gain in using AI to support healthcare in the long run will be how to connect both worlds by finding responsible, safe and efficient ways to utilize AI to have direct clinical impact for patients while at the same time creating efficiencies and better use of resources. Digital pathology is a field that promises to enable just that [30] as it moves towards a tipping point for broader adoption of AI technology in digital healthcare [31] (Fig. 40.7).

New digital health solutions—toward digital therapeutics

A subcategory of digital health are digital medicines, and a subset thereof are digital therapeutics (DTx) (Fig. 40.8). DTx are software-enabled treatment programs tailored to

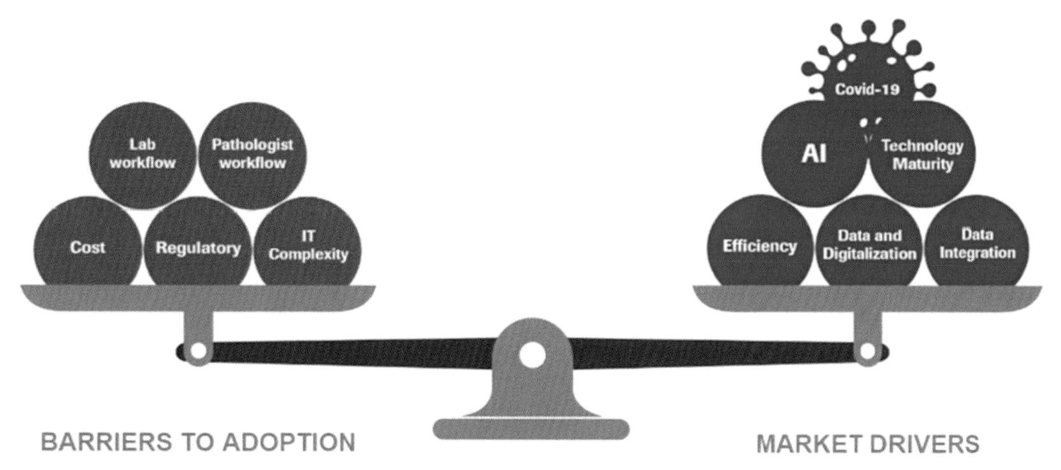

FIGURE 40.7 Digital pathology is one of the tipping points in digital health for broader adoption of AI.

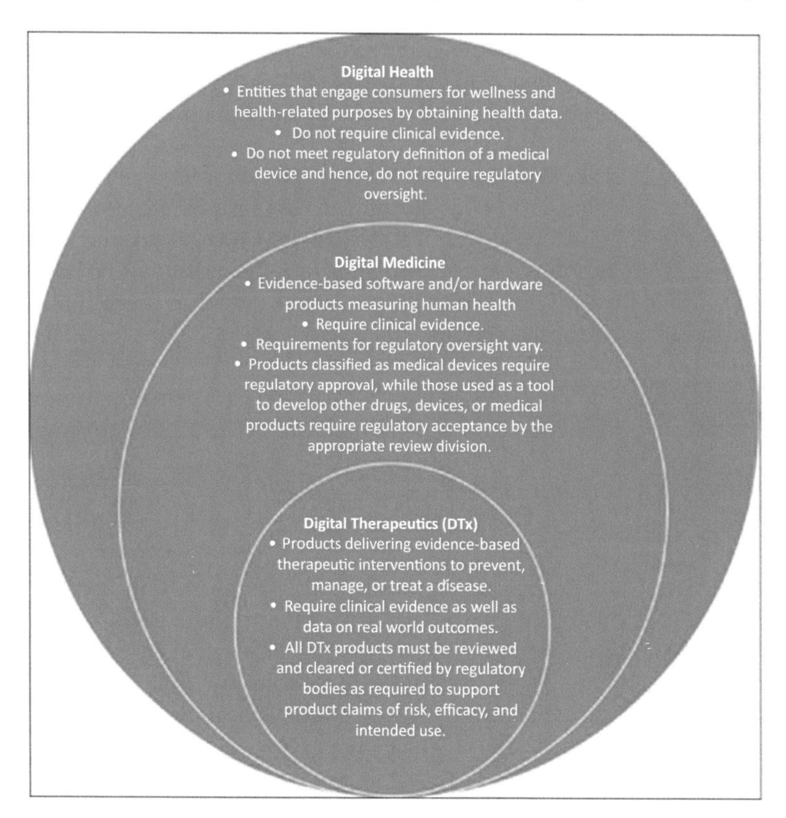

FIGURE 40.8 Differences between digital health, digital medicine, and digital therapeutic products concerning different risks and corresponding levels of necessary evidence and regulatory oversight. Source: *Figure and caption published with permission from* The Lancet. *Original in Dang A, Arora D, Rane P. Role of digital therapeutics and the changing future of healthcare. J Family Med Prim Care. 2020;9 (5):2207–2213. https://doi.org/10.4103/jfmpc.jfmpc_105_20.*

specific conditions, such as for example mental health and neurological disorders, chronic diseases like diabetes, heart disease, high blood pressure, and pulmonary diseases like COPD [32].

The predominant mechanism employed by DTx is the combination of patient monitoring to automatically and in real-time detect patient-specific unhealthy lifestyle patterns or digital biomarkers and to then use this information to induce behavioral changes through intervention or patient coaching. AI, and particularly DL, is uniquely suited to automatically detect signatures in large, noisy, unstructured monitoring data. Another subfield of AI, NLP, allows to personalize human-computer interaction which plays a key role in delivering behavioral guidance and intervention at the point-of-care and in a highly personalized manner [13]. AI-enabled DTx innovation will continue to advance precision medicine and to empower and encourage patients to proactively manage their health and wellness.

While DTx development, design and deployment are governed by strict regulatory frameworks, DTx can generally be brought to patients quicker than traditional pharmaceuticals. This is an appealing value proposition for the pharma industry which explores DTx approaches to complement, enhance and speed up the drug development cycle. In 2022, the FDA approved the marketing of a DTx for ADHD, which opens new possibilities affecting approximately 4 million children at ages 6–11 [33].

Artificial intelligence for drug discovery and design

Drug discovery is a long and complex process, requiring large multidisciplinary teams working at the edge of our current scientific understanding of disease. Repeated cycles of hypothesis construction and confirmation are key features of the discovery process (Fig. 40.9). Nonetheless, there are significant opportunities to utilize ML to increase the probability of success and efficiency of discovery.

The starting point for a new therapeutic is the association of an unmet medical need with a biological mechanism, the regulation of which with a molecule (synthetic organic, peptide, antibody, oligonucleotide, or complex combinations) will alter the disease state to a desired phenotype. Target identification is the process of selecting the target based on the information at hand. With modern biology, data is becoming abundant in both scale and diversity: human genetic information from large and diverse patient populations, functional genomics experiments in relevant cells, and transcriptomic, proteomic, and metabolomic analyses of healthy and diseased tissue. Such scale and diversity of data is ideally suited to ML methods, particularly graph-based neural networks which can directly take advantage of biological networks for input. Having identified an association between a target gene/protein and a disease, it is necessary to prove a causal relationship in the target validation phase. ML methods may assist in prioritizing targets which are most likely to exhibit causality, normally assisted by a knowledge graph which is able to describe and help experts navigate the relationship between targets, disease, symptoms, and drug entities. Knowledge graphs can combine structured data with unstructured data from scientific papers, patents, reports, and real-world evidence. A recent example is the use of a knowledge graph for the repurposing of drugs for use with COVID-19 [34]. This work identified ACE2 as a common biological target in pathways for COVID-19 infection and Clathrin-mediated endocytosis. The medicine baricitinib, licensed for the treatment of rheumatoid arthritis, was identified as a potential treatment to prevent the cytokine storms seen in COVID-19 patients.

With a target identified, a therapeutic molecule must be found. The majority of ML literature is applied to the discovery of "small molecules"—synthetic organic compounds of modest size and suitable for dosing via the familiar oral tablet or pill. A proposed oral medicine needs to bind to its intended target with sufficient potency and duration to effectively regulate the biological pathway and must also be soluble and permeable. It must reach the intended site of action and stay there long enough to elicit the desired response while not causing any side effects or safety concerns for the patient. This is normally summarized as a target compound profile (TCP). It is rare to find an existing molecule which satisfies a TCP, and therefore projects engage in a hit identification process to find promising starting points which have some affinity for the target and some promise that they

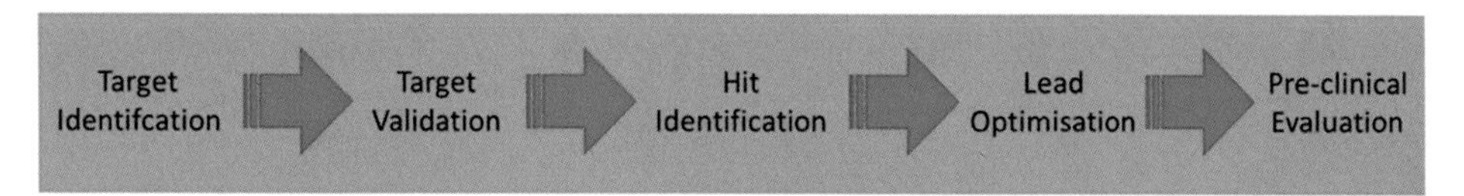

FIGURE 40.9 The drug discovery process.

might be modified to satisfy the TCP. With a hit in hand, chemists will engage in iterations of the classic design–make–test cycle, modifying the compound in a multiparameter optimization process called lead optimization. This process can take several years and has been estimated to cost $10 million per program [35].

Modern target identification processes increase the probability of discovering novel proteins with very little known about them, reducing the immediate opportunities to build ML models. Therefore techniques, such as active learning, may be used to guide experimental data acquisition, which will both discover interesting knowledge and efficiently improve the fledgeling models. Applications of active learning can be found in hit identification [36] where large compound inventories are searched for the most interesting molecules without having to test them all. Active learning models have also been employed in combination with large, accurate, physics-based simulations, such as free energy perturbation for ligand–protein binding affinity [37], where the label is generated by the simulation and the next ligand for simulation decided by the active learning protocol. The outcome of the active learning process is a reliable ML model that can then be used to predict across millions or billions of potential ligands and identify the most potent candidates for synthesis and/or testing.

Many of the other end points for the TCP are not amenable to protein structure-based simulation approaches. In major pharmaceutical companies, there are reasonably sized data sets (thousands to tens of thousands) for end points, such as permeability, clearance, and solubility. ML methods, either traditional (support vector machines, tree-based methods) or DL single- or multitask learners, are used to build so-called "global" models for broad application to any project or chemical series [38]. In drug discovery, these models are known as QSAR (Quantitative Structure Activity Relationships) models. Due to the size of the possible chemical space and a modern discovery process which identifies novel biological mechanisms, it is inevitable that such models will not work for all projects. This is a challenge faced by all QSAR modelers and leads to particular practices for training and validation of models, for example a strong focus on training set selection, temporal validation, and model uncertainty [39]. Where a global model does not perform for a project, a local model is often built. Available data on the specific chemical series will often be minimal (starting with sometimes only 1 hit compound, ending with 100 seconds or maybe 1000 seconds of compounds). New data are slow to acquire (weeks), sparse (some endpoints in the TCP may have data for all compounds, some may only be measured for a handful), biased (some end points may be met for a majority of compounds, some for only a handful) and discontinuous (the phenomenon of the activity cliff [40] where a small change in the molecular structure produces a dramatic fall in affinity). This is a long way from the big data world of images and genetics, and techniques, such as active learning, are often applied to generate the best data for building and improving local models. The frequent need to build models on such limited data sets also drives an interest in "small data" ML, such as one/few-shot [41] and meta-learning [42]. In an attempt to move the industry forward in this regard and generate larger data sets for modeling, pharmaceutical companies have also launched collaborative projects to investigate the use of federated ML [43] and student-teacher models [44]. The improvements seen to date are modest in terms of model predictivity but did show some improvement in domain of applicability.

In an ideal lead optimization process, models would be available for all parts of the TCP. In this scenario, so-called generative models could be used to propose chemical structures predicted to meet the TCP. There are many approaches that have been applied to this challenge: recursive neural networks, variational autoencoders, generative adversarial networks, graph convolutional policy networks, DRL, and many more [45].

This chemical structure generation and prediction process is generally referred to as automated molecular design. The state of the art of this area and a useful framework to classify such systems has recently been proposed [46] (Fig. 40.10). Although no published system meets the highest grade (level 5)—the fully autonomous, AI designer and decision maker—the GSK BRADSHAW system [47] was highlighted as the closest potential level 5 system and has been used to design in vivo active molecules. It should also be noted that there are molecules in the clinic from Ex Scientia and InSilico Medicine that have been discovered using ML-based design processes [48].

While the BRADSHAW system focuses on small molecule generation and optimization, the same concepts and modern ML approaches have significantly impacted and accelerated the design of protein-based therapeutics and tools for target validation, including the lucrative monoclonal antibody drug class. This exploded at the end of 2020 and in 2021 with the breakthrough performance and subsequent release of DeepMind's AlphaFold2

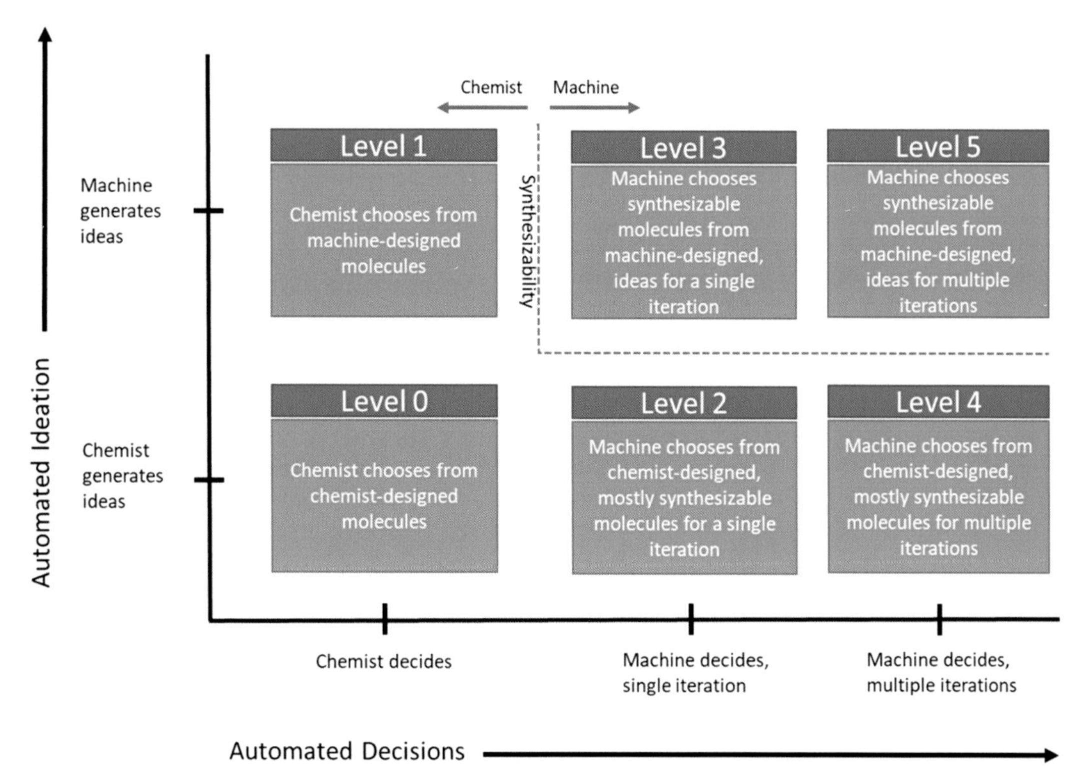

FIGURE 40.10 Classification of approaches to automated molecular design using machine learning models. Source: *Adapted from Goldman B, et al. Defining levels of automated chemical design. J Med Chem 2022 [46].*

model for protein structure prediction [49], opening the floodgates of potential applications in protein modeling and design. The ability to now accurately predict the structures of individual proteins and those in larger complexes and assemblies has inspired further extensions and applications of the AlphaFold2 model in the end-to-end drug discovery process, such as predicting protein conformational changes [50]. These structural models can then feed into a starting point for the structure-based hit identification and lead optimization of synthetic organic molecules and therapeutic proteins, including antibodies [51,52]. Since then, the application of new ML architectures to the general protein structure and design problems, such as generative diffusion models, have only just begun seeing experimental validation of in silico-designed proteins binding to specific targets with significantly improved success rates [53].

Other types of models are being simultaneously developed, relying purely on the sequence of the target protein, and have also recently seen a wide variety of applications. The interface of biology, bioinformatics, and DL has been greatly shaped by exploiting public access to massive genetic sequence databases, containing billions of naturally occurring DNA and amino acid sequences. These resources have facilitated the application of the Transformer model architecture, adapted from the field of NLP, which excels at a range of challenging scientific tasks, ranging from point mutation effect prediction to 3D structure prediction and design [54]. Recent developments have focused on increasing their size and depth [55], rather than iterations or improvements to the fundamental architecture responsible for their groundbreaking performance. Modern models routinely exceed one billion trainable parameters and demonstrate both more effective generalization as well as accurate zero-shot performance in relevant objectives, such as protein function prediction. The size of these models will only increase with in place investments in specialized computational resources, such as TPUs or the Cerebras system.

Due to the availability of a large number of sequences for training, these language models represent an attractive starting point to represent proteins in a BRADSHAW-like framework. In these design–make–test cycles, improving model performance with rapid experimental feedback would require balancing the benefits and tradeoffs of model size when determining which architecture should serve as a foundation for further task-specific fine-tuning [56]. There is still significant effort underway in the field of benchmarking these nascent models, as the promise of continual retraining efforts for updating with evolving experimental data collection, and of coupling with iterative active learning frameworks for further efficient predictive model training, is emerging while larger-scale experimental platforms for hypothesis testing are simultaneously being developed.

As this work was in preparation, the chatGPT large language model (LLM) system was launched by OpenAI [57], an astonishing demonstration of the state of the art in AI applied to language processing and generation. We asked chatGPT the question "What is the future for AI directed drug discovery?." This is the answer, generated in seconds:

"Artificial intelligence (AI) has the potential to revolutionize the drug discovery process by helping to identify new targets for drug development and by enabling the rapid analysis of large amounts of data. In recent years, there has been significant progress in the use of AI in drug discovery, and it is likely that this trend will continue in the future. One promising area of AI in drug discovery is the use of machine learning algorithms to analyze large datasets and identify patterns that may be relevant to drug development. For example, machine learning algorithms can be used to analyze genomic data to identify genetic markers associated with certain diseases, or to analyze chemical compounds to identify those that may have therapeutic potential. Another promising area is the use of AI to design and synthesize new compounds. Machine learning algorithms can be trained to predict the properties of potential drug candidates and to suggest modifications that may improve their effectiveness. This can significantly speed up the drug development process and reduce the cost of bringing new drugs to market. Overall, it is likely that AI will play an increasingly important role in drug discovery in the coming years, and it is possible that it will eventually become an integral part of the process. However, it is important to note that AI is not a replacement for human expertise and judgment, and it will be essential to carefully consider the ethical implications of using AI in drug discovery and to ensure that it is used responsibly."

While there is little to disagree with in this analysis it is not without irony that chatGPT's answer explicitly refers to one of the greatest risks of using LLMs in the scientific domain and particularly in the medical and life sciences sectors: LLMs do not in any way understand the content they create. They are solely exploiting and replicating statistical connections in the training data they have been fed, and this training data so far is sourced essentially unfiltered from the vastness of publicly available data on the internet. Unvalidated and not curated, such data often contains biases, errors, and falsehoods which LLMs may amplify and incorporate in the new content they generate. For LLMs to become useful and trustworthy tools in drug discovery, guardrails and frameworks need to be developed and implemented that ensure that training data selection and LLM output implementation are sound and do not fall victim to the proverbial garbage-in, garbage-out paradigm [58,59].

Artificial intelligence for clinical trial design

After a drug target has been identified and undergone a first phase of preclinical vetting it will be elaborately tested on humans during several clinical trial stages. While clinical trials consume approximately half of the entire drug development budget there is only an approximately 10% probability that a compound entering the clinical trial phase will make it through to regulatory approval and onto the market. A failed clinical trial is not only a write-off of the investment in the trial itself but also sinks the investment costs into previous preclinical phases for that compound which in total can amount to losses in the hundreds of millions of dollars per failed trial (Fig. 40.1). Hence, increasing the success rate of clinical trials constitutes one of the most impactful ways to improve the efficiency of R&D investments into drug development.

There are three main themes of clinical trial design and execution which through leveraging AI technology can be made more efficient: cohort composition, patient recruitment, and patient monitoring (Fig. 40.11) [13]. Targeting suitable and eligible patients is the first hurdle. Motivating and empowering such candidates to enroll in a trial is the second challenge. Then efficiently and reliably monitoring enrolled patients during a trial to assess drop-out risk and their adherence to trial protocol, to retain them in the trial, and for endpoint detection is the third opportunity for AI to make a difference.

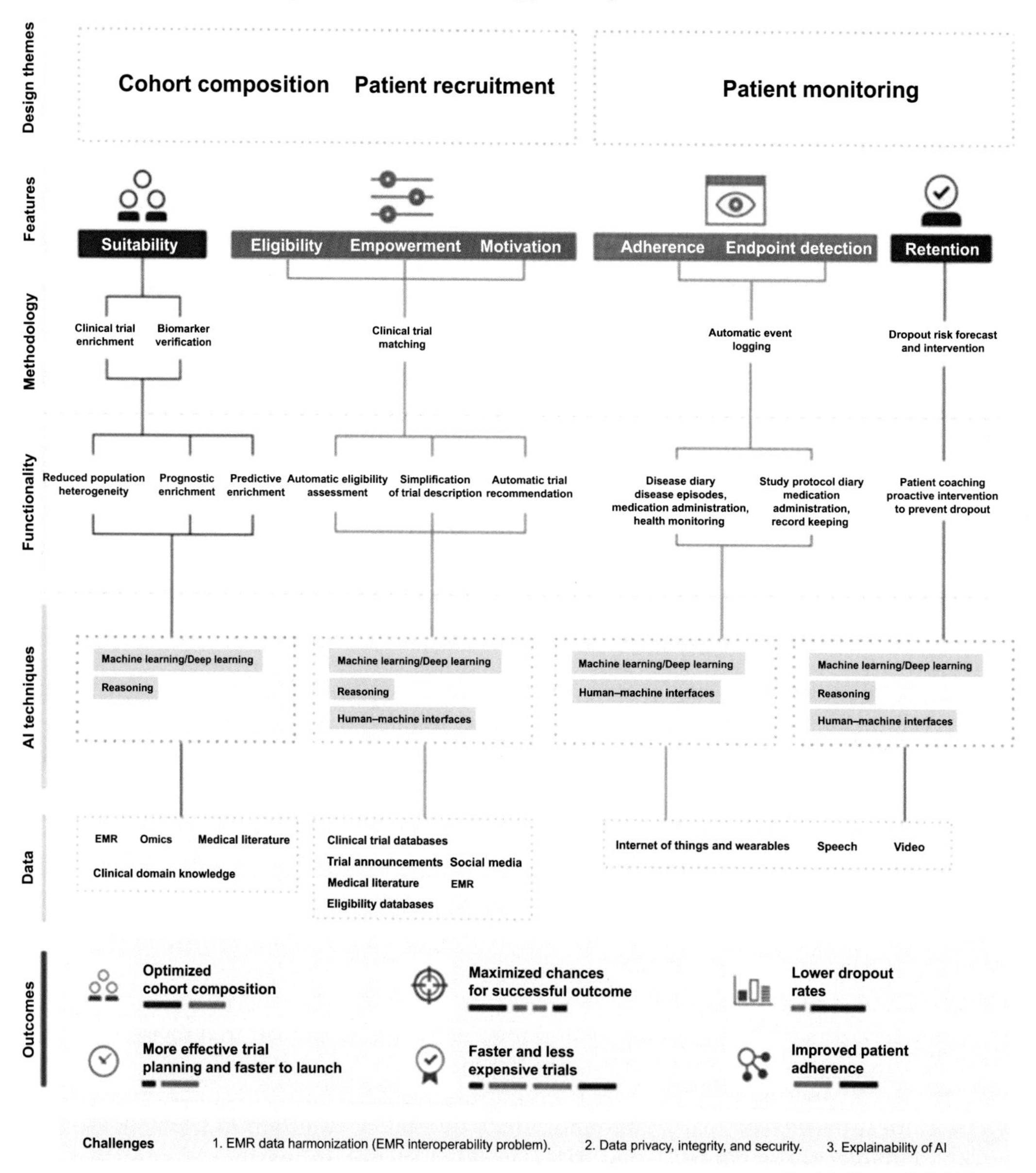

FIGURE 40.11 The schematic visualizes the major ways to infuse AI into the clinical trial design pipeline. The three core design themes—cohort composition, patient recruitment, and patient monitoring (top row)—are based on patient features regarding suitability, eligibility, enrollment empowerment, and motivation, as well as trial features including endpoint detection, adherence control, and patient retention (second row). A variety of design methodologies (third row) are used to implement target functionalities (fourth row). These functionalities are enabled through individual combinations of the three main AI technologies: machine/deep learning, reasoning, and human–machine interfaces (fifth row) which each analyze a specific set of patient- and functionality-specific data sources (sixth row). The relative improvement brought about by such implementations on study outcomes is indicated by the length of the horizontal lines in the color bar code underneath the main outcome aspects (seventh row). Every AI-based study design application is directly dependent on the quality and amount of data it can tap into, and hence faces the same fundamental challenges (bottom row). Abbreviation: EMR, electronic medical record. Source: *Figure and caption published with permission from* Cell Press. *Original in Harrer S, et al. Artificial intelligence for clinical trial design. Trends Pharmacol Sci 2019;40(8):577–591 [13].*

ML, DL, NLP, and reasoning (Box 40.1) can be used to analyze electronic medical records (EMRs), omics data, medical literature and—combined with clinical domain knowledge—enable clinical trial enrichment and biomarker verification for identifying suitable patients.

Adding human-machine interface systems, the same AI techniques can be used to support clinical trial matching approaches for automatically assessing the eligibility of a patient to partake in a trial, for simplifying often convoluted and highly complexly worded trial descriptions, and for producing automated recommendations for prospective patients in search for trials as well as for clinicians looking for patients to enroll in trials. Besides using EMR data, clinical trial matching systems tap into clinical trial databases, trial announcements, eligibility criteria databases, medical literature, and social media information.

Once a patient is enrolled in a trial, protocol-relevant events, such as disease episodes, medication intake, and general health status, need to be logged and documented reliably and efficiently for ensuring adherence to trial protocols and for detecting trial endpoints. This is where the automation and pattern recognition powers of AI come into play allowing to complement and even to replace manual self- or third-party reporting methods. So-called digital disease and study protocol diaries use AI technology to automatically find, label, and record patterns of relevance in patient monitoring data which may be collected by wearable or mobile sensors, video or through speech recognition devices. These techniques also allow to assess drop out risk, to proactively intervene if needed, and to coach patients for retaining them in the trial.

AI-powered digital diaries are also a key building block of what is commonly referred to as digital twin technology. By linking and analyzing a patient's historic and longitudinal health data from genome to exposome and using AI to find personalized patterns of functionality and behavior in such data, digital twins are synthetic models which allow to describe and simulate the personalized biological and environmental functioning of a patient or a patient cohort to a certain extent [60]. Digital twins are not perfect avatars of human beings. Thus their role is not to replace human beings but to complement studies about them, such as clinical trials. In-silico trials for example combine in-vivo patient cohorts with synthetic digital twin ones in virtual external control arms. More variations of clinical trial design patterns which combine synthetic digital twin cohorts with real patient cohorts are possible and being explored [61]. While first examples have emerged where digital twin technology has been cleared for such purposes by regulators in Europe and the United States for example replacing phase III trials lowering recruitment costs and timelines and shortening the approval path [62,63], the use of digital twin technology as part of clinical trials is still in its early stages and classical randomized control trials remain the gold standard of clinical trial design.

Telemedicine and virtual care—paradigms where carer and patient are not interacting face-to-face during consultation or therapy—increasingly incorporate AI elements to enhance user experience and deliver personalized outcomes: automated analysis of video and language streams during a telehealth consultation can assist a carer to make diagnostic and prognostic real-time decisions. A mobile scanner equipped with or remotely connected to AI-driven image recognition software can be used to collect medical imagery data and analyze it in real-time at the point of care substituting the need for the patient to physically check into a clinic [64]. Such emerging applications have enabled so-called decentralized, virtual or digital clinical trials.

Performed predominantly through local healthcare providers or telehealth services these trials remove the need for patients to physically attend a specific trial site during large portions of the trial. Naturally, this can dramatically improve the ability and motivation of patients to enroll in a trial while at the same time offering all the previously described advantages of using digital disease diaries for patient monitoring [65].

Depending on the positioning of AI in clinical trial design and execution, improved outcomes can range from optimized cohort composition, maximized chances for compound success, more effective trial planning and shorter runways to launch to faster and less expensive trials, lower dropout rates, and improved patient adherence to trial protocols. A comprehensive compendium of use cases and more detailed information on technical implementations can be found in open-sourced review [18].

Artificial intelligence for drug manufacturing and distribution

Artificial intelligence for biopharma manufacturing: toward bioprocessing 4.0

Biopharma manufacturing process design and intelligent plant operations are key components of supply chain management that can adopt AI technology for increasing efficiency, quality, and for optimizing stakeholder management [66]. This emerging field is commonly referred to as Bioprocessing 4.0 [67] and built on AI's ability to automatically analyze large unstructured multimodal data in real-time and at scale for detecting and predicting patterns of interest. Such patterns could for example be markers announcing production bottlenecks and disruptions on the plant floor, product features that point to a defect or quality problem or, on a larger scale, market dynamics around demand and supply for the product [68]. AI technology can also be used to build models of production and product design workflows which can then be used to optimize these processes (Fig. 40.12) [69]. These AI-driven supply chain optimization tools are powered with monitoring data from sensors along manufacturing chain and distribution channels which collect information on production performance and capacity, product quality, as well as on its impact and commercial success. As described in the "The digital health value generation cycle" section, the "Five Vs" of big data together with the EHR interoperability problem pose one of the biggest challenges to the effective use of AI tools in biopharma supply chain management.

AI technology can also be used to automate physical manufacturing tasks on the plant floor through robotics technology which can reduce production errors, achieve higher efficiency, and allow more granular control and oversight over production steps [70]. The increasing pervasiveness of AI within the digital health value and biopharma supply chains raises the vulnerability level of critical data and digital infrastructure to cyberattacks. Thereby AI technology acts as both a threat target and a protection system [71].

Artificial intelligence for marketing and distribution

Health care professionals and patients as the consumers of products and services provided by the pharma sector constitute an important source of information on their

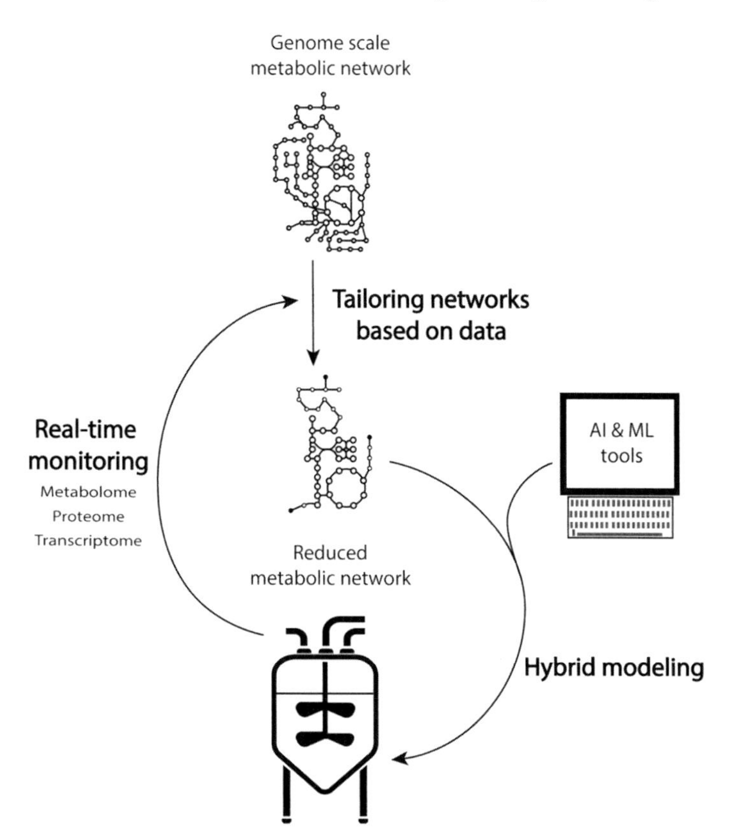

FIGURE 40.12 From experimental data to bioprocess improvement. Systematic workflow using data extracted from real-time monitoring to tailor genome-scale biological networks to core metabolic models that can be combined with artificial intelligence and machine learning tools for an effective implementation of control and optimization strategies. Source: *Figure and caption published with permission from* Springer Nature. *Original in Richelle A, David B, Demaegd D, et al. Towards a widespread adoption of metabolic modeling tools in biopharmaceutical industry: a process systems biology engineering perspective. npj Syst Biol Appl 2020;6:6 [69].*

reception, impact and quality. Such data can be sourced from publicly shared channels, such as social media activity and consumer forums, or it could be collected proactively in a targeted way through focus group surveys. Similar to its use for analyzing social media activity for clinical trial matching purposes, AI—and particularly NLP technology—can be applied to extract, analyze and model product intelligence, and the derived insights can assist the choice of distribution channels, guide personalized marketing and branding narratives and timing, influence the design of future product generations [68], and guide interactions with regulators. Consumer-centric development is not only crucial for achieving optimized user experiences and outcomes when it comes to drugs as the classic product of the pharma sector, but it is also a critical element of designing digital health services along the entire value chain (Fig. 40.2) [20]. To translate and leverage insights from well-established AI-driven marketing and distribution systems in the retail sector, cross-industry alliances between big tech and pharma are emerging which aim to explore novel ways for bringing traditional therapeutics and DTx to patients. For example, Amazon and Microsoft assist Novartis with digitizing its R&D operations, supply chain and distribution systems [72]. Engagements like this cut both ways: on one hand pharma gets access to digital retail infrastructure and logistics operations knowhow in support of its digital transformation. On the other hand, big tech gains sector-specific domain knowledge in the pharma industry, and we see forays of big tech enterprises indicating their interest in developing a pharma business themselves, such as Amazon's acquisition of PillPack and pharmapacks [17]. A digital pharmacy ecosystem, run in the image of AI-powered e-commerce platforms, will face regulatory hurdles but is an increasingly realistic new paradigm for the delivery of medication and health services from prescription or over the counter to the point-of-care.

Summary and trends

The evolution of AI has reached a point at which first commoditized analytical tools can be integrated into real-world workflows for assisting human practitioners in the health and life sciences sector with making better and faster decisions and with managing information more efficiently. This trend has been expedited substantially through the steep rise and fast adoption of generative AI technology, such as for example ChatGPT, in health and medicine [73]. Pharma is embracing this development by exploring the use of AI technology not only for increasing the efficiency and capabilities of drug discovery and development but also for addressing new markets and adopting new business models in digital health. Since 2018 several big pharma enterprises have built substantial in-house capabilities in data science, AI/ML research and development, struck partnerships with big tech, EMR, cloud and analytics vendors, and expanded their data and AI footprint through acquisitions of innovative start- and scale-ups [17,63].

Some of the most promising use cases of AI along the drug development cycle are the discovery of new drug targets, the custom-design of novel drug compounds, the repurposing of existing drugs, the design and execution of more efficient clinical trials, as well as the automation and optimization of drug manufacturing, distribution, and marketing channels. Expanding and transporting these applications into the digital health sector, pharma explores the use of AI, data, and cloud technology to bring new diagnostic, prognostic, personalized treatment and prevention schemes for health, disease, and wellness management to patients and carergivers. DTx, digital diagnostics spearheaded by digital pathology, telehealth and virtual care solutions, patient monitoring systems and interoperable and securely linked electronic health data management platforms are some of the key areas, which the pharma industry is invested into.

The golden yardstick to assess whether the introduction of AI technology has merit or not is to scrutinize evidence—or lack thereof—of whether a quantifiable added value has been created for users and consumers of AI-powered services in an ethical and responsible way. This can for example be measured by means of improved patient care, better outcomes, or increased operational efficiencies. While such value has been demonstrated for digital health related AI applications [74] it is more complicated to assess the impact of AI technology on the overall efficiency of the drug discovery and development cycle. This is predominantly due to the involved timelines: it takes about 10–15 years to bring one single new drug to market but most piloted AI-driven innovations have not been introduced earlier than 3–5 years ago. It will take another 3–5 years of trialing AI-powered modules for drug discovery and development before any direct comparisons to state-of-the-art processes can be made [18][13].

In particular, validation of some of the most innovative and potentially transformative applications of AI in early-stage preclinical discovery phases will take the longest and involve the riskiest investments. A true breakthrough reaching beyond the realms of hype and potential will come from bringing the first drug compound, identified or custom-designed through AI technology, successfully through clinical trials, regulatory stages, and onto the market. This feat has yet to be accomplished, and seeing through such a high risk project -especially given the extremely high costs and risks of running clinical trials—is a commitment which only few pharma organizations are prepared to take. Nevertheless, we are witnessing highly exciting projects, collaborations and new ventures pursuing this path: for example, Google DeepMind's spin-out company Isomorphic Labs uses the AI-powered protein folding predictor AlphaFold for drug discovery and design [75]. Sanofi is partnering with scale-up InSilico Medicine to use AI technology for identifying and developing up to six new drug targets [76], and GSK has formed a dedicated R&D team which uses AI to interpret the human genome in an effort to develop medicines with a higher probability of success [77].

One of the most ambitious scientific stretch-goals and next frontiers in the field of AI-guided drug discovery and design is the creation of a virtual digital cell [78]. The idea is to build an AI model as a digital twin of an entire cell which describes its biological functions with such accuracy that digital experiments performed on the virtual cell would hold true when repeated in a wet lab environment. It is straightforward to imagine the impact such a system would have on the efficiency, speed and scalability of drug discovery and design. However, akin to how Artificial General Intelligence constitutes a far-out vision of where AI might go in the future, AI-based virtual cells are a similarly hypothetical version of what the digital future of drug discovery and development might look like.

At present and for the foreseeable future the pharma industry will focus on integrating, validated, responsible and safe AI technology into scalable electronic medical health data management and analytics workflows. This will render drug development and clinical practice more efficient, which in turn will lead to improved quality of care and better outcomes for both, patients and caregivers.

References

[1] Esteva A, et al. A guide to deep learning in healthcare. Nat Med 2019.
[2] Rajkomar A, Dean J, Kohane I. Machine learning in medicine. N Engl J Med 2019.
[3] Beam AL, Kohane IS. Big data and machine learning in health care. JAMA 2018;319(13):1317–18. Available from: https://doi.org/10.1001/jama.2017.18391.
[4] LeCun Y. A path towards autonomous machine intelligence. OpenReview 2022.
[5] Frey H. The road to AGI. DeepMind Podcast 2022; S2-E5.
[6] Bubeck S, et al. Sparks of artificial general intelligence: Early experiments with gpt-4. arXiv 2023; arXiv:2303.12712.
[7] Kande M, Sonmez M. Don't fear AI. It will lead to long-term job growth. World Econ Forum 2020.
[8] Moor M, et al. Foundation models for generalist medical artificial intelligence. Nature 2023;616(7956):259–65.
[9] Wu E, et al. How medical AI devices are evaluated: limitations and recommendations from an analysis of FDA approvals. Nat Med 2021.
[10] Blueprint for an AI Bill of Rights. The US White House; 2022.
[11] Regulation of the European Parliament and of the council laying down harmonised rules on artificial intelligence (Artificial Intelligence Act) and amending certain union legislative acts. The European Union; 2021.

[12] European Parliament AI Act: a step closer to the first rules on Artificial Intelligence. https://www.europarl.europa.eu/news/en/press-room/20230505IPR84904/ai-act-a-step-closer-to-the-first-rules-on-artificial-intelligence; 2023. [Accessed 05 June 2023].
[13] Harrer S, et al. Artificial intelligence for clinical trial design. Trends Pharmacol Sci 2019.
[14] Hao K. What is AI? We drew you a flowchart to work it out. MIT Technol Rev 2018.
[15] Hao K. What is machine learning? MIT Technol Rev 2018.
[16] Kudumala A. Life sciences artificial intelligence solutions in the age of with – AI and big data services, solutions, and applications for the pharma industry. Deloitte 2021.
[17] The Big Tech in pharma report, CB Insights; 2022.
[18] Harrer S, et al. A new promising way for tackling the pharma dilemma: artificial intelligence for clinical trials. Biochemist 2019.
[19] The dawn of digital medicine, *The Economist*; 2020.
[20] Harrer S. Commercialising digital health: trading on a dynamic data marketplace. Forbes 2021.
[21] Gehde KM, et al. Business model configurations in digital healthcare – a German case study about digital transformation. Intl J Innov Manag 2022.
[22] Topol E. The patient will see you now: the future of medicine is in your hands. New York: Basic Books; 2015.
[23] Topol E. Deep medicine: how artificial intelligence can make healthcare human again. New York: Basic Books; 2019.
[24] Novartis' commitment to the ethical and responsible use of Artificial Intelligence (AI) systems. Novartis Trust and Reputation Committee; 2020.
[25] Reitermann M. The hidden jewel in the healthcare system. Forbes 2018.
[26] US CDC Strengthening Clinical Laboratories—Division of Laboratory Systems. <https://www.cdc.gov/csels/dls/strengthening-clinical-labs.html>.
[27] Baxi V, et al. Digital pathology and artificial intelligence in translational medicine and clinical practice. Mod Pathol 2022;35:23–32.
[28] Harrer S. AI, data and healthcare: buzzword bingo or elevator pitch? Forbes 2022.
[29] Zweig M, Tran D, et al. The AI/ML use cases investors are betting on in healthcare. RockHealth 2018. Available from: https://rockhealth.com/insights/the-ai-mL-use-cases-investors-are-betting-on-in-health-care/.
[30] Retamero JA, et al. Complete digital pathology for routine histopathology diagnosis in a multicenter hospital network. Arch Pathol Lab Med 2020;1:221–8.
[31] Shu C. Sydney-based medtech startup Harrison.ai gets $129M AUD led by Horizon Ventures. TechCrunch 2021.
[32] Natanson E. Digital therapeutics: the future of healthcare will be app-based. Forbes 2017.
[33] FDA permits marketing of first game-based digital therapeutic to improve attention function in children with ADHD, *FDA News Release*. <https://www.fda.gov/news-events/press-announcements/fda-permits-marketing-first-game-based-digital-therapeutic-improve-attention-function-children-adhd>; 2020.
[34] Smith DP, et al. Expert-augmented computational drug repurposing identified baricitinib as a treatment for Covid-19. Front Pharmacol 2021.
[35] Paul SM, et al. How to improve R&D productivity: the pharmaceutical industry's grand challenge. Nat Rev Drug Discovery 2010.
[36] Recker D, Schneider G. Active-learning strategies in computer-assisted drug discovery. Drug Discovery Today 2015.
[37] Konze KD, et al. Reaction-based enumeration, active learning, and free energy calculations to rapidly explore synthetically tractable chemical space and optimize potency of cyclin-dependent kinase 2 inhibitors. J Chem Inf Modeling 2019.
[38] Göller AH, et al. Bayer's in silico ADMET platform: a journey of machine learning over the past two decades. Drug Discovery Today 2020.
[39] Topsha A. Best practices for QSAR model development, validation, and exploitation. Mol Inform 2010.
[40] Vogt M. From activity cliffs to activity ridges: informative data structures for SAR analysis. J Chem Inf Modeling 2011.

[41] StanlyM., et al. A few-shot learning dataset of molecules. In: Thirty-fifth conference on neural information processing systems datasets and benchmarks track (round 2); 2021.
[42] NguyenC.Q., et al. Meta-learning GNN initializations for low-resource molecular property prediction. In: ICML workshop on graph representation learning and beyond; 2020.
[43] Heyndrickx W, et al. MELLODDY: cross pharma federated learning at unprecedented scale unlocks benefits in QSAR without compromising proprietary information. ChemRxiv 2022.
[44] Ponting D. Use of Lhasa Limited products for the in silico prediction of drug toxicity. Silico methods predicting drug toxic. 2022.
[45] Walters WP, Barzilay R. Applications of deep learning in molecule generation and molecular property prediction. ACC Chem Res 2021.
[46] Goldman B, et al. Defining levels of automated chemical design. J Med Chem 2022.
[47] Green DVS, et al. BRADSHAW: a system for automated molecular design. J Comput Mol Des 2020.
[48] The roadmap of drug candidates designed by AI. BiopharmaTrend. <biopharmatrend.com>; 2022.
[49] Jumper J, et al. Highly accurate protein structure prediction with AlphaFold. Nature 2021.
[50] del Alamo D, et al. Predicting alternative conformational states of transporters and receptors with AlphaFold2. eLife 2022.
[51] Zhang Y, et al. Benchmarking refined and unrefined AlphaFold2 structures for hit discovery. ChemRxiv 2022.
[52] Mahajan SP, et al. Hallucinating structure-conditioned antibody libraries for target-specific binders. Front Immunol 2022.
[53] Watson JL, et al. Broadly applicable and accurate protein design by integrating structure prediction networks and diffusion generative models. BioRxiv 2022.
[54] Meier J, et al. Language models enable zero-shot prediction of the effects of mutations on protein function. BioRxiv 2022.
[55] Lin Z, et al. Evolutionary-scale prediction of atomic level protein structure with a language model. BioRxiv 2022.
[56] JeliazkovJ., et al. Agile language transformers for recombinant protein expression optimization. NeurIPS workshop on machine learning for structural biology; 2022.
[57] Lopez G. A Smarter Robot. A new chatbot shows rapid advances in artificial intelligence. N Y Times 2022.
[58] Greene T. Large language models like GPT-3 aren't good enough for pharma and finance. Web 2022.
[59] Harrer S. Attention is not all you need: the complicated case of ethically using large language models in healthcare and medicine. eBiomedicine (The Lancet) 2023;. Available from: https://doi.org/10.1016/j.ebiom.2023.104512.
[60] Lawton G. 10 essential ingredients for digital twins in healthcare. VentureBeat 2022.
[61] Thorlund K, et al. Synthetic and external controls in clinical trials – a primer for researchers. Clin Epidemiol 2020.
[62] Arts D. Debunking top 5 myths about digital twins in clinical trials. Appl Clin Trials 2022.
[63] The future of clinical trials: how technology is making drug trials more efficient, cost effective, and inclusive. *CB Insights*; 2022.
[64] Rothberg JM, et al. Ultrasound-on-chip platform for medical imaging, analysis, and collective intelligence. Proc Natl Acad Sci U S A 2021.
[65] Adams B. 2021 forecast: the rise and rise of the virtual trial model. FierceBiotech 2020.
[66] Owczarek D. Bioprocessing 4.0 and the benefits of introducing AI to biopharmaceutical manufacturing. Nexocode 2021.
[67] Stosch Mv, et al. A roadmap to AI-driven *in silico* process development: bioprocessing 4.0 in practice. Curr OpChem Eng 2021.
[68] Kudumala A. Life sciences artificial intelligence solutions in the age of with: AI and big data services and solutions and capabilities for the pharma industry. Deloitte 2022.
[69] Richelle A, et al. Towards a widespread adoption of metabolic modeling tools in biopharmaceutical industry: a process systems biology engineering perspective. npj Syst Biol Appl 2020.
[70] Nouri S. AI: from drug discovery to robotics. Forbes 2022.
[71] AI in healthcare presents need for security, privacy standards. *HealthITSecurity*; 2022.

[72] Hale C. Novartis' digital transformation continues apace with Amazon supply chain tie-up. FierceBiotech 2019.

[73] Eddy N. Epic, Microsoft partner to use generative AI for better EHRs. https://www.healthcareitnews.com/news/epic-microsoft-partner-use-generative-ai-better-ehrs; 2023. [Accessed 05 June 2023

[74] Siwicki B. AI powered telehealth improves PT care at Essen Health Care. HealthcareITNews 2022.

[75] Method of the year 2021: Protein structure prediction. Nat Methods 2022;19:1. https://doi.org/10.1038/s41592-021-01380-4.

[76] Masson G. Amid 'biotech winter,' Insilico turns up the heat with Sanofi deal worth $1.2B in bio bucks. FierceBiotech 2022.

[77] GSK establishes AI hub in London to discover new drugs. Pharm Technol 2020.

[78] Topol E, Hassabis D. It's not all fun and games: how DeepMind unlocks medicine's secrets. Medscape Med Mach Podcast 2022.

CHAPTER

41

Artificial intelligence in regulatory decision-making for drug and biological products

Tala H. Fakhouri[1], *Qi Liu*[2] *and M. Khair ElZarrad*[1]

[1]Office of Medical Policy, Center for Drug Evaluation and Research, US Food and Drug Administration, Silver Spring, MD, United States [2]Office of Clinical Pharmacology, Office of Translational Sciences, Center for Drug Evaluation and Research, US Food and Drug Administration, Silver Spring, MD, United States

Over the past few decades the volume of data available to support drug development have increased substantially. These increases in data volume were also accompanied by an expansion in data diversity with data originating from disparate sources including biologic data (e.g., genomic, transcriptomic, proteomic, metabolomic data), pharmacological data (e.g., absorption, distribution, metabolism, excretion, and toxicity data), and clinical data (e.g., imaging, electronic health records, and digital health technologies). This growth in data volume and complexity combined with cutting-edge computing power and methodological advancements in artificial intelligence (AI) and machine learning (ML) have the potential to transform how drugs are developed, manufactured, and utilized.

Concurrent with these technological advancements, the Food and Drug Administration (FDA) has seen a significant increase in the number of drug and biologic application submissions using AI and ML components over the past few years, with over 100 submissions reported in 2021 [1]. These submissions traverse the landscape of drug development from drug discovery to postmarket safety monitoring (see Table 41.1) and cut across a range of therapeutic areas (see Fig. 41.1). The application of AI and ML in these submissions aims to improve the efficiency of drug discovery and our understanding of the efficacy and safety of specific treatments. Importantly, the diverse uses of AI in these submissions highlight the need for a careful regulatory assessment of both benefits and risks, and underscores the importance of adopting a risk-based management approach that is proportional with measures commensurate with the level of risk posed by the specific context of use.

Artificial Intelligence in Clinical Practice
DOI: https://doi.org/10.1016/B978-0-443-15688-5.00011-5

TABLE 41.1 Count of regulatory submissions for drug development with key terms "machine learning" or "artificial intelligence" from 2016 to 2021.

	Year					
Submission type (*n*)	**2016**	**2017**	**2018**	**2019**	**2020**	**2021**
IND	1	1	2	5	11	128
NDA, ANDA, BLA	–	–	1	2	2	2
DDT, CPIM	–	–	–	–	1	2
	Year					
Drug development stage (*n*)	2016	2017	2018	2019	2020	2021
Discovery and development	–	–	–	–	1	3
Preclinical research	–	–	–	–	–	8
Clinical research	1	1	3	5	12	118
Postmarket safety monitoring	–	–	–	2	1	3

IND, Investigational New Drug; NDA, New Drug Application; ANDA, Abbreviated New Drug Application; BLA, Biologics License Application; DDT, Drug Development Tool Qualification Programs; CPIM, Critical Path Innovation Meeting.
Internal databases maintained by the FDA Center for Drug Evaluation and Research (CDER).

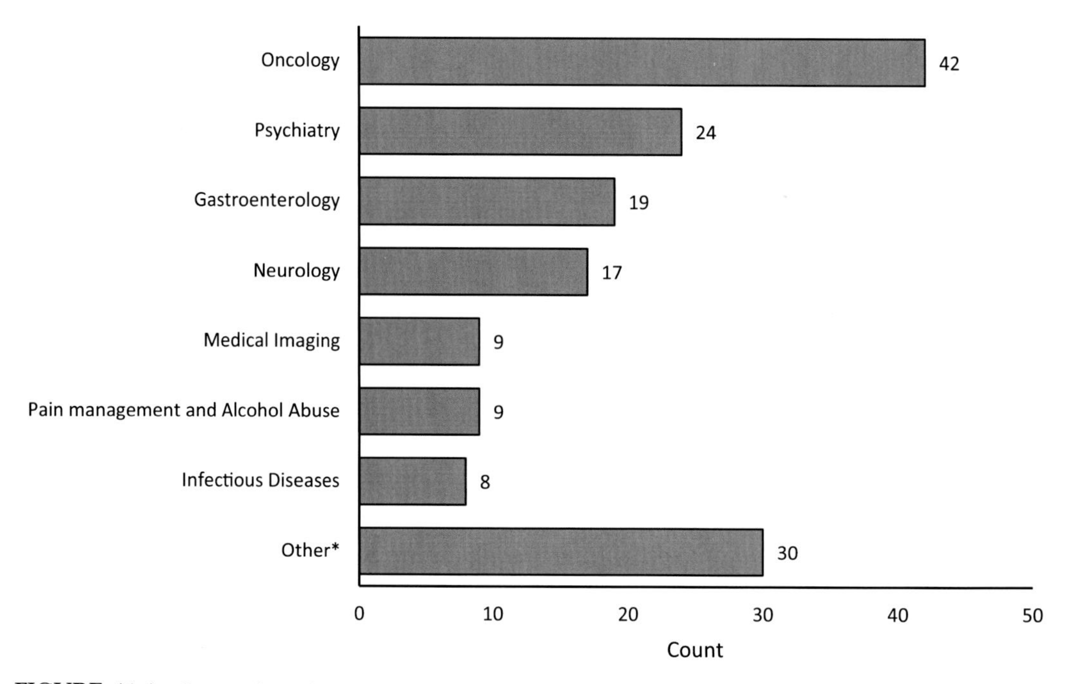

FIGURE 41.1 Count of regulatory submissions for drug development with key terms "machine learning" or "artificial intelligence" by therapeutic area. * The "Other" category includes immunology, pulmonology, endocrinology, dermatology, cardiology, nephrology, and gynecology. Source: *Internal databases maintained by the FDA Center for Drug Evaluation and Research (CDER).*

For any specific AI application in drug development, model risk calculations will be determined by model influence and the decision consequence based on the context of use. For example, high-risk models may require more evidence of credibility than low-risk models, and the regulatory approach may differ accordingly.

As with any innovation, AI and ML creates opportunities, and new and unique challenges. To meet these challenges, the FDA has accelerated its efforts to create an agile regulatory ecosystem that can facilitate innovation and adoption while safeguarding public health [2]. For example, in 2021 the FDA, Health Canada, and the United Kingdom's Medicines and Healthcare products Regulatory Agency jointly published ten guiding principles to inform the development of Good Machine Learning Practices (GMLPs) for medical devices that use AI and ML [3]. The guiding principles touch on a wide range of issues including the importance of bias mitigation by ensuring that the results from ML models are generalizable to the intended patient population, and that the underlying datasets used to train these models are representative of the target population. The principles also underscore the importance of safety and performance monitoring as algorithms evolve after being deployed in the "real world." Specifically, appropriate controls should be in place to manage risks of overfitting, unintended bias, or degradation of the model that may negatively impact the safety and performance of the ML algorithm. Data governance and transparency are also highlighted. Specifically, the GMLPs emphasize that the AI development process should include a human-in-the-loop, where an expert human guides the cyclical development of the ML algorithm, and users (i.e., providers and or patients) are provided access to clear and contextually relevant information. While these GMLPs were not tailored for drug development specifically, their utility and applicability to drug development is being explored to ensure alignment and consistency whenever possible.

As FDA continues to refine the regulatory approaches around the use of AI to facilitate the generation of reliable evidence and to support decision making, the evidentiary standards needed to support drug approvals remain the same regardless of the technological advances involved. AI and ML will undoubtedly play a critical role in drug development, and the FDA remains committed to robust policy development that both protects and promotes public health.

Disclaimer

The views expressed in this chapter are those of the authors and do not necessarily represent the views or policies of the FDA.

References

[1] Liu Q, Huang R, Hsieh J, Zhu H, Tiwari M, Liu G, et al. Landscape analysis of the application of artificial intelligence and machine learning in regulatory submissions for drug development from 2016 to 2021. Clin Pharmacol Ther 2022.

[2] ElZarrad MK, Lee AY, Purcell R, Steele SJ. Advancing an agile regulatory ecosystem to respond to the rapid development of innovative technologies. Clin Transl Sci 2022;. Available from: https://doi.org/10.1111/cts.13267.

[3] Food and Drug Administration. Good machine learning practice for medical device development: guiding principles. <https://www.fda.gov/medical-devices/software-medical-device-samd/good-machine-learning-practice-medical-device-development-guiding-principles>.

CHAPTER

42

Machine learning applications in toxicology

Marc Rigatti[1], *Stephanie Carreiro*[1] *and Edward W. Boyer*[2,3]

[1]Department of Emergency Medicine, University of Massachusetts Chan Medical School, Worcester, MA, United States [2]Department of Emergency Medicine, Harvard Medical School, Boston, MA, United States [3]Emergency Medicine, Ohio State University, Columbus, OH, United States

Potential for artificial intelligence in clinical toxicology

Toxicology is a scientific discipline that seeks to understand the adverse effects of poisons on living organisms. It is a broad field overlapping with biology, chemistry, pharmacology, forensics, and medicine. In practice, clinical toxicology is separated into a multitude of subdisciplines, including occupational toxicology, forensic toxicology, and analytical toxicology. Because of the volume of data from disparate sources that must be integrated in medical toxicology, predicting the clinical course of a given toxin can be difficult. Artificial intelligence (AI) methods of analysis (including machine learning approaches) are of increasing importance in toxicology to mitigate these challenges. Much of the implementation of AI in toxicology to date, however, remains limited to research applications. Accordingly, this chapter will focus on experimental and emerging applications in clinical toxicology, where AI has been applied for the detection and treatment of substance use disorder (SUD), supporting clinical diagnosis for toxidromes, and surveillance for emerging recreational drugs.

Detection and treatment of substance use disorder

Wearable sensors capable of tracking digital biomarkers that enable prediction of a user's state of health [1] have been applied to detect a number of events related to recreational substances, including detecting cocaine relapse, opioid use, opioid dependance, prevention of substance use, and monitoring of substance use of individuals under treatment

Artificial Intelligence in Clinical Practice
DOI: https://doi.org/10.1016/B978-0-443-15688-5.00005-X

for SUD [2–15]. Wrist-worn sensors collect continuous physiologic data streams related to acceleration, skin conductance, skin temperature, pulse and blood oxygen saturation [16–18]. Wrist worn sensors have several advantages including similarity to commonly used smartwatches (to mitigate stigma), comfort and price point [15,19,20]. Disadvantages may include a limited cassette of sensors, variability in performance based on skin tone tones, and access to raw data streams.

The concept of a continuous data monitoring system from wearable sensors to both predict and respond to signals of substance use was first envisioned in 2012 with the iHeal project [12]. iHeal proposed use of wrist sensors to detect subtle changes in physiology associated with events leading to cocaine use including stress or withdrawal and was later extended to detect changes associated with opioid use [15,21]. Early efforts toward such a system demonstrated the ability of wrist-based sensors to detect cocaine use by measuring surrogates of sympathetic nervous system activation; increased electrodermal activity (EDA), locomotion consistent with akathisia observed in stimulant use, and decreased skin temperature [7].

Extension of this work to subjects administered opioids in an emergency department setting demonstrated the ability not only to detect opioid administration but to differentiate administration to naïve vs chronic opioid users [15]. While changes in locomotion, EDA, and temperature were noted following opioid administration, it is difficult to disentangle signals of drug administration from those of drug effect (e.g., sleep, relief of pain), which produce similar changes in locomotion, skin temperature and EDA.

Opioid overdose, one of the most devastating consequences of SUD, is a prime target for AI applications. In 2021 alone opioid overdose claimed 80,816 lives [22]. It presents with a clinical triad of respiratory depression, central nervous system depression, and miosis, and can progress to respiratory arrest and cardiac arrest. The consistency of its pathophysiology and the availability of a highly effective antidote (naloxone) lends itself well to detection using noninvasive sensor technology. Roth et al. deployed a wearable respiratory rate sensor that adheres to clothing close to the chest wall in a population of people who inject drugs in the field and deemed this method insufficient to detect respiratory depression on its own [14]. Nandakumar et al. used a traditional smart phone speaker as a short-range sonar and were able to detect opioid overdose both in clinical and field settings [9]. This system uses the smart phone to transmit frequency-modulated continuous waveforms that reflect off surfaces in the environment and are detected by the phone. Using fast Fourier transform, the frequency shifts of these signals are binned such that the motion corresponding to the subjects breathing is found in a bin corresponding to the distance of the subject from the smart phone. By tracking the corresponding to the breathing signal, the algorithm can also automatically recalibrate to postural changes or changes in positioning of the phone. Smart home sensors attached to walls have also been proposed as adjudicative means to capture respiratory depression [23,24].

Closed-loop systems to pair overdose detection with automatic injection of the opioid reversal agent naloxone are also underway: several proof-of-concept studies have been published to support feasibility [25,26]. These devices detect respiratory rate via accelerometry or ECG and automatically deliver doses of naloxone in response to apnea. Most opioid users surveyed by Kanter et al. report willingness to use a system capable of detecting and reversing opioid overdose [27]. While most were confident that the devices would keep them safer, concerns were expressed regarding potential false positive overdose detection and reversal, as well as maintenance of privacy and confidentiality related to device data and discreetness of the device itself. Similar devices, capable of overdose detection and alert of bystanders rather than naloxone delivery, are overall less acceptable [28]. Concern regarding these systems is primarily related to the vulnerability of users experiencing overdose, particularly for those that are experiencing homelessness [27,28].

Once an individual has initiated treatment and is in recovery from SUD, machine learning fueled digital health systems can support recovery by predicting risk return to drug use and triggering JITA interventions. One example is the Realize, Analyze, Engage System, which leverages algorithms to detect stress and drug craving have in an interactive digital health ecosystem [19,29]. The system includes a wrist-worn wearable sensor which collects continuous physiologic data, a mobile phone application that pushes notifications and microinterventions when craving or stress are detected, and a clinical portal which allows treatment providers to visualize trends in system engagement and risk profiles. Such systems provide patients with just-in-time and just-in-space tools for deescalation during high-risk states, and provide analytics for their healthcare providers to understand trends and triggers.

Clinical diagnostic support for toxicology

The use of AI for diagnostic applications remains challenging, especially within the field of clinical toxicology. Diagnosis of a poisoned patient depends on recognition of patterns of findings on history and physical exam that are suggestive of a particular class of poisons. For example, clinical decision making in toxicology is sometimes reliant on visualization of potential toxins, such as poisonous mushrooms. While AI has been successfully applied to specific plant and mushroom identification, image data are often unavailable or too low in quality to support analysis [30–32]. Machine learning-based image data has not yet proven useful in diagnosing a poisoned patient. Patterns of symptoms indicating a particular class of intoxicant (termed toxidromes) are a more promising area where AI may be able to support clinical decisions. For example, as described above, a patient that overdosed on an opioid is expected to appear somnolent, have pinpoint pupils, and a depressed respiratory rate, while a patient who recently used a sympathomimetic like cocaine or amphetamines may have agitation, dilated pupils, and an elevated heart rate.

The predisposition toward viewing machine learning as a black box reduces trust by bedside clinicians and decreases reliance on the output of such models or systems. Probabilistic logic networks (PLNs) have been proposed as an explainable form of AI that can be applied to toxicological problems to address these issues. For example, Chary et al. developed a PLN based software to classify patient cases into toxidromes; the software performed well on easy and intermediate cases but failed to meet human accuracy on complex or subtle cases that would benefit most from AI assistance [33]. Despite the current state of such models, which are unlikely to outperform a clinician with years of training and experience, the continued development of AI systems that can supplement and support clinician decision making in an explainable way is promising.

Toxicovigilance

Toxicovigilance is defined as the detection and confirmation of clinical adverse events related to toxic exposures in humans. Patterns of drug use evolve at a striking rate of change. Traditional methods of surveillance for emerging drug use trends, such as bedside observation, clinical case reports, survey, or database mining, suffer from dramatic limitations. For example, many current methods rely upon serendipitous outcomes from the convergence of nearly random events in which an individual must first use a new substance, then must become intoxicated, present for clinical care delivered by a practitioner who recognizes a new clinical entity and undergo confirmatory testing that verifies the presence of a new substance. The output of this stochastic approach to toxicovigilance is frequently memorialized in case reports rather than the hypothesis driven literature.

Several features complicate toxicosurveillance for emerging drug of abuse. First, people who use drugs often refer to new substances by new, unexpected names or neologisms. Machine learning algorithms must therefore have the capacity to evaluate an unstable or unknown lexicon. Second, people who use drugs often speak in a vague, indistinct "code" to avoid revealing drug use behaviors. The lack of a stable lexicon requires that machine learning algorithms interpret totally new parts of speech, usage, and contexts to identify new drugs abuse practices. Finally, new classes of recreational substances may have clinical findings that are unknown to practitioners or that have been unreported in the medical literature.

Despite these challenges, machine learning has advanced toxicosurveillance for emerging drug use practices. Web semantic approaches, a variant of natural language processing, has been used on social media, Reddit, and other crowdsourced data to identify new patterns of use of cannabis and opioids [34–36]. Web semantic calculations are precise enough to identify new lexicons of drug use (including existing words used as new drug names), but also patterns of use [37]. Web semantics are also precise enough to identify the location of emerging drug use practices from dialectical variations in language, even when social media posts are not geotagged [38].

Current challenges and future directions

Despite great potential and rapidly evolving technology, numerous barriers confront the widespread implementation of machine learning in toxicology practice. First, stigma associated with drug use have led to concerns regarding privacy and confidentiality of machine learning generated outputs. This fear becomes amplified by the potential for racial, ethnic, or gender bias in AI [39,40]. Second, patients and providers often express fear over legal, social, and employment consequences from inadvertent disclosure of substance use data. We anticipate this will concern be mitigated by recognition of improved encryption/security practices and by the normalization of data sharing with third parties as part of routine digital technology consumption. Finally, physicians and other treatment providers make up a special group of end-users and need to be considered in the design process—both as a consumer of the technology and in many cases the human-in-the-loop.

AI had tremendous potential in the prediction, monitoring and treatment of toxicological diseases using on and off body sensing technology, decision support, and surveillance of crowd- source data. To bridge the current gap between the research laboratory and the clinic, AI enabled technologies must prove to end users (healthcare providers and patients) that they are superior to traditional standards. Additionally, they need to reassure healthcare providers that their goal is to increase efficiency and quality of care by supporting (but not replacing) human healthcare providers. This will likely be achieved through the use of human-in-the-loop design and personalized technologies.

Major take away points

- AI-enabled sensor technology is being widely studied in the detection and treatment of SUD.
- AI can also be used for diagnostic support in toxicology but is likely to be limited by complexity and may benefit from a human-in-the-loop approach.
- Web semantic approaches can be used to accelerate the process of identifying novel substances and patterns of drug use.
- Barriers to AI applications in toxicology include concern for privacy/confidentially, the rapidly evolving landscape of illicit drugs, and acceptability to clinicians.

References

[1] Bhatt P, Liu J, Gong Y, Wang J, Guo Y. Emerging artificial intelligence-empowered mHealth: scoping review. JMIR mHealth uHealth 2022;10:e35053.

[2] Singh R, Lewis B, Chapman B, Carreiro S, Venkatasubramanian K. A machine learning-based approach for collaborative non-adherence detection during opioid abuse surveillance using a wearable biosensor. Proc 12th int jt conf biomed eng syst technol, vol. 5. 2019. p. 310–8.

[3] Mahmud MS, Fang H, Wang H, Carreiro S, Boyer E. Automatic detection of opioid intake using wearable biosensor. 2018 int conf comput netw commun ICNC. 2018. p. 784–8.

[4] Hsu M, Ahern DK, Suzuki J. Digital phenotyping to enhance substance use treatment during the COVID-19 pandemic. JMIR Ment Heal 2020;7:e21814.

[5] Rumbut J, et al. Harmonizing wearable biosensor data streams to test polysubstance detection. 2020 int conf comput netw commun ICNC, 00. 2020. p. 445–9.

[6] Imtiaz MS, Bandoian CV, Santoro TJ. Hypoxia driven opioid targeted automated device for overdose rescue. Sci Rep 2021;11:24513.

[7] Carreiro S, et al. iMStrong: deployment of a biosensor system to detect cocaine use. J Med Syst 2015;39:186.

[8] Scherzer CR, et al. Mobile peer-support for opioid use disorders: refinement of an innovative machine learning tool. J Psychiatry Brain Sci 2020;5:e200001.

[9] Nandakumar R, Gollakota S, Sunshine JE. Opioid overdose detection using smartphones. Sci Transl Med 2019;11.

[10] Gullapalli BT, et al. OpiTrack: a wearable-based clinical opioid use tracker with temporal convolutional attention networks. Proc ACM Interact Mob Wearable Ubiquitous Technol 2021;5:1–29.

[11] Wilson M, Fritz R, Finlay M, Cook DJ. Piloting smart home sensors to detect overnight respiratory and withdrawal symptoms in adults prescribed opioids. Pain Manag Nurs 2022;. Available from: https://doi.org/10.1016/j.pmn.2022.08.011.

[12] Boyer EW, et al. Preliminary efforts directed toward the detection of craving of illicit substances: the iHeal project. J Med Toxicol 2012;8:5–9.

[13] Warren D, et al. Using machine learning to study the effect of medication adherence in opioid use disorder. PLoS One 2022;17:e0278988.
[14] Roth AM, et al. Wearable biosensors have the potential to monitor physiological changes associated with opioid overdose among people who use drugs: a proof-of-concept study in a real-world setting. Drug Alcohol Depen 2021;229:109138.
[15] Carreiro S, et al. Wearable biosensors to detect physiologic change during opioid use. J Med Toxicol 2016;12:255–62.
[16] Empatica | Medical devices, AI and algorithms for remote patient monitoring. <https://www.empatica.com/>.
[17] Fitbit official site for activity trackers & more. <https://www.fitbit.com/global/us/home>.
[18] Garmin International | Home. <https://www.garmin.com/en-US/>.
[19] Carreiro S, et al. Wearable sensor-based detection of stress and craving in patients during treatment for substance use disorder: a mixed methods pilot study. Drug Alcohol Depen 2020;209:107929.
[20] Shrestha S, et al. Towards device agnostic detection of stress and craving in patients with substance use disorder. Proc Annu Hawaii Int Conf Syst Sci Annu Hawaii Int Conf Syst Sci 2023;2023:3156–63.
[21] Carreiro S, et al. Real-time mobile detection of drug use with wearable biosensors: a pilot study. J Med Toxicol 2015;11:73–9.
[22] CDC—Nation Center for Health Statistics—US overdose deaths in 2021 increased half as much in 2020—but are still up 15%. <https://www.cdc.gov/nchs/pressroom/nchs_press_releases/2022/202205.htm#print>; 2022.
[23] Zhang G, et al. Contactless in-home monitoring of the long-term respiratory and behavioral phenotypes in older adults with COVID-19: a case series. Front Psychiatry 2021;12:754169.
[24] Yang Y, et al. Artificial intelligence-enabled detection and assessment of Parkinson's disease using nocturnal breathing signals. Nat Med 2022;28:2207–15.
[25] Chan J, et al. Closed-loop wearable naloxone injector system. Sci Rep-uk 2021;11:22663.
[26] Dhowan B, et al. Simple minimally-invasive automatic antidote delivery device (A2D2) towards closed-loop reversal of opioid overdose. J Control Rel 2019;306:130–7.
[27] Kanter K, et al. Willingness to use a wearable device capable of detecting and reversing overdose among people who use opioids in Philadelphia. Harm Reduct J 2021;18:75.
[28] Ahamad K, et al. Factors associated with willingness to wear an electronic overdose detection device. Addict Sci Clin Pract 2019;14:23.
[29] Carreiro S, et al. Realize, Analyze, Engage (RAE): a digital tool to support recovery from substance use disorder. J Psychiatry Brain Sci 2021;6:e210002.
[30] Prasvita DS, Herdiyeni Y. MedLeaf: mobile application for medicinal plant identification based on leaf image. Int J Adv Sci Eng Inf Technol 2013;3:103–6.
[31] Wibowo A, Rahayu Y, Riyanto A, Hidayatulloh T. Classification algorithm for edible mushroom identification. In: 2018 int conf inf commun technol ICOIACT; 2018, pp. 250–253. Available from: https://doi.org/10.1109/icoiact.2018.8350746.
[32] Wang Z, Cui J, Zhu Y. Review of plant leaf recognition. Artif Intell Rev 2022;1–37. Available from: https://doi.org/10.1007/s10462-022-10278-2.
[33] Chary M, Boyer EW, Burns MM. Diagnosis of acute poisoning using explainable artificial intelligence. aRxiv 2021;. Available from: https://doi.org/10.48550/arxiv.2102.01116.
[34] Tofighi B, et al. Detecting illicit opioid content on Twitter. Drug Alcohol Rev 2020;39:205–8.
[35] Chary M, Genes N, McKenzie A, Manini AF. Leveraging social networks for toxicovigilance. J Med Toxicol 2013;9:184–91.
[36] Ghosh A, Bisaga A, Kaur S, Mahintamani T. Google trends data: a potential new tool for monitoring the opioid crisis. Eur Addict Res 2022;28:33–40.
[37] Lokala U, et al. Drug abuse ontology to harness web-based data for substance use epidemiology research: ontology development study. JMIR Public Heal Surveill 2022;8:e24938.
[38] Zhao F, et al. Computational approaches to detect illicit drug ads and find vendor communities within social media platforms. IEEE ACM Trans Comput Biol Bioinform 2019;19:180–91.
[39] Parikh RB, Teeple S, Navathe AS. Addressing bias in artificial intelligence in health care. JAMA 2019;322:2377–8.
[40] Cirillo D, et al. Sex and gender differences and biases in artificial intelligence for biomedicine and healthcare. npj Digital Med 2020;3:81.

CHAPTER

43

Artificial intelligence in adverse drug events

Ania Syrowatka[1,2,*] *and David W. Bates*[1,2,3,†]

[1]Division of General Internal Medicine, Brigham and Women's Hospital, Boston, MA, United States [2]Harvard Medical School, Boston, MA, United States [3]Department of Health Policy and Management, Harvard T.H. Chan School of Public Health, Boston, MA, United States

Introduction

An adverse drug event is "an injury resulting from medical intervention related to a drug" [1]; this includes adverse drug reactions, such as allergic reactions, and harm caused by medication errors. These events are common in the inpatient and outpatient settings, and there is substantial opportunity to improve prediction and early detection of adverse drug events to prevent or mitigate harm.

Artificial intelligence (AI), defined as "computer applications that can perform tasks that normally require human intelligence" [2], has the potential to support and improve clinical decision-making around ordering and prescribing of medications as well as detection of medication errors or early signs or precursors indicating onset of adverse drug events to improve patient outcomes. This chapter focuses on AI-based algorithms and tools that can be used to inform clinical decision-making for prevention or mitigation of adverse drug events at the point of care. There are several upstream and downstream use cases including the use of AI for development of safer medications, pharmacovigilance, and medication reconciliation that are related but not covered. This overview addresses three types of AI: rule-based models which are currently used in practice, and the next

* Dr. Syrowatka received partial salary support from grants funded by IBM Watson Health, outside the submitted work.

† Dr. Bates reports grants and personal fees from EarlySense, personal fees from CDI Negev, equity from ValeraHealth, equity from Clew, equity from MDClone, equity from Guided Clinical Solutions, personal fees and equity from AESOP, personal fees and equity from FeelBetter, and grants from IBM Watson Health, outside the submitted work.

Artificial Intelligence in Clinical Practice
DOI: https://doi.org/10.1016/B978-0-443-15688-5.00014-0

generation of more complex, modern machine learning, defined as "algorithms and models which machines can use to learn without explicit instructions" [2], and natural language processing-based tools that are under development and in testing.

Current rule-based systems

There are currently multiple rule-based commercial knowledge bases that are widely used to support medication-related decision-making to prevent adverse drug events at the point of care, including MediSpan which is part of Lexicomp (Wolters Kluwer), Cerner Multum, and a suite of solutions available from First Databank PatientFirst. These solutions are integrated within electronic health record (EHR) systems to provide medication error alerts as well as tailored medication recommendations sometimes informed by the patient-level data, such as alerts about medication- or disease-related contraindications, dosing irregularities, and pharmacogenomic interactions. These tools screen medication orders against established "rules" such as clinical guidelines, patient-specific information, such as comorbidities or genomic profiles, and known contraindications from pharmacological databases.

The data leveraged by these models are readily accessible from structured fields in the EHR, such as International Classification of Diseases diagnostic codes documented for comorbidities and medication lists, or medication order information from computerized provider order entry. However, these tools currently only use a small fraction of the data available in the EHR to inform medication-related decision-making. Moreover, clinical impact or performance of these tools has not been widely reported in the literature. A few studies have shown good (91%) to excellent (>99%) coverage for clinically significant drug–drug interactions [3,4]. Although comprehensive and sensitive, these solutions generate a high number of "nuisance" alerts leading to alert fatigue and unsubstantiated overrides and require local implementation rules to overcome alert burden while generating more specific and clinically meaningful alerts [5,6].

Another commercially available but not widely used rule-based system is Seegnal, which has been shown to generate substantially lower rates of alerts (>93% reduction) with substantially higher sensitivity and specificity compared with standard EPIC EHR logic and rules [7]. Machine learning features are under development to further improve the performance of the platform.

Next generation of artificial intelligence tools

More complex forms of AI, such as machine learning and natural language processing, can provide access to substantially more information from unstructured data in the EHR and analyze complex relationships to provide more clinically meaningful alerts with higher accuracy to inform medication-related decision-making. Our team recently conducted a scoping review to identify key use cases for leveraging modern machine learning and natural language processing to reduce the frequency of adverse drug events. We identified four use cases for prediction of: (1) adverse drug events, (2) nonresponse to medications (to prevent adverse drug events from medications that will not benefit the patient), (3) optimal dosing, and (4) most appropriate treatment (to balance risk of adverse drug events and therapeutic benefit) [8].

Modern machine learning algorithms for modeling complex interrelationships

Predictive models were developed using modern machine learning algorithms, which can handle the high-dimensional data available in EHRs to model complex, nonlinear relationships between risk factors and outcomes. Most predictive models relied on structured data, focused on specific adverse drug event or medication classes, and were in the early stages of development and validation. Tree-based algorithms, such as random forests, and neural networks performed well for these use cases. Several models were developed using data from research studies that collected genetic information; coupled with clinical factors, genetic information was shown to substantially improve model performance in some cases [9].

We also identified two use cases for early detection of: (1) adverse drug events and (2) medication errors [8]. These AI solutions generally targeted a broad range of adverse drug event and medication classes, leveraged both structured and unstructured data, and showed improved performance over rule-based systems with higher accuracy and reduced alerting burden.

Three studies evaluated the performance of the MedAware Medication Safety Monitoring Platform which addresses both detection use cases; all three studies demonstrated that the alerts generated by the machine learning-based software were clinically valid (76%–85%) [10–12], and one study conducted a prospective evaluation in a clinical setting and showed a low number of alerts (0.4%) with 43% resulting in changes to medication orders [12]. A follow-up evaluation of MedAware, published after our scoping review, described the recalibration of the machine learning models to fit local contexts, and reported that 41% of the alerts resulted in providers canceling or modifying medication orders [13].

Natural language processing for data extraction

Our review only identified two studies that applied natural language processing to extract information from free-text clinical notes in the EHR. Both focused on improving detection of adverse events documented in the EHR. The first study used natural language processing to identify hypoglycemic events documented in clinical reports that were not captured in structured fields, such as laboratory test results or diagnostic codes, to more accurately measure this outcome [14]. This approach identified an additional 539 events (6.6% of all documented cases) to develop predictive models using modern machine learning to identify patients with diabetes at higher risk of hypoglycemic events. The second study evaluated the performance of the Adverse Drug Effect Recognizer, which leveraged natural language processing to extract information about patient medications, symptoms, abnormal laboratory results, past diagnoses, and other findings from emergency department admission notes to detect preadmission adverse drug reactions related to antihypertensive medications [15]. The extracted information was mapped to standard terminologies and cross-referenced against a database of known adverse events. A prospective evaluation of this tool showed that providers receiving alerts were less likely to order medications that could potentially cause adverse drug reactions (47% vs 58% in the control group).

Current challenges limiting implementation

The EHR data and the technologies to harness these data exist; our review demonstrated that this is an emerging area, with most papers published within the last five years (2016–2020). However, machine learning solutions are still early in the development and validation stages. There are several challenges that will need to be addressed before widespread use of these tools can be realized.

1. *Access to necessary data within the EHR*: Additional data points either not routinely available in the EHR (e.g., genetic profiles) or captured only in free-text clinical notes are not readily accessible for use in clinical models for prediction or early detection of adverse drug events.
2. *Evidence of clinical impact*: There were few prospective evaluations of these tools in clinical settings, and we did not find any that assessed impact on patient outcomes.
3. *Generalizability*: Models developed and validated at one site may not perform as well at other sites or in different healthcare settings, requiring recalibration and further validation to achieve a similar baseline performance.
4. *Confidentiality*: Computationally intensive algorithms, potentially leveraging natural language processing to extract information from free-text clinical notes, must be run within health system firewalls, since most organizations are reluctant to send protected/personal health information outside of the system for processing.
5. *Seamless integration of tools within the EHR*: Most models were developed for specific adverse drug event or medication classes, especially in the prediction domain; each model is independently scalable but implementing separate tools to address each area from various vendors is not feasible, and more comprehensive or integrated solutions are needed.

Future directions

In the short term, well-developed and rigorous prospective evaluations in clinical settings are necessary to demonstrate the real-world impact of these tools on clinician decision-making and patient outcomes. Given the early stages of development in the predictive domain and wide range of use cases, efforts should focus on addressing relatively common adverse drug events associated with higher rates of morbidity or mortality, such as prediction of which patients are likely to develop renal failure or experience severe allergic reactions like anaphylaxis or Stevens-Johnson syndrome. AI solutions should also be developed to avoid use of contraindicated medications in patients with specific genetic issues, including chemotherapeutic drugs. Further applications of natural language processing should be developed to extract information from free-text notes not only about outcomes but also risk factors to develop more comprehensive and accurate clinical prediction models.

Major takeaway points

The use of modern machine learning and natural language processing is an emerging area in the prediction and detection of adverse drug events at the point of care, which has the potential to substantially improve medication-related decision-making compared with rule-based tools. However, most of these models are in the early stages of development or validation and have not yet been tested in real-world clinical settings, particularly in the prediction domain. This is urgently needed. For these to be used in practice, the tools must be implemented and tested, and then adopted into routine care to impact clinician behavior and improve patient outcomes.

References

[1] Kohn LT, Corrigan JM, Donaldson MS. To err is human: building a safer health system. Washington, DC: Institute of Medicine of the National Academy of Sciences; 1999.
[2] Bates DW, Auerbach A, Schulam P, Wright A, Saria S. Reporting and implementing interventions involving machine learning and artificial intelligence. Ann Intern Med 2020;172:S137–44.
[3] Fung KW, Kapusnik-Uner J, Cunningham J, Higby-Baker S, Bodenreider O. Comparison of three commercial knowledge bases for detection of drug-drug interactions in clinical decision support. J Am Med Inf Assoc 2017;24:806–12.
[4] Peters LB, Bahr N, Bodenreider O. Evaluating drug-drug interaction information in NDF-RT and DrugBank. J Biomed Semant 2015;6:19.
[5] Reichley RM, Seaton TL, Resetar E, et al. Implementing a commercial rule base as a medication order safety net. J Am Med Inf Assoc 2005;12:383–9.
[6] Bubp JL, Park MA, Kapusnik-Uner J, et al. Successful deployment of drug-disease interaction clinical decision support across multiple Kaiser Permanente regions. J Am Med Inf Assoc 2019;26:905–10.
[7] Shah SN, Seger DL, Fiskio JM, Horn JR, Bates DW. Comparison of medication alerts from two commercial applications in the USA. Drug Saf 2021;44:661–8.
[8] Syrowatka A, Song W, Amato MG, et al. Key use cases for artificial intelligence to reduce the frequency of adverse drug events: a scoping review. Lancet Digit Health 2022;4:e137–48.
[9] Garcia SL, Lauritsen J, Zhang Z, et al. Prediction of nephrotoxicity associated with cisplatin-based chemotherapy in testicular cancer patients. JNCI Cancer Spectr 2020;4:pkaa032.
[10] Schiff GD, Volk LA, Volodarskaya M, et al. Screening for medication errors using an outlier detection system. J Am Med Inf Assoc 2017;24:281–7.
[11] Rozenblum R, Rodriguez-Monguio R, Volk LA, et al. Using a machine learning system to identify and prevent medication prescribing errors: a clinical and cost analysis evaluation. Jt Comm J Qual Patient Saf 2020;46:3–10.
[12] Segal G, Segev A, Brom A, Lifshitz Y, Wasserstrum Y, Zimlichman E. Reducing drug prescription errors and adverse drug events by application of a probabilistic, machine-learning based clinical decision support system in an inpatient setting. J Am Med Inf Assoc 2019;26:1560–5.
[13] Naor GM, Tocut M, Moalem M, et al. Screening for medication errors and adverse events using outlier detection screening algorithms in an inpatient setting. J Med Syst 2022;46:88.
[14] Li X, Yu S, Zhang Z, et al. Predictive modeling of hypoglycemia for clinical decision support in evaluating outpatients with diabetes mellitus. Curr Med Res Opin 2019;35:1885–91.
[15] Smith JC, Chen Q, Denny JC, Roden DM, Johnson KB, Miller RA. Evaluation of a novel system to enhance clinicians' recognition of preadmission adverse drug reactions. Appl Clin Inf 2018;9:313–25.

CHAPTER

44

Artificial intelligence in mass spectrometry-based proteomics

Wen-Feng Zeng[1], *Matthias Mann*[1,2] *and* *Maximillian T. Strauss*[3]

[1]Department of Proteomics and Signal Transduction, Max Planck Institute of Biochemistry, Martinsried, Germany [2]Proteomics Program, NNF Center for Protein Research, Faculty of Health Sciences, University of Copenhagen, Copenhagen, Denmark [3]Novo Nordisk Foundation Center for Protein Research, University of Copenhagen, Copenhagen, Denmark

Introduction

Dramatic technological advances in mass spectrometry (MS)-based proteomics are increasingly enabling researchers to acquire large-scale datasets with hundreds to thousands of samples. While analyzing proteomics data has also made tremendous progress in recent years, advances in information processing capabilities have not kept up with the expanding data volume and complexity.

Artificial intelligence (AI) is one of the most exciting technologies that is transforming scientific discovery today and can help substantially in this challenge. Here, we give an overview of how AI is used in MS-based proteomics. First, AI technologies, such as machine learning (ML) and deep learning (DL), are now applied across the entire proteomics workflow. Second, insight can be distilled from proteomics data using AI. Third, there are still substantial challenges in applying AI, which are going to be addressed in the near future. For a more in-depth perspective on proteomics in biomarker discovery, we refer to our recent review [1] and for a more technical review, we recommend [2].

Artificial intelligence in the proteomics workflow

The MS-based shotgun proteomics workflow consists of enzymatically digesting proteins into peptides, separating them chromatographically, and ionizing them by electrospray.

Artificial Intelligence in Clinical Practice
DOI: https://doi.org/10.1016/B978-0-443-15688-5.00010-3

The eluting peptides are mass-measured and fragmented. The resulting data is then analyzed to identify and quantify the eluting peptides and assembled into quantified protein groups—protein isoforms that can be distinguished based on their identified peptides.

AI has been applied to almost all steps of this workflow, as shown in the overview Fig. 44.1. In the past decade, ML methods were prevalently used but these are now succeeded by DL methods because they can achieve higher performance. The overall aim is to increase the confidence in and thus the number of peptides and proteins that can be identified and quantified in an experiment. Identification commonly happens by applying a scoring function and comparing experimental data to theoretical data. The premise for using ML and DL is that it can predict the expected behavior of a protein and its resulting peptides during the experiment. This allows better assessment of the likelihood of correct identification.

Starting from the sample preparation, DL models can predict the proteotypic cleavage sites and peptide digestibility [3]. For the subsequent chromatography steps, ML methods have been used to estimate the time when a peptide will elute, starting with support vector machines (SVM) [4–6] and later capsule network DL [7] methods.

Of special interest is the prediction of the intensity pattern of MS/MS spectra from the sequence alone, as this is an excellent parameter to determine the likelihood of correct identification. Initially, models were data-driven and used posterior probabilities [8], which was then succeeded by ensemble learning-based ML methods [9]. Lately, DL methods have been shown to perform much better, with the first DL-based approach being introduced in 2017 [10], and several more following [11,12]. Today, predicting properties has become almost routine, and we have recently provided a generalized frameworks that

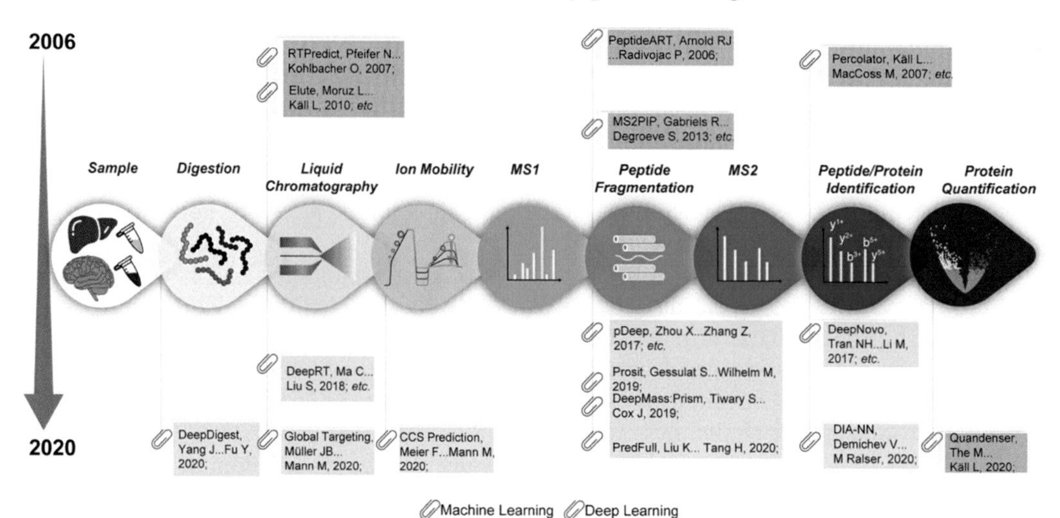

FIGURE 44.1 AI across the proteomics workflow. Each droplet in the scheme depicts a step in the shotgun proteomics workflow. AI can be used to increase the number of peptides and proteins that can be identified and quantified from the raw data. In the last decade, ML methods dominated the field and were subsequently superseded with the advent of DL methods. Source: *Reprinted from Mann M, Kumar C, Zeng W-F, Strauss MT. Artificial intelligence for proteomics and biomarker discovery. Cell Syst 2021;12(8):759–770. https://doi.org/10.1016/j.cels.2021.06.006 [1], with permission from Elsevier.*

can learn any peptide property from their sequences [13]. In this "AlphaPeptDeep" framework, standard models are provided in a Model Shop, which can be refined to the particular task with just a few lines of code.

For the identification task, the state-of-the-art is to generate "nonsense," decoy data and determine a score cutoff based on a false discovery rate reviewed in [14]. While pre-AI scores relied on human-chosen statistical parameters and cut-offs, such as the number of fragments that are present, semisupervised learning automatically determines the best discriminators [15].

DL can also directly predict peptide sequences from the fragmentation spectra [16]. The latest approaches use embeddings to encode spectral data [17] and even jointly embed spectra and sequences for identification [18]. This allows using of distance metrics for scoring and permits clustering.

Artificial intelligence to distill insight from proteomics data

Historically, the results of proteomic experiments have been analyzed with statistical methods to identify key proteins and to extract potential biomarkers. With increasing sizes of study cohorts, ML-based techniques are of ever greater interest, as they are potentially better able to model nonlinear protein regulation and interaction and intrinsically can utilize all the data at hand. Recent studies from our groups show that ML models trained on proteomics data correctly highlight the known proteins as the most decimating features, for example, when studying Alzheimer's disease [19] or Parkinson's disease [20]. Proteomics-based ML classifiers have recently been shown to outperform established assays in early liver diseased detection [21]. We have recently developed OmicLearn, an open source and easy-to-use tool that allows exploration of ML to extract biomarker panels on proteomics data [22]. To further increase insight from proteomics data, it can be integrated with other sources from biomedical data, and AI can assist in this demanding task. This can range from text-mining of literature data [23], fusion of multiomics datasets, or incorporation of very different data types, such as health records, lab tests, or imaging.

Along these lines, AI is playing a key role in an exciting technology termed "Deep Visual Proteomics" that is able to distill spatial insight from AI-assisted segmentation of microscopy images and laser-capture cutting microscopy to cut out samples for subsequent analysis with proteomics [24].

Current challenges of artificial intelligence in proteomics

As with the majority of DL itself, progress in bioinformatic analysis in proteomics is fueled by openly accessible resources. Experimental data is shared on repositories, such as PRIDE [25] and MassIVE [26]. Data accession is facilitated by open-source libraries [27–29] and open analysis software [30,31].

However, this openness can raise complex ethical and philosophical questions, for example, for clinical applications where increased data quality allows (re)identification of patients based on the raw data itself or provide potentially unwanted incidental findings

[32]. This requires careful balancing between data protection and transparency, which could be mitigated by technical innovations like federated learning [33] instead of limiting access to selected researchers.

While ML and DL greatly increase the insight that can be extracted from raw data, it generally comes at the cost of trading human-chosen acceptance criteria for black-box models, making it potentially difficult to develop trust in the findings.

Toward artificial intelligence in proteomics

While ML and DL are commonly listed as AI technologies, the current applications are not yet AI in the sense that there are autonomous decision-making agents. One potential avenue in this direction would be to develop systems that go much beyond presenting the clinician with ever more complex and data to make their decisions. Instead, these systems would incorporate existing biomedical knowledge and they would have models that allow them to make recommendations based on the proteomics and other data at hand for the given patients. Such a system could greatly improve health outcome by providing optimal assistance, while allowing the care giver to return to some of the key virtues of the medical profession that have tended to be lost in recent years [34].

References

[1] Mann M, Kumar C, Zeng W-F, Strauss MT. Artificial intelligence for proteomics and biomarker discovery. Cell Syst 2021;12(8):759–70. Available from: https://doi.org/10.1016/j.cels.2021.06.006.

[2] Wen B, Zeng W-F, Liao Y, Shi Z, Savage SR, Jiang W, et al. Deep learning in proteomics. Proteomics 2020;20 (21–22):1900335. Available from: https://doi.org/10.1002/pmic.201900335.

[3] Yang J, Gao Z, Ren X, Sheng J, Xu P, Chang C, et al. DeepDigest: prediction of protein proteolytic digestion with deep learning. Anal Chem 2021;93(15):6094–103. Available from: https://doi.org/10.1021/acs.analchem.0c04704.

[4] Moruz L, Käll L. Peptide retention time prediction. Mass Spectrometry Rev 2017;36(5):615–23. Available from: https://doi.org/10.1002/mas.21488.

[5] Moruz L, Tomazela D, Käll L. Training, selection, and robust calibration of retention time models for targeted proteomics. J Proteome Res 2010;9(10):5209–16. Available from: https://doi.org/10.1021/pr1005058.

[6] Pfeifer N, Leinenbach A, Huber CG, Kohlbacher O. Statistical learning of peptide retention behavior in chromatographic separations: a new kernel-based approach for computational proteomics. BMC Bioinformatics 2007;8(1):468. Available from: https://doi.org/10.1186/1471-2105-8-468.

[7] Ma C, Ren Y, Yang J, Ren Z, Yang H, Liu S. Improved peptide retention time prediction in liquid chromatography through deep learning. Anal Chem 2018;90(18):10881–8. Available from: https://doi.org/10.1021/acs.analchem.8b02386.

[8] Arnold RJ, Jayasankar N, Aggarwal D, Tang H, Radivojac P. A machine learning approach to predicting peptide fragmentation spectra. Biocomputing, 2006. Maui, Hawaii: World Scientific; 2005. p. 219–30. Available from: https://doi.org/10.1142/9789812701626_0021.

[9] Degroeve S, Martens L. MS2PIP: a tool for MS/MS peak intensity prediction. Bioinformatics 2013;29 (24):3199–203. Available from: https://doi.org/10.1093/bioinformatics/btt544.

[10] Zhou X-X, Zeng W-F, Chi H, Luo C, Liu C, Zhan J, et al. PDeep: predicting MS/MS spectra of peptides with deep learning. Anal Chem 2017;89(23):12690–7. Available from: https://doi.org/10.1021/acs.analchem.7b02566.

[11] Tiwary S, Levy R, Gutenbrunner P, Soto FS, Palaniappan KK, Deming L, et al. High-quality MS/MS spectrum prediction for data-dependent and data-independent acquisition data analysis. Nat Methods 2019;16 (6):519–25. Available from: https://doi.org/10.1038/s41592-019-0427-6.

[12] Gessulat S, Schmidt T, Zolg DP, Samaras P, Schnatbaum K, Zerweck J, et al. Prosit: proteome-wide prediction of peptide tandem mass spectra by deep learning. Nat Methods 2019;16(6):509–18. Available from: https://doi.org/10.1038/s41592-019-0426-7.
[13] Zeng W-F, Zhou X-X, Willems S, Ammar C, Wahle M, Bludau I, et al. AlphaPeptDeep: a modular deep learning framework to predict peptide properties for proteomics. Bioinformatics 2022. Available from: https://doi.org/10.1101/2022.07.14.499992.
[14] Elias JE, Gygi SP. Target-decoy search strategy for increased confidence in large-scale protein identifications by mass spectrometry. Nat Methods 2007;4(3):207–14. Available from: https://doi.org/10.1038/nmeth1019.
[15] Käll L, Canterbury JD, Weston J, Noble WS, MacCoss MJ. Semi-supervised learning for peptide identification from shotgun proteomics datasets. Nat Methods 2007;4(11):923–5. Available from: https://doi.org/10.1038/nmeth1113.
[16] Tran NH, Zhang X, Xin L, Shan B, Li M. De novo peptide sequencing by deep learning. Proc Natl Acad Sci U S A 2017;114(31):8247–52. Available from: https://doi.org/10.1073/pnas.1705691114.
[17] Bittremieux W, May DH, Bilmes J, Noble WS. A learned embedding for efficient joint analysis of millions of mass spectra. Nat Methods 2022;19(6):675–8. Available from: https://doi.org/10.1038/s41592-022-01496-1.
[18] Altenburg T, Muth T, Renard BY. YHydra: deep learning enables an ultra fast open search by jointly embedding MS/MS spectra and peptides of mass spectrometry-based proteomics. Bioinformatics 2021. Available from: https://doi.org/10.1101/2021.12.01.470818.
[19] Bader JM, Geyer PE, Müller JB, Strauss MT, Koch M, Leypoldt F, et al. Proteome profiling in cerebrospinal fluid reveals novel biomarkers of Alzheimer's disease. Mol Syst Biol 2020;16(6). Available from: https://doi.org/10.15252/msb.20199356.
[20] Virreira Winter S, Karayel O, Strauss MT, Padmanabhan S, Surface M, Merchant K, et al. Urinary proteome profiling for stratifying patients with familial Parkinson's disease. EMBO Mol Med 2021. Available from: https://doi.org/10.15252/emmm.202013257.
[21] Niu L, Thiele M, Geyer PE, Rasmussen DN, Webel HE, Santos A, et al. Noninvasive proteomic biomarkers for alcohol-related liver disease. Nat Med 2022;28(6):1277–87. Available from: https://doi.org/10.1038/s41591-022-01850-y.
[22] Torun FM, Winter SV, Doll S, Riese FM, Vorobyev A, Mueller-Reif JB, et al. Transparent exploration of machine learning for biomarker discovery from proteomics and omics data. Biochemistry 2021. Available from: https://doi.org/10.1101/2021.03.05.434053.
[23] Jensen LJ, Saric J, Bork P. Literature mining for the biologist: from information retrieval to biological discovery. Nat Rev Genet 2006;7(2):119–29. Available from: https://doi.org/10.1038/nrg1768.
[24] Mund A, Coscia F, Kriston A, Hollandi R, Kovács F, Brunner A-D, et al. Deep visual proteomics defines single-cell identity and heterogeneity. Nat Biotechnol 2022;40(8):1231–40. Available from: https://doi.org/10.1038/s41587-022-01302-5.
[25] Perez-Riverol Y, Csordas A, Bai J, Bernal-Llinares M, Hewapathirana S, Kundu DJ, et al. The PRIDE database and related tools and resources in 2019: improving support for quantification data. Nucleic Acids Res 2019;47(D1):D442–50. Available from: https://doi.org/10.1093/nar/gky1106.
[26] Wang M, Wang J, Carver J, Pullman BS, Cha SW, Bandeira N. Assembling the community-scale discoverable human proteome. Cell Syst 2018;7(4):412–21. Available from: https://doi.org/10.1016/j.cels.2018.08.004 e5.
[27] Łącki MK, Startek MP, Brehmer S, Distler U, Tenzer S. OpenTIMS, TimsPy, and TimsR: open and easy access to TimsTOF raw data. J Proteome Res 2021;20(4):2122–9. Available from: https://doi.org/10.1021/acs.jproteome.0c00962.
[28] Willems S, Voytik E, Skowronek P, Strauss MT, Mann M. AlphaTims: indexing trapped ion mobility spectrometry–TOF data for fast and easy accession and visualization. Mol & Cell Proteom 2021;20:100149. Available from: https://doi.org/10.1016/j.mcpro.2021.100149.
[29] Goloborodko AA, Levitsky LI, Ivanov MV, Gorshkov MV. Pyteomics—a python framework for exploratory data analysis and rapid software prototyping in proteomics. J Am Soc Mass Spectrometry 2013;24(2):301–4. Available from: https://doi.org/10.1007/s13361-012-0516-6.
[30] Abdrakhimov DA, Bubis JA, Gorshkov V, Kjeldsen F, Gorshkov MV, Ivanov MV. Biosaur: an open-source python software for liquid chromatography–mass spectrometry peptide feature detection with ion mobility support. Rapid Commun Mass Spectrometry 2021. Available from: https://doi.org/10.1002/rcm.9045.
[31] Strauss MT, Bludau I, Zeng W-F, Voytik E, Ammar C, Schessner J, et al. AlphaPept, a modern and open framework for MS-based proteomics. Bioinformatics 2021. Available from: https://doi.org/10.1101/2021.07.23.453379.

[32] Mann SP, Treit PV, Geyer PE, Omenn GS, Mann M. Ethical principles, constraints, and opportunities in clinical proteomics. Mol Cell Proteom 2021;20:100046. Available from: https://doi.org/10.1016/j.mcpro.2021.100046.

[33] Rieke N, Hancox J, Li W, Milletarì F, Roth HR, Albarqouni S, et al. The Future of digital health with federated learning. npj Digital Med 2020;3(1):119. Available from: https://doi.org/10.1038/s41746-020-00323-1.

[34] Topol EJ. Deep medicine: how artificial intelligence can make healthcare human again. 1st ed. New York: Basic Books; 2019.

CHAPTER

45

Artificial intelligence and global health

Jay Vietas

Emerging Technologies Branch, Division of Science Integration, the National Institute for Occupational Safety and Health, Centers for Disease Control and Prevention, Washington DC, United States

Introduction

Artificial intelligence (AI) has the potential to have profound impacts on improvements in global health. Where data are available and of sufficient quality, algorithms can be applied to improve the forecasting of disease outbreaks, assist in identifying pathogens, recommend treatments to address and control outbreaks, and place clinical decision-making tools in the hands of healthcare providers to ensure treatment is optimized and personalized. Many of these tools are in early development, but improvements in predictive tools, such as those to predict malaria disease burden will revolutionize the public health profession and optimize the healthcare response [1]. While there is significant promise for improvements in global health due to the use of AI, the development, implementation, and maintenance of these tools will be critical to ensure the anticipated results are realized.

Disease outbreaks

Instead of waiting for cases to arrive at a local clinic, predictive models can be applied to determine where and when potential disease outbreaks may occur. Since John Snow identified the water at the Broad Street pump as the source of cholera, epidemiologists have been developing relationships between human health outcomes and activities which result in disease [2]. AI builds on these relationships and is able to connect multiple variables in space and time to anticipate where and when disease may occur. Before the

Artificial Intelligence in Clinical Practice
DOI: https://doi.org/10.1016/B978-0-443-15688-5.00034-6

COVID-19 pandemic, a significant amount of effort went into forecasting the early and precise detection of influenza outbreaks by developing artificial intelligent tools from a variety of data sources including from the Centers for Disease Control and Prevention [3], data curated from social media [4], and more complicated systems attempting to combine various data sets [5]. COVID-19 provided ample opportunity to use AI to improve methods of tracking and forecasting the spread of disease [6], including data collected from wastewater [7], the use of mobile phone surveys [8], as well as the exploration of the use of wireless systems and drones [9]. Future use cases of AI will be able to recognize patterns to forecast and monitor disease outbreaks, optimize treatment plans once identified, and assist in the optimization of the use of healthcare supplies and personnel to contain the outbreak [10].

Hazard identification

AI has the capacity to automate and standardize techniques using machine learning and computer vision, which are trained programs able to recognize objects or items from images. These tools, previously reserved for the expertise and training of a limited number of personnel, can now be used in the field and in the laboratory with minimal amounts of training to identify potential hazards or disease vectors. The technology associated with cameras on mobile phones has improved enough to employ AI tools for a wide variety of applications. Additionally, due to the increasing ubiquity of phones with this technology, the use of these tools can occur in ever increasing locations. Photographs and smart phone applications can be used to identify hazardous versus nonhazardous snakes [11], plants [12], and insects [13–15]. Furthermore, this technology typically collects the location of these photographs, too. For entomologists, the combination of tick and mosquito identification with location, especially when combined with molecular techniques, can improve the understanding of relationships between environmental factors and disease spread.

More elaborate AI systems can also be used to protect the public. Computer systems linked to air monitoring sensors can be used to automate pollution warnings [16] and assist first responders to effectively access life-saving information through the use of machine learning [17]. Similar technology is promising for the diagnosis of dental caries using a smartphone and would be useful, particularly in low-income countries, for identifying the most at-risk populations in order to optimize treatment [18].

Clinical decision support systems

AI has the capacity to improve clinical decision-making. Tools have been created to assist surgeons in predicting the need for total knee arthroplasty [19], to guide physicians in breast cancer treatment options [20], and to predict diabetes by noninvasive methods [21]. AI also has the potential to predict the child-birthing location of maternal health program enrollees, and to improve outcomes by distributing support to those not delivering at a healthcare facility [22]. The challenge of relying on these systems is to ensure appropriate use of these tools to minimize misuse or overdiagnosis [23]. Currently clinical decision support tools which provide guidance appear to be the most effective, such as those which assist in prescribing medications [24].

Challenges related to the use of artificial intelligence in global health

To harness the power and promise of AI in global health, adequate investment must be made in communication, computer systems, and supporting personnel to collect, curate, and manage the data necessary to enable benefits and minimize harm from the use of AI-related tools. In developed countries, this has either occurred or is underway. Developing countries run the risk of widening the digital divide if this does not occur [25]. Furthermore, data, models, and forecasts alone do not improve population health. These measures must be paired with appropriately developed policies, logistics, and research efforts to truly make a difference. Training of personnel to understand how these systems work must be accomplished in a manner to develop trust; and the systems themselves must be designed with the user in mind while ensuring data confidentiality and privacy. Finally, the systems must consider the uncertainty of the results, which should include an examination of bias. A review of bias should include the type and representative nature of the data collected, the development and implementation of the algorithm, and the need to continuously examine and validate outcomes to ensure the system is functioning as desired. This will be particularly important for underrepresented populations in the data, and careless use of these tools has the potential to increase health inequities, particularly for those who are already marginalized.

Future directions

Employed properly, AI can improve health systems and health outcomes across the global health landscape. Research, investment, and training will be necessary for programmers, implementers, public health and healthcare workers, and support personnel to ensure that appropriate systems and infrastructure are built to deliver equitable and effective global health outcomes. Finally, the systems must ensure humans are in control of decision-making and that the policies and standards associated with these tools are adequate to fairly safeguard human health.

Disclaimer

The findings and conclusions in this report are those of the author and do not necessarily represent the official position of the National Institute for Occupational Safety and Health, Centers for Disease Control and Prevention.

References

[1] Brown BJ, Manescu P, Przybylski AA, Caccioli F, Oyinloye G, Elmi M, et al. Data-driven malaria prevalence prediction in large densely populated urban holoendemic sub-saharan west africa. Sci Rep 2020;10(1).
[2] Tulchinsky TH. John snow, cholera, the broad street pump; waterborne diseases then and now. Case Stud Public Health 2018;77–99.
[3] Aldhyani THH, Joshi MR, AlMaaytah SA, Alqarni AA, Alsharif N. Using sequence mining to predict complex systems: a case study in influenza epidemics. Complexity 2021;2021:9929013.
[4] Wang F, Wang H, Xu K, Raymond R, Chon J, Fuller S, et al. Regional level influenza study with geo-tagged twitter data. J Med Syst 2016;40(8):189.
[5] Santillana M, Nguyen AT, Dredze M, Paul MJ, Nsoesie EO, Brownstein JS. Combining search, social media, and traditional data sources to improve influenza surveillance. PLoS Comput Biol 2015;11:10.
[6] Atek S, Pesaresi C, Eugeni M, De Vito C, Cardinale V, Mecella M, et al. A geospatial artificial intelligence and satellite-based earth observation cognitive system in response to covid-19. Acta Astronaut 2022; 197:323–35.
[7] Mahmoudi T, Naghdi T, Morales-Narváez E, Golmohammadi H. Toward smart diagnosis of pandemic infectious diseases using wastewater-based epidemiology. TrAC Trends Analyt Chem 2022;153:116635.
[8] Srinivasa Rao ASR, Vazquez JA. Identification of covid-19 can be quicker through artificial intelligence framework using a mobile phone-based survey when cities and towns are under quarantine. Infect Control Hosp Epidemiol 2020;41:826–30.
[9] Alsarhan A, Almalkawi I, Kilani Y. A new covid-19 tracing approach using machine learning and drones enabled wireless network. Int J Interact Mobile Technol 2021;15:111–26.
[10] Davies SE. Artificial intelligence in global health. Ethics Int Affairs 2019;33(2):181–92.
[11] Bolon I, Picek L, Durso AM, Alcoba G, Chappuis F, Ruiz de Castaneda R. An artificial intelligence model to identify snakes from across the world: opportunities and challenges for global health and herpetology. PLoS Negl Trop Dis 2022;16(8):e0010647.
[12] Otter J, Mayer S, Tomaszewski CA. Swipe right: a comparison of accuracy of plant identification apps for toxic plants. J Med Toxicol 2021;17(1):42–7.
[13] Høye TT, Ärje J, Bjerge K, Hansen OLP, Iosifidis A, Leese F, et al. Deep learning and computer vision will transform entomology. Proc Natl Acad Sci U S A 2021;118(2):e2002545117.
[14] Ong S-Q, Nair G, Yusof UK, Ahmad H. Community-based mosquito surveillance: An automatic mosquito-on-human-skin recognition system with a deep learning algorithm. Pest Manag Sci 2022; 78(10):4092–104.
[15] Luo C-Y, Pearson P, Xu G, Rich SM. A computer vision-based approach for tick identification using deep learning models. Insects 2022;13(2):116.
[16] Guo Q, Ren M, Wu S, Sun Y, Wang J, Wang Q, et al. Applications of artificial intelligence in the field of air pollution: a bibliometric analysis. Front Public Health 2022;10.
[17] Do V, Huyen A, Joubert FJ, Gabriel M, Yun K, Lu T, et al. Virtual assistant for first responders using natural language understanding and optical character recognition. In: 12101, proceedings of SPIE—the International Society for Optical Engineering; 2022.
[18] Thanh MT, Van Toan N, Ngoc VT, Tra NT, Giap CN, Nguyen DM. Deep learning application in dental caries detection using intraoral photos taken by smartphones. Appl Sci 2022;12:11.
[19] Lee LS, Chan PK, Wen C, Fung WC, Cheung A, Chan VWK, et al. Artificial intelligence in diagnosis of knee osteoarthritis and prediction of arthroplasty outcomes: a review. Arthroplasty 2022;4(1):16.
[20] Torkey H, Atlam M, El-Fishawy N, Salem H. A novel deep autoencoder based survival analysis approach for microarray dataset. PeerJ Comput Sci 2021;7:e492.
[21] Spänig S, Emberger-Klein A, Sowa JP, Canbay A, Menrad K, Heider D. The virtual doctor: An interactive clinical-decision-support system based on deep learning for non-invasive prediction of diabetes. Artif Intell Med 2019;100.
[22] Fredriksson A, Fulcher IR, Russell AL, Li T, Tsai Y-T, Seif SS, et al. Machine learning for maternal health: Predicting delivery location in a community health worker program in zanzibar. Front Digital Health 2022;4.

[23] Vogt H, Green S, Ekstrøm CT, Brodersen J. How precision medicine and screening with big data could increase overdiagnosis. BMJ 2019;366:5270.

[24] Shahmoradi L, Safdari R, Ahmadi H, Zahmatkeshan M. Clinical decision support systems-based interventions to improve medication outcomes: a systematic literature review on features and effects. Med J Islam Repub Iran 2021;35:27.

[25] Chatterjee J, Nina D. Developing countries are being left behind in the ai race—and that's a problem for all of us. 2022; <https://theconversation.com/developing-countries-are-being-left-behind-in-the-ai-race-and-thats-a-problem-for-all-of-us-180218>.

CHAPTER

46

Legal aspects of artificial intelligence in medical practice

Sarah M.L. Bender and W. Nicholson Price, II

University of Michigan Law School, Ann Arbor, MI, United States

Introduction

As this book illustrates, artificial intelligence (AI) stands to greatly improve medical practice and healthcare outcomes. However, providers' ability to seize the full potential of medical AI is shaped not only by the pace of technological advances, but also by the law.[1]

This chapter provides an overview of the legal frameworks that govern how medical AI is created, marketed, and deployed. We group legal issues into three rough sets based on how close they are to clinical care: "upstream," "midstream," and "downstream" (Fig. 46.1). "Upstream" legal issues influence how AI systems are developed. While they have little immediate bearing on how AI is used in clinical settings, they are important for providers to understand, as they can affect the quality and design of the AI systems they use, and thus also their patients' outcomes. These include questions of privacy and intellectual property, both of which shape what products are developed and how. "Midstream," we focus on regulation of AI technologies, specifically by the US Food and Drug Administration (FDA), familiar to providers using other regulated medical devices or drugs, which determines when AI tools can be marketed for clinical use or otherwise used in patient care. Finally, "downstream" legal issues directly impact the use of medical AI in individual patients' care; here, we consider questions of liability, informed consent, and reimbursement. The legal issues we outline here are limited and incomplete: We cannot give a comprehensive overview of all possible legal questions, and those we address are still in flux. Nevertheless, understanding these key regimes will help providers grasp how law shapes what systems are created, how they make it into healthcare settings, and their role in the clinical encounter.

[1] W. Nicholson Price II, *Black-Box Medicine*, 28 HARV. J.L. & TECH. 419, 421(2015).

Artificial Intelligence in Clinical Practice
DOI: https://doi.org/10.1016/B978-0-443-15688-5.00046-2

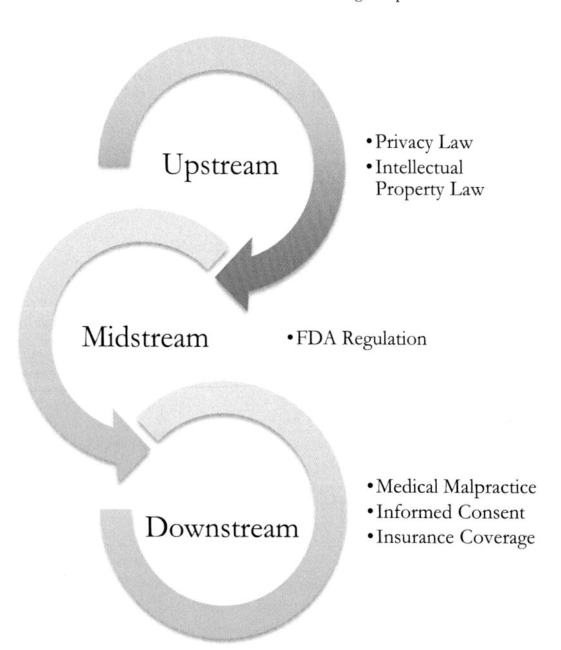

FIGURE 46.1 Schematic representation of legal issues arising in upstream, midstream, and downstream phases of AI implementation.

Upstream

When AI systems are being developed, far upstream from the clinical encounter, two areas of law play a particularly prominent role: privacy and intellectual property. Privacy law shapes the data available for AI training, and carries implications for representativeness, generalizability, and bias. Intellectual property helps shape which entities develop AI in the first place, the process of sharing data for incremental progress, and the ease of validating quality and performance. While providers may not encounter these AI/law interactions in their daily practice, they shape what AI systems providers may find there, and thus merit at least some familiarity.

Privacy law

Healthcare data are governed by a patchwork of state and federal privacy laws, the most important of which are the Health Insurance Portability and Accountability Act (HIPAA)[2] and its implementing regulations, commonly referred to as the HIPAA Privacy Rule.[3] Under the HIPAA Privacy Rule, "covered entities"—which includes most

[2] Health Insurance Portability and Accountability Act of 1996, Pub. L. No. 104–191, 110 Stat. 1936 (1996).

[3] 45 C.F.R. pts. 160, 164 (2022).

healthcare providers, insurance companies, and health information clearinghouses—and their "business associates" may not use or disclose "protected health information" (PHI), except in specific circumstances.[4]

This framework is in many ways underinclusive. Because HIPAA's prohibitions are focused on specific entities, not the data themselves, a large and growing amount of health information is left unprotected.[5] For example, data created and held by other actors—including through the use of Apple Watches, smartphone apps, and Google searches—fall outside the definition of PHI.[6] Health data also loses its protected status when patients request their data from their healthcare provider and then give that information to another, noncovered entity.[7]

Further, while PHI encompasses most individually identifiable health information, it does not include deidentified data.[8] This loophole has some benefits with regard to AI development—namely, that it makes more information available to researchers, including those who may be developing health AIs.[9] However, deidentified data has become increasingly easy to reidentify, which exposes patients to a variety of privacy risks.[10]

In other ways, HIPAA is *over*protective of health data. This, too, can harm patients. Though the Privacy Rule includes a number of exceptions in which covered entities *can* use or disclose PHI, including "quality assessment and improvement activities,"[11] no such exception is made for medical research.[12] To use patient data for research—including AI development—providers must either deidentify the data or obtain patients' authorization.[13] Both approaches impose considerable costs, which low-resource institutions are unlikely to be able to shoulder.[14] As a result, the development of medical AI is highly concentrated in commercial developers or high-resource institutions.[15] These institutions may go on to share their technologies with providers in low-resource settings. However, these AI systems will have been trained using data from the high-resource institution's patient

[4] 45 C.F.R. pts. 160 (2022); Roger Allan Ford & W. Nicholson Price II, *Privacy & Accountability in Black-Box Medicine*, 23 MICH. TELECOMM. & TECH L. REV. 1, 23 (2016); W. Nicholson Price II & I. Glenn Cohen, *Privacy in the Age of Medical Big Data*, 25 NATURE MED. 37, 38–39 (2019).

[5] Price & Cohen, *supra* note 4, at 38.

[6] Price & Cohen, *supra* note 4, at 38.

[7] Price & Cohen, *supra* note 4, at 38.

[8] Price & Cohen, *supra* note 4, at 38.

[9] Ford & Price, *supra* note 4, at 23; *see also* Charlotte A. Tschider, *AI's Legitimate Interest: Towards a Public Benefit Privacy Model*, 21 HOUS. J. HEALTH L. & POL'Y 121, 161 (2021).

[10] Price & Cohen, *supra* note 4, at 40; Ford & Price, *supra* note 4, at 23.

[11] 45 C.F.R. § 164.501 (2022).

[12] W. Nicholson Price II, *Medical AI & Contextual Bias*, 33 HARV. J.L. & TECH. 66, 82 (2019) [hereinafter Price, *Contextual Bias*].

[13] Price, *Contextual Bias*, *supra* note 12; W. Nicholson Price II, *Problematic Interactions Between AI & Health Privacy*, 4 UTAH L. REV. 925, 930 (2021) [hereinafter Price, *Problematic Interactions*].

[14] Price, *Contextual Bias*, *supra* note 12, at 82–83.

[15] Price, *Contextual Bias*, *supra* note 12, at 82–83.

population, which may be unrepresentative of the patients it will be used to help treat in these new contexts, and may mesh poorly with lower-resource workflows.[16] These technologies may thus render less accurate or helpful results in low-resource settings,[17] exacerbating health disparities and putting patients' health at risk.[18]

Intellectual property law

Intellectual property law creates important incentives for technological innovation, without which developers might underinvest in the research needed to develop cutting-edge technologies. Though patents have long been a key tool for promoting medical innovation,[19] US law erects a number of barriers to patent availability for medical AI.

AI involves two types of innovative information—datasets and algorithms.[20] Data do not qualify as "process[es], machine[s], manufacture[s], or composition[s] of matter" and thus are not patentable.[21] However, for many years, at least some types of algorithms were thought to be. Until recently, developers could secure method patents for diagnostic algorithms, so long as they involved a machine or a transformation of matter—such as performing a blood test.[22] This changed in 2012 when the Supreme Court issued the first of a series of decisions curtailing the availability of patents to such technologies.[23] Though inventors are still getting patents on some medical-AI-related inventions, the landscape has gotten less hospitable and remains quite uncertain.[24]

Today, trade secrecy laws provide the primary form of intellectual property protections for both the algorithms and datasets that drive black-box medicine.[25] Trade secrecy protects

[16] Price, *Contextual Bias*, *supra* note 12, at 82–83; Price, *Problematic Interactions*, *supra* note 13, at 932–34.

[17] Price, *Problematic Interactions*, *supra* note 13, at 932–34.

[18] Ana Bracic, Shawneequa L. Callier & W. Nicholson Price II, *Exclusion Cycles: Reinforcing Disparities in Medicine*, 377 SCIENCE 1158 (2022); Mark P. Sendak, Nicholson Price & Suresh Balu, *A Market Failure is Preventing Efficient Diffusion of Health Care AI Software*, STAT (May 24, 2022), https://www.statnews.com/2022/05/24/market-failure-preventing-efficient-diffusion-health-care-ai-software/.

[19] W. Nicholson Price II, *Big Data, Patents, and the Future of Medicine*, 37 CARDOZO L. REV. 1401, 1419 (2016) [hereinafter Price, *Patents*].

[20] Price, *Patents*, *supra* note 19, at 1433.

[21] 35 U.S.C. § 101 (2022).

[22] Price, *Patents*, *supra* note 19, at 1406, 1421.

[23] Mayo Collaborative Servs. v. Prometheus Lab'ys, 132 S. Ct. 1289 (2012); *see also* Ass'n for Molecular Pathology v. Myriad Genetics, Inc., 133 S. Ct. 2107 (2013); Alice Corp. v. CLS Bank Int'l, 134 S. Ct. 2347 (2014).

[24] Mateo Aboy et al., *How Does Emerging Patent Case Law in the US and Europe Affect Precision Medicine?*, 37 NATURE BIOTECHNOLOGY 1118, 1118–21 (2019); *see also* Rebecca S. Eisenberg, *Diagnostics Need Not Apply*, B.U. J. SCI. & TECH. L. 256 (2015). Mateo Aboy et al., Mapping the Patent Landscape of Medical Machine Learning, 41 NATURE BIOTECHNOLOGY 461 (2023).

[25] *See* Price, *Patents*, *supra* note 19, at 1432; Meghan J. Ryan, *Algorithms, IP Rights, & the Public Interest*, 21 NEV. L.J. 61, 86–87 (2020).

such information from misappropriation and thus can help firms to recoup their investments. However, these protections only last as long as the information in question is kept secret.[26]

This framework creates troubling incentives for AI development. It encourages developers to keep their datasets and algorithms to themselves and thus results in AI systems that are created using siloed, often institution-specific data.[27] It also stymies future innovation. Rather than learning and building off of what has already been found to work, firms must make duplicative research investments to acquire similar underlying data for themselves.

Secret algorithms are also hard to validate and oversee. Regulators receive some information to validate some medical AI products, but their oversight is limited, as described below. Evaluation by others, such as health systems or third parties, is significantly challenged by secrecy. This can likewise make it difficult for providers and patients to trust these new tools, hampering their adoption in clinical settings.[28] Those outside evaluation efforts which have succeeded—often identifying significant flaws—have come at the cost of significant investment and often real creativity in reconstructing otherwise-unavailable data.[29]

Midstream

Between upstream questions of development and downstream questions of use in clinical encounters comes an intermediate question: which AI systems make it into the hands of providers in the first place? Many factors influence this choice—price, availability, the business case, compatibility with existing information technology, and the like. But the law is also prominent in helping to validate new technology before use. We turn next to the role of FDA in this process.

FDA regulation

Patients and providers have good reason to be skeptical of medical AI's quality. Though medical AI holds great promise, these systems can also be deeply flawed.[30] While some AI tools simply fail to achieve their stated goals, others are riddled with biases and inaccuracies that put patients' lives at risk.[31] This is true even for widely adopted systems that may appear embedded in familiar electronic health record systems, for instance.[32]

[26] Price, *Patents, supra* note 19, at 1432–33.

[27] Price, *Patents, supra* note 19, at 1434–36.

[28] Price, *Patents, supra* note 19, at 1435–36.

[29] Andrew Wong et al., *External Validation of a Widely Implemented Proprietary Sepsis Prediction Model in Hospitalized Patients*, 181 JAMA INTERNAL MED. 1065 (2021); Ziad Obermeyer et al., *Dissecting Racial Bias in an Algorithm Used to Manage the Health of Populations*, 366 SCIENCE 447 (2019).

[30] W. Nicholson Price II, *Distributed Governance of Medical AI*, 25 SMU SCI. & TECH. L. REV. 3, 4 (2022) [hereinafter Price, *Distributed Governance*].

[31] Obermeyer et al., *supra* note 29; Bracic et al., *supra* note 18.

[32] Wong et al., *supra* note 29.

These quality concerns highlight the need for some way to ensure safety and effectiveness. Typically, providers look to FDA to fulfill this role as regulator of new biomedical technologies. For most biomedical technologies, including drugs and medical devices, FDA approval or clearance serves as a baseline marker of quality, assuaging worries that the technology simply won't work. FDA assigns medical devices (a broad category which includes many AI systems) to one of three categories—class I, II, or III—depending on their level of risk.[33] Class I devices are those deemed the least dangerous, and that are thus less closely scrutinized, while class III products are subject to additional regulatory hurdles.

Many medical AI systems, however, will never go through any form of FDA review.[34] One reason for this is that FDA only regulates commercial products, but many medical AI systems are developed exclusively for institutions' in-house use.[35] Because these in-house AIs are never sold to others or otherwise made available for commerce, they are generally exempted from FDA review, despite the fact that they too are prone to quality issues.[36] Further, many medical AI systems do not meet the statutory definition of a "medical device" for various reasons, and thus also fall outside the scope of FDA's oversight authority.[37] FDA is actively interrogating these boundary questions, which remain unsettled.[38]

Even the AI systems that *are* subject to FDA review are arguably underregulated.[39] Of the medical AIs that have received some kind of FDA authorization, the vast majority have gone through the 510(k) clearance process, as opposed to full FDA approval.[40] Developers can obtain 510(k) clearance by demonstrating that their medical AI is substantially equivalent to an already-approved product—which need not also be an AI tool.[41] Experts have raised concerns about the 510(k) process's lack of rigor, and raised questions about whether FDA is equipped to evaluate AI tools under this framework.[42] For instance, in a study of 130 FDA-evaluated devices, almost all were evaluated based only on retrospective studies, most with data from only a single site.[43]

[33] *Classify Your Medical Device*, FDA (Feb. 7, 2020), https://www.fda.gov/medical-devices/overview-device-regulation/classify-your-medical-device.

[34] Price, *Distributed Governance*, *supra* note 30, at 3–4.

[35] W. Nicholson Price II, Rachel E. Sachs, & Rebecca S. Eisenberg, *New Innovation Models in Medical AI*, 99 WASH. U. L. REV. 1121, 1144 (2022); Price, *Distributed Governance*, *supra* note 30, at 4–5.

[36] Price, *Distributed Governance*, *supra* note 30, at 4–5.

[37] Price, Sachs, & Eisenberg, *supra* note 35, at 1144–45; Price, *Distributed Governance*, *supra* note 30, at 3–4.

[38] U.S. Food & Drug Admin., Clinical Decision Support Software: Guidance For Industry and Food and Drug Administration Staff (2022), https://www.fda.gov/media/109618/download.

[39] Price, *Distributed Governance*, *supra* note 30, at 7.

[40] Charlotte Tschider, *Medical Device Artificial Intelligence: The New Tort Frontier*, 46 BYU L. REV. 1551, 1597 (2021).

[41] Price, *Distributed Governance*, *supra* note 30, at 8–9.

[42] Ravi B. Parikh, Ziad Obermeyer, & Amol S. Navathe, *Regulation of Predictive Analytics in Medicine*, 363 SCIENCE 810 (2019); Price, *Distributed Governance*, *supra* note 30, at 8–9; Eric Wu et al., *How Medical AI Devices Are Evaluated: Limitations and Recommendations from an Analysis of FDA Approvals*, 27 NATURE MED. 582, 583 (2021).

[43] Wu et al., *supra* note 42.

More rigorous forms of FDA review are still limited in their ability to evaluate and monitor AI as providers are likely to encounter it.[44] As described above, where and how AI systems are used can have a significant impact on the quality of their outcomes.[45] Unlike other medical devices, AI systems can learn and adapt their performance with use.[46] They can also be tuned to their environment (with different patients and different workflows), which is often essential as initial performance frequently fails to generalize. Though AI's adaptive, malleable nature is one of its key advantages, these qualities also frustrate FDA's regulatory design, because central evaluation does not account for local variation in training and performance.[47] To adequately validate black-box systems, oversight must include localized and ongoing elements.[48] Large, high-resource health systems should and do invest in governance procedures that allow them to validate their AI tools on a regular basis.[49] However, doing so may be difficult for institutions with fewer resources, which could further exacerbate disparities in AI access and quality.[50]

Providers should thus be aware: AI systems are unlikely to have the same type of generalizable validation from FDA that they may be accustomed to seeing in other medical devices. In the absence of that quality signal, localized governance (such as committees overseeing health system adoption of new technologies) might play a larger role, or evaluation may simply be absent. Even where products have undergone FDA review, that provides relatively little assurance that the system will function effectively in a new setting.

Downstream

Medical AI also raises new questions regarding the point of care. AI will not fulfill its promise of revolutionizing health care unless it actually changes how things are done—for example, by getting doctors to follow AI-generated treatment recommendations in lieu of the traditional treatment plans they would previously have followed. Whether and how that adoption happens, including how AI is used in the clinical encounter itself, will be shaped to some extent by the law of liability, informed consent, and reimbursement.

Medical malpractice liability

AI systems are not perfect. Even highly accurate medical AI systems will render erroneous recommendations that go on to cause patient injuries.[51] Whether hospitals and

[44] Boris Babic et al., *Algorithms on Regulatory Lockdown in Medicine*, 366 SCIENCE 1202, 1202 (2019).

[45] Price, *Contextual Bias*, *supra* note 12, at 82–83; Price, *Distributed Governance*, *supra* note 30, at 9–10; Trishan Panch et al., *A Distributed Approach to the Regulation of Clinical AI.*, 1 PLOS DIGITAL HEALTH e0000040 (2022).

[46] Babic et al., *supra* note 44, at 1202.

[47] Babic et al., *supra* note 44, at 1202.

[48] Price, *Distributed Governance*, *supra* note 30, at 11–12.

[49] Price, *Distributed Governance*, *supra* note 30, at 12–14.

[50] Price, *Distributed Governance*, *supra* note 30, at 14–15.

[51] Hannah R. Sullivan & Scott J. Schweikart, *Are Current Tort Liability Doctrines Adequate for Addressing Injury Caused by AI?*, 21 AMA J. ETHICS E160 (2019).

providers can be held liable for such injuries will have a significant impact on how they incorporate AI into patient care.[52]

Medical providers

Generally speaking, medical providers are shielded from malpractice liability so long as they follow the standard of care.[53] Though AI is already outpacing doctors at performing certain medical tasks,[54] its use is not yet the medical standard of care for any procedure or practice.[55] So long as this remains the case, malpractice liability will largely incentivize physicians to stick with the treatment plan they would have followed sans AI, regardless of what the AI system recommends.[56]

Of course, the standard of care is not set in stone. As more physicians rely on medical AI systems, following AI recommendations will likely become part of the standard of care in certain fields.[57] When this occurs, incentives will shift, such that providers who followed incorrect AI recommendations will be able to shield themselves from liability.[58] Conversely, those who *reject* correct AI recommendations may eventually be on the hook for resulting patient harms. However, the latter outcome is unlikely to arise in the near future, in part because providers can typically invoke the "respectable minority" or "two schools of thought" defenses so long as there remains a variation of clinical judgment regarding the use of medical AI, arguing that differences of opinion render either course within the standard of care.

Health systems

Healthcare institutions could also face liability related to the use of medical AI. This could happen in one of two ways. First, institutions may face derivative liability for the negligence of providers who are their actual or apparent agents.[59] Thus in cases where a patient has a valid medical malpractice claim against a hospital employee relating to their use of medical AI, and the AI's use was within the scope of the employee's employment, both the institution and provider may be on the hook.

[52] Dhruv Khullar et al., *Public vs. Physician Views of Liability for Artificial Intelligence in Health Care*, 28 J. AM. MED. INFORMATICS ASSOC. 1574 (2021); Kevin Tobia, Aileen Nielsen & Alexander Stremitzer, *When Does Physician Use of AI Increase Liability?* 62 J. NUCL. MED. 17 (2021).

[53] W. Nicholson Price II, Sara Gerke & I. Glenn Cohen, *Potential Liability for Physicians Using Artificial Intelligence*, 322 JAMA 1765, 1765 (2019) [hereinafter Price, Gerke & Cohen, *Potential Liability*].

[54] I. Glenn Cohen, *Informed Consent and Medical Artificial Intelligence: What to Tell the Patient?*, 108 GEO. L.J. 1425, 1430 (2020); A. Michael Froomkin, Ian Kerr & Joelle Pineau, *When AIs Outperform Doctors: Confronting the Challenges of a Tort-Induced Over-Reliance on Machine Learning*, 61 ARIZ. L. REV. 33, 39–40 (2019).

[55] W. Nicholson Price II, Sara Gerke & I. Glenn Cohen, *Liability for Use of Artificial Intelligence in Medicine, in* RESEARCH HANDBOOK ON HEALTH, AI, AND THE LAW (Barry Solaiman & I. Glenn Cohen eds., 2023) [hereinafter Price, Gerke & Cohen, *Liability for Use of A.I.*]; Froomkin, Kerr & Pineau, *supra* note 56.

[56] Price, Cohen & Gerke, *Potential Liability*, *supra* note 53, at 1765.

[57] Price, Gerke & Cohen, *Liability for Use of A.I.*, *supra* note 55, at 5–7.

[58] Price, Gerke & Cohen, *Liability for Use of A.I.*, *supra* note 55, at 5.

[59] Price, Gerke & Cohen, *Liability for Use of A.I.*, *supra* note 55, at 8.

Second, hospitals and health systems owe various duties to their patients, the breach of which can result in direct liability for the institution.[60] These obligations may extend to institutions' use of medical AI. For instance, in most jurisdictions, institutions have a duty to exercise reasonable care in the selection or retention of its medical staff, and in the acquisition of equipment.[61] It is plausible that these duties might be extended by court to cover hospitals' selection and retention of medical AI systems, in which case institutions may be liable for negligently choosing, implementing, and using medical AI.[62]

Artificial intelligence liability insurance

As with more traditional forms of malpractice, liability insurance can mitigate the risks associated with medical AI.[63] This in turn can promote the adoption of such tools. However, malpractice plans and other healthcare liability insurance products will not necessarily cover the use of such new technologies.[64] Institutions and providers seeking to adopt medical AI systems should assess whether their malpractice coverage extends to these technologies, or whether additional AI liability insurance may be needed.

Alternatively, health systems and physicians may be able to shield themselves from liability via contract.[65] AI developers may agree to indemnify their products' users for harms resulting from the use of their systems and purchase AI liability insurance themselves.[66] This was the approach taken by the firm that developed IDx-Dr, the first autonomous AI diagnostic approved for marketing by FDA.[67] The American Medical Association has also advocated for this type of arrangement, on the basis that AI developers "are in the best position to manage issues of liability arising directly from system failure or misdiagnosis."[68]

[60] Price, Gerke & Cohen, *Liability for Use of A.I.*, *supra* note 55, at 10; W. Nicholson Price II, *Medical Malpractice and Black-Box Medicine*, *in* BIG DATA, HEALTH LAW, AND BIOETHICS 295, 303 (I. Glenn Cohen et al. eds., 2018) [hereinafter Price, *Medical Malpractice*].

[61] Price, Gerke & Cohen, *Liability for Use of A.I.*, *supra* note 53, at 10.

[62] Price, Gerke & Cohen, *Liability for Use of A.I.*, *supra* note 55, at 10; Price, *Medical Malpractice*, *supra* note 60, at 303.

[63] Ariel Dora Stern et al., *AI Insurance: How Liability Insurance Can Drive the Responsible Adoption of Artificial Intelligence in Health Care*, NEJM CATALYST, Apr. 2022, at 1, 1, 3.

[64] *See* Stern et al., *supra* note 63, at 5–6.

[65] Stern et al., *supra* note 63, at 3.

[66] Stern et al., *supra* note 63, at 3.

[67] Stern et al., *supra* note 63, at 3; Jessica Kim Cohen, *Liability Concerns May Pose Roadblock for Hospital AI*, MODERN HEALTHCARE (Oct. 26, 2019, 1:00 AM), https://www.modernhealthcare.com/technology/liability-concerns-may-pose-roadblock-hospital-ai.

[68] AM. MED. ASS'N, AUGMENTED INTELLIGENCE IN HEALTH CARE (2019), https://www.ama-assn.org/system/files/2019-08/ai-2018-board-policy-summary.pdf.

Informed consent

Outside of malpractice, medical providers may also face liability if they fail to secure patients' informed consent. Informed consent standards vary by jurisdiction: in states with a patient-based framework, physicians must disclose any information that a reasonable patient would find material, while in physician-based states, physicians are held to the standard of what a reasonable medical practitioner would disclose under similar circumstances.[69]

What remains unclear under either framework is whether providers must inform their patients that an AI shaped their treatment decisions. On the one hand, AI recommendations may be viewed as just another input into physicians' decision-making processes—analogous to reading a journal article or consulting a colleague.[70] Neither the physician- nor the patient-based standard obliges physicians to disclose the different steps in their reasoning processes.[71] Why would physicians' use of AI systems be any different?

On the other hand, research has consistently shown that AIs can be deeply biased and render less accurate results for certain populations.[72] As described above, AIs trained in high-resource settings may perform poorly in other contexts. Because AI systems are often trained on data that reflect racial disparities in health care, they also often render worse outcomes for patients of color.[73] As a result, some scholars contend that physicians have a duty not only to tell patients when their medical recommendations are influenced by AI, but also to explain that these technologies may be affected by such biases.[74] As it stands, whether providers have a duty to disclose their use of medical AI remains an open question, the answer to which could have a significant impact on how such tools are used.

Payment

Last but not least is the question of reimbursement: who will pay for the use of medical AI, and how? Reimbursement is crucial for the adoption of new medical technologies. The US government has taken some steps in establishing payment models for medical AI, with the Centers for Medicare & Medicaid Services approving reimbursement through two of its programs in recent years.[75] But many questions remain, particularly around reimbursement for the implementation and use of background monitoring and risk-prediction

[69] Cohen, *supra* note 54, at 1442–43; Khiara M. Bridges, *Artificial Equality: Race, Prenatal Care, and AI* (forthcoming; p. 32 of draft).

[70] Cohen, *supra* note 54, at 1442.

[71] Cohen, *supra* note 54, at 1442.

[72] Bridges, *supra* note 69, at 18; Sharona Hoffman & Andy Podgurski, *Artificial Intelligence and Discrimination in Health Care*, 19 YALE J. HEALTH POL'Y, L., & ETHICS 1 (2020).

[73] Bridges, *supra* note 69, at 7–9, 18.

[74] Bridges, *supra* note 69, at 1, 29.

[75] Melissa M. Chen, Lauren Parks Golding & Gregory N. Nicola, *Who Will Pay for AI?*, 3 RADIOLOGY: A.I. 1, 1 (2021).

systems, rather than single-event diagnostics or analytics. There is no standard reimbursement framework for medical AI.[76] Some costs may be recouped through increased quality and efficiency, but in the absence of reimbursement structures specifically tailored for the use of AI and its careful monitoring, additional barriers to best-practices adoption may arise, especially in small and low-resource institutions.[77]

Conclusion

The law regarding the use of medical AI is very much in flux. Particularly, at the midstream and downstream levels, there remain a number of unresolved legal questions, which could have a significant bearing on when and how medical AI systems are adopted and used in patient care. Medical providers should monitor the evolving legal landscape surrounding these technologies. Doing so will allow them to not only better protect themselves against unexpected liability, but also better understand the quality and design of the AI systems being used in their patients' care.

[76] Sendak, Price & Balu, *supra* note 18.

[77] Sendak, Price & Balu, *supra* note 18.

CHAPTER

47

Socioeconomic bias in applying artificial intelligence models to health care

Young J. Juhn[1], *Momin M. Malik*[2], *Euijung Ryu*[3], *Chung-Il Wi*[1] *and John D. Halamka*[4]

[1]Department of Pediatric and Adolescent Medicine, Mayo Clinic, Rochester, MN, United States [2]The Center for Digital Health, Mayo Clinic, Rochester, MN, United States [3]Department of Quantitative Health Sciences, Mayo Clinic, Rochester, MN, United States [4]Mayo Clinic Platform, Mayo Clinic, Rochester, MN, United States

The excitement about the potential benefits of artificial intelligence (AI) models in health care has come with accompanying concerns about the potential for these models to exacerbate health care inequities. Indeed, there is an urgent need for national guidelines to address AI bias and inequities in health care. Such bias can take various forms and falls into several distinct categories, including bias related to race, ethnic group, gender, and socioeconomic status (SES); these inequities are impacting millions of lives. Long-standing concerns about the potential of AI models to perpetuate social problems ("compounding injustice") [1] like long-standing health care inequities [2] were dramatically demonstrated in the now ubiquitously cited explicit example of Obermeyer et al. [3]. They analyzed a large, commercially available dataset used to determine which patients have complex health needs and require priority attention. In conjunction with a large academic hospital, the investigators identified 43,539 white and 6,059 Black primary care patients who were part of risk-based contracts. The analysis revealed that at any given risk score Black patients were considerably sicker than white patients based on the number and severity of chronic diseases. However, the commercial dataset did not recognize the greater disease burden in Black patients because it was designed to assign risk scores based on total health care costs accrued in one year. In this case, the bias took the form of *differential accuracy*, where calculated risk scores of Black

Artificial Intelligence in Clinical Practice
DOI: https://doi.org/10.1016/B978-0-443-15688-5.00044-9

patients reflected the severity of health conditions less accurately than those of white patients. They found that this disparity could be explained by how risk scores were calculated not based on direct (but more difficult to gather) measures of health, but on the more easily available proxy of *total annual health care costs* per patient. This proxy captured something about health, but it also captured systemic underinvestment in the health care of Black patients [2,4]. In this way, the compounded interpersonal and institutional biases that Black patients face [5–7] were exacerbated, not alleviated, by the use of data and AI.

This was one dramatic, explicit example, but there are others, potentially with different types of and/or mechanisms behind different types of bias. Gender bias has been documented in medical imaging datasets used to train and test AI systems used for computer-assisted diagnosis. Larrazabal et al. [8] studied the performance of deep neural networks used to diagnose 14 thoracic diseases using X-rays. When they compared gender-imbalanced datasets with datasets in which male and female candidates were equally represented, they found that "with a 25%/75% imbalance ratio, the average performance across all diseases in the minority class is significantly lower than a model trained with a perfectly balanced dataset." Here, for a similar type of bias (differential accuracy), the mechanism was more technical in nature: their analysis concluded that datasets that underrepresent one gender result in biased classifiers, which in turn may lead to greater misclassification of pathology in the minority group. The classifier for one group being underpowered compared to the classifier for the other group can be addressed by adequately representative data. Still, the historic underrepresentation of women in clinical trials and analyses and the ways in which women are less likely to receive high-quality care and are more likely to die when they receive suboptimal care [9] may show up at multiple levels of data used to develop AI.

In a systematic look at models to help clinicians identify patients most likely to be readmitted within 30 days, Johns Hopkins University investigators [10] examined the four most commonly used models (all were validated, evaluated, and applied in at least two external settings) and found widespread problems. The researchers developed an 11-question checklist for AI model evaluation, such as, were the parameters being evaluated by the readmission model realistic proxies for patients' health care outcomes or needs?; were any important features left out of each model?; and, were validation studies conducted to evaluate differences among various subpopulations? While some models passed some of the 11 items, they found that *none* of the models considered label bias, measurement bias from incompleteness, or "model bias" (a large category including aggregation bias, errors from models lacking appropriate control variables, and not considering the true target is what *action* would cause a desired change rather than just what inputs correlate [11]). One problem is that, similar to the case investigated by Obermeyer et al., these models all also had health care *utilization* as their modeling target, rather than some more direct (but harder to gather) measure of health care *needs*. These problems, they argue, could lead to bias in model performance between Black and white patients, and between patients of lower and higher socioeconomic status (they did not define or measure socioeconomic status but made the argument by citing studies using some of the proxies and indexes that we will discuss below). As they pointed out, these tools are currently being used "to direct care to high-readmission-risk patients, standardize readmissions-based quality metrics across hospitals, and forecast all-cause and condition-specific readmissions."

One of the team's 11 items is population bias, where the data do not represent the population; while they found that the four models they examined found limited potential for concern, this is a widespread problem elsewhere and describes the problem studied by Larrazabal et al. There is wider evidence to suggest that machine learning models that rely on electronic health record data in general under-represent patients in lower socioeconomic groups [12], which would then lead to parallel problems as those around gender bias.

Attempts to comprehensively understand types and sources of bias are only beginning. Inequities from the use of AI are already impacting millions of lives and will potentially impact billions more if and as the use of AI expands [13]. If we choose to continue to develop and deploy AI in hopes of unlocking its full promised benefits, it is essential that we also study the sources, dimensions, and extent of differential performance of AI systems and take steps to mitigate disparities in performance that will exacerbate health care inequities. At the same time, others hope that careful and proper implementation of AI models could potentially redress health disparities [14].

In this chapter, we focus on bias along the dimension of SES. This is a construct representing one's relative position within society and one's ability to access desired resources [15], and disadvantages that come from being of low socioeconomic status act alongside processes like racism and gender bias because SES is a key element of the social determinants of health (SDH). We first review how to conceptualize and define AI bias. Next, we discuss the theoretical role of SES alongside the practicalities of how to measure it, a prerequisite to incorporating analysis of SES in AI development. Then, we present our work looking at how measures of SES indeed reveal bias in the performance of AI models and exploring what the underlying mechanisms might be. Lastly, we look at possibilities for mitigation and future strategies.

Artificial intelligence bias

Today, "AI" refers almost exclusively to *statistical machine learning*, and we adopt this usage. Statistical machine learning, in turn, was conceptualized by statistician Leo Breiman in 2001 [16] as fundamentally a style of statistical modeling, distinguished from traditional statistics in not being concerned with modeling the data-generating process but rather only with making reliable input-output mappings, which Baiocchi and Radu propose calling *outcome reasoning* [17]. So long as new data comes from the same process as previously observed data, higher quality input-output mappings are usually achieved by highly flexible nonparametric and/or ensemble models, which are also much more applicable to high-dimensional data. Such models have a weakness in not necessarily corresponding to causal process and, in fact, potentially disguising causal mechanisms behind model success that can explain how and why these models fail. One consequence is that, unlike traditional statistical modeling, machine learning is not concerned with finding appropriate control variables since the goal is not to make a model that accurately captures underlying population parameters. While models that are poor reflections of the world may be effective at minimizing loss over a population, this success may come from prioritizing accuracy in "easy to model" individuals and at the cost of larger errors in "harder to model" individuals, which can be at odds with the goals of equity.

While there are many examples of potential or observed undesirable outcomes that are referred to as "AI bias," defining this more precisely (and identifying the source of problems) proves to be tricky. Many proposed definitions of AI model bias remain vague, and when used without definition, the word is used in ways that often unhelpfully collapse multiple unrelated phenomena. As Danks and London [18] point out, "bias" is an overloaded word, used in vastly different ways across fields. For example, *interpersonal* bias is a microscale psychological process; *institutional* bias is a macro-scale structural process; and measurement bias and *sampling* bias are technical concepts resulting from the specifics of a process of data collection. They also point out that, insofar as AI models are fundamentally statistical estimators, perhaps the most natural meaning for AI bias would be *estimator* bias, the statistical idea of whether or not a modeling process is able to, across multiple hypothetical realizations and with no problems with the data, on average have zero difference with what is considered as underlying "true" parameter values that causally link the inputs to the output. However, this is almost certainly not what most people mean by AI bias; and furthermore, AI models are by this definition usually "biased," as most AI models sacrifice the property of being asymptotically unbiased for having lower variance, which (based on how performance is defined) can result in overall superior performance. Sampling bias and measurement bias are analogs of estimator bias in terms of imagining an underlying "truth," but around input data rather than of the mathematical functions that process those data. For sampling bias, we imagine a population parameter as being the "true" underlying parameter of interest, and if we do not have a representative sample, then descriptive statistics from the sample and the parameters of any model will be biased estimators of population parameters. Measurement bias similarly assumes a "true" underlying value for quantity of interest (which is often metaphysical and subjective), and the way that measurements systematically deviate from this in nonrandom ways are known as measurement bias.

Instead, we hold that what most people mean by "AI model bias" is none of the above notions of bias but rather *differential model performance*, where the AI model performs differently on different subgroups. Specifically, the bias is of a form that gives further benefits to already-privileged subgroups and/or places further burdens on already marginalized subgroups, like how the system examined by Obermeyer et al. made it easier for white patients than for Black patients to be recommended into or defaulted into programs providing additional support. The *sources* of these differential performances may related to biases discussed above, [19–21] but differential performance is the standard that ultimately leads us to say that some AI model is biased.

Under this definition, the notion of AI bias ties into philosophical debates about how we should define fairness: for example, as equality of opportunity or equality of outcomes or some other criteria. One persistent problem is that the unjust status quo means that acting on what is empirically accurate can be unfair. Legal theorist Deborah Hellman [22] gives the example of how in the late 1980s, US health insurance companies began charging the victims of domestic violence more (higher premiums) for health insurance or denying coverage altogether. Such people had empirically higher health care costs and higher mortality (especially if they tried to leave their abusers), but Hellman argues that to make use of this empirical fact is to "pile on" further punishment to those who are already "victims of injustice and cruelty," and should not be done regardless of its accuracy. Indeed, many social movements have opposed the use of data to further individualize risk, instead advocating for collectivizing risk through social programs [23].

This is not to say that forms of bias other than differential performance are not relevant for AI model development [19,24]. [19–21] Sampling bias, for example, is a major source of differential model performance as described above [8], insofar as the generalizability of an AI model depends entirely on the representativeness of the sample on which it is fit, and the performance of an AI model on one subgroup in a population may depend on whether the sample size for that group is sufficiently large. Personal biases of developers and modelers in choosing what to optimize (such as in the choice of health care costs as a proxy for health described above) and what is important, can similarly lead to differential performance. If AI model developers are from more privileged social backgrounds, they may not appreciate the number of unmeasured factors that are key for correctly modeling a phenomenon or for people to adhere to recommended/"optimal" courses of action (e.g., transportation or housing insecurity making it difficult to come to appointments). But even in these cases, the ultimate concern is that AI models may not perform as well on those from marginalized populations, potentially excluding them from beneficial programs and treatment or subjecting them to suboptimal care.

Given the ways in which those of low SES are already marginalized, using models with such biases in their performance would further health care disparities ("compounding injustice") [1]. If a sample does not have a representative count of those of lower SES, the covariance with the target variable will have a noisier estimate, leading to worse performance (whether SES is included in a model directly or not). Lower SES may make interventions more likely to fail, such that if AI models are optimized to whether interventions are likely to succeed or not (versus if they are *needed* or not), those of lower SES may receive less resources. Unmeasured processes leading to differential data quality along those of lower SES may result in differential model performance; but this will be impossible to see if we do not measure SES or capture inaccurate SES.

Socioeconomic status as a key social determinant of health

Decades of research have firmly established how variation in social context is often as or more impactful on health than variation in biological factors alone, and that these social variations are more associated with health inequities [25–39]. These are known as social determinants of health (SDH): health effects from the "conditions in which people are born, grow, live, work and age" [40]. This covers community context, economic stability, education access and quality, health care access and quality, and neighborhood and built environment [41] as well as the structural forces that cause differences in these social factors that lead to worse health outcomes for certain individuals and groups. A 2017 report from the National Academies of Medicine (NAM), *Communities in Action: Pathways to Health Equity*, conceptualizes the pathways of SDH to include differential access to health care resources, differential health knowledge, and differential literacy and behavior [41].

SES is a closely related construct from social science, referring generally to relative social position and how society is stratified into varying levels of privilege and disadvantage. Conceptually, Oakes and Rossi define it as one's ability to access desired resources (human, materialistic, and social) [15]. Practically, SES encompasses economic factors but also human and social factors like "amount and kind of education, type and prestige of occupation, place of residence, social network and—in some societies or parts of society—ethnic origin or religious background" [42]. However, having a measure that effectively captures this construct is difficult, a theme we will return to later.

SES does not necessarily include race/ethnicity, which the US Preventive Services Task Force defines as a social construct and not biological fact [43]; that is, being *perceived* and *treated* as, for example, Black rather than anything related to genetics or phenotype is what leads to being affected by systemic racism and suffering negative health outcomes (nor do perceptions of and treatments as a given race/ethnicity map consistently onto genetics or phenotype) [44,45]. In various frameworks for SDH, race (or rather, rac*ism*) may or may not be included. For example, Dahlgren-Whitehead recently argued racism should not be included as a determinant of health but rather be conceptualized as an important "driving force" influencing almost all determinants of health in the model and driving the social and ethnic patterning of determinants [29]. Overall, the literature suggests that the associations of SES with access to health care resources exist across all groups of race/ethnicity, from preventive screening services to implantation and cancer therapy, meaning that we can consider these factors as separate but interacting [46–49]. That is, it is not sufficient to consider race alone. The difference is that self-reported race is usually a readily available measure (even if the more relevant construct might be the *racialization* of how others differentially perceive and treat an individual; and even if accepted categories like "Asian" may be too coarse to capture, for example, vastly different social situations and health outcomes based on different circumstances of migration from different countries of origin in different periods of history, like between Southeast Asian refugee populations resettled in deprived areas in the 1970s and recent South and East Asian highly-educated professionals). In contrast, often even poor proxies for SES are not readily available.

The most straightforward proxy for SES is income, even though SES is explicitly broader than income. Income is generally not available in electronic health records (EHR), but when collected, even this one dimension shows stark disparities. Chetty et al. assessed life-expectancy of 40-year-old men and women in the United States by household income between 2001 and 2014 [31], and found 10.1 years for women and 15 years for men in differences in life expectancy between populations in the top and bottom 1 percentiles. During their study period, populations in the top 5 percentile gained an extra 2.3–2.9 years of life expectancy while those in bottom 5 percentile gained only 0.3–0.04 years of life expectancy. This life-expectancy gap was worsened during the COVID-19 pandemic among populations at different percentile incomes [50]. Cross-sectional analyses using the 2008 National Health Interview Survey ($n = 17{,}337$ nonelderly adults) showed that compared to adults with higher income, those with lower income had significantly lower health care access for a broad range of preventive and medical care services including medical homes (odds ratio [OR] = 0.41–0.5), health maintenance visits (OR = 0.55–0.70), dental visits (OR = 0.32–0.43), colorectal screening in the past 2 years (OR = 0.56–0.60), Pap smear screening in the past 3 years (OR = 0.44–0.57), mammogram screening in the past 2 years (OR = 0.35–0.47), blood pressure tests (OR = 0.44–0.56), and cholesterol tests (OR = 0.35–0.48), with further variation explained by race/ethnicity [46]. A 2019 report based on a cohort study comprised of 85,000 participants in 12 states of the United States between 2002 and 2012 showed that, while colonoscopies are significantly associated with reduced risk of colorectal cancer and mortality, endoscopy screening rates were significantly lower among those with household incomes below $15,000 compared to those with household incomes over $50,000 (about 30% vs. 55%, respectively) [51]. These study results suggest that disparities in access to preventive services for cancer remain unchanged over time.

While SES is comparatively fixed (changes in SES are theorized in terms of social mobility), the relationship between SES and health impacts is much more variable, with public health policy playing a major role. An analysis from Gorey et al. found that income was directly associated with lymph node evaluation, chemotherapy, and survival among patients residing in San Francisco but not among patients in Toronto. Similarly, high-income persons had better survival rates in San Francisco compared to low-income persons, but this same difference was not present in Toronto [47]. The difference is the Canadian health care system makes the same quality of health care accessible regardless of ability to pay, removing much of the differential impact of SES on health outcomes.

Education is another major component of SES that, once collected, is easy to include in analysis and explains more than income alone. Braveman et al. conducted a national study on five child health indicators (infant mortality, health status, activity limitation, healthy eating, sedentary adolescents) and six adult health indicators (life expectancy, health status, activity limitation, heart disease, diabetes, obesity) by both income and education [39]. They found that those with the lowest income and who were least educated were consistently the least healthy. They suggested that health in the United States is patterned strongly along both socioeconomic and racial/ethnic lines, arguing for policies prioritizing those groups and for routine health reporting to examine various socioeconomic and racial/ethnic disparity patterns and their overlaps [39].

Just as policies can alleviate the mechanisms that lead from SES to health disparities, they can also magnify these mechanisms. If performance metrics of health care organizations (HCOs) used for accountability applications (e.g., public report and payment) do not factor in the risks from SDH, accountability measures can lead to feedback loops of not giving adequate resources to HCOs serving riskier patients. A recent study assessed 30-day preventable hospital readmission rates among 43 Children's Hospitals in the United States adjusting for race, ethnicity, the payer, and compared it before and after adjusting for median household income of the patient's postal code (a proxy discussed further below) [52]. About 12% of Children's Hospitals had changes in ranking for a penalty threshold after adjusting for ZIP code-level median household income [52]. However, the proxy of the median household income of ZIP code area does not capture individual-level variability, and thus this finding is likely an under-estimate the impact of SES on the study outcome. Similar findings have been reported on Medicare beneficiaries for adults [53]. For these reasons, the State of Massachusetts began including SDH in the payment formula for Medicaid managed care organizations in 2016 [54]. Given the significant implications of performance measurement of HCOs on accountability, both the NAM and the National Quality Forum (NQF) recently issued their policy positions recommending adjustment of performance measures of HCOs by social risk factors [55,56]. While both NAM and NQF suggested that data on social risk factors should be as granular as possible to ensure accuracy, they highlighted the *lack of* a *suitable measure for individual-level SDH* as the key challenge for accurate risk-adjustment of HCOs' performance to stakeholders including payers, government, HCOs, and patients.

Measuring socioeconomic status

Going beyond lone aspects of SES like income or education have proven difficult, with little work in health care to develop or validate new measures [15]. Largely, either lone aspects are used, or aggregate-level indexes are imported from sociology and demography. While a full review of available measures is beyond the scope of this book chapter [15,57–64] one important distinction is between individual-level measures of SES and area-level aggregate measures of SES.

Area-level aggregate measures

Among the foremost challenges of individual-level data is simply collecting it. However, the US Census readily provides rich data aggregated at the level of geographic units, and so has been the source of sociological and demographic research on SES, particularly in the Area Deprivation Index [65] and Social Vulnerability Index [66]. These are dimension reductions of various properties of geographic units, are widely used in health research as well, and correlate with health outcomes [52,54,67–69]. But they have several limitations, much more pronounced for health research than for other disciplines.

First, neighborhood environment or larger geographic environment is its own epidemiological construct which is distinct from individual-level socioeconomic environment, as shown in studies employing multilevel analysis to distinguish between the two [70–75]. For example, our previous study demonstrated that children residing in Census tracts with high traffic volume had higher incidence of asthma compared to those living in Census tracts without high traffic volume, and that this was independent of a measure of one aspect of SES, maternal educational level [70]. That is, traffic volume at the Census tract level influences the risk of childhood asthma, but this does not capture all the effects of individual-level SES, as there can be substantial variability between individuals living in the same Census tracts. Then, geographic access to specialists is an important factor for health outcomes (diagnosis and management of chronic diseases) but does not account for a potentially large amount of variability in access for the people who live in different parts of a geographic unit, for example, differential access to transportation [76–78]. Area-level and individual-level processes are conceptually distinct and will suggest entirely different interventions (e.g., an area-level intervention would be to incentivize specialists to practice in under-served areas, whereas an individual-level intervention would be to improve an individual's access to transportation), which makes it essential to distinguish the extent to which health outcomes are driven by area-level processes or by individual-level processes.

Second, area-level SES measures collapse individual-level heterogeneity, which leads to substantial inaccuracy in assigning individual-level SES. In a case where both individual-level (education and income) and area-level SES measures were available, we assessed discordance between the two; using the area-level measures resulted in a misclassification rate of quartiles of income and education of 20%–35% (Cohen's weighted kappa [κ_w] = 0.15–0.22) in Olmsted County, Minnesota, and Kansas City, Missouri [79,80]. A previous study compared 1999 income data from the National Birth Defects Prevention Study with 2000 Census block group income data for the residence location of these same mothers [81], finding only poor to fair concordance between the two data sources (control mothers, $\kappa_w = 0.28$; 95% CI = 0.19–0.37 and case mothers, $\kappa_w = 0.18$; 95% CI = 0.13–0.24). The authors suggested that "caution should be used if block-level data is used as a proxy for individual-level household incomes in population-based birth defects surveillance and research" [81].

Third, there is also the danger of the *ecological fallacy*: [65] conclusions made at the aggregate level do not necessarily hold at the individual level. For example, one aggregate analysis found that lung cancer rates are *negatively* correlated with average indoor radon concentration on a per-county analysis, but "a protective effect from breathing radioactive gas seems unlikely" [82]. This can be due to individual-level variability, how aggregation (which requires the choice of summary statistics) can disguise nonlinearities, or the lack of availability of key controls at the aggregate level, but all with the effect that using area-level aggregate measures of SES may significantly underestimate the impact of SES on health outcomes.

Lastly, much of area-level data also comes with sampling error: American Community Survey data, commonly used as sources of neighborhood-level SES, uses a random sample of about 1%/year [83], and now that the US Census Bureau has adopted differential privacy for the 2020 Census, even decennial Census figures contain area-level noise.

Individual-level measures

Despite the limitations of area-level aggregate ways of measuring SES, attempts at getting individual-level measures (beyond simple proxies of only one dimension of SES like income or education) are even more difficult. Simple (unidimensional) measures, such as income, wealth, education, and occupation [63], are relatively easy to gather but do not capture the multidimensional nature of SES. Composite SES measure such as the Hollingshead and Nakao-Treas indices [59–61], which attempt to capture multiple dimensions of SES, often do not have good validity properties. For example, the CAPSES index from Oakes et al. made up of measures of material capital, human capital, and social capital, did not outperform simple measures at higher criterion-related validity [15], which is where a measure of a construct is considered valid if it correlates with other quantities that the construct is theorized to cause.

In a systematic review of the use of measures related to SES, Braverman et al. [84] found that in most studies, greater wealth was associated with better health, even after adjusting for other SES measures. The findings appeared most consistent when using detailed wealth measures on specific assets and debts rather than a single question, for example, about income. Adjusting for wealth also generally decreased, but did not eliminate, observed racial/ethnic disparities in health. Thus *simple measures of wealth may be the most effective default for an individual-level measure* [85].

However, wealth is not the only contributor to health outcomes, as shown by associations of education with health outcomes after controlling for income, so an ideal individual-level index for SES would go beyond wealth, but there are challenges to achieving these. From a measurement theory perspective, both indexes for wealth and composite indexes have problems related to: the lack of precision and reliability of measures; the difficulty of collecting individual-level data for patients; the dynamic nature of SES over a lifetime; the classification of women, children, retired, and unemployed persons, whose access to resources may not be reflected in their individual wealth; inaccurate or misleading interpretations of study results; and, particularly important for constructing a single-number index that combines several dimensions to represent SES, a lack of or poor correlation between individual SES measures [63]. For example, at the level of individuals, there are only weak correlations between measures of traditional components of SES measures, such as education, income, and occupation ($r = 0.33$ for education and income, $r = 0.40$ for occupation and income, and $r = 0.61$ for occupation and education) [62].

Data availability

The other major barrier to incorporating individual-level measures of SES into health research is that income, or wealth, is not a part of EHR. Thus, many studies seeking measures of SES will link a patient's provided address to area-level SES measures or use payer data (specifically, Medicaid and Medicare eligibility or enrollment, including dual eligibility) as a proxy for SES. This is because, of the 17 SDH factors included in the National Academy of Medicine report [55], only 3 variables (dual eligibility [Medicaid and Medicare], nativity, and urbanity/rurality) were readily and consistently available in EHR. Medicaid eligibility as a binary variable lacks granularity and still disguises significant within-group heterogeneity of SES similar to area-level SES measures. In addition, change of payer information is frequent enough to make it difficult to reliably use insurance information as an SES measure, and the state policy for Medicaid eligibility varies significantly (i.e., expansion of Medicaid eligibility under the Affordable Care Act varies among the states) [86].

For these reasons, SES is often understandably but unfortunately passed over. A previous study assessing the reporting of baseline SES in reports of randomized controlled trials of stroke and transient ischemic attack published in 12 major journals in the disciplines of general medicine, general neurology, cerebrovascular disease, and rehabilitation [87] found that *only 12% of studies* reported any SES measure. Journal categories did not differ in rate of SES reporting, and SES reporting did not increase over time [87].

The HOUsing-based SocioEconomic Status index

Our research has sought to address this national challenge by developing and applying a measure of SES suitable for assessing and monitoring health equity in health care settings. HOUsing-based SocioEconomic Status (HOUSES) index has several desirable properties [88]: it is an index that is individual level; it does not require patient contact (which is burdensome and can result in nonresponse bias, especially since a survey study found 67% of patients uncomfortable with disclosing household income, 40% uncomfortable with disclosing sexual orientation, and 38% uncomfortable with disclosing educational background [89]); it does not rely on self-report (which can suffer from recall bias and social desirability bias even among those who do respond); it is based on a measure that plausibly captures multiple dimensions of the underlying construct of SES (housing); it is scalable and cost-effective; and it is dynamic, allowing us to capture changes in SES over time.

HOUSES is a single number, resulting from a standardized (z-score) dimension reduction from four real property data (number of bedrooms, number of bathrooms, square footage of the unit, and estimated value of the unit) from the county Assessor's office [88]. The theoretical justification is that an individual's housing features reflect one's current household wealth and income status in a way that also overcomes the limitations of using income alone (since people without a formal income but living in high-value housing will likely have access to resources). They even reflect past (e.g., credit) and future potential household wealth (e.g., occupation) that are evaluated by mortgage companies and lenders when purchasing a house. More directly, housing is a crucial site in the day-to-day life of most individuals that reflects the distribution of wealth and income, control over life circumstances, and access to social resources. Housing is an important factor in processes of social identity formation and the establishment and maintenance of social relationships [90]. Housing markets are especially significant in the distribution of wealth [91–93], and housing circumstance is crucial in the production and reproduction of social identity and social status [94–99]. Furthermore, the number of bedrooms and housing values is proportional to the income of families of occupied units [100]. As the number of units for apartment buildings can be captured, HOUSES can still be calculated for those living in apartment buildings, by which these four factors also capture dimensions of housing relevant to SES like tenure, renting versus owning, etc., not explicitly included in the index.

For obtaining one's HOUSES index, their address at the time of index date of interest (e.g., incidence date of disease) is geocoded to link real property data of the housing unit. Each property item corresponding to the individual's address is standardized into z-scores and are aggregated into an overall z-score for the four items such that a higher HOUSES score indicates higher SES. HOUSES is standardized within the county based on available real property data at a given year, as real property data is ascertained and updated from the county (or city) Assessor's office on a regular basis. Then, z-scores can be converted to groups or categories of interests for HOUSES by quartile, decile, percentile, etc.

The HOUSES index has shown strong validity by criteria used in psychometrics. Criterion validity that HOUSES indeed measures SES is shown by moderate to good correlations with other measures of SES including education, income, Hollingshead Index (HS), and Nakao-Treas Index (NT) in Olmsted County, MN ($R = 0.29–0.54$, $p < .001$) and Jackson County, MO ($r = 0.39–0.59$, $p < .001$), respectively [88]. Since Jackson County, MO—an urban setting encompassing Kansas City—is much more socioeconomically diverse than Olmsted County, MN (e.g., 75.9% vs. 88.7% white, 17.5% vs. 6.7% without any college education, and 7.8% vs. 1.8% with a family income less than $24,999, respectively), the HOUSES index demonstrated its applicability to both urban and mixed urban-rural settings.

Even more importantly than concordance with other SES indexes, given their limitations, HOUSES has demonstrated stronger construct validity in correlating more than other SES measures [88,101–104] to a broad range of health behaviors (e.g., smoking exposure) and outcomes (e.g., obesity, preterm birth, and general health) theorized to relate to SES and correlating with other unidimensional aspects of SES, as well as with poorly controlled childhood asthma by Asthma Control Test (ACT) in Sioux Falls, South Dakota [88,101–110]. HOUSES correlates with incidence of graft failure among kidney transplant recipients, while both alternative operationalizations of individual-level SES (educational level) and area-level aggregate SES measure *failed to* correlate with incidence of graft failure [111].

Since the original validation of the HOUSES index, it has been further shown to correlate with a broad range of health outcomes for both adults and children (about 40 different health outcomes), which are known to be inversely associated with SES, including: acute conditions (e.g., myocardial infarction [101], all-cause hospitalizations and readmission [105,112], accidental falls [106], and critical care outcomes [113]); chronic conditions (e.g., rheumatoid arthritis, coronary heart disease, asthma, mood disorder, hypertension, diabetes [102,108], Vit D status [114], transplantation outcomes like postkidney transplantation graft failure [111], and multiple complex chronic conditions [115]); behavioral health (e.g., smoking status [107], advance care planning [116]); childhood conditions (e.g., adverse self-rated health [109], poorly controlled asthma per the ACT score [103], invasive pneumococcal disease [104], pertussis or human papillomavirus vaccine compliance [110,117], and mood disorder [118]); [118,115] and mortality [102,119].

The HOUSES Cloud system through which HOUSES index is provided to users enables a scalable solution for formulating and managing HOUSES index data in the *entire United States* that is compliant with relevant data privacy regulations (e.g., Health Insurance Portability and Accountability Act [HIPAA]), and is typically available for more than 95% of study populations. Through this cloud-based system, HOUSES index data can be automatically formulated and provided to users who upload relevant parcel data (e.g., address information), thereby preserving data privacy. All data in transit and at rest is encrypted. The secure, privacy-protecting cloud-based platform hosts large publicly available real property data which is the basis for real property tax assessments available in all counties of the United States. It formulates the HOUSES index by linking patient's address information from various datasets to the stored HOUSES index in the cloud system.

Considering that the median duration of residence in the United States was only 4.7 years and 1.9 years for people aged 25–34 [120], HOUSES can capture changes in individual SES over time. Indeed, housing consumption often reflects longitudinal change in SES. For example, changes in homeownership or a move up or down in housing amenities are commonly associated with altered social position [90]. It also covers traditionally vulnerable subpopulations, such as racial minorities, rural populations, and women, children, retired, and unemployed persons, although we acknowledge that greater bias and/or variance in public assessment of housing value for such populations may affect the estimate of HOUSES [121,122].

Being formulated at the population level introduces the high initial burden of needing to carry out the dimension reduction methodology for each new area of application, but standardizing one's SES in the community context enables fair comparisons of study results across different geographic areas with different community wealth. For example, an annual income of $100,000 may reflect a different level of SES in Olmsted County, MN (median household income: $83,070 in 2021), compared to Loudoun county, VA (median household income: $153,506 in 2021) [123]. As standardizing one's other SES measures (e.g., income) to enable fair comparisons across different geographic regions requires data collection from all community residents, it may be logistically challenging for other SES measures, if not infeasible, unlike for the HOUSES index which is normalized within a given county as tax jurisdiction area.

Socioeconomic status and artificial intelligence bias

Just as racial bias and gender bias in health care is mirrored by AI models exhibiting racial and gender bias, so too do we expect the ways in which lower SES, as a social determinant of health, leads to worse health outcomes, to be mirrored in the performance of AI models. Having a suitable measure of SES in HOUSES, we can examine differential performance of AI models along this metric; and having identified a dimension along which AI models may be biased, we can also examine further to see what underlying process may explain how lower SES is associated with AI bias.

In recent work we applied the HOUSES index to "audit" the fairness of two AI models for anticipating asthma exacerbation (AE) episodes for children with asthma within 1 year [124]. This was a core component of an AI-assisted clinical decision support system named A-GPS (Asthma-Guidance and Prediction System). A-GPS is a tool that summarizes asthma-related information (both known risk factors and a history of asthma-related visits) extracted from a patient's EHR alongside a prediction score for AE based on a naïve Bayes (NB) model and gradient boosting machine (GBM) model fitted to historic cases of AE with an overall area under the ROC curve of 0.78 for NB and 0.74 for GBM [125].

To assess potential socioeconomic bias in the performance of these models, we "audited" the model by comparing the model's performance on children with persistent asthma who received routine medical care at Mayo Clinic in different quartiles of the HOUSES index [124]. Auditing may happen for the same metric used for model success (e.g., AUROC, accuracy, precision, recall), or there may be other metrics used for the purpose of determining model fairness [126]. There are forms of "impossibility theorems" that show that parity within certain sets of metrics cannot be simultaneously achieved, like equal positive/negative predictive values, and equal false positive/negative rates, unless we have a perfect classifier or unless subgroups have equal base rates, both of which in practice never happen. In particular, marginalized groups often have a different base rate (e.g., higher incidence of AE).

Using a suggestion from the fairness literature, for an audit metric, we used balanced error rate (BER), defined as the unweighted average of two error rates (false positive rate and false negative rate), as the primary fairness metric, while also considering other measures, such as accuracy (representing overall agreement), sensitivity (representing equal opportunity), and positive predictive value (PPV or precision, representing predictive parity). Note that some recent work has proposed a more principled system for choosing a single audit metric, which we plan to apply in the future [127]. For each fairness metric, the ratio comparing the least privileged group (i.e., HOUSES Q1: the lowest SES) with all others (HOUSES Q2–Q4) was calculated. A ratio of 1 means there is no difference, with a ratio greater than 1 implying inferior model performance in those with low SES for BER, and a ratio less than 1 implying inferior model performance for accuracy, sensitivity, and PPV.

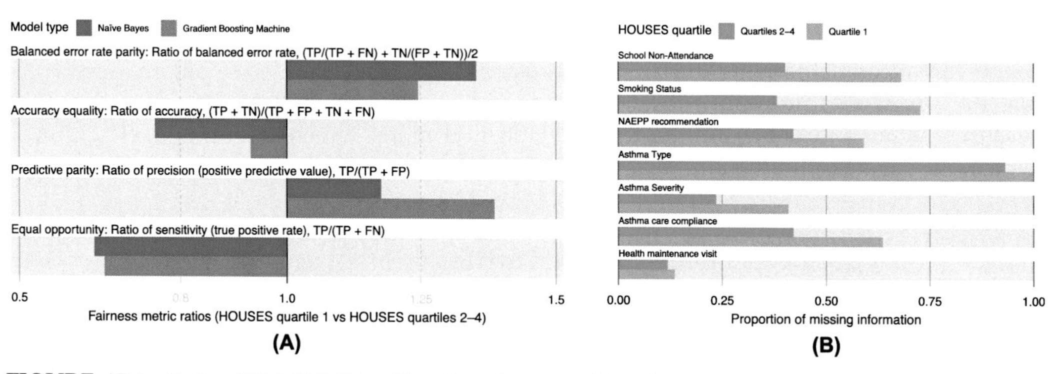

FIGURE 47.1 Ratio of HOUSES Q1 to Q2–Q4 performance (A) and missingness of features in EHRs (B).

In the study cohort, roughly 20% of subjects had low SES by HOUSES index (Q1). Subjects with low SES tend to have higher frequency of AE compared to those with higher SES, which aligns with epidemiological findings in asthma literature demonstrating the importance of considering SES in asthma outcomes. For all fairness metrics considered except PPV, performance of both NB and GBM models was inferior in subjects with low SES (e.g., ratio of BER was greater than 1; Fig. 47.1A).

While we did this audit manually, there are a number of tools that do so automatically, including testing with concept activation vectors, Audit-AI, and AI Fairness 360. In addition, Mayo Clinic Platform recently launched Validate, a digital product to measure model sensitivity, specificity, area under the curve, and bias by subgroups of race, gender, and other available sociodemographic variables [128]. In the future, it will be increasingly easier to carry out such basic audits, although finding the source of disparities will still require further investigation.

We hypothesized that lower SES corresponds to lower levels of health care access, which would affect data in producing lower-quality data within the EHR of patients of lower SES [124]. To investigate this, we compared the availability of variables relevant for asthma care like asthma severity. In addition to clinical importance, the availability of these features also affects performance of ML models as these features are used for model building. Fig. 47.1B shows how children with low SES tend to have higher frequency of missing data for these features compared to those with higher SES.

Lower SES might also result in measurement bias in the variable of asthma itself. We compared the frequency of subjects with recurrent asthma symptoms documented in EHR who fulfilled asthma criteria, but without a diagnosis of asthma, indicating undiagnosed asthma [124]. Specifically, we classified children as having undiagnosed asthma if they met Predetermined Asthma Criteria but did not have a physician diagnosis of asthma. In the study (training) cohort, roughly 12% of subjects with low SES had undiagnosed asthma compared to 9.8% with higher SES. Given that a physician diagnosis is often required to have an adequate asthma care plan, higher frequency of undiagnosed asthma in subjects with low SES is aligned with higher frequency of missing data relevant to asthma care, which in turn affects model performance.

This supports our hypothesized mechanism: disparities correlate with the quality of signals, resulting in models that perform worse for those experiencing disparities. If variables representing those disparities are explicitly included, it may serve as a control, but equally likely (if many of those of lower SES are incorrectly labeled as not having asthma) is an AI model will assign a negative weight to them for the estimated probability of a negative outcome, potentially leading to health care resources being withheld.

Mitigation

There are a number of proposals and guidance for model-based mitigation of differential model performance [14,21,129–131]. Many of these are agnostic to the sources of model bias, focusing instead on producing post hoc corrections to equalize performance and/or audit metrics across subgroups. In a follow-up to the original work documenting the bias by SES, we explored some existing tools for mitigation but found them ineffective.

Although both NB and GBM models were assessed in our prior work for developing and testing Asthma-Guidance and Prediction System (A-GPS) [125], we considered only a GBM model in this exercise as it included socio-demographic (including HOUSES) as influential features along with asthma-related features, unlike the NB model. The GBM model had asthma symptom frequency (relative influence [RI] = 19.7), previous AE counts (RI = 14%), HOUSES index (RI = 11.4%), race/ethnicity (RI = 11.1%), and asthma visit frequencies (RI = 7%) as top features, implying the importance of socioeconomic factors in AE.

We applied some preprocessing and postprocessing approaches implemented in AI Fairness 360 (AIF360) [130], which includes "disparate impact remover," a reweighting approach, uniform resampling, preferential resampling, Reject Option-based Classification pivot, and cutoff manipulation. Some mitigation approaches were able to reduce bias by SES in a fairness metric at the expense of other metrics (e.g., the disparate impact remover reduced bias in accuracy but increased bias in sensitivity). However, none of the approaches were able to consistently produce the results that could be considered as a "fair model" (a ratio between Q1 and Q2–Q4 between 0.8 and 1.25, a standard effect size based on the 80% rule of thumb from legal work around "disparate impact," with 1/0.8 = 1.25) [130]. This is in line with other work finding that there are generally trade-offs in post hoc attempts to balance metrics of fairness and model performance (Fig. 47.2).

While we found current tools for mitigating model bias to be empirically ineffective for investigating the underlying sources, our group is planning to generate new methods relating to the quality of EHR [132–136].

Differentially inaccurate labels of asthma are a major problem as such errors will propagate through AI models. One approach is to use other data (free text) in EHR as an alternative way to extract asthma information (improve quality of EHRs). We developed natural language process (NLP) systems to apply predetermined asthma criteria (PAC) [137–139] and Asthma Predictive Index (API) [139,140], sets of asthma criteria, to EHR and validated these against manual chart review for each asthma criteria. While this may not solve the problem of fewer doctor visits from those of lower SES, leading to less data in which to find PAC and API and other potential sources of variability (seeing different doctors on different visits, potentially leading to lower quality from lack of continuity of care and difficulty of capturing concept from free text embedded in EHRs [given that 80% of health care data is in free text form] [141]), it is one approach to fixing one of the underlying problems of the

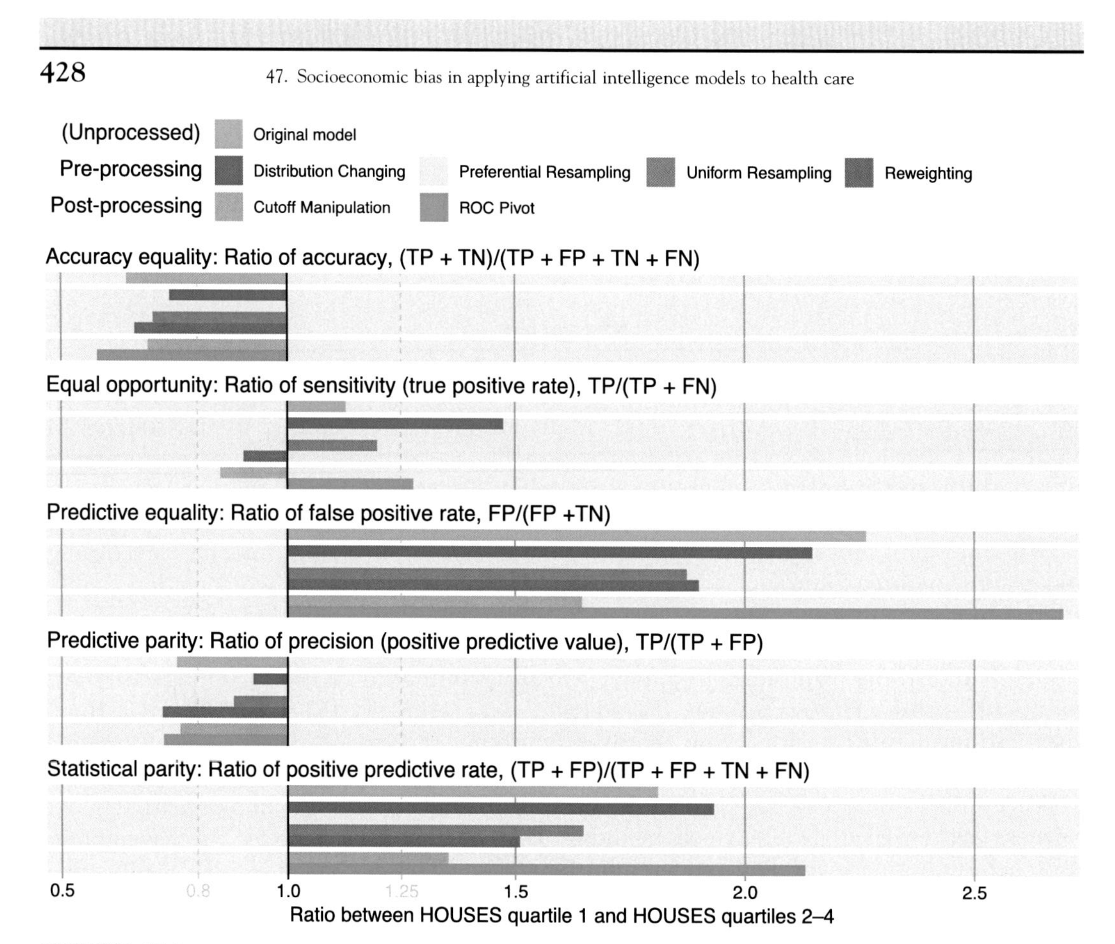

FIGURE 47.2 Preprocessing and postprocessing approaches for model-based mitigation of differential model performance.

data itself. Other fixes we are exploring include addressing missingness through imputation. Lastly, the ultimate solution may not be a modeling one at all: it might be to address health equity issues by specifically mitigating issues around SDH. As the National Academy of Medicine suggested [20], perhaps, achieving health equity in health care systems should be considered to be one of the quintuple aims. We have found that the AI fairness literature largely takes the quality of data for granted, and this is especially evident for health care applications. We argue that the root causes of differential health care access and quality act through data quality, and we believe AI modeling is unlikely to solve health care disparities through modeling approaches alone that do not address this mechanism. Nor do we think that simply being more careful about the data that we currently have, or collecting more data from the same measures, will be sufficient [142]. It is possible that further technical developments will find ways of balancing fairness with performance, such as balancing the two as part of model selection [143] or by including reasoning about the underlying social processes producing the observed patterns and using hypothesized causal relationships as a guide for building AI models [144]. Still, because AI is a form of statistical modeling even more dependent on data than traditional statistics, without reasoning about the quality of data and sources of differential quality in order to address them, we are unlikely to completely address the potential of AI to worsen disparities. The concern may not be differential performance itself but rather when AI models are implemented in clinical practice [19]. The potential problematic interface between AI models and clinical triage systems (which are prone to cause disparities described above) may play a greater role in causing health inequity. From this standpoint, some have argued that postdeployment monitoring and surveillance will be crucially important for AI stewardship [145]. At present, AI bias is largely realized in the context of mathematical validation, but we argue for the great need for clinical validation (clinical effectiveness, usability, and safety) at the time of implementation and postimplementation, recognizing that only a handful of AI models demonstrated clinical effectiveness in health care [146].

Conclusion

The 2019 report on "AI in Health Care" by the National Academy of Medicine suggested "algorithmic fairness" as a crucial step toward health equity as disparate model performance can exacerbate inequity on a large scale ("The impact of a single biased human is far less than that of a global or national AI") [20]. This concern is largely shared by the White House and other health policy and government agencies, such as the Organization for Economic Co-operation and Development [147,148], World Health Organization [149], FDA [150], National Institute of Health (NIH) [21], Federal Trade Commission [151], American Medical Association [152], health care executives [153], and the academia-industry Coalition for Healthcare AI [154].

In this work, we have argued that a meaningful definition of AI bias is around differential model performance; that it is crucial to consider SES as a dimension along which AI models can be biased; and that the HOUSES index provides a promising way to meaningfully measure individual-level SES, including for the purposes of measuring AI model bias. We also argue that it is not enough to measure such bias or to try to apply automated tools that provide post hoc model corrections regardless of the underlying sources of bias, perhaps contrary to some formulations of "algorithmic fairness." We argue that there is no shortcut to reasoning about and investigating possible underlying mechanisms that produce differential performance. Since AI models are dependent entirely on data, reasoning through ways in which low SES, along with other structural processes like racism and gender bias, can result in less accurate data can give us important insight into how to pursue reliable mitigation approaches. While research continues into how to best mitigate AI bias, auditing remains a clear step for anticipating when proposed applications might worsen these disparities. Beyond mathematical validation, it is crucially important to assess clinical validation (effectiveness, usability, and safety) at the time of implementation and post-implementation. It is also important to determine when to "de-adopt" and discard the biased, harmful or even low-value AI models through postimplementation surveillance. Doing this lets us avoid propagate biased models through a health care system already susceptible to health inequity, and instead work to redesign both AI models and care delivery systems towards being safe, equitable, effective, and valuable.

References

[1] Hellman D. Big data and compounding injustice. J Moral Philos 2021. Virginia Public Law and Legal Theory Research Paper No. 2021-27. <https://ssrn.com/abstract = 3840175>.

[2] Benjamin R. Assessing risk, automating racism. Science 2019;366:421–2.

[3] Obermeyer Z, Powers B, Vogeli C, et al. Dissecting racial bias in an algorithm used to manage the health of populations. Science 2019;366:447–53.

[4] Ledford H. Millions of black people affected by racial bias in health-care algorithms. Nature 2019;574:608–9.

[5] Beach MC, Saha S, Park J, et al. Testimonial injustice: linguistic bias in the medical records of Black patients and women. J Gen Intern Med 2021;36:1708–14.

[6] Sun M, Oliwa T, Peek ME, Tung EL. Negative patient descriptors: documenting racial bias in the electronic health record. Health Aff 2022;41:203–11.

[7] Hoffman KM, Trawalter S, Axt JR, et al. Racial bias in pain assessment and treatment recommendations, and false beliefs about biological differences between Blacks and whites. Proc Natl Acad Sci USA 2016; 113:4296–301.

[8] Larrazabal AJ, Nieto N, Peterson V, et al. Gender imbalance in medical imaging datasets produces biased classifiers for computer-aided diagnosis. Proc Natl Acad Sci USA 2020;117:12592–4.

[9] Li S, Fonarow GC, Mukamal KJ, et al. Sex and race/ethnicity-related disparities in care and outcomes after hospitalization for coronary artery disease among older adults. Circ Cardiovasc Qual Outcomes 2016;9:S36–44.

[10] Wang HE, Landers M, Adams R, et al. A bias evaluation checklist for predictive models and its pilot application for 30-day hospital readmission models. J Am Med Inf Assoc 2022;29:1323–33.

[11] Schulman P, Saria S. Reliable decision support using counterfactual models. In: Proceedings of the 31st International Conference on Neural Information Processing Systems; 2017.

[12] Gianfrancesco MA, Tamang S, Yazdany J, et al. Potential biases in machine learning algorithms using electronic health record data. JAMA Intern Med 2018;178:1544–7.

[13] Cerrato P, Halamka J, Pencina M. A proposal for developing a platform that evaluates algorithmic equity and accuracy. BMJ Health Care Inf 2022;29.

[14] Pierson E, Cutler DM, Leskovec J, et al. An algorithmic approach to reducing unexplained pain disparities in underserved populations. Nat Med 2021;27:136–40.

[15] Oakes JM, Rossi PH. The measurement of SES in health research: current practice and steps toward a new approach. Soc Sci Med 2003;56:769–84.

[16] Breiman L. Statistical modeling: the two cultures (with comments and a rejoinder by the author). Stat Sci 2001;16:199–231.

[17] Baiocchi M, Rodu J. Reasoning using data: two old ways and one new. Observ Stud 2021;7.

[18] Danks D, London AJ. Algorithmic bias in autonomous systems. In: Proceedings of the 26th International Joint Conference on Artificial Intelligence (IJCAI); 2017.

[19] Vayena E, Blasimme A, Cohen IG. Machine learning in medicine: addressing ethical challenges. PLoS Med 2018;15:e1002689.

[20] Matheny ME, Whicher D, Thadaney, Israni S. Artificial intelligence in health care: a report from the National Academy of Medicine. JAMA 2020;323:509–10.

[21] Cutillo CM, Sharma KR, Foschini L, et al. Machine intelligence in healthcare—perspectives on trustworthiness, explainability, usability, and transparency. NPJ Digit Med 2020;3:47.

[22] Hellman D. When is discrimination wrong? Harvard University Press; 2011.

[23] Ochigame R. The long history of algorithmic fairness. Phenomonal World; 2020.

[24] d'Alessandro B, O'Neil C, LaGatta T. Conscientious classification: a data scientist's guide to discrimination-aware classification. Big Data 2017;5:120–34.

[25] Bach PB, Pham HH, Schrag D, et al. Primary care physicians who treat Blacks and whites. N Engl J Med 2004;351:575–84.

[26] Warnecke RB, Oh A, Breen N, et al. Approaching health disparities from a population perspective: the National Institutes of Health Centers for Population Health and Health Disparities. Am J Public Health 2008;98:1608–15.

[27] Adler NE, Newman K. Socioeconomic disparities in health: pathways and policies. Health Aff 2002;21:60–76.

[28] Bernheim SM, Ross JS, Krumholz HM, et al. Influence of patients' socioeconomic status on clinical management decisions: a qualitative study. Ann Family Med 2008;6:53–9.

[29] Dahlgren G, Whitehead M. The Dahlgren-Whitehead model of health determinants: 30 years on and still chasing rainbows. Public Health 2021;199:20–4.
[30] Snyder-Mackler N, Burger JR, Gaydosh L, et al. Social determinants of health and survival in humans and other animals. Science 2020;368:eaax9553.
[31] Chetty R, Stepner M, Abraham S, et al. The association between income and life expectancy in the United States, 2001–2014. JAMA 2016;315:1750–66.
[32] Venkataramani A, Daza S, Emanuel E. Association of social mobility with the income-related longevity gap in the United States: a cross-sectional, county-level study. JAMA Intern Med 2020;180:429–36.
[33] Smith D, Shipley M, Rose G. The magnitude and causes of socio-economic differentials in mortality; further evidence from the Whitehall Study. J Epidemiol Commun Health 1990;44:265–70.
[34] Steptoe A, Hamer M, O'Donnell K, et al. Socioeconomic status and subclinical coronary disease in the Whitehall II epidemiological study. PLoS One 2010;5:e8874.
[35] Stringhini S, Sabia S, Shipley M, et al. Association of socioeconomic position with health behaviors and mortality. JAMA 2010;303:1159–66.
[36] Elovainio M, Ferrie JE, Singh-Manoux A, et al. Socioeconomic differences in cardiometabolic factors: social causation or health-related selection? Evidence from the Whitehall II cohort study, 1991-2004. Am J Epidemiol 2011;174:779–89.
[37] Forde I, Chandola T, Raine R, et al. Socioeconomic and ethnic differences in use of lipid-lowering drugs after deregulation of simvastatin in the UK: the Whitehall II prospective cohort study. Atherosclerosis 2011;215:223–8.
[38] Stringhini S, Berkman L, Dugravot A, et al. Socioeconomic status, structural and functional measures of social support, and mortality. Am J Epidemiol 2012;175:1275–83.
[39] Braveman PA, Cubbin C, Egerter S, et al. Socioeconomic disparities in health in the United States: what the patterns tell us. Am J Public Health 2010.
[40] Solar O, Irwin A. A conceptual framework for action on the social determinants of health. Social determinants of health discussion paper 2 (policy and practice). World Health Organization; 2010.
[41] National Academies of Sciences, Engineering, and Medicine; Health and Medicine Division; Board on Population Health and Public Health Practice; Committee on Community-Based Solutions to Promote Health Equity in the United States. Communities in action: pathways to health equity. Baciu A, Negussie Y, Geller A, Weinstein JN, editors. Washington, DC: National Academies Press; 2017.
[42] American Psychological Association. Socioeconomic status (SES). APA Dictionary of Psychology, 2nd ed.; 2015.
[43] Doubeni CA, Simon M, Krist AH. Addressing systemic racism through clinical preventive service recommendations from the US Preventive Services Task Force. JAMA 2021;325:627–8.
[44] Chokshi DA, Foote MMK, Morse ME. How to act upon racism—not race—as a risk factor. JAMA Health Forum 2022;3:e220548.
[45] Keeys M, Baca J, Maybank A. Race, racism, and the policy of 21st Century medicine. Yale J Biol Med 2021;94:153–7.
[46] Dubay LC, Lebrun LA. Health, behavior, and health care disparities: disentangling the effects of income and race in the United States. Int J Health Serv 2012;42:607–25.
[47] Gorey KM, Luginaah IN, Bartfay E, et al. Effects of socioeconomic status on colon cancer treatment accessibility and survival in Toronto, Ontario, and San Francisco, California, 1996–2006. Am J Public Health 2011;101:112–19.
[48] Fujiwara RJT, Ishiyama G, Ishiyama A. Association of socioeconomic characteristics with receipt of pediatric cochlear implantations in California. JAMA Netw Open 2022;5 e2143132-e2143132.
[49] Silver ER, Truong HQ, Ostvar S, et al. Association of neighborhood deprivation index with success in cancer care crowdfunding. JAMA Netw Open 2020;3 e2026946-e2026946.
[50] Schwandt H, Currie J, von Wachter T, et al. Changes in the relationship between income and life expectancy before and during the COVID-19 pandemic, California, 2015–2021. JAMA 2022.
[51] Warren Andersen S, Blot WJ, Lipworth L, et al. Association of race and socioeconomic status with colorectal cancer screening, colorectal cancer risk, and mortality in southern US adults. JAMA Netw Open 2019;2:e1917995.
[52] Sills MR, Hall M, Colvin JD, et al. Association of social determinants with children's hospitals' preventable readmissions performance. JAMA Pediatr. 2016;170:350–8.

[53] Roberts ET, Zaslavsky AM, Barnett ML, et al. Assessment of the effect of adjustment for patient characteristics on hospital readmission rates: implications for pay for performance. JAMA Intern Med 2018.
[54] Ash AS, Mick EO, Ellis RP, et al. Social determinants of health in managed care payment formulas. JAMA Intern Med 2017;177:1424–30.
[55] Committee on Accounting for Socioeconomic Status in Medicare Payment Programs; Board on Population Health and Public Health Practice; Board on Health Care Services; Institute of Medicine; National Academies of Sciences, Engineering, and Medicine. Accounting for social risk factors in Medicare payment: identifying social risk factors. Washington, DC: National Academies Press; 2016.
[56] National Quality Forum. Evaluation of NQF trial period for risk adjustment for social risk factors final report. Washington, DC: National Quality Forum; 2017.
[57] Braveman P. Health disparities and health equity: concepts and measurement. Annu Rev Public Health 2006;27:167–94.
[58] Galobardes B, Shaw M, Lawlor DA, et al. Indicators of socioeconomic position (part 1). J Epidemiol Commu Health 2006;60:7–12.
[59] Rieppi R, Greenhill LL, Ford RE, Chuang S, Wu M, et al. Socieconomic status as a moderator of ADHD treatment outcomes. J Am Acad Child Adolesc Psychiatry 2002;41:269–77.
[60] Hollingshead A. Four factor index of social status. New Haven, CT: Yale University Department of Psychology; 1975.
[61] Nakao KTJ. The 1989 socioeconomic index of occupations: construction from the 1989 occupational prestige scores (General Social Survey methodological report no 74). Chicago: University of Chicago, National Opinion Research Center; 1992.
[62] Liberatos P, Link BG, Kelsey JL. The measurement of social class in epidemiology. Epidemiol Rev 1988;10:87–121.
[63] Shavers VL. Measurement of socioeconomic status in health disparities research. J Natl Med Assoc 2007;99:1013–23.
[64] Berkman L, Kawachi I. Social epidemiology. Oxford: Oxford University Press; 2000.
[65] Geronimus AT. Invited commentary: using area-based socioeconomic measures—think conceptually, act cautiously. Am J Epidemiol 2006;164:835–40 discussion 841–3.
[66] Flanagan BE, Gregory EW, Hallisey EJ, et al. A social vulnerability index for disaster management. J Homel Secur Emerg Manag 2011;8.
[67] Krieger N. Overcoming the absence of socioeconomic data in medical records: validation and application of a Census-based methodology. Am J Public Health 1992;82:703–10.
[68] Chien AT, Wroblewski K, Damberg C, et al. Do physician organizations located in lower socioeconomic status areas score lower on pay-for-performance measures? J Gen Intern Med 2012;27:548–54.
[69] Franks P, Fiscella K. Effect of patient socioeconomic status on physician profiles for prevention, disease management, and diagnostic testing costs. Med Care 2002;40:717–24.
[70] Juhn YJ, Sauver JS, Katusic S, et al. The influence of neighborhood environment on the incidence of childhood asthma: a multilevel approach. Soc Sci & Med 2005;60:2453–64.
[71] Wong GY, Mason WM. The hierarchial logistic regression model for multilevel analysis. J Am Stat Assoc 1985;80:513–24.
[72] Diez-Roux AV, Nieto J, Muntaner C, et al. Neighborhood environments and coronary heart disease: a multilevel analysis. Am J Epidemiol 1997;146:48–59.
[73] O'Campo P, Xue X, Wang M-C, et al. Neighborhood risk factors for low birthweight in Baltimore: a multilevel analysis. Am J Public Health 1997;87:1113–18.
[74] Li Q, Kirby RS, Sigler RT, et al. A multilevel analysis of individual, household, and neighborhood correlates of intimate partner violence among low-income pregnant women in Jefferson County, Alabama. Am J Public Health 2010;100:531–9.
[75] Dundas R, Leyland AH, Macintyre S. Early-life school, neighborhood, and family influences on adult health: a multilevel cross-classified analysis of the Aberdeen children of the 1950s study. Am J Epidemiol 2014;180:197–207.
[76] Xu WY, Jung J, Retchin SM, et al. Rural-urban disparities in diagnosis of early-onset dementia. JAMA Netw Open 2022;5:e2225805.
[77] Flores G, Snowden-Bridon C, Torres S, et al. Urban minority children with asthma: substantial morbidity, compromised quality and access to specialists, and the importance of poverty and specialty care. J Asthma 2009;46:392–8.

[78] Walker AF, Hu H, Cuttriss N, et al. The neighborhood deprivation index and provider geocoding identify critical catchment areas for diabetes outreach. J Clin Endocrinol Metab 2020;105(9):3069–75.
[79] Narla PN, Pardo-Crespo MR, Beebe TJ, et al. Concordance between individual vs. area-level socioeconomic measures in an urban setting. J Health Care Poor Underserved 2015;26:1157–72.
[80] Pardo-Crespo MR, Narla NP, Williams AR, et al. Comparison of individual-level versus area-level socioeconomic measures in assessing health outcomes of children in Olmsted County, Minnesota. J Epidemiol Commun Health 2013;67:305–10.
[81] Marengo L, Ramadhani T, Farag NH, et al. Should aggregate US Census data be used as a proxy for individual household income in a birth defects registry? J Regist Manag 2011;38:9–14.
[82] Gelman A, Ansolabehere S, Price PN, et al. Models, assumptions, and model checking in ecological regressions. J R Stat Soc 2001;164:101–18.
[83] United States Census Bureau. American Community Survey: sample size definitions, vol. 2022. Available from: https://www.census.gov; 2022.
[84] Pollack CE, Chideya S, Cubbin C, et al. Should health studies measure wealth? A systematic review. Am J Prev Med 2007;33:250–64.
[85] Cubbin C, Pollack C, Flaherty B, et al. Assessing alternative measures of wealth in health research. Am J Public Health 2011;101:939–47.
[86] Griffith K, Evans L, Bor J. The Affordable Care Act reduced socioeconomic disparities in health care access. Health Aff 2017;36:1503–10.
[87] Magin P, Victoire A, Zhen XM, et al. Under-reporting of socioeconomic status of patients in stroke trials: adherence to CONSORT principles. Stroke 2013;44:2920–2.
[88] Juhn YJ, Beebe TJ, Finnie DM, et al. Development and initial testing of a new socioeconomic status measure based on housing data. J Urban Health 2011;88:933–44.
[89] Kirst M, Shankardass K, Bomze S, et al. Sociodemographic data collection for health equity measurement: a mixed methods study examining public opinions. Int J Equity Health 2013;12:75.
[90] Dunn JR. Housing and health inequalities: review and prospects for research. Hous Stud 2000;15:341–66.
[91] Harvey D. Social justice and the city. Oxford: Blackwell; 1973.
[92] Duncan J. Housing and identity. London: Croom Helm; 1981.
[93] Badcock B. Unfairly structured cities. Oxford: Basil Blackwell; 1984.
[94] Despres C. The meaning of home: literature review and directions for future research and theoretical development. J Archit Plan Res 1991;8:96–115.
[95] Harvey D. The condition of postmodernity. Oxford: Blackwell; 1989.
[96] Harris R, Pratt G. The meaning of home, home ownership, and public policy. Montreal and Kingston: McGill-Queen's Press; 1993.
[97] Smith S. The essential qualities of a home. J Environ Psychol 1994;14:31–46.
[98] Marcus C. House as a mirror of self. Berkley: Conari Press; 1995.
[99] Dunn JR, Hayes MV. Social inequality, population health, and housing: a study of two Vancouver neighborhoods. Soc Sci Med 2000;51:563–87.
[100] Department of Housing and Urban Development, US Department of Commerce. American Housing Survey for the United States: 1999. Washington, DC: US Department of Housing and Urban Development, US Department of Commerce; 2003.
[101] Bang DW, Manemann SM, Gerber Y, et al. A novel socioeconomic measure using individual housing data in cardiovascular outcome research. Int J Environ Res Public Health 2014;11:11597–615.
[102] Ghawi H, Crowson CS, Rand-Weaver J, et al. A novel measure of socioeconomic status using individual housing data to assess the association of SES with rheumatoid arthritis and its mortality: a population-based case-control study. BMJ Open 2015;5:e006469.
[103] Harris MN, Lundien MC, Finnie DM, et al. Application of a novel socioeconomic measure using individual housing data in asthma research: an exploratory study. NPJ Prim Care Respir Med 2014; 24:14018.
[104] Johnson MD, Urm SH, Jung JA, et al. Housing data-based socioeconomic index and risk of invasive pneumococcal disease: an exploratory study. Epidemiol Infect 2013;141:880–7.
[105] Takahashi PY, Ryu E, Hathcock MA, et al. A novel housing-based socioeconomic measure predicts hospitalisation and multiple chronic conditions in a community population. J Epidemiol Commun Health 2016; 70:286–91.

[106] Ryu E, Juhn YJ, Wheeler PH, et al. Individual housing-based socioeconomic status predicts risk of accidental falls among adults. Ann Epidemiol 2017;27:415–20 e2.
[107] Wi CI, Gauger J, Bachman M, et al. Role of individual-housing-based socioeconomic status measure in relation to smoking status among late adolescents with asthma. Ann Epidemiol 2016;26:455–60.
[108] Wi CI, St Sauver JL, Jacobson DJ, et al. Ethnicity, socioeconomic status, and health disparities in a mixed rural-urban US community-Olmsted County, Minnesota. Mayo Clin Proc 2016;91:612–22.
[109] Butterfield MC, Williams AR, Beebe T, et al. A two-county comparison of the HOUSES index on predicting self-rated health. J Epidemiol Commun Health 2011;65:254–9.
[110] Hammer R, Capili C, Wi C-I, et al. A new socioeconomic status measure for vaccine research in children using individual housing data: a population-based case-control study. BMC Public Health 2016;16:1–9.
[111] Stevens MA, Beebe TJ, Wi CI, et al. HOUSES index as an innovative socioeconomic measure predicts graft failure among kidney transplant recipients. Transplantation 2020;104:2383–92.
[112] Zurek KI, Boswell CL, Miller NE, et al. Association of early and late hospital readmissions with a novel housing-based socioeconomic measure. Health Serv Res Manag Epidemiol 2022;9.
[113] Barwise A, Wi CI, Frank R, et al. An innovative individual-level socioeconomic measure predicts critical care outcomes in older adults: a population-based study. J Intensive Care Med 2020;36(7):828–37.
[114] Thacher TD, Dudenkov DV, Mara KC, et al. The relationship of 25-hydroxyvitamin D concentrations and individual-level socioeconomic status. J Steroid Biochem Mol Biol 2019;197:105545.
[115] Bjur KA, Wi CI, Ryu E, et al. Epidemiology of children with multiple complex chronic conditions in a mixed urban-rural US community. Hosp Pediatr 2019;9:281–90.
[116] Barwise A, Juhn YJ, Wi CI, et al. An individual housing-based socioeconomic status measure predicts advance care planning and nursing home utilization. Am J Hosp Palliat Care 2019;36(5):362–9.
[117] MacLaughlin KL, Jacobson RM, Sauver JLS, et al. An innovative housing-related measure for individual socioeconomic status and human papillomavirus vaccination coverage: a population-based cross-sectional study. Vaccine 2020;38:6112–19.
[118] Bjur KA, Wi CI, Ryu E, et al. Socioeconomic status, race/ethnicity, and health disparities in children and adolescents in a mixed rural-urban community-Olmsted County, Minnesota. Mayo Clin Proc 2019;94:44–53.
[119] Kaur H, Lachance DH, Ryan CS, et al. Asthma and risk of glioma: a population-based case-control study. BMJ Open 2019;9:e025746.
[120] Schachter JP, Kuenzi JJ. Seasonality of moves and the duration and tenure of residence: 1996. Washington, DC: United States Census Bureau; 2002.
[121] Dam AV. Black families pay significantly higher property taxes than white families, new analysis shows. *The Washington Post*; 2020.
[122] Avenancio-Léon CHT. The assessment gap: racial inequalities in property taxation. The Washington Center for Equitable Growth; 2020.
[123] United States Census Bureau. Income in the past 12 months (in 2021 inflation-adjusted dollars); 2021.
[124] Juhn YJ, Ryu E, Wi CI, et al. Assessing socioeconomic bias in machine learning algorithms in health care: a case study of the HOUSES index. J Am Med Inform Assoc 2022;ocac052.
[125] Seol HY, Shrestha P, Muth JF, et al. Artificial intelligence-assisted clinical decision support for childhood asthma management: a randomized clinical trial. PLoS One 2021;16:e0255261.
[126] Corbett-Davies S., Goel S. The measure and mismeasure of fairness: a critical review of fair machine learning. arXiv preprint arXiv:1808.00023, 2018.
[127] Rodolfa K, Saleiro P, Ghani R. Bias and fairness. In: Big Data and social science. 2nd ed. Chapman and Hall/CRC; 2021. p. 281–312.
[128] Mayo Clinic Platform. Mayo Clinic Platform_Validate; 2022.
[129] Park Y, Hu J, Singh M, et al. Comparison of methods to reduce bias from clinical prediction models of postpartum depression. JAMA Netw Open 2021;4:e213909.
[130] Bellamy RKE, Dey K, Hind M, et al. AI Fairness 360: an extensible toolkit for detecting and mitigating algorithmic bias. IBM J Res Dev 2019;63(4):1–4 15.
[131] Forno E, Celedón JC. Health disparities in asthma. Am J Respir Crit Care Med 2012;185:1033–5.
[132] Weiskopf NG, Hripcsak G, Swaminathan S, et al. Defining and measuring completeness of electronic health records for secondary use. J Biomed Inf 2013;46:830–6.
[133] Weiskopf NG, Rusanov A, Weng C. Sick patients have more data: the non-random completeness of electronic health records. AMIA Annu Symp Proc 2013;2013:1472–7.

[134] Weiskopf NG, Weng C. Methods and dimensions of electronic health record data quality assessment: enabling reuse for clinical research. J Am Med Inf Assoc 2013;20:144–51.

[135] Kahn MG, Callahan TJ, Barnard J, et al. A harmonized data quality assessment terminology and framework for the secondary use of electronic health record data. EGEMS 2016;4:1244.

[136] Yan M, Pencina MJ, Boulware LE, et al. Observability and its impact on differential bias for clinical prediction models. J Am Med Inf Assoc 2022;29:937–43.

[137] Wi CI, Sohn S, Rolfes MC, et al. Application of a natural language processing algorithm to asthma ascertainment: an automated chart review. Am J Respir Crit Care Med 2017;196:430–7.

[138] Wi CI, Sohn S, Ali M, et al. Natural language processing for asthma ascertainment in different practice settings. J Allergy Clin Immunol Pract 2018;6:126–31.

[139] Seol HY, Rolfes MC, Chung W, et al. Expert artificial intelligence-based natural language processing characterises childhood asthma. BMJ Open Respir Res 2020;7(1):e000524.

[140] Kaur H, Sohn S, Wi CI, et al. Automated chart review utilizing natural language processing algorithm for asthma predictive index. BMC Pulm Med 2018;18:34.

[141] Juhn Y, Liu H. Artificial intelligence approaches using natural language processing to advance EHR-based clinical research. J Allergy Clin Immunol 2020;145:463–9.

[142] Obermeyer Z, Nissan R, Stern M, et al. Algorithmic bias playbook. Cent Appl AI Chic Booth 2021.

[143] Rodolfa KT, Lamba H, Ghani R. Empirical observation of negligible fairness–accuracy trade-offs in machine learning for public policy. Nat Mach Intell 2021;3:896–904.

[144] Nabi R, Malinsky D, Shpitser I. Learning optimal fair policies. In: Proceedings of the 36th International Conference on Machine Learning, vol. 97. 2019. p. 4674–82.

[145] Eaneff S, Obermeyer Z, Butte AJ. The case for algorithmic stewardship for artificial intelligence and machine learning technologies. JAMA 2020;324:1397–8.

[146] Angus DC. Randomized clinical trials of artificial intelligence. JAMA 2020;323:1043–5.

[147] OECD—Organisation for Economic Co-operation and Development. OECD principles on AI; 2019.

[148] White House Office of Science and Technology Policy. Blueprint for an AI Bill of Rights: making automated systems work for the American people. White House Office of Science and Technology Policy; 2017.

[149] World Health Organization. Ethics and governance of artificial intelligence for health. World Health Organization; 2021.

[150] US Food and Drug Administration. Executive summary for the patient engagement advisory committee meeting: artificial intelligence (AI) and machine learning (ML) in medical devices. In: Services DoHaH, FDA; 2020.

[151] Jercich K. Hold yourself accountable on keeping AI unbiased, or the FTC might do it for you. Healthcare IT News; 2021.

[152] American Medical Association. AMA passes first policy recommendations on augmented intelligence. American Medical Association; 2018.

[153] Intel Corporation. Overcoming barriers in AI adoption in healthcare. Intel Corporation; 2018.

[154] Coalition for Heathcare AI. Blueprint for trustworthy AI implementation guidance and assurance for healthcare. 2023.

CHAPTER

48

Artificial intelligence in dermatology

Samuel Yeroushalmi, Alexander Ildardashty, Mimi Chung, Erin Bartholomew, Marwa Hakimi, Tina Bhutani and Wilson Liao

Department of Dermatology, University of California, San Francisco, San Francisco, CA, United States

Introduction

The use of artificial intelligence (AI) has the potential to greatly impact the field of dermatology through applications, such as aiding in disease diagnosis and advancement of personalized medicine approaches. Machine learning (ML) algorithms can be developed from an abundance of database information in the form of electronic medical records, clinical and histopathologic images, as well as translational data (e.g., genomic, transcriptomic, epigenetic) [1]. Here, we briefly summarize key trials as well as recent research which may be of interest to both dermatologists and other medical providers.

Clinical disease classification

Melanoma

One of the most important roles of the dermatologist is to accurately identify cutaneous malignancies like melanoma. Distant-site malignant melanoma at diagnosis only has a 30% 5-year survival rate compared to local-site which has a 99% 5-year survival, highlighting the importance of accurate skin cancer screening [2]. Nasr-Esfahani et al. were among the first to use a convolutional neural network (CNN) to detect melanoma. Their algorithm was trained on over 6000 clinical images and was able to correctly diagnose melanoma with a sensitivity and specificity of 0.81 and 0.80, respectively [3]. Esteva et al. utilized the Google Inception V3 CNN to develop a diagnostic algorithm using nearly 130,000 clinical images to detect skin cancer. The authors reported that their CNN achieved 72.1%

Artificial Intelligence in Clinical Practice
DOI: https://doi.org/10.1016/B978-0-443-15688-5.00047-4

accuracy in diagnosing skin cancer and its performance was comparable to dermatologists [4]. Brinker et al. showed that CNNs can even outperform dermatologists, as was shown by their pretrained ResNet50 CNN which achieved a sensitivity and specificity of 82.3% and 67.2% respectively in classifying dermatoscopic images of melanoma [5].

Nonmelanoma skin cancer

Nonmelanoma skin cancers (NMSCs), which include basal cell carcinomas (BCCs) and squamous cell carcinomas (SCCs), are the most common types of cancer worldwide [6]. Like melanoma, AI technology which can accurately detect these types of cancers is of major interest to the clinician and is currently being researched. Tschandl et al. trained a combined CNN which classified both clinical and dermatoscopic images. Though the CNN had an area under the receiver operatic characteristic (AUROC) of 0.742 compared to 0.695 of human ratings, the CNN did not outperform humans with regards to the percentage of correct diagnoses [7]. Cho et al. showed that deep convoluted neural networks (DCNNs) have the potential to diagnose skin cancer as well as dermatologists. They evaluated a DCNN which categorized benign and malignant disease and found that the algorithm was equivalent to dermatologists and outperformed nondermatologists with sensitivity ranging from 0.803 to 0.827 and specificity ranging from 0.702 to 0.759 depending on the image set used [4].

Other skin disease

A deep learning system (DLS) developed by Liu et al. was able to primary predictions for 26 common skin conditions including psoriasis, eczema, acne, melanocytic nevi, seborrheic keratosis, seborrheic dermatitis, alopecia areata, verruca vulgaris, folliculitis, and keratinocyte carcinomas. This DLS also showed superiority to primary care physicians and nurse practitioners, and was noninferior to dermatologists [8]. These diagnostic algorithms can serve as a useful aid for nondermatologist clinicians in appropriately diagnosing, triaging, and managing their patients.

Dermatopathology

AI has the potential to diagnose skin disease with the use of histopathological images in addition to gross images. The digitalization of whole slide images allows for the capture of high-resolution data which can be analyzed by an AI algorithm. Arevalo et al. showed that an unsupervised model, which does not require input of a clinical to train the model, was able to identify features of BCCs with an AUROC of 98.1% [9]. Hekler et al. reported the results of a CNN which was superior to 11 dermatopathologists in classifying randomly cropped histopathologic images of melanoma. The algorithm had a discordance of 18% with a single histopathologist, though a limitation of this study was the stringent time restrictions placed on histopathologists to review slides (12 second per image) [10].

Point-of-care diagnosis and telehealth

Point-of-care AI diagnostic tools are currently available for healthcare practitioners to use as a diagnostic aid at bedside with the use of their smartphones. VisualDx's DermExpert is an AI system that can identify and categorize skin lesions' morphologies with an accuracy of 68%, compared to a 36% diagnostic accuracy of primary care physicians without any aid [11]. Another smartphone application which utilized ML developed by Skinvision was 95.1% sensitive in detecting premalignant conditions for malignant melanoma and other NMSCs with a specificity of 78.3% [12]. The convenience of AI-integrated smartphone technology can dramatically improve the real-time diagnostic capabilities of primary care practitioners who lack the training an experience of a dermatologist.

Dermatologists may also need to rely on digital images of patients' skin via live telehealth or store-and-forward technology. This can be challenging if image quality is poor, so Vodrahalli et al. developed an ML algorithm known as TrueImage which was able to detect low quality dermatology photos. Their algorithm was able to reject 50% of poor quality images while retaining 80% of high quality ones, allowing for more efficient care through telehealth [13].

Precision medicine

The use of AI combined in combination with big data stores that are currently available has propelled advancing in precision medicine, otherwise known as personalized or individualized medicine. These technologies can be used to predict treatment response or even disease progression and have numerous applications in dermatology.

Psoriasis

Correa da Rosa et al. aimed to address the lag time between when a psoriasis patient's clinical response to treatment can be determined. Using gene-expression profile from biopsy samples, the authors were able to develop ML prediction models that could accurately predict a 12-week clinical endpoint from a sample taken within the first 4 weeks of treatment with systemic medications [14]. Foulkes et al. was able to predict psoriasis patients response to etanercept using a multiomics model based on gene expression and biochemical pathways associated with TNF-alpha and major histocompatibility complex [15]. Continuing research in this field can help reduce trial-and-error treatment selection for psoriasis patients which would improve patient outcomes as well as reduce healthcare spending.

Psoriatic arthritis

A major question which has yet to be answered is which patient with psoriasis will go on to develop psoriatic arthritis? Research by Patrick et al. used ML to find nine new loci for psoriasis or its subtypes using data from over 7000 genotyped psoriasis and psoriatic arthritis patients. The ML prediction was also able to distinguish psoriatic arthritis from cutaneous-only psoriasis with an AUROC of 0.82 using genetic markers [16]. Liu et al. used ML of single cell RNA-seq and proteomic data from peripheral blood to distinguish psoriatic arthritis patients from psoriasis patients and healthy individuals with an AUROC of 0.87 [17]. Given that there is currently no diagnostic or predictive test for psoriatic arthritis, further development in this field can lead to early treatment or even prevention of permanent joint disease in these patients.

Melanoma

Genetic and data have also be used to predict whether pigmented lesions are malignant or not. Torres et al. used an ML-based pipeline that used a dataset which included next-generation microRNA sequencing and developed a model which could distinguish melanomas from nevi with a sensitivity of 81% and specificity of 88% with an AUROC of 0.98 [18].

Current limitations and future directions

AI has numerous potential applications within the field of dermatology; however, it is not without its limitations. First, as discussed by Chan et al. AI algorithms can be seen as a black box: when used without physician input, it may not be possible to determine why the algorithm made a certain decision [19]. The accuracy of the algorithm is also a reflection of the dataset, and poor quality data can detriment its outputs. Ultimately, the results of AI should be used within a clinical context and should supplement the clinician's decision making, rather than replace it. Many studies included data from white patients and these trained algorithms may not be accurate when used with patients from racial and ethnic minorities [20].

The future of AI should continue to focus on making these technologies inclusive of and accessible to all patients of varying backgrounds. Further clinical trials are needed to determine the effectiveness of AI in dermatology before it is routinely incorporated in clinical practice.

Major takeaway points

1. AI algorithms can diagnose melanoma, NMSC, and other dermatologic diseases potentially more accurately than nondermatologist practitioners and in some cases, on par with dermatologists or dermatopathologists.
2. Point-of-care mobile technology which utilizes AI can improve real-time diagnostic accuracy.
3. Precision medicine AI can be used to predict treatment response, diagnose skin diseases including skin cancer, and possibly predict disease prognosis and disease onset.
4. Though further research on AI in dermatology which includes prospective clinical trials is needed, AI has a strong potential to supplement clinical decision making both in the dermatology and primary care practice setting.

Conflicts of interest

Tina Bhutani is a principal investigator for trials sponsored by Abbvie, Castle, CorEvitas, Dermavant, Galderma, Mindera, and Pfizer. She has received research grant funding from Novartis and Regeneron. She has been an advisor for Abbvie, Arcutis, Boehringer-Ingelheim, Bristol Myers Squibb, Janssen, Leo, Lilly, Novartis, Pfizer, Sun, and UCB. Wilson Liao has received research grant funding from Abbvie, Amgen, Janssen, Leo, Novartis, Pfizer, Regeneron, and TRex Bio. The remaining authors have nothing to disclose.

References

1. Wehner MR, Levandoski KA, Kulldorff M, Asgari MM. Research techniques made simple: an introduction to use and analysis of big data in dermatology. J Invest Dermatol 2017;137(8):e153–8. Available from: https://doi.org/10.1016/j.jid.2017.04.019.
2. SEER Cancer Statistics Review, 1975–2016. SEER. <https://seer.cancer.gov/csr/1975_2016/index.html> [accessed 05.05.22].
3. Nasr-Esfahani E, Samavi S, Karimi N, et al. Melanoma detection by analysis of clinical images using convolutional neural network. In: 2016 38th annual international conference of the IEEE Engineering in Medicine and Biology Society (EMBC); 2016. p. 1373–1376. Available from: https://doi.org/10.1109/EMBC.2016.7590963.
4. Esteva A, Kuprel B, Novoa RA, et al. Dermatologist-level classification of skin cancer with deep neural networks. Nature. 2017;542(7639):115–18. Available from: https://doi.org/10.1038/nature21056.
5. Brinker TJ, Hekler A, Enk AH, et al. Deep learning outperformed 136 of 157 dermatologists in a head-to-head dermoscopic melanoma image classification task. Eur J Cancer 2019;113:47–54. Available from: https://doi.org/10.1016/j.ejca.2019.04.001.
6. The incidence and clinical analysis of non-melanoma skin cancer. Sci Rep. <https://www.nature.com/articles/s41598-021-83502-8> [accessed 10.05.22].
7. Tschandl P, Rosendahl C, Akay BN, et al. Expert-level diagnosis of nonpigmented skin cancer by combined convolutional neural networks. JAMA Dermatol 2019;155(1):58–65. Available from: https://doi.org/10.1001/jamadermatol.2018.4378.
8. Liu Y, Jain A, Eng C, et al. A deep learning system for differential diagnosis of skin diseases. Nat Med 2020;26(6):900–8. Available from: https://doi.org/10.1038/s41591-020-0842-3.
9. Arevalo J, Cruz-Roa A, Arias V, Romero E, González FA. An unsupervised feature learning framework for basal cell carcinoma image analysis. Artif Intell Med 2015;64(2):131–45. Available from: https://doi.org/10.1016/j.artmed.2015.04.004.
10. Hekler A, Utikal JS, Enk AH, et al. Pathologist-level classification of histopathological melanoma images with deep neural networks. Eur J Cancer 2019;115:79–83. Available from: https://doi.org/10.1016/j.ejca.2019.04.021.
11. Dulmage B, Tegtmeyer K, Zhang MZ, Colavincenzo M, Xu S. A point-of-care, real-time artificial intelligence system to support clinician diagnosis of a wide range of skin diseases. J Invest Dermatol 2021;141(5):1230–5. Available from: https://doi.org/10.1016/j.jid.2020.08.027.
12. Udrea A, Mitra GD, Costea D, et al. Accuracy of a smartphone application for triage of skin lesions based on machine learning algorithms. J Eur Acad Dermatol Venereol 2020;34(3):648–55. Available from: https://doi.org/10.1111/jdv.15935.
13. Vodrahalli K, Daneshjou R, Novoa RA, Chiou A, Ko JM, Zou J. TrueImage: a machine learning algorithm to improve the quality of telehealth photos. arXiv:201002086 [cs, eess]. <http://arxiv.org/abs/2010.02086> [accessed 10.05.22].
14. Correa da Rosa J, Kim J, Tian S, Tomalin LE, Krueger JG, Suárez-Fariñas M. Shrinking the psoriasis assessment gap: early gene-expression profiling accurately predicts response to long-term treatment. J Invest Dermatol 2017;137(2):305–12. Available from: https://doi.org/10.1016/j.jid.2016.09.015.
15. Foulkes AC, Watson DS, Carr DF, et al. A framework for multi-omic prediction of treatment response to biologic therapy for psoriasis. J Invest Dermatol 2019;139(1):100–7. Available from: https://doi.org/10.1016/j.jid.2018.04.041.

16. Patrick MT, Stuart PE, Raja K, et al. Genetic signature to provide robust risk assessment of psoriatic arthritis development in psoriasis patients. Nat Commun 2018;9(1):4178. Available from: https://doi.org/10.1038/s41467-018-06672-6.
17. Liu J, Kumar S, Hong J, et al. Combined single cell transcriptome and surface epitope profiling identifies potential biomarkers of psoriatic arthritis and facilitates diagnosis via machine learning. Front Immunol 2022;13:835760. Available from: https://doi.org/10.3389/fimmu.2022.835760.
18. Torres R, Lang UE, Hejna M, et al. MicroRNA ratios distinguish melanomas from nevi. J Invest Dermatol 2020;140(1):164–73. Available from: https://doi.org/10.1016/j.jid.2019.06.126.
19. Chan S, Reddy V, Myers B, Thibodeaux Q, Brownstone N, Liao W. Machine learning in dermatology: current applications, opportunities, and limitations. Dermatol Ther (Heidelb) 2020;10(3):365–86. Available from: https://doi.org/10.1007/s13555-020-00372-0.
20. Navarrete-Dechent C, Dusza SW, Liopyris K, Marghoob AA, Halpern AC, Marchetti MA. Automated dermatological diagnosis: hype or reality? J Invest Dermatol 2018;138(10):2277–9. Available from: https://doi.org/10.1016/j.jid.2018.04.040.

CHAPTER

49

Artificial intelligence in gastroenterology and hepatology

Joseph C. Ahn and Vijay H. Shah

Division of Gastroenterology and Hepatology, Mayo Clinic, Rochester, MN, United States

Introduction

The field of gastroenterology and hepatology (GI) is a vast discipline that encompasses the functioning and health of the complex human digestive system. The digestive system consists of the alimentary tract, including the esophagus, stomach, small intestine, and colon, as well as accessory organs, such as the liver, gallbladder, and pancreas. Each of these organs can be affected by various acute and chronic pathologic processes, resulting in bothersome symptoms, impaired organ function, development of neoplasia, and, in severe cases, death. GI disorders are among the leading causes of healthcare burden in the United States. According to recent survey data, annual GI healthcare expenditures in the United States exceed $100 billion, with over 40 million ambulatory visits and nearly 4 million hospital admissions per year [1]. In 2018 GI diseases and cancers caused 255,407 deaths in the United States, while globally, GI illnesses lead to over 8 million deaths annually [2].

Patients with GI disorders generate an abundance of health data, including documented history and physical examination, laboratory tests, radiologic studies, endoscopic images, and tissue samples. Historically, various scoring systems and prediction models have been developed using those data to help with diagnosis and clinical decision making. Some popular examples include the Glasgow-Blatchford score [3] for upper gastrointestinal hemorrhage, Ranson's criteria [4] for acute pancreatitis, Maddrey's discriminant function [5] for alcohol-associated hepatitis, and the model for end-stage liver disease (MELD) score [6] for cirrhosis. Most of the currently used clinical prediction tools have been built using traditional modeling techniques, such as linear regression or logistic regression, relying on a small number of structured clinical variables (e.g., age, sex, vital signs, laboratory values). While these models have been widely adopted with success, they have limited ability to capture complex, nonlinear relationships between variables.

Artificial Intelligence in Clinical Practice
DOI: https://doi.org/10.1016/B978-0-443-15688-5.00016-4

The emergence of artificial intelligence (AI) presents exciting opportunities to advance the care of patients with GI disorders. Advanced machine learning (ML) methods can analyze nonlinear relationships between hundreds and even thousands of clinical variables, yielding novel insights on disease risk factors and phenotypes. Moreover, deep learning algorithms can rapidly and accurately process unstructured, high-dimensional data, such as texts, images, and waveforms. AI is poised to revolutionize the field of diagnostic endoscopy in the near future. This chapter provides an overview and highlights some notable examples of the many potential applications of AI in the field of GI (Table 49.1).

TABLE 49.1 Application of artificial intelligence in gastroenterology and hepatology.

Organ	Common disorders	Types of data generated	Potential role for artificial intelligence
Esophagus	GERD Esophagitis Motility disorders Barrett's esophagus Malignancy	History, vitals, exam, labs EGD images Histopathology Esophagram Esophageal manometry pH monitoring studies	Automated diagnosis of esophageal pathologies on EGD Automated interpretation of manometry and pH studies Prediction of dysplasia in BE Endoscopic staging of esophageal cancer
Stomach	Peptic ulcer disease *Helicobacter pylori* infection Bleeding Malignancy	History, vitals, exam, labs EGD images Histopathology Radiologic studies	Diagnosis of PUD and *H. pylori* on endoscopy Prediction of clinical outcomes in PUD/GI bleeds Gastric cancer diagnosis, staging, treatment response
Small intestine	Celiac disease and other malabsorptive disorders Bleeding Malignancy	History, vitals, exam, labs EGD images Wireless capsule endoscopy Histopathology Radiologic studies	Automated reading of WCE for detection of bleeding and masses Assessment of celiac disease and response to therapy
Colon	Colon polyps Colorectal cancer Bleeding IBD	History, vitals, exam, labs Colonoscopy images Histopathology Radiologic studies	Automated colon polyp and colon cancer detection on colonoscopy Standardized assessment of endoscopic and histologic severity in IBD
Liver	Steatosis, inflammation, fibrosis, cirrhosis HCC/other lesions Ascites, VH, HE, HRS	History, vitals, exam, labs Liver ultrasound Multiphase CT/MRI Histopathology	Automated grading of steatosis and fibrosis on imaging/pathology Prediction of decompensation and death in cirrhosis Diagnosis/staging of HCC and other liver lesions
Pancreas	Pancreatitis PDAC/other lesions	History, vitals, exam, labs Endoscopic ultrasound CT/MRI Histopathology	Prediction of severity and complications in acute pancreatitis Risk stratification of cystic lesions Diagnosis/staging of PDAC

Esophagus

The esophagus is a muscular tube that transports swallowed oral contents to the stomach. Some major esophageal disorders include gastroesophageal reflux disease (GERD), esophagitis, motility disorders of the esophageal, Barrett's esophagus, and esophageal cancer.

Gastroesophageal reflux disease

GERD is characterized by the reflux of gastric contents into the esophagus, causing troublesome symptoms and/or complications [7]. Accurately diagnosing GERD is often challenging as many other conditions can mimic its symptoms, and patients with GERD may present with atypical symptoms [8]. In the early 21st century several studies applied ML algorithms to symptom-based patient surveys to aid with diagnosis of GERD. Pace et al. conducted a survey of 159 patients presenting with symptoms of suspected GERD and trained an artificial neural network (ANN) using 45 demographic and clinical variables, which achieved a 100% accuracy rate in correctly identifying GERD compared to 78% with traditional discriminant analysis [9]. The same group also trained another questionnaire-based ANN to classify patients with GERD into erosive reflux disease versus nonerosive reflux disease, but with an unsatisfactory accuracy of 62.2% [10]. Horowitz et al. also used survey data from 132 patients to build an ANN that helped diagnose GERD with an AUROC of 0.787 [11]. More recently, a deep learning algorithm trained on the 24-hour pH impedance tests of 106 patients was able to identify actual reflux events and postreflux swallow-induced peristaltic wave index with excellent correlation with manual interpretations [12]. Lastly, AI has shown promise in grading the severity of reflux esophagitis according to the Los Angeles classification system. A deep learning model trained on 2081 endoscopic images of gastroesophageal junction was able to correctly classify the severity of reflux esophagitis with 86.7% accuracy, outperforming junior (71.5%) and experienced (77.4%) endoscopists [13].

Esophageal motility disorders

Disorders of esophageal motility, such as achalasia and esophageal spasm, are caused by abnormalities in esophageal peristalsis and relaxation mechanism at the gastroesophageal junction. High-resolution esophageal manometry (HREM) [14] is considered the gold standard for diagnosing esophageal motility disorders [15]. Kou et al. used 1,741 HREM studies with 26,115 swallows to train a deep learning model that classified swallow types into normal, hypercontractile, weak-fragmented, failed, and premature with a 88% accuracy at the study level [16]. The same group subsequently developed a multistage diagnostic system combining deep learning and feature-based models to predict swallow type and pressurization with 88% and 93% accuracies, respectively, and the study-level diagnosis with 81% accuracy for the top-1 prediction and 92% accuracy for the top-2 predictions [17]. Popa et al. also built a convolutional neural network (CNN) on 1570 HREM images that was able to classify the HREM images into 10 different categories of esophageal motility disorders with an overall precision greater than 93% [18].

Barrett's esophagus and esophageal adenocarcinoma

Barrett's esophagus refers to intestinal metaplasia that occurs in patients with chronic GERD and is a precursor to esophageal adenocarcinoma [19]. CNN-based computer-aided detection (CAD) systems are emerging as powerful new tools to facilitate the endoscopic diagnosis and staging of malignant and premalignant esophageal lesions. After several early-phase pilot studies with promising results [20–22], investigators from Amsterdam University Medical Centers developed a hybrid ResNet-UNet model CAD system that classified neoplastic versus nondysplastic Barrett's esophagus on endoscopy with nearly 90% accuracy and significantly outperformed human endoscopists [23] (Fig. 49.1).

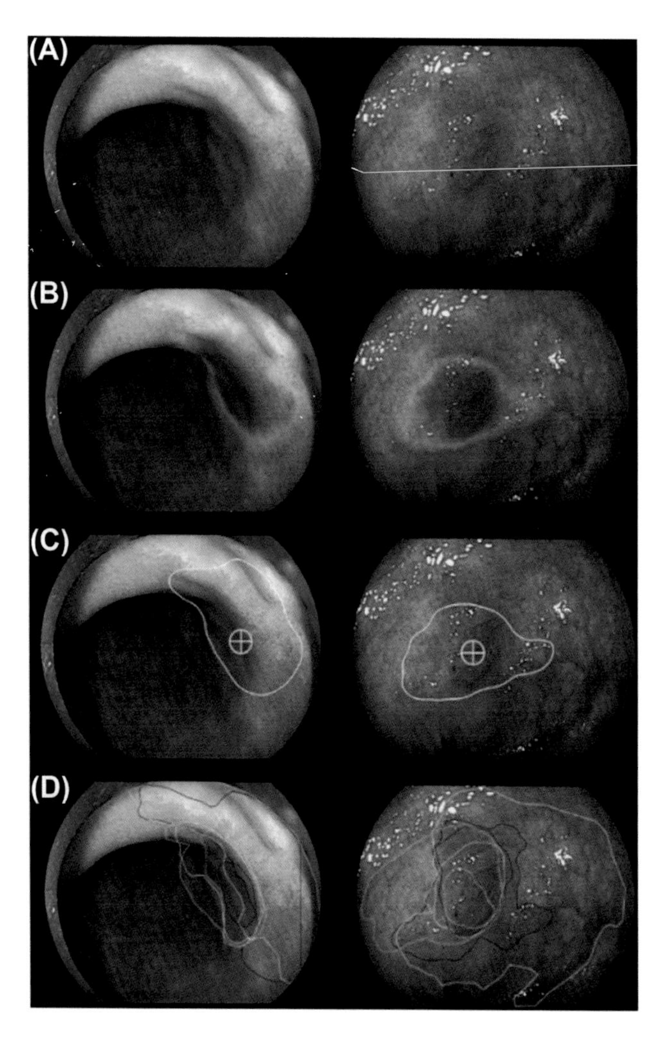

FIGURE 49.1 Detection of neoplasia in Barrett's esophagus using deep learning. Example of (A) 2 neoplastic lesions with (B) the heatmap visualization by the CAD system and (C) its corresponding delineation and biopsy site indicator. (D) Ground truth is established by expert delineations. Source: *Fig. 1 from https://doi.org/10.1053/j.gastro.2019.11.030.*

Additional studies have shown encouraging results for deep learning-based detection of early esophageal adenocarcinoma in patients with Barrett's esophagus [24–26]. A meta-analysis of AI-based CADs for Barrett's esophagus and esophageal adenocarcinoma showed an AUROC of 0.90 and 0.97, respectively with performances that were either comparable or superior to human endoscopists [27].

Stomach

Peptic ulcer disease

Peptic ulcer disease is one of the most common digestive disorders and can cause dyspepsia, abdominal pain, bleeding, gastric outlet obstruction, penetration into adjacent structures, and perforation leading to peritonitis and death [28]. CNN models have demonstrated outstanding performances for detecting peptic ulcers and other gastric lesions on endoscopic images [29–31]. *Helicobacter pylori* (*H. pylori*) infection is the most common cause of peptic ulcer disease worldwide. The presence or absence of *H. pylori* cannot be distinguished by human eyes during endoscopy, and diagnosis is typically made through endoscopic biopsies, breath tests, or stool antigen tests. In 2017 a 22-layer CNN trained on a dataset of 32,208 endoscopic images was able to predict the presence of *H. pylori* infection with a 87.7% accuracy compared to 82.4% by six board-certified endoscopists [32]. Additional studies from Japan and China have continued to show excellent performances of deep learning models in predicting *H. pylori* infection on endoscopic images [33–36]. In 2020 a deep learning-based CAD was able to classify endoscopic images of patients into three categories of *H. pylori* infection: uninfected, currently infected, and posteradication with over 80% accuracy [37]. Furthermore, AI-based prediction models have also demonstrated satisfactory performances for prediction of clinical outcomes, such as recurrent bleeding and mortality, among patients presenting with complications of peptic ulcer disease [38–40].

Gastric cancer

Gastric adenocarcinoma is a common and often deadly form of cancer worldwide, and early detection on upper endoscopy is crucial for ensuring curative therapy and favorable long-term survival [41]. Recent studies have demonstrated the effectiveness of CAD systems for detecting gastric cancer on endoscopic images. A pooled analysis of studies from Asia [42–46] found that CAD systems had an overall sensitivity, specificity, positive predictive value, and negative predictive value of 88%, 89%, 88%, and 89%, respectively for detecting gastric cancer detection on endoscopic images [47]. Additionally, researchers from Korea have developed a cutting-edge clinical decision support system (CDSS) that uses deep learning to automate the diagnosis and invasion-depth prediction of gastric neoplasms in real-time endoscopy. In a prospective multicenter external validation study, the CDSS achieved high accuracy for both tasks, with a lesion detection rate of 95.6% [48]. The system was also able to accurately classify four stages of gastric cancer (no cancer, dysplasia, early gastric cancer, and advanced gastric cancer) with 89.7% accuracy in the training cohort and 81.5% accuracy in external validation.

In addition to diagnosing and grading gastric cancer using endoscopic images, deep learning-based radiomics models built from abdominal CT scans have effectively predicted response to therapy, recurrence, and prognosis in patients with gastric cancer [49–53]. Jin et al. developed a deep learning model that predicts pathologically confirmed lymph node metastasis using preoperative CT images from gastric cancer patients undergoing gastrectomy with lymph node dissection. In external validation, the deep learning system demonstrated excellent predictive performance that was superior to clinicopathologic variables (median AUROC 0.876 vs. 0.652) [52]. Li et al. developed a CT-based AI model that diagnosed signet-ring cell carcinoma-subtype of gastric cancer on CT scan with an AUROC of 0.786 and classified patients into high- and low-risk categories [53]. Although molecular subtypes of gastric cancer are important predictors of response to chemotherapy or immunotherapy, such information is not readily available in every patient due to the lack of molecular testing capacities. Flinner et al. developed an ensemble CNN that predicts different molecular subtypes of gastric cancer on routine hematoxylin-eosin stained pathology slides [54] (Fig. 49.2).

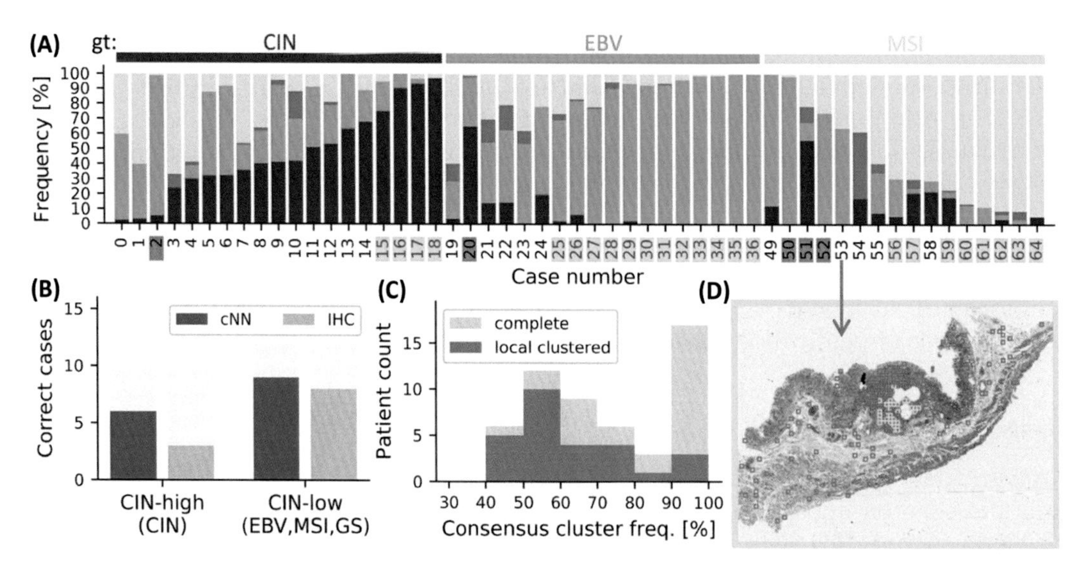

FIGURE 49.2 Gastric cancer and deep learning prediction of molecular subtypes on H&E stained pathology slides. (A) Four class ensemble cNN models trained with TCGA data. The frequency of consensus prediction for all tiles of individual patients [UKC(− GS) test data] is shown. The ground truth (gt) is indicated in the top row. Green/red squares below indicate patients with secure correct/incorrect consensus prediction using ensemble cNNs. (B) Comparison of ensemble cNN-based deep learning using the staining approach in predicting CIN-high (molecular subclass designation: CIN) and CIN-low (molecular subclass designation: EBV, GS, MSI). CIN ratios above 0.5 are usually considered chromosomally unstable. (C) Histogram for the frequencies of the class with the consensus prediction (whether correct or not) for the UKC(− GS) test dataset (light grey). Dark grey indicates the number of patients in whom at least one molecular subclass was found at a given location and is counted as locally clustered. (D) Example of a whole-slide image for patient #53 where more than one subclass was found, each at a defined region. Source: *Fig. 3 from https://doi.org/10.1002/path.5879.*

Small intestine

The standard endoscopic examinations, such as the esophagogastroduodenoscopy and colonoscopy, are limited in their ability to visualize most of the small intestine whose length varies between 600 and 1500 cm [55]. Wireless capsule endoscopy (WCE) is widely used as a noninvasive and effective method to visualize the small intestinal lumen to look for abnormalities, such as bleeding, ulcerations, or mass lesions [56]. However, the manual reading of WCE is very time intensive and prone to human error as the reader has to go through thousands of images in which relevant findings may be present only in a few frames [57]. Therefore AI has a great potential to improve the accuracy and time-efficiency of reading WCE. Multiple groups have applied deep learning for automated interpretation of WCE with successful outcomes. According to a meta-analysis by Soffer et al. [58], deep learning models demonstrated outstanding sensitivity of 93%–98% and specificity of 94%–99% for detection of bleeding [59–63] or mucosal ulcers [64–68]. Celiac disease is an increasingly common small intestinal disorder characterized by villous atrophy and intestinal malabsorption. The monitoring of celiac disease and response to treatment requires endoscopic assessment of small bowel mucosa. Recently, a deep learning model for assessing the severity of villous damage on WCE in patients with celiac disease showed excellent agreement with expert gastroenterologists [69] (Fig. 49.3).

Colon

Colon polyp detection

Colorectal cancer is the third most common and second most deadly cancer globally [70]. Colonoscopy is the gold standard procedure for both screening and prevention of colorectal cancer, as it enables gastroenterologists to detect and remove precancerous polyps. However, colonoscopy is a complex procedure and the quality of colonoscopy varies depending on the technical expertise of the endoscopist. Detecting polyps during a colonoscopy can be difficult especially if they are small or flat. Inadequate quality of bowel preparation, tortuosity of the colon, and patient discomfort during the procedure further compromise gastroenterologists' ability to successfully detect and remove polyps. Therefore it is hoped that deep learning-based CAD systems will help improve real-time adenoma detection rates during colonoscopy.

Deep learning models trained on large, retrospective datasets of images and videos during colonoscopy have demonstrated great performances at detecting polyps. In 2018, Misawa et al. used colonoscopy videos from 73 patients to develop an original AI-based CAD system, which showed sensitivity, specificity, and accuracy of 90%, 63.3%, and 76.5%, respectively for identifying polyps [71]. Urban et al. trained another CNN-based CAD system using a set of 8,641 hand-labeled images from over 2,000 screening colonoscopies. This model achieved an AUROC of 0.991 and accuracy of 96.4%, and was able to be applied to real-time endoscopy videos to help improve polyp detection by human endoscopists [72]. Subsequent studies from other centers have produced similar performances, confirming the feasibility of using AI for polyp detection during colonoscopy [73,74] (Fig. 49.4).

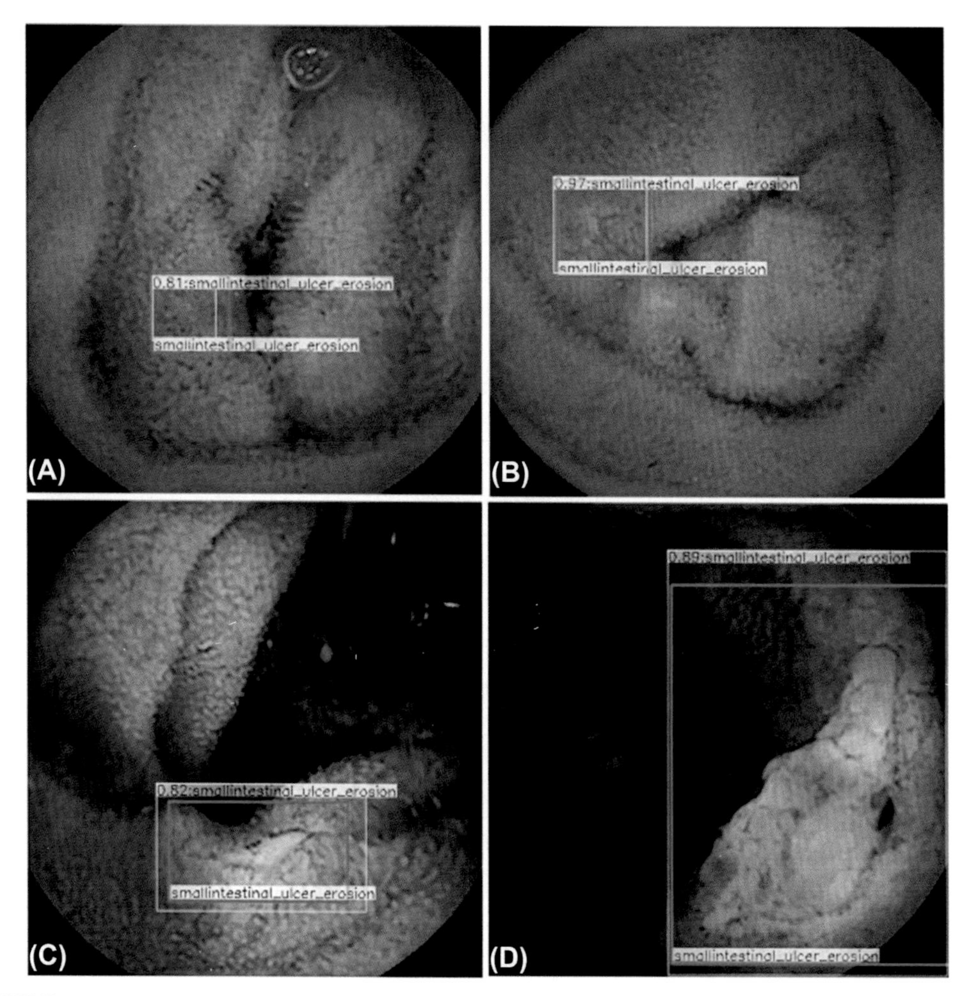

FIGURE 49.3 AI detection of erosions and ulcerations on wireless capsule endoscopy. here, the ABCD do not mean much and they are just four different examples. the original image does not contain specific descriptions for different parts, either. Representative images of erosions and ulcerations correctly detected by the convolutional neural network (CNN) in the validation set (green box, true lesion; yellow box, region identified as an erosion or ulceration by the CNN; number, the probability score by the CNN). Source: *Fig. 2 from https://doi.org/10.1016/j.gie.2018.10.027.*

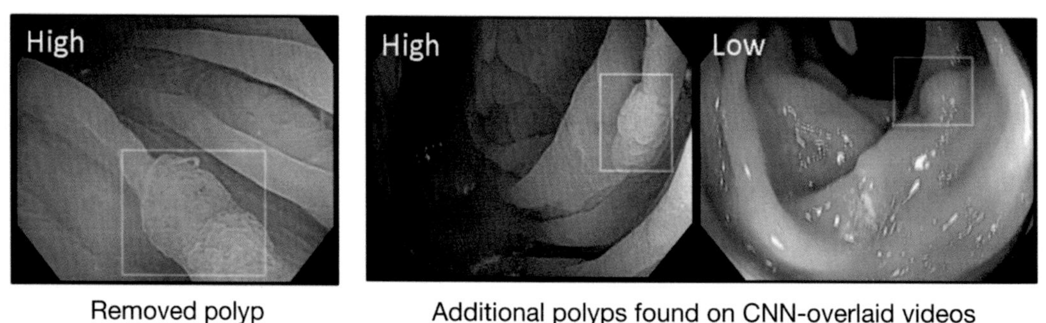

FIGURE 49.4 Real-time AI detection of colon polyps during colonoscopy. Source: *Fig. 2 from https://doi.org/10.1053/j.gastro.2018.06.037.*

Recently several prospective randomized controlled trials (RCTs) have been conducted to evaluate the effectiveness of using AI-based CAD systems for colon polyp detection. Wang et al. conducted an open, nonblinded RCT of 1058 patients where the addition of CAD led to significantly higher adenoma detection rate (29.1% vs. 20.3%, $P < .001$) and mean number of adenomas per patient (0.53 vs. 0.31, $P < .001$). Of note, however, this was primarily due to detection of a higher number of diminutive adenomas and hyperplastic polyps, and no significant difference was seen in detection of larger adenomas [75]. Another RCT by Liu et al. showed similar results where CAD improved the overall adenoma detection rate but did not improve the detection of larger adenomas [76]. In 2019 Su et al. developed an AI-based automatic quality control system (AQCS) for colonoscopy which not only helped with polyp detection but also with timing of the withdrawal, withdrawal stability, and evaluation of bowel preparation [77]. In an RCT of 659 patients, the AQCS increased adenoma detection rate for both small (0–5 mm) and larger (>5 mm) polyps (0.289 vs. 0.165, $P < .001$) and also led to superior withdrawal time (7.03 vs. 5.68 minutes, $P < .001$) and developed bowel prep rate (87.34% vs. 80.63%, $P = .023$) [77]. In 2020 Gong et al. reported another deep learning-based real-time colonoscopy quality improvement system called ENDOANGEL which helps monitor real-time withdrawal speed and endoscope slipping [78]. In an RCT of 704 patients, the ENDOANGEL system led to significantly improved adenoma detection rate as well as improved detection of large adenomas greater than 10 mm in size [78]. Most of the RCTs described above had been single-center studies in Asian countries. In 2022 Wallace et al. reported the results of an international, multicenter RCT of AI-assisted colonoscopy conducted across eight centers in Europe and the United States [79]. In this study, the use of a CNN-based CAD system was associated with significantly lower adenoma miss rate for polyps smaller than 5 mm (15.85% vs. 35.75%, $P < .001$) and lower number of adenomas found at second colonoscopy (0.33 vs. 0.70, $P < .001$) [79]. Despite the highly encouraging results of RCTs, the uptake of CAD systems by clinicians in real-world practice settings is moving slowly, likely due to the lack of familiarity and the additional add-on costs. It is also not yet proven that the improved adenoma detection with AI-assisted colonoscopy will lead to decreased risk of colorectal cancer and death after colonoscopy [80].

Inflammatory bowel disease

Inflammatory bowel disease (IBD) refers to ulcerative colitis (UC) and Crohn's disease, both of which are chronic and relapsing autoimmune conditions with increasing morbidity and mortality worldwide [81,82]. IBD is diagnosed on the basis of compatible symptoms, endoscopic appearance, and histopathologic confirmation. As the assessment and grading of disease severity on endoscopy can be highly subjective, several endoscopic scoring systems have been developed to reduce interobserver variability [83]. AI is expected to help further establish reproducibility and homogeneity of the endoscopic findings [84]. In 2019 Stidham et al. used a large number of endoscopic images from over 3000 patients with UC to train a 159-layer CNN which was excellent at distinguishing endoscopic remission from moderate-to-severe disease with an AUROC of 0.97 [85]. Ozawa et al. also developed a CNN-based CAD system based on GoogLeNet architecture which was excellent at distinguishing Mayo score 0–1 images from Mayo score 2–3 images with AUROC of 0.98 [86].

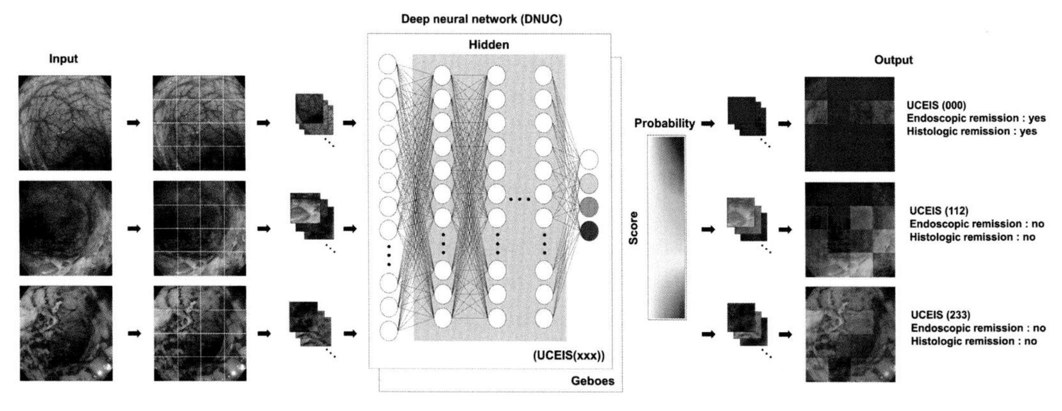

FIGURE 49.5 AI prediction of IBD activity on real-time endoscopy. Source: *Fig. 1 from https://doi.org/10.1053/j.gastro.2020.02.012.*

Additional studies have shown outstanding performances by AI-based CAD systems for automated determination of endoscopic and histologic disease activity [87–90]. In 2020 Gottlieb et al. prospectively collected 795 full-length colonoscopy videos from an international multicenter clinical trial of 249 patients from 14 countries, and demonstrated that a deep learning model could accurately predict UC severity from full-length endoscopy videos with excellent agreement with standardized endoscopic scores [91]. In 2022 Maeda et al. conducted an open-label prospective cohort study of 135 patients undergoing real-time AI-assisted colonoscopy. At 12-month follow-up, patients classified as "Active" by the AI model had a significantly higher relapse rate compared to patients classified as "Healing" (28.4% vs. 4.9%, $P < .001$) [92] (Fig. 49.5).

In addition to endoscopic assessments, AI is expected to play an important role in standardized assessment of histologic disease activity in patients with IBD. In several recent single-center studies, CNN-based AI models trained on colonic biopsy images of patients with UC were capable of accurately assessing histologic activity and remission with excellent agreements with expert pathologists and standardized indices [93–95]. Finally, AI models trained on clinical data from electronic medical records may have a role in prediction of prognosis and response to therapy [96–99].

Liver

Liver is the largest solid organ in human body and has many critical roles including production of bile, metabolism of macronutrients, synthesis of major proteins, clearance of drugs and toxins, and immunity. Liver diseases, such as acute liver failure, cirrhosis, and hepatocellular carcinoma (HCC), account for approximately 2 million deaths per year globally [100]. Despite the development of effective antiviral agents against hepatitis B and C viruses, the overall burden of liver diseases continue to rise due to marked increase in the prevalence of alcohol-related liver disease and nonalcoholic fatty liver disease [101]. There are many unmet needs in the field of liver diseases including new tools for early detection of liver disease, accurate estimation of prognosis, prediction of decompensating events (ascites, variceal hemorrhage, hepatic encephalopathy, hepatorenal syndrome), and prediction of HCC risk. It is hoped that state-of-the-art AI models will rise to the challenge to address such unmet needs [102].

Assessment of liver fibrosis and steatosis

Abdominal ultrasound is widely used as the first imaging modality to visualize the liver, but its accuracy is highly operator dependent. Several studies showed promising results in applying AI for accurate assessment of hepatic steatosis and fibrosis on ultrasound images [103–105]. In 2018 Wang et al. conducted a prospective multicenter study applying deep-learning radiomics for staging liver fibrosis on 2D-shear wave elastography which demonstrated excellent AUROCs of 0.97 for F4, 0.98 for ≥F3, and 0.85 for ≥F2 fibrosis [106]. In addition to fibrosis staging, AI algorithms have shown high accuracy for detection of hepatic steatosis on ultrasound [107–110]. Cross-sectional imaging modalities, such as CT and MRI, are also useful for evaluation of liver diseases and portal hypertension. Choi et al. used a large single-center dataset of CT images from 7461 patients with various stages of liver fibrosis to develop a CNN which was able to stage liver fibrosis with AUROC of 0.96, 0.97, and 0.95 for significant fibrosis (F2–4), advanced fibrosis (F3–4), and cirrhosis (F4), respectively [111]. MRI-based AI radiomics models also showed encouraging performances for predicting liver stiffness and fibrosis [111,112].

While radiologic studies provide valuable information on the liver contour and features of portal hypertension, liver biopsy remains the gold standard for diagnosis of various liver diseases. However, significant intra- and interobserver variations among pathologists have been reported [113]. Deep learning-based AI models are emerging as effective tools for automated and standardized interpretation of digitized liver biopsy slides in patients with liver diseases [114–116]. In 2021 Taylor-Weiner et al. used over 1000 liver biopsy slides from three RCTs of therapies for patients with NAFLD to develop a CNN for quantitative histologic grading and assessment of disease severity [117]. The deep learning model's predictions of NAFLD activity score and fibrosis exhibited significant correlations with expert pathologists.

Prediction of disease severity and prognosis

Currently the MELD score [6] and its variations, such as MELD-Na [118], are used to estimate the severity of liver disease and rank patients awaiting for liver transplantation. However, the MELD score is imperfect and requires ongoing revisions to improve its predictive performance. Various AI models utilizing clinical variables have been developed to more accurately predict clinical outcomes in specific conditions, such as chronic hepatitis C infection [119], primary sclerosing cholangitis [120], acute liver failure [121], and posttransplant settings [122]. However, in an ambitious study of the multicenter North American Consortium for the Study of End-Stage Liver Disease cohort, use of complex AI techniques, such as random forest and support vector machine, failed to show any improvements in prediction of hospital readmissions and death compared to using MELD-Na alone [123].

The severity of portal hypertension is the best predictor of hepatic decompensation and poor outcomes in patients with cirrhosis, but the gold standard technique for estimating portal venous pressure is an invasive measurement of hepatic venous pressure gradient (HVPG). Qi et al. developed a deep learning model for noninvasive estimation of HVPG using computational fluid dynamics from CT angiography images [124]. In a prospective validation study, the "virtual HVPG" showed a good correlation with invasive HVPG measures with an AUROC of 0.89, and inter- and intraobserver agreement of 0.88 and 0.96, respectively [124].

Bosch et al. attempted to predict the HVPG from findings seen on liver histology. The authors used HVPG measurements and liver biopsy images from 218 patients with compensated cirrhosis to train a CNN which was able to predict the presence of clinically significant portal hypertension with an AUROC of 0.76 in the test set [125]. In a creative study by Ahn et al., the researchers used 12-lead electrocardiograms (ECGs) from over 5000 patients with cirrhosis and 20,000 age- and sex-matched controls to train a CNN that detects the presence of cirrhosis on standard, 12-lead ECGs [126]. The CNN had an AUROC of 0.908 for predicting the presence of cirrhosis on ECGs and also produced a score which appears to be correlated with the severity of liver disease.

Liver lesions and cancers

Liver lesions are extremely common findings on abdominal imaging. While most incidentally discovered liver lesions are benign findings, malignancy must be excluded. Patients with cirrhosis or chronic hepatitis B infection must be monitored for HCC, a highly aggressive primary liver cancer associated with poor outcomes [127]. Ultrasound is the most commonly used screening modality for HCC. In 2012 an ANN trained on contrast-enhanced ultrasound images from 112 patients was able to classify five different types of liver lesions (HCC, hypervascular and hypovascular liver metastases, hemangiomas, or focal fatty changes) with 87% accuracy in the test set [128]. Additional studies employing CNN showed even better performances at differentiating between benign and malignant lesions on ultrasound [129–132]. Beyond classifying the lesions that are already present on ultrasound images, Jin et al. created a deep learning radiomics model from a cohort of 434 chronic HBV patients which was able to predict 5-year HCC development with an AUROC of 0.90 [133].

Multiphase CT or MRI are highly sensitive for detection of liver lesions and can in fact establish the diagnosis of HCC without a biopsy if certain imaging characteristics are present [134]. In 2018 Yasaka et al. used CT images from 460 patients with liver lesions to train a CNN that classifies the liver lesions into one of five categories (HCC, other malignancy, indeterminate, hemangiomas, and cysts) with median AUROC of 0.92 for differentiating malignant versus nonmalignant lesions [135]. In 2020 Shi et al. demonstrated that combining a deep learning model with a three-phase CT achieved a diagnostic accuracy comparable to that of a four-phase CT, which reduce the amount of radiation exposure for patients [136]. Hamm et al. used multiphasic MRI images from 434 patients with liver lesions to train a CNN which demonstrated an AUC of 0.992 for classification of HCC and greatly outperformed human radiologists [137]. In order to make the CNN interpretable, the same group performed a post hoc analysis using

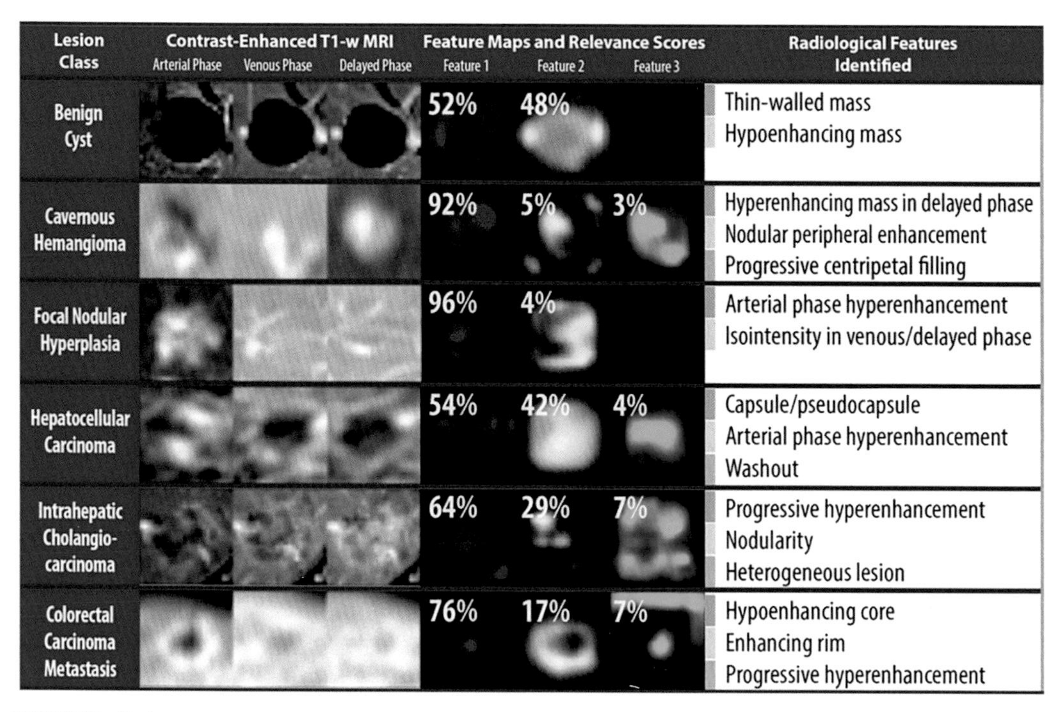

Lesion Class
Contrast-Enhanced T1-w MRI
Arterial Phase
Venous Phase
Delayed Phase
Feature Maps and Relevance Scores
Feature 1
Feature 2
Feature 3
Radiological Features Identified
Benign Cyst
52%
48%
Thin-walled mass
Hypoenhancing mass
Cavernous Hemangioma
92%
5%
3%
Hyperenhancing mass in delayed phase
Nodular peripheral enhancement
Progressive centripetal filling
Focal Nodular Hyperplasia
96%
4%
Arterial phase hyperenhancement
Isointensity in venous/delayed phase
Hepatocellular Carcinoma
54%
42%
4%
Capsule/pseudocapsule
Arterial phase hyperenhancement
Washout
Intrahepatic Cholangio-carcinoma
64%
29%
7%
Progressive hyperenhancement
Nodularity
Heterogeneous lesion
Colorectal Carcinoma Metastasis
76%
17%
7%
Hypoenhancing core
Enhancing rim
Progressive hyperenhancement

FIGURE 49.6 AI classification of liver lesions on MRI using feature maps for explainability. Source: *Fig. 4 from https://doi.org/10.1007/s00330-019-06214-8.*

feature maps whose activation pattern correlated well with the radiologic features human radiologists use to make their decisions [138] (Fig. 49.6).

In addition to classification of lesions, imaging-based AI models can be useful in accurate staging of liver cancer, treatment planning, and predicting response to therapy. In patients with HCC, imaging-based CNN models have been able to successfully detect presence of microvascular invasion [139–141], assess ablative margins following microwave ablation [142], and predict response to trans-arterial chemoembolization [143–145].

Pancreas

Pancreatitis

Acute pancreatitis is one of the most common GI causes of hospital admission in the United States and is a significant cause of morbidity and mortality. One out of every five patients develop severe pancreatitis characterized by local and systemic complications and have a mortality rate approaching 20% [146]. Therefore in patients presenting to the emergency department with acute pancreatitis, it is important to predict the likelihood of progression to severe pancreatitis and development complications. Multiple neural network-based prediction models have been developed over the past decade and outperformed traditional scoring systems, such as Glasgow and APACHE-II [147–153]. Recently, a large, multicenter cohort of more than 5000 patients from 30 centers across 13 European countries was used to develop two XGBoost-based ML-based scores for prediction of pancreatic necrosis (NECRO-APP) and severe pancreatitis (EASY-APP) with AUROCs of 0.76 and 0.81, respectively [154,155].

Autoimmune pancreatitis is a challenging diagnosis whose imaging characteristics can closely mimic those of pancreatic ductal adenocarcinoma (PDAC). In 2021, Marya et al. developed an endoscopic ultrasound-based CNN model trained to differentiate autoimmune pancreatitis from PDAC, chronic pancreatitis, and normal pancreas. Trained on a large number of EUS images and videos from 583 patients who underwent endoscopic ultrasound, the CNN demonstrated over 90% specificity and specificity for distinguishing autoimmune pancreatitis from the other pancreatic conditions [156].

Pancreatic cancer

Prediction and early detection of pancreatic cancer is perhaps the biggest unmet need in the field of pancreatology. PDAC is projected to become the second-leading cancer of cancer death in the United States by 2030, and continues to have a high rate of recurrence and dismal long-term survival rate [157]. Multiple studies have shown the ability of AI algorithms to correctly diagnose PDAC on radiologic and EUS images [158–161]. In 2021 a deep learning-based pancreatic segmentation analysis on CT scan showed that compared to controls without PDAC, mean whole gland pancreatic attenuations in Hounsfield Unit were significantly lower in prediagnostic CT scans from patients who later developed PDAC, potentially suggesting fatty pancreas as a precursor of PDAC [162]. In addition, the same group also developed radiomics-based ML models that could detect PDAC on prediagnostic CT scans of 155 patients who were later diagnosed with PDAC. All of the ML models had very high AUROCs ranging above 0.95 for detection of PDAC in prediagnostic CT scans previously reported as negative for PDAC [163]. Additional studies have shown promises in using AI for categorization and risk stratification of pancreatic cystic lesions [164–167].

Limitations and future directions

There are several limitations, not necessarily unique to the field of GI, that need to be overcome before universal implementation of AI for patientcare. Complex ML models, such as deep learning, are considered to be "black-box" models meaning humans cannot understand their decision making processes. Interpretability of the deep learning models will be crucial for physicians and regulating authorities to accept and trust them, and also for troubleshooting and improving the models. Another limitation is the lack of generalizability. Most published AI algorithms have been developed using retrospective, single-center data from large academic centers and may not accurately represent the real-world population of patients seen at community hospitals. As the algorithms are so heterogeneous and not easily accessible to outside providers, it is difficult for them to be incorporated into practice guidelines. Lastly, there is currently a lack of high quality evidence that incorporation of AI is cost-effective and actually improves patient outcomes in GI—this issue will likely be resolved with time and publication of more large-scale, prospective multicenter studies. Despite the limitations, we are witnessing an explosive growth in the application of AI across all domains of GI which only continues to accelerate. Incorporation of AI into clinical practice in GI is no longer a futuristic idea and will certainly become a reality for most GI providers.

References

[1] Peery AF, Crockett SD, Murphy CC, Jensen ET, Kim HP, Egberg MD, et al. Burden and cost of gastrointestinal, liver, and pancreatic diseases in the United States: update 2021. Gastroenterology 2022;162:621−44.

[2] Milivojevic V, Milosavljevic T. Burden of gastroduodenal diseases from the global perspective. Curr Treat Opt Gastroenterol 2020.

[3] Blatchford O, Murray WR, Blatchford M. A risk score to predict need for treatment for uppergastrointestinal haemorrhage. Lancet 2000;356:1318−21.

[4] Ranson JH, Rifkind KM, Roses DF, Fink SD, Eng K, Spencer FC. Prognostic signs and the role of operative management in acute pancreatitis. Surg Gynecol Obstet 1974;139:69−81.

[5] Maddrey WC, Boitnott JK, Bedine MS, Weber Jr. FL, Mezey E, White Jr. RI. Corticosteroid therapy of alcoholic hepatitis. Gastroenterology 1978;75:193−9.

[6] Kamath PS, Wiesner RH, Malinchoc M, Kremers W, Therneau TM, Kosberg CL, et al. A model to predict survival in patients with end-stage liver disease. Hepatology 2001;33:464−70.

[7] Vakil N, van Zanten SV, Kahrilas P, Dent J, Jones R. The Montreal definition and classification of gastroesophageal reflux disease: a global evidence-based consensus. Am J Gastroenterol 2006;101:1900−20 quiz 1943.

[8] Gyawali CP, Kahrilas PJ, Savarino E, Zerbib F, Mion F, Smout A, et al. Modern diagnosis of GERD: the Lyon Consensus. Gut 2018;67:1351−62.

[9] Pace F, Buscema M, Dominici P, Intraligi M, Baldi F, Cestari R, et al. Artificial neural networks are able to recognize gastro-oesophageal reflux disease patients solely on the basis of clinical data. Eur J Gastroenterol Hepatol 2005;17:605−10.

[10] Pace F, Riegler G, de Leone A, Pace M, Cestari R, Dominici P, et al. Is it possible to clinically differentiate erosive from nonerosive reflux disease patients? A study using an artificial neural networks-assisted algorithm. Eur J Gastroenterol Hepatol 2010;22:1163−8.

[11] Horowitz N, Moshkowitz M, Halpern Z, Leshno M. Applying data mining techniques in the development of a diagnostics questionnaire for GERD. Dig Dis Sci 2007;52:1871−8.

[12] Wong MW, Liu MX, Lei WY, Liu TT, Yi CH, Hung JS, et al. Artificial intelligence facilitates measuring reflux episodes and postreflux swallow-induced peristaltic wave index from impedance-pH studies in patients with reflux disease. Neurogastroenterol Motil 2022;e14506.

[13] Ge Z, Wang B, Chang J, Yu Z, Zhou Z, Zhang J, et al. Using deep learning and explainable artificial intelligence to assess the severity of gastroesophageal reflux disease according to the Los Angeles Classification System. Scand J Gastroenterol 2023;1−9.

[14] Pandolfino JE, Kahrilas PJ. AGA technical review on the clinical use of esophageal manometry. Gastroenterology 2005;128:209−24.

[15] Ravi K. The role of endoscopic impedance planimetry in esophageal disease. Gastroenterol Hepatol (N Y) 2021;17:282−4.

[16] Kou W, Galal GO, Klug MW, Mukhin V, Carlson DA, Etemadi M, et al. Deep learning-based artificial intelligence model for identifying swallow types in esophageal high-resolution manometry. Neurogastroenterol Motil 2022;34:e14290.

[17] Kou W, Carlson DA, Baumann AJ, Donnan EN, Schauer JM, Etemadi M, et al. A multi-stage machine learning model for diagnosis of esophageal manometry. Artif Intell Med 2022;124:102233.

[18] Popa SL, Surdea-Blaga T, Dumitrascu DL, Chiarioni G, Savarino E, David L, et al. Automatic diagnosis of high-resolution esophageal manometry using artificial intelligence. J Gastrointestin Liver Dis 2022;31:383−9.

[19] Hirota WK, Loughney TM, Lazas DJ, Maydonovitch CL, Rholl V, Wong RK. Specialized intestinal metaplasia, dysplasia, and cancer of the esophagus and esophagogastric junction: prevalence and clinical data. Gastroenterology 1999;116:277−85.

[20] van der Sommen F, Zinger S, Curvers WL, Bisschops R, Pech O, Weusten BL, et al. Computer-aided detection of early neoplastic lesions in Barrett's esophagus. Endoscopy 2016;48:617–24.
[21] de Groof J, van der Sommen F, van der Putten J, Struyvenberg MR, Zinger S, Curvers WL, et al. The Argos project: the development of a computer-aided detection system to improve detection of Barrett's neoplasia on white light endoscopy. United Eur Gastroenterol J 2019;7:538–47.
[22] de Groof AJ, Struyvenberg MR, Fockens KN, van der Putten J, van der Sommen F, Boers TG, et al. Deep learning algorithm detection of Barrett's neoplasia with high accuracy during live endoscopic procedures: a pilot study (with video). Gastrointest Endosc 2020;91:1242–50.
[23] de Groof AJ, Struyvenberg MR, van der Putten J, van der Sommen F, Fockens KN, Curvers WL, et al. Deep-learning system detects neoplasia in patients with Barrett's esophagus with higher accuracy than endoscopists in a multistep training and validation study with benchmarking. Gastroenterology 2020;158 915–929.e914.
[24] Ebigbo A, Mendel R, Probst A, Manzeneder J, Souza Jr. LA, Papa JP, et al. Computer-aided diagnosis using deep learning in the evaluation of early oesophageal adenocarcinoma. Gut 2019;68:1143–5.
[25] Hashimoto R, Requa J, Dao T, Ninh A, Tran E, Mai D, et al. Artificial intelligence using convolutional neural networks for real-time detection of early esophageal neoplasia in Barrett's esophagus (with video). Gastrointest Endosc 2020;91 1264–1271.e1261.
[26] Ghatwary N, Zolgharni M, Ye X. Early esophageal adenocarcinoma detection using deep learning methods. Int J Comput Assist Radiol Surg 2019;14:611–21.
[27] Visaggi P, Barberio B, Gregori D, Azzolina D, Martinato M, Hassan C, et al. Systematic review with meta-analysis: artificial intelligence in the diagnosis of oesophageal diseases. Aliment Pharmacol Ther 2022;55:528–40.
[28] Peery AF, Crockett SD, Murphy CC, Lund JL, Dellon ES, Williams JL, et al. Burden and cost of gastrointestinal, liver, and pancreatic diseases in the United States: update 2018. Gastroenterology 2019;156 254–272.e211.
[29] Alaskar H, Hussain A, Al-Aseem N, Liatsis P, Al-Jumeily D. Application of convolutional neural networks for automated ulcer detection in wireless capsule endoscopy images. Sens (Basel) 2019;19.
[30] Wang S, Xing Y, Zhang L, Gao H, Zhang H. Deep convolutional neural network for ulcer recognition in wireless capsule endoscopy: experimental feasibility and optimization. Comput Math Methods Med 2019;2019:7546215.
[31] Zhang L, Zhang Y, Wang L, Wang J, Liu Y. Diagnosis of gastric lesions through a deep convolutional neural network. Dig Endosc 2021;33:788–96.
[32] Shichijo S, Nomura S, Aoyama K, Nishikawa Y, Miura M, Shinagawa T, et al. Application of convolutional neural networks in the diagnosis of Helicobacter pylori infection based on endoscopic images. EBioMedicine 2017;25:106–11.
[33] Itoh T, Kawahira H, Nakashima H, Yata N. Deep learning analyzes Helicobacter pylori infection by upper gastrointestinal endoscopy images. Endosc Int Open 2018;6:E139–e144.
[34] Nakashima H, Kawahira H, Kawachi H, Sakaki N. Artificial intelligence diagnosis of Helicobacter pylori infection using blue laser imaging-bright and linked color imaging: a single-center prospective study. Ann Gastroenterol 2018;31:462–8.
[35] Shichijo S, Endo Y, Aoyama K, Takeuchi Y, Ozawa T, Takiyama H, et al. Application of convolutional neural networks for evaluating Helicobacter pylori infection status on the basis of endoscopic images. Scand J Gastroenterol 2019;54:158–63.
[36] Zheng W, Zhang X, Kim JJ, Zhu X, Ye G, Ye B, et al. High accuracy of convolutional neural network for evaluation of helicobacter pylori infection based on endoscopic images: preliminary experience. Clin Transl Gastroenterol 2019;10:e00109.
[37] Nakashima H, Kawahira H, Kawachi H, Sakaki N. Endoscopic three-categorical diagnosis of Helicobacter pylori infection using linked color imaging and deep learning: a single-center prospective study (with video). Gastric Cancer 2020;23:1033–40.
[38] Wong GL, Ma AJ, Deng H, Ching JY, Wong VW, Tse YK, et al. Machine learning model to predict recurrent ulcer bleeding in patients with history of idiopathic gastroduodenal ulcer bleeding. Aliment Pharmacol Ther 2019;49:912–18.
[39] Yen H-H, Wu P-Y, Su P-Y, Yang C-W, Chen Y-Y, Chen M-F, et al. Performance comparison of the deep learning and the human endoscopist for bleeding peptic ulcer disease. J Med Biol Eng 2021;41:504–13.
[40] Søreide K, Thorsen K, Søreide JA. Predicting outcomes in patients with perforated gastroduodenal ulcers: artificial neural network modelling indicates a highly complex disease. Eur J Trauma Emerg Surg 2015;41:91–8.

[41] Rawla P, Barsouk A. Epidemiology of gastric cancer: global trends, risk factors and prevention. Prz Gastroenterol 2019;14:26–38.

[42] Wu L, Zhou W, Wan X, Zhang J, Shen L, Hu S, et al. A deep neural network improves endoscopic detection of early gastric cancer without blind spots. Endoscopy 2019;51:522–31.

[43] Horiuchi Y, Aoyama K, Tokai Y, Hirasawa T, Yoshimizu S, Ishiyama A, et al. Convolutional neural network for differentiating gastric cancer from gastritis using magnified endoscopy with narrow band imaging. Dig Dis Sci 2020;65:1355–63.

[44] Li L, Chen Y, Shen Z, Zhang X, Sang J, Ding Y, et al. Convolutional neural network for the diagnosis of early gastric cancer based on magnifying narrow band imaging. Gastric Cancer 2020;23:126–32.

[45] Kanesaka T, Lee T-C, Uedo N, Lin K-P, Chen H-Z, Lee J-Y, et al. Computer-aided diagnosis for identifying and delineating early gastric cancers in magnifying narrow-band imaging. Gastrointest Endosc 2018;87:1339–44.

[46] Ikenoyama Y, Hirasawa T, Ishioka M, Namikawa K, Yoshimizu S, Horiuchi Y, et al. Detecting early gastric cancer: comparison between the diagnostic ability of convolutional neural networks and endoscopists. Dig Endosc 2021;33:141–50.

[47] Sharma P, Hassan C. Artificial intelligence and deep learning for upper gastrointestinal neoplasia. Gastroenterology 2022;162:1056–66.

[48] Gong EJ, Bang CS, Lee JJ, Baik GH, Lim H, Jeong JH, et al. Deep-learning-based clinical decision support system for gastric neoplasms in real-time endoscopy: development and validation study. Endoscopy 2023;.

[49] Cui Y, Zhang J, Li Z, Wei K, Lei Y, Ren J, et al. A CT-based deep learning radiomics nomogram for predicting the response to neoadjuvant chemotherapy in patients with locally advanced gastric cancer: a multicenter cohort study. EClinicalMedicine 2022;46:101348.

[50] Hao D, Li Q, Feng QX, Qi L, Liu XS, Arefan D, et al. Identifying prognostic markers from clinical, radiomics, and deep learning imaging features for gastric cancer survival prediction. Front Oncol 2021;11:725889.

[51] Guan X, Lu N, Zhang J. Computed tomography-based deep learning nomogram can accurately predict lymph node metastasis in gastric cancer. Dig Dis Sci 2022;.

[52] Jin C, Jiang Y, Yu H, Wang W, Li B, Chen C, et al. Deep learning analysis of the primary tumour and the prediction of lymph node metastases in gastric cancer. Br J Surg 2021;108:542–9.

[53] Li C, Qin Y, Zhang WH, Jiang H, Song B, Bashir MR, et al. Deep learning-based AI model for signet-ring cell carcinoma diagnosis and chemotherapy response prediction in gastric cancer. Med Phys 2022;49:1535–46.

[54] Flinner N, Gretser S, Quaas A, Bankov K, Stoll A, Heckmann LE, et al. Deep learning based on hematoxylin–eosin staining outperforms immunohistochemistry in predicting molecular subtypes of gastric adenocarcinoma. J Pathol 2022;257:218–26.

[55] Raines D, Arbour A, Thompson HW, Figueroa-Bodine J, Joseph S. Variation in small bowel length: factor in achieving total enteroscopy? Dig Endosc 2015;27:67–72.

[56] Iddan G, Meron G, Glukhovsky A, Swain P. Wireless capsule endoscopy. Nature 2000;405:417.

[57] Lee NM, Eisen GM. 10 years of capsule endoscopy: an update. Expert Rev Gastroenterol Hepatol 2010;4:503–12.

[58] Soffer S, Klang E, Shimon O, Nachmias N, Eliakim R, Ben-Horin S, et al. Deep learning for wireless capsule endoscopy: a systematic review and meta-analysis. Gastrointest Endosc 2020;92 831–839.e838.

[59] Jia X, Meng M.Q.H. A deep convolutional neural network for bleeding detection in Wireless Capsule Endoscopy images. In: 2016 38th annual international conference of the IEEE Engineering in Medicine and Biology Society (EMBC), 16–20 August 2016; 2016. p. 639–642.

[60] Jia X, Meng M.Q.H. Gastrointestinal bleeding detection in wireless capsule endoscopy images using handcrafted and CNN features. In: Proceedings of the annual international conference of the IEEE engineering in medicine and biology society, EMBS; 2017. p. 3154–3157.

[61] Leenhardt R, Vasseur P, Li C, Saurin JC, Rahmi G, Cholet F, et al. A neural network algorithm for detection of GI angiectasia during small-bowel capsule endoscopy. Gastrointest Endosc 2019;89:189–94.

[62] Aoki T, Yamada A, Kato Y, Saito H, Tsuboi A, Nakada A, et al. Automatic detection of blood content in capsule endoscopy images based on a deep convolutional neural network. J Gastroenterolo Hepatol 2020;35:1196–200.

[63] Tsuboi A, Oka S, Aoyama K, Saito H, Aoki T, Yamada A, et al. Artificial intelligence using a convolutional neural network for automatic detection of small-bowel angioectasia in capsule endoscopy images. Dig Endosc 2020;32:382–90.

[64] Fan S, Xu L, Fan Y, Wei K, Li L. Computer-aided detection of small intestinal ulcer and erosion in wireless capsule endoscopy images. Phys Med Biol 2018;63:165001.

[65] Aoki T, Yamada A, Aoyama K, Saito H, Tsuboi A, Nakada A, et al. Automatic detection of erosions and ulcerations in wireless capsule endoscopy images based on a deep convolutional neural network. Gastrointest Endosc 2019;89 357–363.e352.

[66] Alaskar H, Hussain A, Al-Aseem N, Liatsis P, Al-Jumeily D. Application of convolutional neural networks for automated ulcer detection in wireless capsule endoscopy images. Sensors 2019.

[67] Klang E, Barash Y, Margalit RY, Soffer S, Shimon O, Albshesh A, et al. Deep learning algorithms for automated detection of Crohn's disease ulcers by video capsule endoscopy. Gastrointest Endosc 2020;91 606–613.e602.

[68] Wang S, Xing Y, Zhang L, Gao H, Zhang H. A systematic evaluation and optimization of automatic detection of ulcers in wireless capsule endoscopy on a large dataset using deep convolutional neural networks. Phys Med Biol 2019;64:235014.

[69] Chetcuti Zammit S, McAlindon ME, Greenblatt E, Maker M, Siegelman J, Leffler DA, et al. Quantification of celiac disease severity using video capsule endoscopy: a comparison of human experts and machine learning algorithms. Curr Med Imaging 2023;.

[70] Xi Y, Xu P. Global colorectal cancer burden in 2020 and projections to 2040. Transl Oncol 2021;14:101174.

[71] Misawa M, Kudo S-e, Mori Y, Cho T, Kataoka S, Yamauchi A, et al. Artificial intelligence-assisted polyp detection for colonoscopy: initial experience. Gastroenterology 2018;154 2027–2029.e2023.

[72] Urban G, Tripathi P, Alkayali T, Mittal M, Jalali F, Karnes W, et al. Deep learning localizes and identifies polyps in real time with 96% accuracy in screening colonoscopy. Gastroenterology 2018;155 1069–1078.e1068.

[73] Wang P, Xiao X, Glissen Brown JR, Berzin TM, Tu M, Xiong F, et al. Development and validation of a deep-learning algorithm for the detection of polyps during colonoscopy. Nat Biomed Eng 2018;2:741–8.

[74] Yamada M, Saito Y, Imaoka H, Saiko M, Yamada S, Kondo H, et al. Development of a real-time endoscopic image diagnosis support system using deep learning technology in colonoscopy. Sci Rep 2019;9:14465.

[75] Wang P, Berzin TM, Glissen Brown JR, Bharadwaj S, Becq A, Xiao X, et al. Real-time automatic detection system increases colonoscopic polyp and adenoma detection rates: a prospective randomised controlled study. Gut 2019;68:1813–19.

[76] Liu WN, Zhang YY, Bian XQ, Wang LJ, Yang Q, Zhang XD, et al. Study on detection rate of polyps and adenomas in artificial-intelligence-aided colonoscopy. Saudi J Gastroenterol 2020;26:13–19.

[77] Su J-R, Li Z, Shao X-J, Ji C-R, Ji R, Zhou R-C, et al. Impact of a real-time automatic quality control system on colorectal polyp and adenoma detection: a prospective randomized controlled study (with videos). Gastrointest Endosc 2020;91 415–424.e414.

[78] Gong D, Wu L, Zhang J, Mu G, Shen L, Liu J, et al. Detection of colorectal adenomas with a real-time computer-aided system (ENDOANGEL): a randomised controlled study. Lancet Gastroenterol Hepatol 2020;5:352–61.

[79] Wallace MB, Sharma P, Bhandari P, East J, Antonelli G, Lorenzetti R, et al. Impact of artificial intelligence on miss rate of colorectal neoplasia. Gastroenterology 2022;163 295–304.e295.

[80] Rex DK, Berzin TM, Mori Y. Artificial intelligence improves detection at colonoscopy: why aren't we all already using it? Gastroenterology 2022;163:35–7.

[81] Ungaro R, Mehandru S, Allen PB, Peyrin-Biroulet L, Colombel JF. Ulcerative colitis. Lancet 2017;389:1756–70.

[82] Torres J, Mehandru S, Colombel JF, Peyrin-Biroulet L. Crohn's disease. Lancet 2017;389:1741–55.

[83] Lee JS, Kim ES, Moon W. Chronological review of endoscopic indices in inflammatory bowel disease. Clin Endosc 2019;52:129–36.

[84] Da Rio L, Spadaccini M, Parigi TL, Gabbiadini R, Dal Buono A, Busacca A, et al. Artificial intelligence and inflammatory bowel disease: Where are we going? World J Gastroenterol 2023;29:508–20.

[85] Stidham RW, Liu W, Bishu S, Rice MD, Higgins PDR, Zhu J, et al. Performance of a deep learning model vs human reviewers in grading endoscopic disease severity of patients with ulcerative colitis. JAMA Netw Open 2019;2:e193963.

[86] Ozawa T, Ishihara S, Fujishiro M, Saito H, Kumagai Y, Shichijo S, et al. Novel computer-assisted diagnosis system for endoscopic disease activity in patients with ulcerative colitis. Gastrointest Endosc 2019;89 416–421.e411.

[87] Maeda Y, Kudo SE, Mori Y, Misawa M, Ogata N, Sasanuma S, et al. Fully automated diagnostic system with artificial intelligence using endocytoscopy to identify the presence of histologic inflammation associated with ulcerative colitis (with video). Gastrointest Endosc 2019;89:408–15.

[88] Takenaka K, Ohtsuka K, Fujii T, Negi M, Suzuki K, Shimizu H, et al. Development and validation of a deep neural network for accurate evaluation of endoscopic images from patients with ulcerative colitis. Gastroenterology 2020;158:2150–7.

[89] Byrne M, East J, Iacucci M, Panaccione R, Kalapala R, Duvvur N, et al. DOP13 artificial intelligence (AI) in endoscopy—deep learning for detection and scoring of ulcerative colitis (UC) disease activity under multiple scoring systems. J Crohns Colitis 2021;15:S051–2.

[90] Yao H, Najarian K, Gryak J, Bishu S, Rice MD, Waljee AK, et al. Fully automated endoscopic disease activity assessment in ulcerative colitis. Gastrointest Endosc 2021;93 728–736.e721.

[91] Gottlieb K, Requa J, Karnes W, Chandra Gudivada R, Shen J, Rael E, et al. Central reading of ulcerative colitis clinical trial videos using neural networks. Gastroenterology 2021;160 710–719.e712.

[92] Maeda Y, Kudo SE, Ogata N, Misawa M, Iacucci M, Homma M, et al. Evaluation in real-time use of artificial intelligence during colonoscopy to predict relapse of ulcerative colitis: a prospective study. Gastrointest Endosc 2022;95 747–756.e742.

[93] Vande Casteele N, Leighton JA, Pasha SF, Cusimano F, Mookhoek A, Hagen CE, et al. Utilizing deep learning to analyze whole slide images of colonic biopsies for associations between eosinophil density and clinicopathologic features in active ulcerative colitis. Inflamm Bowel Dis 2022;28:539–46.

[94] Peyrin-Biroulet L, Adsul S, Dehmeshki J, Kubassova O. DOP58—an artificial intelligence-driven scoring system to measure histological disease activity in ulcerative colitis. J Crohns Colitis 2022;16 i105.

[95] Villanacci V, Parigi TL, Del Amor R, Mesguer Esbrì P, Gui X, Bazarova A, et al. OP15 A new simplified histology artificial intelligence system for accurate assessment of remission in ulcerative colitis. J Crohns Colitis 2022;16:i015–17.

[96] Stidham RW, Yu D, Zhao X, Bishu S, Rice M, Bourque C, et al. Identifying the presence, activity, and status of extraintestinal manifestations of inflammatory bowel disease using natural language processing of clinical notes. Inflamm Bowel Dis 2022.

[97] Reddy BK, Delen D, Agrawal RK. Predicting and explaining inflammation in Crohn's disease patients using predictive analytics methods and electronic medical record data. Health Inform J 2019;25:1201–18.

[98] Li Y, Pan J, Zhou N, Fu D, Lian G, Yi J, et al. A random forest model predicts responses to infliximab in Crohn's disease based on clinical and serological parameters. Scand J Gastroenterol 2021;56:1030–9.

[99] Waljee AK, Liu B, Sauder K, Zhu J, Govani SM, Stidham RW, et al. Predicting Corticosteroid-free biologic remission with vedolizumab in Crohn's disease. Inflamm Bowel Dis 2018;24:1185–92.

[100] Asrani SK, Devarbhavi H, Eaton J, Kamath PS. Burden of liver diseases in the world. J Hepatol 2019;70:151–71.

[101] Paik JM, Golabi P, Younossi Y, Mishra A, Younossi ZM. Changes in the global burden of chronic liver diseases from 2012 to 2017: the growing impact of NAFLD. Hepatology 2020;72:1605–16.

[102] Ahn JC, Connell A, Simonetto DA, Hughes C, Shah VH. Application of artificial intelligence for the diagnosis and treatment of liver diseases. Hepatology 2021;73:2546–63.

[103] Li W, Huang Y, Zhuang BW, Liu GJ, Hu HT, Li X, et al. Multiparametric ultrasomics of significant liver fibrosis: a machine learning-based analysis. Eur Radiol 2019;29:1496–506.

[104] Gatos I, Tsantis S, Spiliopoulos S, Karnabatidis D, Theotokas I, Zoumpoulis P, et al. Temporal stability assessment in shear wave elasticity images validated by deep learning neural network for chronic liver disease fibrosis stage assessment. Med Phys 2019;46:2298–309.

[105] Chen Y, Luo Y, Huang W, Hu D, Zheng R-q, Cong S-z, et al. Machine-learning-based classification of real-time tissue elastography for hepatic fibrosis in patients with chronic hepatitis B. Comput Biol Med 2017;89:18–23.

[106] Wang K, Lu X, Zhou H, Gao Y, Zheng J, Tong M, et al. Deep learning Radiomics of shear wave elastography significantly improved diagnostic performance for assessing liver fibrosis in chronic hepatitis B: a prospective multicentre study. Gut 2019;68:729.

[107] Kuppili V, Biswas M, Sreekumar A, Suri HS, Saba L, Edla DR, et al. Extreme learning machine framework for risk stratification of fatty liver disease using ultrasound tissue characterization. J Med Syst 2017;41:1–20.

[108] Byra M, Styczynski G, Szmigielski C, Kalinowski P, Michałowski Ł, Paluszkiewicz R, et al. Transfer learning with deep convolutional neural network for liver steatosis assessment in ultrasound images. Int J Comput Assist Radiol Surg 2018;13:1895–903.

[109] Biswas M, Kuppili V, Edla DR, Suri HS, Saba L, Marinhoe RT, et al. Symtosis: a liver ultrasound tissue characterization and risk stratification in optimized deep learning paradigm. Comput Methods Prog Biomed 2018;155:165–77.

[110] Alshagathrh FM, Househ MS. Artificial intelligence for detecting and quantifying fatty liver in ultrasound images: a systematic review. Bioengineering 2022;9:748.

[111] Choi KJ, Jang JK, Lee SS, Sung YS, Shim WH, Kim HS, et al. Development and validation of a deep learning system for staging liver fibrosis by using contrast agent-enhanced CT images in the liver. Radiology 2018;289:688–97.

[112] Ahmed Y, Hussein RS, Basha TA, Khalifa AM, Ibrahim AS, Abdelmoaty AS, et al. Detecting liver fibrosis using a machine learning-based approach to the quantification of the heart-induced deformation in tagged MR images. NMR Biomed 2020;33:e4215.

[113] Gawrieh S, Knoedler DM, Saeian K, Wallace JR, Komorowski RA. Effects of interventions on intra- and interobserver agreement on interpretation of nonalcoholic fatty liver disease histology. Ann Diagn Pathol 2011;15:19–24.

[114] Gawrieh S, Sethunath D, Cummings OW, Kleiner DE, Vuppalanchi R, Chalasani N, et al. Automated quantification and architectural pattern detection of hepatic fibrosis in NAFLD. Ann Diagn Pathol 2020; 47:151518.

[115] Forlano R, Mullish BH, Giannakeas N, Maurice JB, Angkathunyakul N, Lloyd J, et al. High-throughput, machine learning-based quantification of steatosis, inflammation, ballooning, and fibrosis in biopsies from patients with nonalcoholic fatty liver disease. Clin Gastroenterol Hepatol 2020;18 2081–2090.e2089.

[116] Roy M, Wang F, Vo H, Teng D, Teodoro G, Farris AB, et al. Deep-learning-based accurate hepatic steatosis quantification for histological assessment of liver biopsies. Lab Invest 2020;100:1367–83.

[117] Taylor-Weiner A, Pokkalla H, Han L, Jia C, Huss R, Chung C, et al. A machine learning approach enables quantitative measurement of liver histology and disease monitoring in NASH. Hepatology 2021;74.

[118] Kim WR, Biggins SW, Kremers WK, Wiesner RH, Kamath PS, Benson JT, et al. Hyponatremia and mortality among patients on the liver-transplant waiting list. N Engl J Med 2008;359:1018–26.

[119] Konerman MA, Beste LA, Van T, Liu B, Zhang X, Zhu J, et al. Machine learning models to predict disease progression among veterans with hepatitis C virus. PLoS One 2019;14:e0208141.

[120] Eaton JE, Vesterhus M, McCauley BM, Atkinson EJ, Schlicht EM, Juran BD, et al. Primary sclerosing cholangitis risk estimate tool (PREsTo) predicts outcomes of the disease: a derivation and validation study using machine learning. Hepatology 2020;71:214–24.

[121] Speiser JL, Lee WM, Karvellas CJ, Group UALFS. Predicting outcome on admission and post-admission for acetaminophen-induced acute liver failure using classification and regression tree models. PLoS One 2015;10:e0122929.

[122] Briceño J, Cruz-Ramírez M, Prieto M, Navasa M, De Urbina JO, Orti R, et al. Use of artificial intelligence as an innovative donor-recipient matching model for liver transplantation: results from a multicenter Spanish study. J Hepatol 2014;61:1020–8.

[123] Hu C, Anjur V, Saboo K, Reddy KR, O'Leary J, Tandon P, et al. Low predictability of readmissions and death using machine learning in cirrhosis. Am J Gastroenterol 2021;116:336–46.

[124] Qi X, An W, Liu F, Qi R, Wang L, Liu Y, et al. Virtual hepatic venous pressure gradient with CT angiography (CHESS 1601): a prospective multicenter study for the noninvasive diagnosis of portal hypertension. Radiology 2018;290:370–7.

[125] Bosch J, Chung C, Carrasco-Zevallos OM, Harrison SA, Abdelmalek MF, Shiffman ML, et al. A machine learning approach to liver histological evaluation predicts clinically significant portal hypertension in NASH cirrhosis. Hepatology 2021;74:3146–60.

[126] Ahn JC, Attia ZI, Rattan P, Mullan AF, Buryska S, Allen AM, et al. Development of the AI-cirrhosis-ECG score: an electrocardiogram-based deep learning model in cirrhosis. Am J Gastroenterol 2022;117:424–32.

[127] Yang JD, Hainaut P, Gores GJ, Amadou A, Plymoth A, Roberts LR. A global view of hepatocellular carcinoma: trends, risk, prevention and management. Nat Rev Gastroenterol Hepatol 2019;16:589–604.

[128] Streba CT, Ionescu M, Gheonea DI, Sandulescu L, Ciurea T, Saftoiu A, et al. Contrast-enhanced ultrasonography parameters in neural network diagnosis of liver tumors. World J Gastroenterol 2012;18:4427–34.

[129] Hassan TM, Elmogy M, Sallam E-S. Diagnosis of focal liver diseases based on deep learning technique for ultrasound images. Arab J Sci Eng 2017;42:3127–40.

[130] Bharti P, Mittal D, Ananthasivan R. Preliminary study of chronic liver classification on ultrasound images using an ensemble model. Ultrason Imaging 2018;40:357–79.

[131] Schmauch B, Herent P, Jehanno P, Dehaene O, Saillard C, Aubé C, et al. Diagnosis of focal liver lesions from ultrasound using deep learning. Diagn Interv Imaging 2019;100:227–33.

[132] Brehar R, Mitrea DA, Vancea F, Marita T, Nedevschi S, Lupsor-Platon M, et al. Comparison of deep-learning and conventional machine-learning methods for the automatic recognition of the hepatocellular carcinoma areas from ultrasound images. Sens (Basel) 2020;20.

[133] Jin J, Yao Z, Zhang T, Zeng J, Wu L, Wu M, et al. Deep learning radiomics model accurately predicts hepatocellular carcinoma occurrence in chronic hepatitis B patients: a five-year follow-up. Am J Cancer Res 2021;11:576–89.

[134] Lee Y-T, Wang JJ, Zhu Y, Agopian VG, Tseng H-R, Yang JD. Diagnostic criteria and LI-RADS for hepatocellular carcinoma. Clin Liver Dis 2021;17:409–13.

[135] Yasaka K, Akai H, Abe O, Kiryu S. Deep learning with convolutional neural network for differentiation of liver masses at dynamic contrast-enhanced CT: a preliminary study. Radiology 2018;286:887–96.

[136] Shi W, Kuang S, Cao S, Hu B, Xie S, Chen S, et al. Deep learning assisted differentiation of hepatocellular carcinoma from focal liver lesions: choice of four-phase and three-phase CT imaging protocol. Abdom Radiol (NY) 2020;45:2688–97.

[137] Hamm CA, Wang CJ, Savic LJ, Ferrante M, Schobert I, Schlachter T, et al. Deep learning for liver tumor diagnosis part I: development of a convolutional neural network classifier for multi-phasic MRI. Eur Radiol 2019;29:3338–47.

[138] Wang CJ, Hamm CA, Savic LJ, Ferrante M, Schobert I, Schlachter T, et al. Deep learning for liver tumor diagnosis part II: convolutional neural network interpretation using radiologic imaging features. Eur Radiol 2019;29:3348–57.

[139] Zhang Y, Lv X, Qiu J, Zhang B, Zhang L, Fang J, et al. Deep learning with 3D convolutional neural network for noninvasive prediction of microvascular invasion in hepatocellular carcinoma. J Magn Reson Imaging 2021;54:134–43.

[140] Jiang YQ, Cao SE, Cao S, Chen JN, Wang GY, Shi WQ, et al. Preoperative identification of microvascular invasion in hepatocellular carcinoma by XGBoost and deep learning. J Cancer Res Clin Oncol 2021;147:821–33.

[141] Wang G, Jian W, Cen X, Zhang L, Guo H, Liu Z, et al. Prediction of microvascular invasion of hepatocellular carcinoma based on preoperative diffusion-weighted MR using deep learning. Acad Radiol 2021;28 (Suppl 1):S118–s127.

[142] An C, Jiang Y, Huang Z, Gu Y, Zhang T, Ma L, et al. Assessment of ablative margin after microwave ablation for hepatocellular carcinoma using deep learning-based deformable image registration. Front Oncol 2020;10:573316.

[143] Peng J, Kang S, Ning Z, Deng H, Shen J, Xu Y, et al. Residual convolutional neural network for predicting response of transarterial chemoembolization in hepatocellular carcinoma from CT imaging. Eur Radiol 2020;30:413–24.

[144] Liu QP, Xu X, Zhu FP, Zhang YD, Liu XS. Prediction of prognostic risk factors in hepatocellular carcinoma with transarterial chemoembolization using multi-modal multi-task deep learning. EClinicalMedicine 2020;23:100379.

[145] Zhang L, Xia W, Yan ZP, Sun JH, Zhong BY, Hou ZH, et al. Deep learning predicts overall survival of patients with unresectable hepatocellular carcinoma treated by transarterial chemoembolization plus sorafenib. Front Oncol 2020;10:593292.

[146] Mederos MA, Reber HA, Girgis MD. Acute pancreatitis: a review. JAMA 2021;325:382–90.

[147] Andersson B, Andersson R, Ohlsson M, Nilsson J. Prediction of severe acute pancreatitis at admission to hospital using artificial neural networks. Pancreatology 2011;11:328–35.

[148] Fei Y, Gao K, Li WQ. Artificial neural network algorithm model as powerful tool to predict acute lung injury following to severe acute pancreatitis. Pancreatology 2018;18:892–9.

[149] Mofidi R, Duff MD, Madhavan KK, Garden OJ, Parks RW. Identification of severe acute pancreatitis using an artificial neural network. Surgery 2007;141:59–66.

[150] Fei Y, Hu J, Li WQ, Wang W, Zong GQ. Artificial neural networks predict the incidence of portosplenomesenteric venous thrombosis in patients with acute pancreatitis. J Thromb Haemost 2017;15:439–45.

[151] Qiu Q, Nian YJ, Guo Y, Tang L, Lu N, Wen LZ, et al. Development and validation of three machine-learning models for predicting multiple organ failure in moderately severe and severe acute pancreatitis. BMC Gastroenterol 2019;19:118.

[152] Qiu Q, Nian YJ, Tang L, Guo Y, Wen LZ, Wang B, et al. Artificial neural networks accurately predict intra-abdominal infection in moderately severe and severe acute pancreatitis. J Dig Dis 2019;20:486–94.

[153] Cheng Y, Yang J, Wu Q, Cao L, Wang B, Jin X, et al. Machine learning for the prediction of acute kidney injury in patients with acute pancreatitis admitted to the intensive care unit. Chin Med J (Engl) 2022;135:2886–7.

[154] Kiss S, Pintér J, Molontay R, Nagy M, Farkas N, Sipos Z, et al. Early prediction of acute necrotizing pancreatitis by artificial intelligence: a prospective cohort-analysis of 2387 cases. Sci Rep 2022;12:7827.

[155] Kui B, Pintér J, Molontay R, Nagy M, Farkas N, Gede N, et al. EASY-APP: an artificial intelligence model and application for early and easy prediction of severity in acute pancreatitis. Clin Transl Med 2022;12:e842.

[156] Marya NB, Powers PD, Chari ST, Gleeson FC, Leggett CL, Abu Dayyeh BK, et al. Utilisation of artificial intelligence for the development of an EUS-convolutional neural network model trained to enhance the diagnosis of autoimmune pancreatitis. Gut 2021;70:1335–44.

[157] Park W, Chawla A, O'Reilly EM. Pancreatic cancer: a review. JAMA 2021;326:851–62.

[158] Chu LC, Park S, Kawamoto S, Fouladi DF, Shayesteh S, Zinreich ES, et al. Utility of CT radiomics features in differentiation of pancreatic ductal adenocarcinoma from normal pancreatic tissue. AJR Am J Roentgenol 2019;213:349–57.

[159] Das A, Nguyen CC, Li F, Li B. Digital image analysis of EUS images accurately differentiates pancreatic cancer from chronic pancreatitis and normal tissue. Gastrointest Endosc 2008;67:861–7.

[160] Gao X, Wang X. Deep learning for World Health Organization grades of pancreatic neuroendocrine tumors on contrast-enhanced magnetic resonance images: a preliminary study. Int J Comput Assist Radiol Surg 2019;14:1981–91.

[161] Săftoiu A, Vilmann P, Dietrich CF, Iglesias-Garcia J, Hocke M, Seicean A, et al. Quantitative contrast-enhanced harmonic EUS in differential diagnosis of focal pancreatic masses (with videos). Gastrointest Endosc 2015;82:59–69.

[162] Janssens LP, Weston AD, Singh D, Spears G, Harmsen WS, Takahashi N, et al. Determining age and sex-specific distribution of pancreatic whole-gland CT attenuation using artificial intelligence aided image segmentation: associations with body composition and pancreatic cancer risk. Pancreatology 2021;21:1524–30.

[163] Mukherjee S, Patra A, Khasawneh H, Korfiatis P, Rajamohan N, Suman G, et al. Radiomics-based machine-learning models can detect pancreatic cancer on prediagnostic computed tomography scans at a substantial lead time before clinical diagnosis. Gastroenterology 2022;163 1435–1446.e1433.

[164] Dalal V, Carmicheal J, Dhaliwal A, Jain M, Kaur S, Batra SK. Radiomics in stratification of pancreatic cystic lesions: machine learning in action. Cancer Lett 2020;469:228–37.

[165] Machicado JD, Koay EJ, Krishna SG. Radiomics for the diagnosis and differentiation of pancreatic cystic lesions. Diagnostics (Basel) 2020;10.

[166] Wei R, Lin K, Yan W, Guo Y, Wang Y, Li J, et al. Computer-aided diagnosis of pancreas serous cystic neoplasms: a radiomics method on preoperative MDCT images. Technol Cancer Res Treat 2019;18 1533033818824339.

[167] Yang J, Guo X, Ou X, Zhang W, Ma X. Discrimination of pancreatic serous cystadenomas from mucinous cystadenomas with CT textural features: based on machine learning. Front Oncol 2019;9:494.

CHAPTER

50

Artificial intelligence in nutrition research

Mélina Côté[1,2] *and Benoît Lamarche*[1,2]

[1]Centre Nutrition, santé et société (NUTRISS), Institut sur la nutrition et les aliments fonctionnels (INAF), Laval University, Québec, QC, Canada [2]School of Nutrition, Laval University, Québec, QC, Canada

Artificial intelligence (AI), a field undergoing rapid development, has led to numerous advancements in the field of nutrition research. Indeed, AI-based approaches, including machine learning (ML), deep learning (DL) and natural language processing (NLP), have been employed to improve **dietary assessment** tools, to personalize nutritional advice at an individual level (**precision nutrition**) and to improve nutrition and health at a populational level (**public health in nutrition**). This short chapter provides an overview of the applications of AI in nutrition research and discusses current challenges and limitations to address.

Dietary assessment

Traditional dietary assessment tools bear inherent limitations, including that they measure food intakes with systematic and random errors. They may also be subjective, affected by response biases, costly and resource demanding [1]. AI-based methods, such as ML and DL, have thus been employed to attempt to address such limitations and improve dietary intake data collection and analysis. Three main AI-based dietary assessment tools have been developed: audio-based, motion-based and image-based tools.

First, audio-based dietary assessment tools have been developed by training ML and DL algorithms to detect eating, drinking, swallowing or ambient noise sounds from frequency distribution data of foods eaten recorded through microphones [2–4]. For example, a ML algorithm was trained to differentiate the sound frequencies of eating apples, eating chips, drinking water, speaking and ambient noise [2]. Capacity to estimate energy and nutrient intakes using such audio-based approaches has yet to be developed.

Artificial Intelligence in Clinical Practice
DOI: https://doi.org/10.1016/B978-0-443-15688-5.00031-0

Other audio-based dietary assessment tools have been developed using NLP. Indeed, NLP has been employed to extract food names and serving size from food intake data dictated by participants and recorded using microphones [5–10]. This information is then linked to food and nutrient databases to estimate energy and nutrient intakes. These NLP-based algorithms have been incorporated into smartphone applications to facilitate reporting food intakes for users [10].

Second, ML and DL have been developed to detect eating behaviors using motion sensors [11–15]. For example, sensors like accelerometers and gyroscopes have been placed on upper limbs, eyeglasses and meal plates to record motion. ML classification algorithms, like decision trees, random forests and support vector machines, or DL algorithms, like recurrent neural networks, are then employed to recognize eating and noneating actions, utensil used, speaking motion, head movements and walking motion. However, such motion-based ML tools do not allow a comprehensive assessment of dietary intakes, providing limited information on specific foods consumed, serving size and energy and nutrient intakes. Motion-based tools remain to date complementary to other dietary assessment tools.

Finally, AI-based methods have been by far most employed to develop image-based dietary assessment tools [16–21]. Indeed, ML algorithms have been developed to collect, analyze, and classify food images into categories, such as fruits, vegetables, dairy products, grain products and more. Complex DL algorithms, like convolutional neural networks and recurrent neural networks, have been developed to not only recognize foods but also estimate nutrient and calorie intakes. For example, Fang et al. [19] developed a DL algorithm to estimate the energy content of foods in images. The algorithm predicted energy values of the food images with an average (and relatively large) error of 209 kilocalories on an average eating occasion of 546 kilocalories (39% average estimation error). The performance of such food recognition algorithms depends greatly on gathering voluminous datasets to adequately train the algorithms. Therefore large datasets, often ranging from 5000 to 250,000 images of foods, have been and continue to be generated to improve predictive performance of the algorithms [22]. AI-based dietary assessment algorithms are developed to ultimately be incorporated into dietary assessment applications on mobile devices, thereby facilitating the recording of food intakes by users and the acquisition of more objective and precise food and nutrient intakes by dietitians and research teams. Examples of such applications include Snap-n-eat [23], Keenoa [24], and goFOOD [25]. However, the validity and usability of such applications must still be further evaluated [24].

Precision nutrition

Following the coinage of the term *precision medicine*, interest has greatly flourished in regard to personalizing nutritional recommendations according to the numerous factors that can impact one's response to diet, a paradigm described as *precision nutrition*. Precision nutrition aims to obtain a finer understanding of the interindividual variability in metabolic responses to diet and to consider this information when administering nutritional advice to patients, ultimately aiming to achieve better health. AI has played an important role in advancing the field of precision nutrition.

To better understand the interindividual variability in responses to diet, important emphasis has been placed on -omics data, including genomic, metabolomic, and microbiomic data. The capacity of ML algorithms to harness voluminous and multidimensional datasets and to detect nonlinear relationships among variables have made such algorithms salient for the process and the analysis of -omics data. Indeed, supervised and unsupervised ML algorithms have been employed to study genetic variants [26–28], metabolic phenotypes [29–31], and microbiota composition [32,33] in the context of nutrition-related diseases. For instance, genetic variants predicting susceptibility to developing obesity and risk of developing type 2 diabetes were identified using ML algorithms, such as support vector machines and random forests [26,27]. Variants in gut microbiota composition were also identified by ML algorithms as predictors of diet-related health outcomes, such as obesity [32] and ulcerative colitis [33].

A better understanding of the factors that influence one's response to diet has led to the development of personalized dietary recommendation systems. Such systems have been mostly developed to help better predict or manage type 2 diabetes, a disease that can be greatly prevented or managed through nutritional advice. Personalized dietary recommendation systems have been developed to provide personalized advice to patients according to multiple patient characteristics, such as dietary intakes, anthropometrics, sociodemographic characteristics, as well as gut microbiota, genetic and metabolic data [34–36]. For example, a 6-month randomized controlled trial was conducted to compare the clinical effect of a personalized diet with a Mediterranean diet on glycemic control in patients with prediabetes [35]. The personalized diet was determined using a ML algorithm that integrated clinical and microbiomic data to predict daily postprandial glucose response. The results showed significantly greater reductions of daily glucose levels among patients on the personalized diet compared to patients on the Mediterranean diet [35]. Thus, with further research, AI-based applications may assist dieticians and other health professionals in offering personalized nutrition counseling to patients and possibly further improve the health of their patients.

Public health in nutrition

The multiple advantages of AI have led to the use of, ML, DL, and NLP algorithms in the field of public health in nutrition. Indeed, AI-based algorithms have been employed to improve outcome prediction performance in nutritional epidemiology as well as to enhance the collection of geotagged data to better monitor spatial and temporal distributions of food environments and nutrition-related behaviors and diseases.

Nutritional epidemiology

Traditional statistical models in epidemiology are based on a set of rules and assumptions that do not allow for much flexibility in developing prediction models [37–39]. Moreover, the number of variables that can be incorporated in such models are limited.

In contrast, AI includes more flexible algorithms that make few assumptions and that can include a larger set of variables. Hence, several ML algorithms have been employed to predict disease outcomes based on diet-related data. For example, decision tree, random forest, support vector machine and k-nearest neighbor algorithms using dietary intake data have been employed to predict risk of type 2 diabetes [40], obesity [41–44], depression [45], malnutrition [46], as well as cardiometabolic or cardiovascular risk [47–51]. Several studies comparing traditional statistical models to ML algorithms have revealed better prediction performances of the latter compared to the former, but this has not always been the case [52–55]. Moreover, most applications of ML in public health nutrition are oriented toward prediction purposes, but interest has grown regarding the use of ML algorithms for causal inference purposes. Nonparametric algorithms can be used along with doubly robust estimation techniques, such as targeted likelihood estimation (TMLE), to estimate causal effects [56,57]. For instance, by combining a ML algorithm with TMLE, Bodnar et al. [58] demonstrated an inverse and significative association between the consumption of a diet rich in vegetables and fruits and the risk of adverse pregnancy outcomes, an association that was not revealed by a logistic regression model. Hence, AI algorithms could lead to the development of high-performance prediction models, and possibly of causal inference models. However, the added-value of AI algorithms over traditional statistical models has yet to be further evaluated and demonstrated.

Public health monitoring

ML algorithms have been used to collect thousands of georeferenced nutrition-related data. Indeed, ML techniques have been used to extract information about lifestyle habits and food environments from social media, such as *Twitter*, *Instagram*, and *Reddit*, providing additional information to better monitor dietary habits of the population. For example, georeferenced data retrieved and analyzed by ML algorithms have been used to map the quality of food supplies in food deserts [59], to represent the availability of healthy and unhealthy foods in Canada [60], as well as to map out a population's dietary habits and physical activity practices [61]. Other studies have also used ML to analyze social media posts to better understand specific eating behaviors. Indeed, many studies have taken advantage of the abundance of posts publicly available on social media, particularly *Twitter* and *Reddit*, to gather quantitative and qualitative data regarding specific nutrition topics, such as perceptions and thoughts related to a particular topic, and to study correlations among those topics. In these cases, NLP algorithms are largely used to perform topic modeling and content analyses. For example, NLP algorithms have been employed to better characterize perceptions and opinions regarding emotional eating [62], weight loss [63], as well as obesity, diet, diabetes, and exercise [64–67]. Thus, on account of AI, social media has become a new source of information in public health, enhancing the monitoring of a population's health by georeferencing multiple nutrition-related variables. Analysis of social media content using ML can also help better understand beliefs and perceptions related to behavioral dietary patterns.

Current challenges and limitations

Although AI has contributed to many advancements in the fields of dietary assessment, precision nutrition, and public health nutrition, multiple challenges remain and must be addressed to lead to real-life applications in a foreseeable future.

In the field of dietary assessment, AI-based tools remain somewhat subjective and prone to response bias when participants are aware that they are taking a picture or recording their intakes [1]. Also, many tools do not yet calculate with high accuracy calorie and nutrient intakes. Moreover, most AI-based dietary assessment tools are developed to be used in smartphone applications, which requires owning a smartphone, having access to the internet, and having a certain level of digital literacy. Such requirements may represent barriers for some, especially in equity deserving groups. Finally, many AI-based dietary assessment tools have yet to be rigorously validated in epidemiological or clinical settings [1,24,68]. Even if the use of AI-based tools for dietary assessment will likely lead to greater accuracy and precision in dietary assessment compared to relying solely on self-reported data, AI-based tools still face many challenges and limitations, suggesting that such tools will remain complementary to traditional methods in the near future.

In the fields of precision nutrition and public health nutrition, the quantity and quality of data available is an important limitation to real-life applications of AI in nutrition. Indeed, since AI algorithms are training-based, an important quantity of data is essential to adequately develop and test the algorithms. More importantly, these data must be representative of the population to develop algorithms that are equitable and unbiased. Recruiting and retaining participants, especially from equity deserving populations, is an important challenge in nutrition research. Therefore, efforts must be put into collecting more important sample sizes that best represent the target population to adequately develop AI algorithms. Moreover, the quality of data is also a major challenge when collecting data from social media platforms. Indeed, social media data is often incomplete, incorrect and highly biased [69,70]. For instance, there is a notable underrepresentation of equity deserving groups on social media [60,62,69–71], interpretation of results based on social media data is limited since sociodemographic data in unavailable, and data is often collected over a short period of time, reflecting results over only this period of time [64,65]. Hence, although social media can be a relevant source of voluminous and current data, it should be used as a complement to other data sources [72]. Finally, the choice of ML algorithms may also be a challenge to the application of AI in nutrition. Indeed, some ML and DL algorithms are uninterpretable and unexplainable, often referred to as "black-box models," which means that the design of the algorithm as well as the reasoning behind a prediction cannot be observed or explained [73,74]. Such algorithms are problematic in terms of applicability in real-life settings, because of the impossibility to properly translate recommendations to patients and to assure their equitable treatment [75,76]. Thus, further research must be conducted on developing high-performance AI algorithms that are interpretable and explainable.

Conclusion

In conclusion, AI has generated new data sources and increased analytical power in nutrition research, leading to important advancements in the fields of dietary assessment, precision nutrition and public health nutrition. The promising applications of AI in nutrition suggest that such approaches will continue to contribute to nutrition research, ultimately yielding real-life applications. Nonetheless, the hype around AI must not overshadow its inherent challenges, including verifying acceptability among populations, validity and applicability of AI-based tools in real-life settings and addressing multiple ethical concerns associated with AI-based algorithms and applications.

Major takeaway points

- The emergence of AI has contributed to the field of dietary assessment, precision nutrition, and public health nutrition by generating new data sources and increasing analytical power.
- For dietary assessment, AI has led to the development of algorithms and applications capable of recognizing dietary intakes through audio recordings, motion sensors and images, and, in some cases, measuring serving and nutrient intakes.
- For precision nutrition, AI has helped to better understand the interindividual variability in metabolic responses to diet, ultimately contributing to personalizing nutritional advice to patients.
- For public health nutrition, AI has led to the development of more flexible prediction models in nutrition-outcome studies, as well as to the collection and analysis of geotagged data to better monitor food environments and nutrition-related behaviors and diseases.
- However, important challenges and limitations remain and must be addressed to yield real-life applications of AI-based approaches and methods in nutrition research.

References

[1] Zhao X, Xu X, Li X, He X, Yang Y, Zhu S. Emerging trends of technology-based dietary assessment: a perspective study. Eur J Clin Nutr 2021;75(4):582–7.
[2] Kalantarian H, Sarrafzadeh M. Audio-based detection and evaluation of eating behavior using the smartwatch platform. Comput Biol Med 2015;65:1–9.
[3] Makeyev O, Lopez-Meyer P, Schuckers S, Besio W, Sazonov E. Automatic food intake detection based on swallowing sounds. Biomed Signal Process Control 2012;7(6):649–56.
[4] Detection of food intake events from throat microphone recordings using convolutional neural networks. In: Tugtekin Turan MA, Erzin E, editors. International conference on multimedia & expo workshops, 2018. IEEE; 2018.
[5] Hezarjaribi N, Reynolds CA, Miller DT, Chaytor N, Ghasemzadeh H. S2NI: a mobile platform for nutrition monitoring from spoken data. IEEE Eng Med Biol Soc 2016;1991–4.
[6] Hezarjaribi N, Mazrouee S, Ghasemzadeh H. Speech2Health: a mobile framework for monitoring dietary composition from spoken data. IEEE J Biomed Health Inform 2018;22(1):252–64.
[7] Korpusik M, Huang C, Price M, Glass J. Distributional semantics for understanding spoken meal descriptions. In: International conference on acoustics, speech and signal processing. IEEE; 2016. p. 6070–4.

[8] Korpusik M, Glass J. Spoken language understanding for a nutrition dialogue system. IEEE ACM Trans Audio Speech Lang Process 2017;25(7):1450–61.
[9] Dodd CT, Adam MTP, Rollo ME. Speech recording for dietary assessment: a systematic literature review. IEEE Access 2022;10:37658–69.
[10] Taylor S, Korpusik M, Das S, Gilhooly C, Simpson R, Glass J, et al. Use of natural spoken language with automated mapping of self-reported food intake to food composition data for low-burden real-time dietary assessment: method comparison study. J Med Internet Res 2021;23(12):e26988.
[11] Heydarian H, Adam M, Burrows T, Collins C, Rollo ME. Assessing eating behaviour using upper limb mounted motion sensors: a systematic review. Nutrients. 2019;11(5):1168.
[12] Chung J, Chung J, Oh W, Yoo Y, Lee WG, Bang H. A glasses-type wearable device for monitoring the patterns of food intake and facial activity. Sci Rep 2017;7:41690.
[13] Farooq M, Sazonov E. A novel wearable device for food intake and physical activity recognition. Sensors. 2016;16(7):1067.
[14] Farooq M, Doulah A, Parton J, McCrory MA, Higgins JA, Sazonov E. Validation of sensor-based food intake detection by multicamera video observation in an unconstrained environment. Nutrients. 2019;11(3):609.
[15] Mertes G, Ding L, Chen W, Hallez H, Jia J, Vanrumste B. Measuring and localizing individual bites using a sensor augmented plate during unrestricted eating for the aging population. IEEE J Biomed Health Inform 2020;24(5):1509–18.
[16] Jia W, Li Y, Qu R, Baranowski T, Burke LE, Zhang H, et al. Automatic food detection in egocentric images using artificial intelligence technology. Public Health Nutr 2019;22(7):1168–79.
[17] Wang W, Min W, Li T, Dong X, Li H, Jiang S. A review on vision-based analysis for automatic dietary assessment. Trends Food Sci Technol 2022;122:223–37.
[18] Silva B, Juan C. A survey on automated food monitoring and dietary management systems. J Health Med Inform 2017;8(3):272.
[19] Fang S, Shao Z, Kerr DA, Boushey CJ, Zhu F. An end-to-end image-based automatic food energy estimation technique based on learned energy distribution images: protocol and methodology. Nutrients. 2019;11(4):877.
[20] Lo FPW, Sun Y, Qiu J, Lo B. Image-based food classification and volume estimation for dietary assessment: a review. IEEE J Biomed Health Inform 2020;24(7):1926–39.
[21] Mezgec S, Korousic Seljak B. NutriNet: a deep learning food and drink image recognition system for dietary assessment. Nutrients. 2017;9(7):657.
[22] Tahir G, Loo CK. A review of the vision-based approaches for dietary assessment. arXiv pre-print, 2021.
[23] Zhang W, Yu Q, Siddiquie B, Divakaran A, Sawhney H. "Snap-n-eat": food recognition and nutrition estimation on a smartphone. J Diabetes Sci Technol 2015;9(3):525–33.
[24] Ji Y, Plourde H, Bouzo V, Kilgour RD, Cohen TR. Validity and usability of a smartphone image-based dietary assessment app compared to 3-day food diaries in assessing dietary intake among canadian adults: randomized controlled trial. JMIR Mhealth Uhealth 2020;8(9):e16953.
[25] Lu Y, Stathopoulou T, Vasiloglou MF, Pinault LF, Kiley C, Spanakis EK, et al. goFOOD(TM): an artificial intelligence system for dietary assessment. Sens (Basel) 2020;20(15):4283.
[26] Curbelo Montañez CA, Fergus P, Hussain A, Al-Jumeily D, Dorak MT, Abdullah R. Evaluation of phenotype classification methods for obesity using direct to consumer genetic data. In: Huang D-S, Jo K-H, Figueroa-Garcia JC, editors. Intelligent conference on intelligent computing. Cham: Springer; 2017. p. 350–62.
[27] Lopez B, Torrent-Fontbona F, Vinas R, Fernandez-Real JM. Single nucleotide polymorphism relevance learning with random forests for type 2 diabetes risk prediction. Artif Intell Med 2018;85:43–9.
[28] Wang Y, Zhang L, Niu M, Li R, Tu R, Liu X, et al. Genetic risk score increased discriminant efficiency of predictive models for type 2 diabetes mellitus using machine learning: cohort study. Front Public Health 2021;9:606711.
[29] Hillesheim E, Brennan L. Metabotyping and its role in nutrition research. Nutr Res Rev 2020;33(1):33–42.
[30] Urpi-Sarda M, Almanza-Aguilera E, Llorach R, Vázquez-Fresno R, Estruch R, Corella D, et al. Non-targeted metabolomic biomarkers and metabotypes of type 2 diabetes: a cross-sectional study of PREDIMED trial participants. Diabetes Metab 2019;45(2):167–74.
[31] Hall H, Perelman D, Breschi A, Limcaoco P, Kellogg R, McLaughlin T, et al. Glucotypes reveal new patterns of glucose dysregulation. PLoS Biol 2018;16(7):e2005143.

[32] Fernandez-Navarro T, Diaz I, Gutierrez-Diaz I, Rodriguez-Carrio J, Suarez A, de Los Reyes-Gavilan CG, et al. Exploring the interactions between serum free fatty acids and fecal microbiota in obesity through a machine learning algorithm. Food Res Int 2019;121:533–41.
[33] Barberio B, Facchin S, Patuzzi I, Ford AC, Massimi D, Valle G, et al. A specific microbiota signature is associated to various degrees of ulcerative colitis as assessed by a machine learning approach. Gut Microbes 2022;14(1):2028366.
[34] Zeevi D, Korem T, Zmora N, Israeli D, Rothschild D, Weinberger A, et al. Personalized nutrition by prediction of glycemic responses. Cell. 2015;163(5):1079–94.
[35] Ben-Yacov O, Godneva A, Rein M, Shilo S, Kolobkov D, Koren N, et al. Personalized postprandial glucose response-targeting diet versus mediterranean diet for glycemic control in prediabetes. Diabetes Care 2021;44(9):1980–91.
[36] Fazakis N, Kocsis O, Dritsas E, Alexiou S, Fakotakis N, Moustakas K. Machine learning tools for long-term type 2 diabetes risk prediction. IEEE Access 2021;9:103737–57.
[37] Côté M, Lamarche B. Artificial intelligence in nutrition research: perspectives on current and future applications. Appl Physiol Nutr Metab 2021;1–8.
[38] Lavigne M, Mussa F, Creatore MI, Hoffman SJ, Buckeridge DL. A population health perspective on artificial intelligence. Healthc Manag Forum 2019;32(4):173–7.
[39] Morgenstern JD, Rosella LC, Costa AP, de Souza RJ, Anderson LN. Perspective: big data and machine learning could help advance nutritional epidemiology. Adv Nutr 2021;12(3):621–31.
[40] Xiong X-l, Zhang R-x, Bi Y, Zhou W-h, Yu Y, Zhu D-l. Machine learning models in type 2 diabetes risk prediction: results from a cross-sectional retrospective study in Chinese adults. Curr Med Sci 2019;39(4):582–8.
[41] Fu Y, Gou W, Hu W, Mao Y, Tian Y, Liang X, et al. Integration of an interpretable machine learning algorithm to identify early life risk factors of childhood obesity among preterm infants: a prospective birth cohort. BMC Med 2020;18(1):184.
[42] Selya AS, Anshutz D. Machine learning for the classification of obesity from dietary and physical activity patterns. In: Giabbanelli P, Mago V, Papageorgiou E, editors. Advanced data analytics in health. smart innovation, systems and technologies, vol. 93. Cham: Springer; 2018. p. 77–97.
[43] Using machine learning to predict obesity in high school students. In: Zheng Z, Ruggiero K, editors. International conference on bioinformatics and biomedicine, 2017. IEEE; 2017.
[44] Dunstan J, Aguirre M, Bastías M, Nau C, Glass TA, Tobar F. Predicting nationwide obesity from food sales using machine learning. Health Inform J 2020;26(1):652–63.
[45] Oh J, Yun K, Maoz U, Kim TS, Chae JH. Identifying depression in the National Health and Nutrition Examination Survey data using a deep learning algorithm. J Affect Disord 2019;257:623–31.
[46] Talukder A, Ahammed B. Machine learning algorithms for predicting malnutrition among under-five children in Bangladesh. Nutrition. 2020;78:110861.
[47] Panaretos D, Koloverou E, Dimopoulos AC, Kouli GM, Vamvakari M, Tzavelas G, et al. A comparison of statistical and machine-learning techniques in evaluating the association between dietary patterns and 10-year cardiometabolic risk (2002–2012): the ATTICA study. Br J Nutr 2018;120(3):326–34.
[48] Zhao Y, Naumova EN, Bobb JF, Claus Henn B, Singh GM. Joint associations of multiple dietary components with cardiovascular disease risk: a machine-learning approach. Am J Epidemiol 2021;190(7):1353–65.
[49] Rigdon J, Basu S. Machine learning with sparse nutrition data to improve cardiovascular mortality risk prediction in the USA using nationally randomly sampled data. BMJ Open 2019;9(11):e032703.
[50] Morgenstern JD, Rosella LC, Costa AP, Anderson LN. Development of machine learning prediction models to explore nutrients predictive of cardiovascular disease using Canadian linked population-based data. Appl Physiol Nutr Metab 2022;47(5):529–46.
[51] Alaa AM, Bolton T, Di Angelantonio E, Rudd JHF, van der Schaar M. Cardiovascular disease risk prediction using automated machine learning: a prospective study of 423,604 UK Biobank participants. PLoS One 2019;14(5):e0213653.
[52] Côté M, Osseni MA, Brassard D, Carbonneau E, Robitaille J, Vohl MC, et al. Are machine learning algorithms more accurate in predicting vegetable and fruit consumption than traditional statistical models? An exploratory analysis. Front Nutr 2022;9:740898.
[53] Christodoulou E, Ma J, Collins GS, Steyerberg EW, Verbakel JY, Van Calster B. A systematic review shows no performance benefit of machine learning over logistic regression for clinical prediction models. J Clin Epidemiol 2019;110:12–22.

[54] Lynam AL, Dennis JM, Owen KR, Oram RA, Jones AG, Shields BM, et al. Logistic regression has similar performance to optimised machine learning algorithms in a clinical setting: application to the discrimination between type 1 and type 2 diabetes in young adults. Diagn Progn Res 2020;4:6.

[55] Nusinovici S, Tham YC, Chak Yan MY, Wei Ting DS, Li J, Sabanayagam C, et al. Logistic regression was as good as machine learning for predicting major chronic diseases. J Clin Epidemiol 2020;122:56–69.

[56] Naimi AI, Balzer LB. Stacked generalization: an introduction to super learning. Eur J Epidemiol 2018;33(5):459–64.

[57] Schuler MS, Rose S. Targeted maximum likelihood estimation for causal inference in observational studies. Am J Epidemiol 2017;185(1):65–73.

[58] Bodnar LM, Cartus AR, Kirkpatrick SI, Himes KP, Kennedy EH, Simhan HN, et al. Machine learning as a strategy to account for dietary synergy: an illustration based on dietary intake and adverse pregnancy outcomes. Am J Clin Nutr 2020;111(6):1235–43.

[59] De Choudhury M, Sharma S, Kiciman E. Characterizing dietary choices, nutrition, and language in food deserts via social media. In: Proceedings of the 19th ACM conference on computer-supported cooperative work & social computing: association for computing machinery; 2016. p. 1157–70.

[60] Widener MJ, Li W. Using geolocated Twitter data to monitor the prevalence of healthy and unhealthy food references across the US. Appl Geogr 2014;54:189–97.

[61] Shah N, Srivastava G, Savage DW, Mago V. Assessing Canadians health activity and nutritional habits through social media. Front Public Health 2019;7:400.

[62] Hwang Y, Kim HJ, Choi HJ, Lee J. Exploring abnormal behavior patterns of online users with emotional eating behavior: topic modeling study. J Med Internet Res 2020;22(3):e15700.

[63] Liu Y, Yin Z. Understanding weight loss via online discussions: content analysis of reddit posts using topic modeling and word clustering techniques. J Med Internet Res 2020;22(6):e13745.

[64] Shaw G, Karami A. Computational content analysis of negative tweets for obesity, diet, diabetes, and exercise. Proc Assoc Inf Sci Technol 2017;54(1):357–65.

[65] Karami A, Dahl AA, Turner-McGrievy G, Kharrazi H, Shaw G. Characterizing diabetes, diet, exercise, and obesity comments on Twitter. Int J Inf Manag 2018;38(1):1–6.

[66] Money V, Karami A, Turner-McGrievy B, Kharrazi H. Seasonal characterization of diet discussions on Reddit. Proc Assoc Inf Sci Technol 2020;57(1):e320.

[67] Yeruva VK, Junaid S, Lee Y. Contextual word embeddings and topic modeling in healthy dieting and obesity. J Healthc Inform Res 2019;3(2):159–83.

[68] Vasiloglou MF, Marcano I, Lizama S, Papathanail I, Spanakis EK, Mougiakakou S. Multimedia data-based mobile applications for dietary assessment. J Diabetes Sci Technol 2022;17(4):1056–65. Available from: http://doi.org/10.1177/19322968221085026.

[69] Lanfranchi V. Machine learning and social media in crisis management: agility vs ethics. In: Proceedings of the international conference on information systems for crisis response and management IMT mines Albi-Carmaux (École Mines-Télécome); 2017.

[70] Matheny ME, Whicher D, Thadaney Israni S. Artificial intelligence in health care: a report from the national academy of medicine. JAMA. 2020;323(6):509–10.

[71] Safdar NM, Banja JD, Meltzer CC. Ethical considerations in artificial intelligence. Eur J Radiol 2020;122:108768.

[72] Lynn T, Rosati P, Leoni Santos G, Endo PT. Sorting the healthy diet signal from the social media expert noise: preliminary evidence from the healthy diet discourse on Twitter. Int J Environ Res Public Health 2020;17(22):8557.

[73] Marcinkevics R, Vogt JE. Interpretability and explainability: a machine learning zoo mini-tour. arXiv pre-print, 2020.

[74] Amann J, Blasimme A, Vayena E, Frey D, Madai VI, Precise Qc. Explainability for artificial intelligence in healthcare: a multidisciplinary perspective. BMC Med Inform Decis Mak 2020;20(1):310.

[75] Landry LG, Ali N, Williams DR, Rehm HL, Bonham VL. Lack of diversity in genomic databases is a barrier to translating precision medicine research into practice. Health Aff 2018;37(5):780–5.

[76] Reel PS, Reel S, Pearson E, Trucco E, Jefferson E. Using machine learning approaches for multi-omics data analysis: a review. Biotechnol Adv 2021;49:107739.

CHAPTER

51

Artificial intelligence in cardiac electrophysiology

Sulaiman S. Somani[1], *Sanjiv M. Narayan*[1,2] *and Albert J. Rogers*[1,2]

[1]Department of Medicine, Stanford University School of Medicine, Stanford, CA, United States
[2]Cardiovascular Institute, Stanford University School of Medicine, Stanford, CA, United States

Introduction

Cardiac electrophysiology (EP) is a specialty involving the study and treatment of heart rhythm disorders. These disorders are common and may have severe health consequences. Important information about heart rhythm disorders often lies within large collections of data, which must be properly collected, organized, analyzed, and acted upon to select the appropriate treatment course. However, as the number of patients with heart rhythm disorders increases, and the volume of data collected from each patient increases, there has been an explosion in the data that is accessible for interpretation. As datasets grow larger and more diverse, from varying data sources ranging from wearable monitors to three-dimensional imaging studies, the capacity of experts to manually integrate the information into a clinical workflow becomes a bottleneck. In the wake of these challenges, artificial intelligence (AI) has found fertile ground in implementation. EP data, including patient characteristics, vital signs, electrical signals, biomarkers, drug administration, imaging, and genomes, are increasingly stored centrally. In central repositories, they can be accessed *en masse* and federated between institutions. Additionally, these data elements are increasingly structured, which helps more easily provide the resources necessary for training AI models. In this chapter, we will discuss the milestones achieved, unmet needs in the specialty, and future directions in the implementation of AI for patients with atrial arrhythmias, ventricular arrhythmias (VAs), and sudden cardiac death (SCD). We will review the barriers to achieving further implementation of AI in EP and possible pathways to overcome them.

Artificial Intelligence in Clinical Practice
DOI: https://doi.org/10.1016/B978-0-443-15688-5.00021-8

Artificial intelligence to guide management of atrial fibrillation

Atrial fibrillation (AF) is the most common sustained arrhythmia, and its incidence is rising rapidly in both developed and developing nations [1]. AF is characterized by disorganized electrical activity from enhanced automaticity, re-entry circuits, and triggered activity in the background of structural, electrical, epigenetic, and genomic predispositions [2]. Patients with AF experience a reduced quality of life and are at increased risk for stroke and heart failure. In this section, we will review how AI has been applied to guide medical management across the continuum of care for patients with AF. Fig. 51.1 provides an overview of the patient journey with AF and the points at which AI has been or is poised to be employed. Large populations at risk for AF may be screened for the presence of occult disease or be assessed for the risk of developing the arrhythmia. Once a diagnosis of AF has been made, identification of the underlying cause of the disease and the curation of a personalized plan for AF management are topics of current research. Decisions for treatment of AF with drugs or interventions are made in the clinical context and may eventually benefit from AI prediction models or clinical decision support tools. Finally, implementation of AI to enable robust and efficient follow-up monitoring during continued care can provide feedback to optimize these upstream decisions.

Large-scale atrial fibrillation screening

AF is associated with a range of physiological effects, symptoms, and associations with other diseases. AF may be a sign that there is a problem with thyroid function, such as

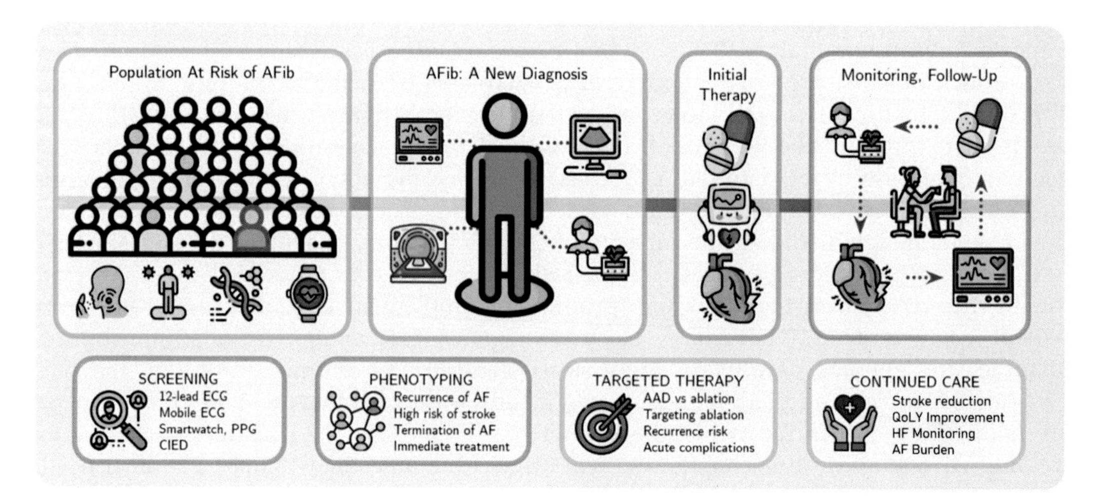

FIGURE 51.1 The journey of care of patients with AF. Large populations (left panel) at risk of developing AF are screened or risk is predicted. New diagnoses of AF (middle left) require phenotyping to understand treatment decisions, prognosis, and pathophysiology of disease. Targeted therapy (middle right) is provided using AI-enabled tools. Monitoring and follow-up (right panel) during continued care provide feedback on the upstream process.

hyperthyroidism, pulmonary disease, such as chronic bronchitis or obstructive sleep apnea, or other cardiac disease, such as heart failure, valvular disease, or inflammation. Serious downstream effects of AF may include stroke, systemic embolization, or arrhythmia-related cardiomyopathy. While symptomatic AF may cause palpitations, chest pain, shortness of breath, loss of consciousness, or fatigue, many patients with AF may be asymptomatic or remain undiagnosed due to a lack of access to care. Several clinical risk scores have been validated with modest accuracy for the risk of undiagnosed or incident AF, including EHR-AF [3], CHARGE-AF [4,5], and C2HEST [6]. Large-scale AF screening has become possible through development of new wearable technologies that have reduced the cost and effort for assessment [7,8]. Large-scale AF screening has been enabled by AI in a number of preexisting modalities including 12-lead electrocardiogram (ECG), consumer device ECG, and photoplethysmography (PPG) devices. Each modality has been evaluated for accuracy of rhythm identification or risk prediction (Table 51.1). Yet, it is still unknown whether treatment of asymptomatic AF detected by screening methods reduces stroke risk [17].

12-Lead electrocardiogram

The standard 12-lead ECG is a noninvasive recording of the heart's electrical activity from the skin of the thorax and limbs. The standard recording is taken with a patient in the supine position for 10 seconds. The recording is generated by measuring the differences across the electrodes, filtering, and digitally recording them for review by a clinician. Modern sampling rates for a standard ECG are >1000 Hz and filtering of signals generally employs a low-frequency cutoff of 0.05–0.5 Hz and a high-frequency cutoff of >150 Hz in adults and higher in children [18]. Convolutional neural network-based models to detect patients currently in sinus rhythm who may have AF (or will develop it within the coming month) have been developed using retrospective cohorts [19] and later were prospectively validated [16]. Longer-term prediction (up to 5 years) of AF incidence has also recently been reported and validated in external cohorts using ECG alone and in conjunction with clinical comorbidity information [20].

Wearable and consumer electrocardiogram devices

Advances in high fidelity amplifiers and communication with smart phone technology have enabled several consumer ECG devices. These devices allow the ECG recording of 1–6 signals representative of the frontal leads of the standard ECG configuration, most commonly lead I (between the left and right arms). These devices are generally consumer activated and record for several seconds at frequencies and filtering settings somewhat more restrictive than the standard 12-lead ECG to reduce the influence of noise in less-controlled settings. The devices have been made in different form factors including watch bands [14,21], watches, adhesive patches [9], and handheld devices [15,22]. Companies developing these devices provide automated and FDA-cleared assessments of recorded rhythms including bradycardia, tachycardia, normal rhythm, and AF. The performance of these algorithms was first evaluated on patients undergoing cardioversion and compared against physician interpretation with a high level of agreement. Prescription-grade patch-based recordings have been assessed for AI-enabled automated rhythm diagnoses including AF and other important rhythm diagnoses with excellent performance in comparison to consensus of clinician interpreters [22].

TABLE 51.1 Machine learning-based models addressing needs in cardiac electrophysiology.

Study	Data	Outcome	Ground truth	Architecture	Performance and limitations
Hannun et al. [9]	91,232 Ziopatch recordings (30 seconds) from 53,545 patients (2013–2017), sampled based on a specific rhythm demonstration (below). Training set labeled by ECG technicians as part of the iRhythm workflow.	Classification of arrhythmia (12 categories: AF/AFL, AVB, Bigeminy, EAR, IVR, Junctional, Noise, NSR, SVT, Trigeminy, VT, Wenkebach); performance with a separate panel of individual Cardiologists ($N = 6$) who labeled the test set as well	Labeled by expert consensus panel ($N = 9$; 8 EP, 1 non-EP) of cardiologists	CNN	AUC 0.97. F1-score 0.84 exceeded that of independent Cardiologists (0.780, on average). VT was one rhythm that the model performed worse on w.r.t. F1 score (thrown off by IVR vs VT (rate >100) and SVT with aberrancy) though with increased sensitivity (makes sense, catching these things); when accounted for these, the model again improved performance.
Apple Heart Study Perez et al. [10]	419,297 participants over 8 months, from an Apple Watch (11/2017–8/2018), age > 22, English proficiency.	Over a median of 117 days of monitoring, 2161 participants (0.52%) received notifications of irregular pulse	ECG patch worn over 7 days	Proprietary	0.5% (2161) notified of irregular pulse. 450 returned ECG patches. 34% of these had AF. Misclassification bias given structure of enrollment (pragmatic, open label, siteless). Younger demographic—more likely those with smartwatch finesse
Huawei Heart Study Guo et al. [11]	187,912 individuals between 10/2018–5/2019	AF	Secondary evaluation via telecare center, clinical evaluation, ECG, 24 h Holter	N/A	of whom 0.23% received "suspected AF" notification. ~ one half followed up (262/424) and of them 87% confirmed as having AF. 25% of these were high-risk and started on A/C.
Torres-Soto et al. [12]	>500 K labeled PPG signals from >100 individuals undergoing elective cardioversions and stress test, transfer learning from synthetic data.	predict AF from NSR.	Chart history	CDAE autoencoder, with transfer learning	F1 score 0.93 on prospectively collected dataset. Novel approach that leverages CDAE to improve performance (F1-score 0.54→0.96 on validation data). Performs both quality assessment and rhythm classification. Validated on IEEE open dataset.

Han et al. [13]	3185 veterans from 2004–2009 with remote ICD monitoring including raw electrograms that detected AF and not on oral A/C with and without stroke event.	Predict stroke risk	VA EHR	CNN, RF	AUC RF 0.662 > CHADS-VASC AUC 0.5. Combining CHADS-VASC + RF + CNN improved performance to 0.634 (test).
Bumgarner et al. [14]	100 patients undergoing elective DCCV for AF with pre-DCCV and post-DCCV KB (Kardioband) and ECG recordings.	How good the KB interpretation algorithm works (comparison with physician interpretation) and how well the KB device captures data (comparison with ECG)	ECG diagnosis of AF (simultaneous ECG, KB recordings).	N/A	Nearly 1/3 of recordings were noninterpretable. Comparable performance of KB algo and physicians (93% vs 99% sens, 84% vs 83% spec), K = 0.88 on head-to-head interrater comparison. Limitations: prespecified cohort of patients undergoing DCCV
Wegner et al. [15]	9 patients, 296 ECGs	Kardia ECG monitor (standard placement to give lead I, novel placement on sternum to give novel parasternal lead).	12-lead ECG diagnosis of AF by cardiac electrophysiologist.	N/A	High NPV (100%), low sensitivity, low specificity, low accuracy. Validate both recording of NPL and algorithm performance. Unknown appropriateness for screening. Readings: 100%, 96% sensitivity for AF (94%, 97% spec).
Noseworthy et al. [16]	1003 patients, ECGs while in NSR	Prediction of AF (or rather high-risk and low-risk AF groups)	Holter monitor, up to 30 days, after flagged	CNN	Increased risk of detection of AF compared to standard of care (10.6% vs 3.6%) in high risk and (2.4% vs 0.9%) in low-risk groups, which improves effectiveness of picking those who should be screened. Yield ~1 in 13 patients chosen.

Photoplethysmography

Perhaps the most nonobtrusive wearable rhythm assessment method is the PPG built into a wristwatch. Although the signal derived from this sensor is the pulse rather than the ECG, the pattern of the pulse is related to the heart rhythm and can be used for AF screening. The benefit of this modality is that it can be worn continuously and does not require activation by the patient. However, the pulse rate is limited in accuracy compared with the gold-standard ECG measurement [23]. Several studies have looked at using this as a tool, combined with a method of confirmation of the ECG, for large-scale screening of AF. Because of the widespread availability of these devices, the trials are amenable to rapid recruitment in a pragmatic study design [10,11]. The Apple Heart Study successfully recruited 419,297 participants over 8 months in partnership with Stanford University and assessed accuracy and frequency of pulse irregularity notifications delivered by the device [10]. Specialized methods of PPG signal processing have been developed to filter noise from signals and improve detection of arrhythmias with AI methods, including convolutional autoencoding [12].

Advances in wearable technology, data collection, and implementation of AI in AF rhythm analysis have established the feasibility of large-scale AF screening using several modalities and sensors. However, future work is required to address several remaining issues. Clinical trials are required to show that screening of asymptomatic individuals results in treatment that prevents cardiovascular morbidity or mortality (e.g., prevention of stroke with administration of anticoagulants). Further work is required to determine whether early intervention to AF therapy in the screened population can reduce arrhythmia progression. Some evidence shows that early intervention with catheter ablation may reduce adverse outcomes in early symptomatic patients with AF [24,25], but it is unclear if this extends to the large-scale asymptomatic population. Finally, prevention of AF, through techniques, such as lifestyle modification or treatment of conditions that predispose to AF, requires further work before application of predictive models.

Prediction of stroke in atrial fibrillation

Thromboembolic phenomena, including cerebrovascular accident (stroke) and systemic embolization, are some of the most feared complications of AF. While not all embolization in patients with AF is due to the heart rhythm, proper identification of patients at risk and mitigation with anticoagulation has high clinical utility [26]. Systemic anticoagulation is an effective method of reducing these events, with meta-analyses demonstrating a near 60% reduction in stroke risk in those with AF [27]. However, long-term anticoagulation with vitamin K antagonists, low-molecular weight heparin, and direct oral anticoagulants also comes with an increased risk of bleeding. The tradeoff between thromboembolic phenomena and bleeding events functions in a dose-dependent manner, with studies demonstrating increased stroke risk and decreased bleeding risk in lower-dose DOACs compared to normal dose [28]. Clinical scores, such as CHA_2DS_2-VASc [29], have been applied in detecting the risk of stroke in patients with AF and are used to guide therapy with anticoagulants; however, the risk of embolization also correlates with risk of bleeding, as demonstrated by the shared features between CHA_2DS_2-VASc and HAS-BLED [30,31]. For these reasons, AI holds tremendous promise to leverage novel data streams and improved discrimination between thrombotic and bleeding risks to better guide anticoagulant initiation.

In current literature, AI has been applied to raw electrocardiographic data streams from either body surface ECGs or remote telemonitoring data. One approach sought to etiologically identify AF in patients with cryptogenic stroke by using body surface ECGs. Researchers used a previously designed AI model to predict AF from ECG waveforms in which the patient was in a normal sinus rhythm [30] and were unable to predict incident stroke *per se*. However, the team did reconfirm strong correlation between the risk predicted from this model and the likelihood of detecting AF on routine ambulatory monitoring after the incident stroke event (e.g., establishing the most likely etiology of stroke) [32]. Further work by the same group refined the output of this AF model by categorizing patients at low- and high-risk of AF from normal ECGs, which demonstrated that 62% of those patients in the high-AF-risk group had an AF-related stroke within 3 years [33], thus identifying a cohort more likely to benefit from anticoagulation. Separately, signatures of daily AF burden from cardiac implantable electronic devices were used to train both independent and ensemble machine learning models and demonstrated an improvement in stroke risk prediction compared with the CHA2DS2-VASc score [13], highlighting how patterns of arrhythmia recording by devices may be an additionally valuable feature in determining stroke risk.

Imaging modalities, such as echocardiography, computed tomography (CT), and cardiac magnetic resonance imaging, may also carry unique signatures to help prognosticate stroke risk in patients with AF, though use of AI in this space currently remains limited. Historically, research evaluating atrial myopathy from imaging has focused on imaging features that correlate with risk for stroke. For instance, left atrial size, presence of left ventricular dysfunction, and left atrial appendage emptying volume from transthoracic echocardiograms are predictive of thromboembolic risk in patients with nonvalvular AF. These markers have been used to identify those at low-risk who may not need anticoagulation using traditional statistical methods [34,35]. Left atrial strain studies have also been shown to correlate with incident stroke [36]. Other transthoracic echocardiogram approaches have also been hypothesized to play a role, such as speckle-tracking strain analysis and three-dimensional echocardiography [37], but none have been used within an AI framework directly. Cardiac MRI also extends these dynamic parameter measurements by adding another spatial dimension compared with echocardiography to further improve stroke risk prediction [38–40]. Combining local tissue properties on MRI with deterministic computational tools defines substrate arrhythmogenic capacity and has also been implicated in stroke risk, particularly with a focus around cryptogenic stroke patients [41]. The use of deep learning to better characterize early atrial electrophysiological changes, such as fibrosis and minor atrial mechanical dysfunction is an area of unmet need. Application of ML in this area has the potential to critically augment the risk-benefit tradeoff to initiate anticoagulation, especially in those patients in whom high bleeding risks exist.

Revealing novel phenotypes of atrial fibrillation

AF is a complex pathophysiologic and spatiotemporal disease that manifests as various clinical phenotypes. Different phenotypes of AF may have different responses to treatments, such as recurrence of the arrhythmia after drug administration or catheter ablation strategies. New phenotypes have emerged from analysis of patients from large cohorts and registries [42–45]. These studies use cluster analyses of various clinical, demographic, and electrophysiological features. Ultimately, these phenotypes may share a heterogenous response to therapies, contain variable aspects of socioeconomic or comorbidity factors, and graded adverse clinical outcomes. AI offers multiple avenues for tackling this problem, both with its unsupervised machine learning techniques that are more robust than simple cluster analyses and with the ability to leverage complex data modalities, such as electrical signals from the catheter procedures and implanted devices, remote monitoring, and surface ECGs, imaging, text documents, and genomic profiles.

Predicting recurrence

Various modalities have thus been combined to investigate success after catheter ablation for AF, as defined by recurrence of AF. Most models use structured patient data in combination with features manually or semiautomatically detected from conventional signal and image processing features [46–50]. Some studies have approached this problem using a multimodal learning approach with promising initial results. One proof-of-concept study developed a machine learning model for predicting recurrence after PVI using features derived from an AF induction simulator and extracted tissue quality features from preprocedural cardiac MRIs [51]. Compared with only features from the raw images, the model performance was significantly greater (AUC 0.47–0.81) when combining features from the simulation of AF induction. Another study confirmed the benefit of this complementary use of data-driven ML models with biophysical simulations, with significantly improved performance using combined models, with the greatest predictive value coming from AF electrical patterns followed by imaging parameters then history (AUC range: 0.85–0.66) [52]. A study applying a multimodal approach including intracardiac electrograms and surface ECGs, along with clinical demographics and imaging features to predict risk of AF recurrence after catheter ablation, found a modest improvement in recurrent risk compared to other existing clinical scores [53]. Other efforts have also utilized noninvasive data modalities tracking atrial signatures, such as one study demonstrating the correlation of surface ECG atrial activity with both complexity of underlying AF and long-term recurrence risk [54]. Further directions for disease phenotyping include optimization of models with high density data from patients undergoing detailed examination with electrograms, body surface ECG, and imaging modalities, such as CT and MRI, and attempts to better understand the pathophysiology of AF. Once underlying mechanisms are elucidated using these data-hungry methods, perhaps more simple clinical identification methods may scale new knowledge to meet the widespread unmet need for clinical decision support and new therapies.

Targeting ablation

The mainstay of catheter ablation for AF is to target the pulmonary veins, from which AF triggers arise. Electrical isolation of these anatomic structures is performed using radiofrequency energy, laser, or cryotherapy but does not result in uniform elimination of AF.

Therefore operators may target ablation at additional anatomic structures or at sites of interest as determined by analysis of the electrical signals in AF, or AF *mapping*. There is great interest in targeting functional substrate for AF, especially when isolation of the pulmonary veins does not result in durable freedom from AF. Multiple studies have investigated targeting these intracardiac electrograms with machine learning to identify sites of ablation. One study investigated how CNNs classified AF image grids into those with or without rotational sites and demonstrated how the model predicted flagged areas using the same clinical decision rules used by expert clinicians [55]. Additionally, operator and center dependence can greatly influence the reproducibility across ablations due to variability in the assessment of intracardiac electrical signals. To address this problem, another study developed real-time ML models to analyze intracardiac electrograms and guide operators to sites of high spatiotemporal dispersion. Ablation of these sites resulted in elimination of AF during the procedure and in follow up [56,57]. Another study attempted to preoperatively predict AF trigger origins from CT scans, as being around the pulmonary veins or outside, to determine the role of pulmonary vein isolation in ablation procedural success and AF recurrence [58]. Interestingly, this model's test characteristics suggest that there may be a role in phenotyping patients who may not benefit from vein isolation and require alternative strategies. Similarly, another study investigated the role of cardiac MRI to identify areas of fibrosis and develop segmented atrial models to feed into biophysical simulation studies and identify possible ablation studies that can terminate AF, thus lowering intracardiac electrical mapping time and potentially improve targeted AF ablation outcomes [59].

Pharmaceutical therapy

Medical therapy for maintaining sinus rhythm in patients with AF is determined mostly by restrictions of each medicine's use rather than the likelihood that it will achieve rhythm control in each patient. For example, antiarrhythmic medications may be contraindicated in renal dysfunction or heart failure. Some medications for AF require increased monitoring or even admission to the hospital for initiation. After medications are filtered out based on patient factors, the medications are trialed empirically, without available *a priori* evidence for personalized performance. Understanding the distinct groups of patients that will or will not respond to antiarrhythmic therapy is a critically important question and a major unmet need in AF. Similarly, refining patients who may experience significant side effects, including liver, thyroid, lung, or cardiac toxicities before initiating a medication would be highly valuable. One study investigated the response to a common class III antiarrhythmic, dofetilide, using a mixed statistical model approach to identify key factors contributing to successful loading and an unsupervised learning approach to identify distinct phenotypes [60]. Interestingly, they identified that dose adjustment during loading was a strong negative predictor of a successful load. The study provided the backbone for future applications of ML in which prediction of dose adjustment or final dosage could be achieved with multimodal inputs. However, few studies leverage AI to identify patients who may benefit from AADs, which remains a major unmet need.

Artificial intelligence to guide management of ventricular arrhythmias and sudden cardiac death

SCD remains the leading cause of death, accounting for up to 300,000 deaths per year in the United States. The primary cause of sudden death in most of these cases is VAs [61,62]. VAs often arise in the setting of cardiomyopathies, or failing ventricular muscle, that may be a result of genetic abnormalities, immune system dysfunction, blockages of the coronary arteries, infections, toxins, such as chemotherapy, or radiation. Although VAs are often the cause of death for patients with cardiomyopathy, the mechanism (or pathophysiology) of the arrhythmia may be different depending on the type of cardiomyopathy. VA with a duration longer than 30 seconds is defined as "sustained" and includes ventricular tachycardia (VT) and ventricular fibrillation (VF). Both rhythms have a high mortality rate because of the effect on the ventricles, the large pumping chamber of the heart. There is a lower prevalence of VAs due to the high mortality rate of the disease, which reduces the size of available datasets significantly compared with atrial arrhythmias. Further, VA intervention is limited by the increased thickness of the ventricle compared with the atrium, which increases the impact of three-dimensional complexity. Because of its unstable and complex nature, VF has especially limited therapeutic options. Despite these challenges, treatment of VA is an important direction for AI implementation because of the critical clinical impact of treating the disorder and the complex data streams that elucidate their nature. In this section, we discuss efforts to apply AI in sudden cardiac arrest diagnosis, early detection, and response, tools for guiding VA therapy with medications or interventions, better understanding the rhythm pathophysiology and phenotypes of VAs, and to study arrhythmia specific cardiomyopathies.

Sudden cardiac death

SCD is defined as sudden unexpected death either within 1 hour of symptom onset if the event was witnessed, or within 24 hours of having been witnessed as symptom free if the event was unwitnessed by the WHO [63]. Although SCD is a heterogeneous disease in which the underlying cause may be VA or other severe process, cardiac arrhythmias occur in over half of these patients [64]. Although patients with preexisting cardiac disease history are at higher risk for SCD, more than half of patients who meet the definition of SCD do not have known history of cardiac disease, although it may be occult. AI has been employed to help predict, detect, and respond to SCD.

Sudden cardiac death detection and response

Early detection of SCD and response can mean the difference between life and death. In arrhythmic sudden cardiac arrest, rapid response to cardiopulmonary resuscitation and defibrillation with an automated external defibrillator (AED) are key for regaining spontaneous circulation of blood and oxygen in the body. Each minute that passes during pulselessness greatly decreases the chance that the patient will make a full neurologic recovery. AI has found footing in early detection by enabling detection of SCD in wearable devices and in contactless monitoring situations. Models for early response based on SCD detection are under development and may begin saving lives in the next 5 years. Contactless vital sign monitoring has been developed using AI tools for feeds from video cameras, such as security cameras and webcams which are increasingly present in residential and commercial areas [65]. Other contactless biomarkers include audible monitoring for detection of agonal breathing, a breathing pattern that occurs as a response to early SCD. A support vector machines model based on Google's VGGish feature platform was shown to discriminate this sound via smart speakers and cellular phone recordings [66]. The benefits of contactless options for detection are that no action is required by patients prior to detection and perhaps multiple persons in an area could be monitored simultaneously. An end-to-end solution is proposed for an SCD response system based on smartwatches. The protocol paper for the HEART-SAFE project describes a user controllable mobile application that links a user's smartwatch with a first responder and emergency services dispatch center via a real-time AI analysis of PPG, movement, and impact detection from the wrist [67]. Further improvements in the response to SCD could be gained from detection of the *onset* of SCD with short-term predictive features of ECG, respiration, or oxygenation signs. Prior work has been developed robust prediction models for impending cardiac arrest using heart rate variability (HRV), respiratory rate variability (RRV), or ECG features. HRV and RRV were shown in one study to predict VA events in an ICU population up to 1 hour prior to the onset of the event [68]. Similarly, the QRS morphology, or the shape of the ventricular depolarization signal on the ECG, in the minute before the VF event has been shown to accurately predict the onset of the event [69]. Finally, AI enhances response systems for SCD include delivery of AED, telephone call triage, and even detection of a heart rhythm that is treatable by defibrillation during chest compressions [70].

Sudden cardiac death and ventricular arrhythmia risk prediction

Risk prediction of individuals who are at higher risk of VAs and SCD is an important problem to prevent events or quickly reverse them when they occur. Patients with a high enough risk of VAs, such as those with severe ventricular dysfunction, or those with certain specific cardiomyopathies, warrant an implant of an implantable cardioverter defibrillator (ICD). An ICD is a device that is implanted in the body that detects VAs and provides an electrical shock across the chest (cardioversion or defibrillation) to restore normal rhythm. Because ICD implantation has a risk of procedural and long-term complications, they are only recommended to those at highest risk. AI has been employed to improve clinical risk prediction of VAs and mortality, with the promise that the decision of ICD implantation, other therapy, and prognosis may be more tailored to the individual. The 12-lead ECG has been used with convolutional neural networks to predict overall mortality over the next 1 year [71]. Prediction of sudden cardiac arrest in patients with heart failure using multimodality datasets including baseline demographic, clinical, biological, electrophysiological, social and psychological variables is also underway [72]. Beyond vitals and ECG data, other modalities, such as imaging, have been employed to predict VAs. For example, in hypertrophic cardiomyopathy, magnetic resonance imaging (MRI) texture analysis based on k-nearest neighbors and synthetic minority oversampling improved prediction of VT events over standard clinical models [73].

Guiding ventricular arrhythmia therapy

Sustained VAs include VT and VF, which may have different underlying causes and options for therapy. Generally, VT involves rapid activation of the ventricle where each activation has the same pattern of activity. VT pathophysiology may include an area of diseased tissue that fires repeatedly, or it may involve a circuit that meanders through the ventricular target. In these scenarios, the result is rapid ventricular beating that may cause death, syncope, or chest pain because of inefficient pumping of blood to the body. Because VT has a specific site of origin or circuit that is coordinated and repeated, it is more amenable to study and to intervention therapy compared with VF. VF, on the other hand, does not have a clear repeatable and organized pattern and has little to no pumping action for blood flow to the body. Therefore it is difficult to study and intervene on VF, where few therapeutic options are available for patients with recurrent VF. Preprocessing signals from ECG can be used to decrease the complexity of input data and make learning more efficient. Time-frequency representations of rhythms have successfully been used to improve classification of VF and VT [74]. Other signal preprocessing techniques have shown similar success, such as variational mode decomposition [75].

Ventricular tachycardia mapping and ablation

Mapping is the process of studying a rhythm to determine its site of origin or circuit of activation. It involves collecting signals from the heart during sinus rhythm, cardiac pacing, or during VT to determine targets for ablation. Ablation of a tachycardia may involve severing the circuit or eliminating the disordered cells at the site of origin using heat, cold, or electrical energy. VT therapy requires understanding of the 3D location of the circuit responsible for the rhythm. AI has allowed rapid determination of VAs with a high degree of accuracy and repeatability. For example, early AI-based models have shown strong accuracy in locating the circuits involved in VT in patients with a history of myocardial infarct [76]. More recently, a model developed to distinguish VAs coming from the right side versus the left side of the heart achieved accuracy of >97% using features extracted from the 12-lead surface ECG [77]. Work aimed at addressing the same question showed very high accuracy using a random forest model in a prospective validation cohort [78]. Other recent models in patients with structurally normal hearts were trained in 18,612 ECG recordings was trained against the localization based on invasive catheter ablation and the model achieved >98% accuracy in a testing cohort [79]. These localization models based on noninvasive testing improves procedural efficiency, procedural planning, and patient counseling prior to the VT ablation.

With current mapping technology, thousands of electrical recordings can be taken across the three-dimensional surface of the heart in a few minutes. Signal characteristics are analyzed to provide the operator with regional information about the heart. For example, the amplitude of the electrical signal recorded between two electrodes (bipolar voltage) is related to the presence of electrically active tissue. Higher voltages are recorded in "healthy" areas of the ventricle while lower voltages are recorded in areas of "scar," or replacement by fibrotic scar. Fibrosis has been shown to create substrate for VAs, and as such, has been a common target for ablation therapy. Advances using AI have been made in image texture analysis of MRI images to define where the fibrosis exists and its quality

[73]. Similarly, analysis of CT imaging has shown that areas of myocardial thinning relate to fibrosis [80]. Analysis of these segments is time consuming; however, pipelines using AI segmentation is under development [81]. Finally, the authors of this chapter joined machine learning methods to predict clinical outcomes based on the shape of ventricular action potentials and inferred specific ion current alterations that might explain the pathophysiologic changes [82]. This bench to bedside framework outperformed traditional clinical predictors and forms the basis of a more mechanism-guided management of VA.

Future work in mapping of VAs could be aimed at streamlining the procedure planning and assessing improvement in clinical outcomes. AI models with implementation along the process may provide an efficient setup for the electrophysiologist as displayed in Fig. 51.2. First, wearable devices and ECG recording devices record the clinical rhythm and distinguish the type of VA. A 12-lead ECG recording could be then analyzed for localization of the circuit to indicate the region of the arrhythmia. Imaging studies performed would further elucidate areas of fibrosis and ventricular dysfunction that assist in determining the source of the arrhythmia. Similarly, AI-enabled segmentation models assist in identifying ventricular access points and areas that should be avoided for safety, such as the coronary arteries, and phrenic nerves. Finally, AI could be applied to intracardiac signals in VAs for real-time ablation guidance and mapping. The challenge in VA therapy remains that the information required to fully understand the rhythm problem is acquired in a time-consuming and invasive manner, which limits the size of datasets in any single institution.

Phenotyping ventricular arrhythmias

VAs have been classified in several ways: appearance and duration of time on ECG; disease state of the heart from which they arise; and the mechanism of the tachycardia, based

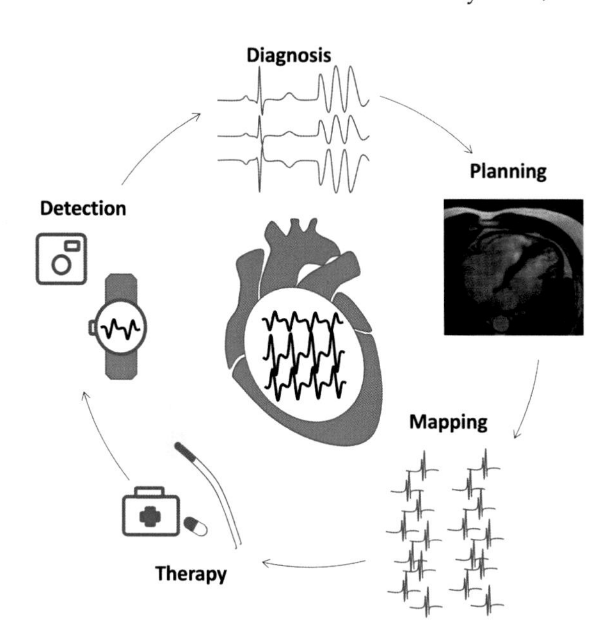

FIGURE 51.2 Artificial intelligence implementation in VAs and sudden cardiac death. AI integration with wearable devices enables detection, 12-lead ECG enables improved diagnosis, implementation in cardiac imaging for procedural planning and VA phenotyping, future direction to intracardiac mapping signals for guiding personalized therapy (clockwise).

in its electrophysiologic underpinnings. However, there still lacks an overarching classification that defines arrhythmias based on an individual person's characteristics that also determines the most appropriate course of therapy. There is much interest in using AI to describe the intersection of the electrical signals, electrophysiologic state of the cells, and disease of the heart. This description, or phenotype, would be most useful if it indicated a course of treatment, whether it be with drugs or intervention, and if it provided a clinically useful prognosis.

AI has been used to analyze electrophysiologic data to understand and detect diseases or dysfunction of the heart. The 12-lead ECG has been used to detect worsening ejection fraction from a large cohort of patients using both transthoracic echocardiogram images and 12-lead ECG recordings [83]. Further application of this technology may assist in identifying patients with other diseases of the heart muscle that would be associated with VAs. In patients with hypertrophic cardiomyopathy, a disease in which the heart muscle thickens due to improper growth of the heart muscle cells, unsupervised learning was used to determine clusters, or possible phenotypes, based on ECG biomarkers [84]. The phenotypes created related to specific anatomic findings and a clinical tool that could be applied by a clinician evaluating the 12-lead ECG. Prognosis of patients with cardiomyopathy using electrophysiologic data from cardiac resynchronization therapy has also been established for mortality [85,86] and response to treatment [87]. Connections between systems in the body may be used for multiorgan interaction, such as the immune system activation in patients with pulmonary hypertension [88]. Such discoveries may provide new inroads and a wholistic approach to arrhythmia therapy.

Arrhythmia-specific cardiomyopathies and channelopathies

Channelopathies are diseases of the ion channels which create the action potential in the heart cells. Disorders of these ion channels may occur due to genetic abnormalities, such as with the long QT syndromes, short QT syndromes, Brugada syndrome, or catecholaminergic polymorphic VT. Changes in ion channels, or electrophysiological remodeling, may also be induced due to acquired disorders, such as drug exposure, ischemia of the heart, infections, such as Chagas disease, or in structural disorders like myotonic dystrophy and arrhythmogenic right ventricular cardiomyopathy. Disorders of channelopathies have been studied at multiple scales, including genetics, single cells, cell sheets, tissue models, and in vivo. AI has been employed to link these multiscale cardiac EP to track changes in channelopathies to the clinical arrhythmias [89]. Drug induced VAs, most commonly torsade de points, is a frequent challenge for development of medical therapies for seizures, pain, arrhythmias, depression, and nausea, among others. Testing of these new therapies for safety is costly. Efforts to use high throughput models that are closer to humans and avoid animal testing are promising to improve efficiency, reduce cost and increase safety. Models spanning multiple scales may link drug targets and molecules to clinical outcomes using a computational pipeline [90]. In addition to computational modeling, incorporation of large numbers of cells with variable genetic expression with novel bench tools, such as iPSC and single-cell techniques [91].

Challenges and future directions for the use of artificial intelligence in electrophysiology

AI is poised to make significant additional contributions to the field of EP. Future applications that would be amenable to enhancement by AI, include prediction for arrhythmia response to drugs, enhancements to EP mapping, incorporation of imaging with clinical EP markers, and efficient management of ambulatory ECG and implantable cardiac device data, among others. Recent, single-center, data shows that ML methods may be used to combine information from multiple streams of data. In a recent study by Tang et al., electrical signals, imaging, and biomarkers were incorporated using a fusion model to predict long-term recurrence of AF [53]. Extension of this technique for other heart rhythm problems and cardiac diseases is readily feasible. Illustrative pseudocode outlining the process for loading and fusing pretrained limbs of a multimodal framework is provided in Fig. 51.3. Automation of EP data analysis and even navigation of intracardiac catheters

```
from data import DataGenerator
from models import MultimodalAFibModel
from trainer import ModelTrainer
from utils import CONFIG

# Load data generator.
data_gen = DataGenerator()

# Set up the data.
data_gen.set_ehr_files(CONFIG.PATH_TO_EHR_DATA) # X1
data_gen.set_ecg_files(CONFIG.PATH_TO_ECG_DATA) # X2
data_gen.set_ct_files(CONFIG.PATH_TO_ECG_DATA) # X3
data_gen.set_outcome_file(CONFIG.PATH_TO_OUTCOME_DATA) # Y

# Splits dataset into training, validation, and testing.
data_gen.split_dataset(CONFIG.DATA_SPLIT_PARAMS)

# Load the multimodal model class.
mm_afib_model = MultimodalAFibModel(CONFIG.MM_MODEL_PARAMS)

# Load all pretrained models.
mm_afib_model.load_ehr_model(CONFIG.PATH_TO_EHR_MODEL, output_layer=CONFIG.EHR_MM_OL)
mm_afib_model.load_ecg_model(CONFIG.PATH_TO_ECG_MODEL, output_layer=CONFIG.ECG_MM_OL)
mm_afib_model.load_ct_model(CONFIG.PATH_TO_CT_MODEL, output_layer=CONFIG.CT_MM_OL)

# Load our model trainer and train to clinical outcome.
trainer = ModelTrainer(mm_afib_model, data_gen)
trainer.train(CONFIG.TRAIN_PARAMETERS)
trainer.save(CONFIG.TRAIN_PARAMETERS)
```

FIGURE 51.3 Pseudocode outlining one process for combining multimodal models to predict a clinical outcome. A data generator provides the input (X) and output (Y) matrices to the trainer. Pretrained models are loaded, and output layers are identified. Finally, the trainer begins the training process on the multimodal model and saves the resulting model for evaluation.

may be incorporated to assist interventional procedures in real time [92]. AI has not been used in guiding treatment for VA as much as it has in AF, likely due to reduced size of datasets and reduce prevalence in the population. However, its success in AF implementation portends a good outlook for improvements in VA therapy by AI. Finally, using AI as a discovery tool is still in its early development. Interpretation of models that are traditionally thought of as a black box would benefit our understanding of the mechanisms of disease and improve confidence in the output of the models. Promising foundational work to link clinical outcomes with cellular EP available in patients, such as shape, duration, or other features of the action potential, could prove useful for defining new disease phenotypes or develop drug therapy [82]. Future efforts at understanding model decisions using attention models, blanking, or simulation studies may help in this way. Finally, AI tools applied at the basic science, histologic, genomic, and bioengineering levels, may provide answers or new insights that deliver clinical advances.

To incorporate new AI models into medical care, careful attention must be applied to validation and translation. First, structured data sets must be curated, and quality checked for accurate labels. Data is often stored in proprietary formats, which requires translation and preprocessing to be incorporated into model development or application of the model. There may be differences in acquisition techniques for imaging or electrocardiographic recordings that may become apparent after development. Semiautomated and automated tools are being developed to train models and label data, including complex ones for imaging resources. For example, an interactive medical imaging application for processing and labeling data is the CemrgApp [93]. There may be an opportunity for AI to be applied in the labeling process itself for some of these repetitive tasks. Some standardization and best practices for interpreting certain types of electrophysiologic data and outputs may be required, as currently several different model architectures have been applied at different institutions with varying results. Fig. 51.4 shows several published methods of interpreting the voltage time series of electrocardiographic or intracardiac electrogram recordings. It is unclear whether the optimal model architecture is different for different applications of the same data or whether these differences arise from technique or specifics of those datasets. In general, the safe deployment of a new clinical model requires careful development of the model, followed by external validation, prospective validation, and finally clinical deployment including regulatory and reimbursement assessment [94].

The generalizability of AI models across disparate data may be limited due to lack of data in minority populations, including racial or gender minorities or due to rare diseases and events. This challenge is not unique to EP but remains a significant area for improvement. Interinstitutional data sharing and training of models across institutions may provide one way of improving generalizability of models [95]. Possible options for data sharing across institutions include sharing of the data itself, which is limited by privacy and security concerns, sharing of the weights of the model, or sharing of model architectures. Institutional sharing improves, but does not eliminate systemic biases and the availability of data, which requires special attention to ensure inclusive and universally beneficial datasets [96].

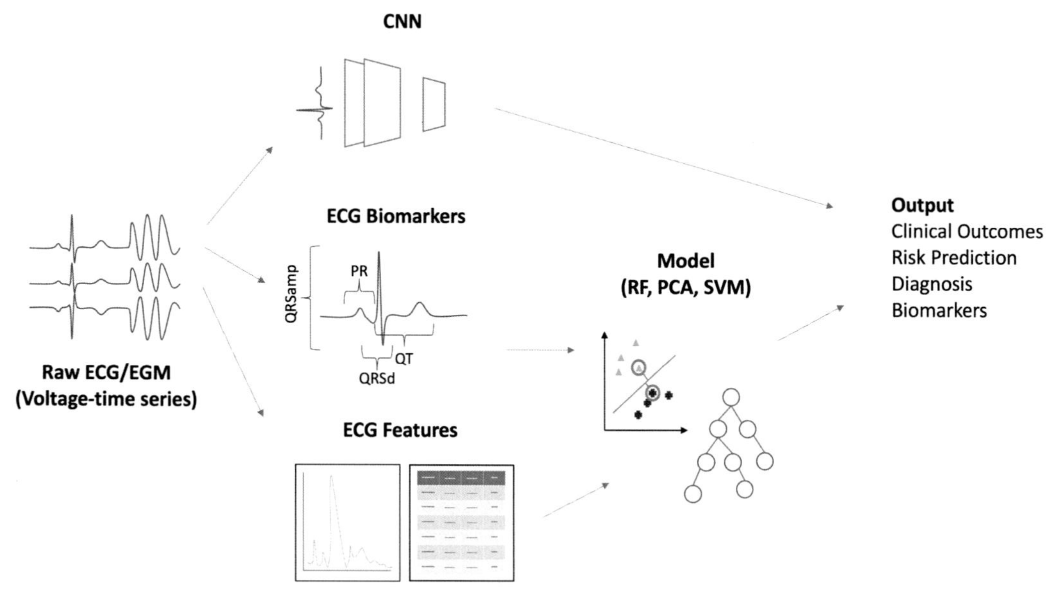

FIGURE 51.4 Common methods of interpreting surface ECG or intracardiac electrogram signals using neural networks, signal biomarkers, or mathematical features. ECG, Electrocardiography; EGM, electrogram; CNN, convolutional neural network; RF, random forest; PCA, principle component analysis; SVM, support vector machines.

Summary

1. Cardiac EP is fertile ground for AI implementation.
2. Landmark trials in AI have been performed on EP datasets.
3. AI applications span prediction, diagnosis, automation, and discovery, which may require different techniques.
4. New methods are needed to address diverse datasets, interinstitutional datasets, and bias in datasets.
5. AI is poised to address the most challenging unmet needs in cardiac EP.

References

[1] Chugh SS, Havmoeller R, Narayanan K, Singh D, Rienstra M, Benjamin EJ, et al. Worldwide epidemiology of atrial fibrillation: a Global Burden of Disease 2010 Study. Circulation. 2014;129:837–47.
[2] Issa ZF, Miller JM, Zipes DP. Clinical arrhythmology and electrophysiology: a companion to Braunwald's heart disease. 3rd ed. Elsevier; 2019.
[3] Hulme OL, Khurshid S, Weng L-C, Anderson CD, Wang EY, Ashburner JM, et al. Development and validation of a prediction model for atrial fibrillation using electronic health records. JACC Clin Electrophysiol 2019;5:1331–41.

[4] Sinner MF, Stepas KA, Moser CB, Krijthe BP, Aspelund T, Sotoodehnia N, et al. B-type natriuretic peptide and C-reactive protein in the prediction of atrial fibrillation risk: the CHARGE-AF Consortium of community-based cohort studies. Europace. 2014;16:1426–33.
[5] Alonso A, Roetker NS, Soliman EZ, Chen LY, Greenland P, Heckbert SR. Prediction of atrial fibrillation in a racially diverse cohort: the Multi-Ethnic Study of Atherosclerosis (MESA). J Am Heart Assoc 2016;5. Available from: https://doi.org/10.1161/JAHA.115.003077.
[6] Li Y-G, Bisson A, Bodin A, Herbert J, Grammatico-Guillon L, Joung B, et al. C2 HEST score and prediction of incident atrial fibrillation in poststroke patients: a French nationwide study. J Am Heart Assoc 2019;8:e012546.
[7] Gladstone DJ, Wachter R, Schmalstieg-Bahr K, Quinn FR, Hummers E, Ivers N, et al.SCREEN-AF Investigators and Coordinators Screening for atrial fibrillation in the older population: a randomized clinical trial. JAMA Cardiol 2021;6:558–67.
[8] Svennberg E, Friberg L, Frykman V, Al-Khalili F, Engdahl J, Rosenqvist M. Clinical outcomes in systematic screening for atrial fibrillation (STROKESTOP): a multicentre, parallel group, unmasked, randomised controlled trial. Lancet. 2021;398:1498–506.
[9] Hannun AY, Rajpurkar P, Haghpanahi M, Tison GH, Bourn C, Turakhia MP, et al. Cardiologist-level arrhythmia detection and classification in ambulatory electrocardiograms using a deep neural network. Nat Med 2019;25:65–9.
[10] Perez MV, Mahaffey KW, Hedlin H, Rumsfeld JS, Garcia A, Ferris T, et al. Large-scale assessment of a smartwatch to identify atrial fibrillation. N Engl J Med 2019;381:1909–17.
[11] Guo Y, Wang H, Zhang H, Liu T, Liang Z, Xia Y, et al. Mobile photoplethysmographic technology to detect atrial fibrillation. J Am Coll Cardiol 2019;74:2365–75.
[12] Torres-Soto J, Ashley EA. Multi-task deep learning for cardiac rhythm detection in wearable devices. npj Digit Med 2020;3:116.
[13] Han L, Askari M, Altman RB, Schmitt SK, Fan J, Bentley JP, et al. Atrial fibrillation burden signature and near-term prediction of stroke: a machine learning analysis. Circ Cardiovasc Qual Outcomes 2019;12:e005595.
[14] Bumgarner JM, Lambert CT, Hussein AA, Cantillon DJ, Baranowski B, Wolski K, et al. Smartwatch algorithm for automated detection of atrial fibrillation. J Am Coll Cardiol 2018;71:2381–8.
[15] Wegner FK, Kochhäuser S, Ellermann C, Lange PS, Frommeyer G, Leitz P, et al. Prospective blinded evaluation of the smartphone-based AliveCor Kardia ECG monitor for atrial fibrillation detection: the PEAK-AF study. Eur J Intern Med 2020;73:72–5.
[16] Noseworthy PA, Attia ZI, Behnken EM, Giblon RE, Bews KA, Liu S, et al. Artificial intelligence-guided screening for atrial fibrillation using electrocardiogram during sinus rhythm: a prospective non-randomised interventional trial. Lancet. 2022;400:1206–12.
[17] US Preventive Services Task Force, Davidson KW, Barry MJ, Mangione CM, Cabana M, Caughey AB, et al. Screening for atrial fibrillation: US Preventive Services Task Force recommendation statement. JAMA. 2022;327:360–7.
[18] Kligfield P, Gettes LS, Bailey JJ, Childers R, Deal BJ, Hancock EW, et al. Recommendations for the standardization and interpretation of the electrocardiogram: part I: the electrocardiogram and its technology: a scientific statement from the American Heart Association Electrocardiography and Arrhythmias Committee, Council on Clinical Cardiology; the American College of Cardiology Foundation; and the Heart Rhythm Society: endorsed by the International Society for Computerized Electrocardiology. Circulation. 2007;115:1306–24.
[19] Attia ZI, Noseworthy PA, Lopez-Jimenez F, Asirvatham SJ, Deshmukh AJ, Gersh BJ, et al. An artificial intelligence-enabled ECG algorithm for the identification of patients with atrial fibrillation during sinus rhythm: a retrospective analysis of outcome prediction. Lancet. 2019;394:861–7.
[20] Khurshid S, Friedman S, Reeder C, Di Achille P, Diamant N, Singh P, et al. ECG-based deep learning and clinical risk factors to predict atrial fibrillation. Circulation. 2022;145:122–33.
[21] Rajakariar K, Koshy AN, Sajeev JK, Nair S, Roberts L, Teh AW. Accuracy of a smartwatch based single-lead electrocardiogram device in detection of atrial fibrillation. Heart. 2020;106:665–70.
[22] William AD, Kanbour M, Callahan T, Bhargava M, Varma N, Rickard J, et al. Assessing the accuracy of an automated atrial fibrillation detection algorithm using smartphone technology: the iREAD study. Heart Rhythm 2018;15:1561–5.

[23] Shcherbina A, Mattsson CM, Waggott D, Salisbury H, Christle JW, Hastie T, et al. Accuracy in wrist-worn, sensor-based measurements of heart rate and energy expenditure in a diverse cohort. J Pers Med 2017;7:3.
[24] Andrade JG, Wells GA, Deyell MW, Bennett M, Essebag V, Champagne J, et al. Cryoablation or drug therapy for initial treatment of atrial fibrillation. N Engl J Med 2021;384:305–15.
[25] Andrade JG, Deyell MW, Macle L, Wells GA, Bennett M, Essebag V, et al. Progression of atrial fibrillation after cryoablation or drug therapy. N Engl J Med 2023;388:105–16.
[26] Freedman B, Potpara TS, Lip GYH. Stroke prevention in atrial fibrillation. Lancet. 2016;388:806–17.
[27] Hart RG, Pearce LA, Aguilar MI. Meta-analysis: antithrombotic therapy to prevent stroke in patients who have nonvalvular atrial fibrillation. Ann Intern Med 2007;146:857–67.
[28] Ruff CT, Giugliano RP, Braunwald E, Hoffman EB, Deenadayalu N, Ezekowitz MD, et al. Comparison of the efficacy and safety of new oral anticoagulants with warfarin in patients with atrial fibrillation: a meta-analysis of randomised trials. Lancet. 2014;383:955–62.
[29] Lip GYH, Nieuwlaat R, Pisters R, Lane DA, Crijns HJGM. Refining clinical risk stratification for predicting stroke and thromboembolism in atrial fibrillation using a novel risk factor-based approach: the euro heart survey on atrial fibrillation. Chest. 2010;137:263–72.
[30] Pisters R, Lane DA, Nieuwlaat R, de Vos CB, Crijns HJGM, Lip GYH. A novel user-friendly score (HAS-BLED) to assess 1-year risk of major bleeding in patients with atrial fibrillation: the Euro Heart Survey. Chest. 2010;138:1093–100.
[31] Lip GYH, Frison L, Halperin JL, Lane DA. Comparative validation of a novel risk score for predicting bleeding risk in anticoagulated patients with atrial fibrillation: the HAS-BLED (Hypertension, Abnormal Renal/Liver Function, Stroke, Bleeding History or Predisposition, Labile INR, Elderly, Drugs/Alcohol Concomitantly) score. J Am Coll Cardiol 2011;57:173–80.
[32] Rabinstein AA, Yost MD, Faust L, Kashou AH, Latif OS, Graff-Radford J, et al. Artificial intelligence-enabled ECG to identify silent atrial fibrillation in embolic stroke of unknown source. J Stroke Cerebrovasc Dis 2021;30:105998.
[33] Raghunath S, Pfeifer JM, Ulloa-Cerna AE, Nemani A, Carbonati T, Jing L, et al. Deep neural networks can predict new-onset atrial fibrillation from the 12-lead ECG and help identify those at risk of atrial fibrillation-related stroke. Circulation. 2021;143:1287–98.
[34] Predictors of thromboembolism in atrial fibrillation: II. Echocardiographic features of patients at risk. The Stroke Prevention in Atrial Fibrillation Investigators. Ann Intern Med 1992;116:6–12.
[35] Uzięblo-Życzkowska B, Kapłon-Cieślicka A, Gawałko M, Budnik M, Starzyk K, Wożakowska-Kapłon B, et al. Risk factors for left atrial thrombus in younger patients (aged < 65 years) with atrial fibrillation or atrial flutter: Data from the multicenter left atrial thrombus on transesophageal echocardiography (LATTEE) registry. Front Cardiovasc Med 2022;9:973043.
[36] Alhakak AS, Biering-Sørensen SR, Møgelvang R, Modin D, Jensen GB, Schnohr P, et al. Usefulness of left atrial strain for predicting incident atrial fibrillation and ischaemic stroke in the general population. Eur Heart J Cardiovasc Imaging 2022;23:363–71.
[37] Goldberger JJ, Arora R, Green D, Greenland P, Lee DC, Lloyd-Jones DM, et al. Evaluating the atrial myopathy underlying atrial fibrillation: identifying the arrhythmogenic and thrombogenic substrate. Circulation. 2015;132:278–91.
[38] Inoue YY, Alissa A, Khurram IM, Fukumoto K, Habibi M, Venkatesh BA, et al. Quantitative tissue-tracking cardiac magnetic resonance (CMR) of left atrial deformation and the risk of stroke in patients with atrial fibrillation. J Am Heart Assoc 2015;4. Available from: https://doi.org/10.1161/JAHA.115.001844.
[39] Lee DC, Markl M, Ng J, Carr M, Benefield B, Carr JC, et al. Three-dimensional left atrial blood flow characteristics in patients with atrial fibrillation assessed by 4D flow CMR. Eur Heart J Cardiovasc Imaging 2016;17:1259–68.
[40] Markl M, Lee DC, Ng J, Carr M, Carr J, Goldberger JJ. Left atrial 4-dimensional flow magnetic resonance imaging. Invest Radiol 2016;51:147–54.
[41] Bifulco SF, Scott GD, Sarairah S, Birjandian Z, Roney CH, Niederer SA, et al. Computational modeling identifies embolic stroke of undetermined source patients with potential arrhythmic substrate. eLife 2021;10. Available from: https://doi.org/10.7554/eLife.64213.
[42] Inohara T, Shrader P, Pieper K, Blanco RG, Thomas L, Singer DE, et al. Association of of atrial fibrillation clinical phenotypes with treatment patterns and outcomes: a multicenter registry study. JAMA Cardiol 2018;3:54–63.

[43] Proietti M, Vitolo M, Harrison SL, Lane DA, Fauchier L, Marin F, et al. Impact of clinical phenotypes on management and outcomes in European atrial fibrillation patients: a report from the ESC-EHRA EURObservational Research Programme in AF (EORP-AF) General Long-Term Registry. BMC Med 2021;19:256.
[44] Ogawa H, An Y, Nishi H, Fukuda S, Ishigami K, Ikeda S, et al. Characteristics and clinical outcomes in atrial fibrillation patients classified using cluster analysis: the Fushimi AF Registry. Europace. 2021;23:1369–79.
[45] Pastori D, Antonucci E, Milanese A, Menichelli D, Palareti G, Farcomeni A, et al. Clinical phenotypes of atrial fibrillation and mortality risk—a cluster analysis from the nationwide Italian START registry. J Pers Med 2022;12:785.
[46] Park J-W, Kwon O-S, Shim J, Hwang I, Kim YG, Yu HT, et al. Machine learning-predicted progression to permanent atrial fibrillation after catheter ablation. Front Cardiovasc Med 2022;9:813914.
[47] Firouznia M, Feeny AK, LaBarbera MA, McHale M, Cantlay C, Kalfas N, et al. Machine learning–derived fractal features of shape and texture of the left atrium and pulmonary veins from cardiac computed tomography scans are associated with risk of recurrence of atrial fibrillation postablation. Circ Arrhythm Electrophysiol 2021;14:e009265.
[48] Budzianowski J, Hiczkiewicz J, Burchardt P, Pieszko K, Rzeźniczak J, Budzianowski P, et al. Predictors of atrial fibrillation early recurrence following cryoballoon ablation of pulmonary veins using statistical assessment and machine learning algorithms. Heart Vessel 2019;34:352–9.
[49] Furui K, Morishima I, Morita Y, Kanzaki Y, Takagi K, Yoshida R, et al. Predicting long-term freedom from atrial fibrillation after catheter ablation by a machine learning algorithm: validation of the CAAP-AF score. J Arrhythm 2020;36:297–303.
[50] Hung M, Lauren E, Hon E, Xu J, Ruiz-Negrón B, Rosales M, et al. Using machine learning to predict 30-day hospital readmissions in patients with atrial fibrillation undergoing catheter ablation. J Pers Med 2020;10:82.
[51] Shade JK, Ali RL, Basile D, Popescu D, Akhtar T, Marine JE, et al. Preprocedure application of machine learning and mechanistic simulations predicts likelihood of paroxysmal atrial fibrillation recurrence following pulmonary vein isolation. Circ Arrhythm Electrophysiol 2020;13:e008213.
[52] Roney CH, Sim I, Yu J, Beach M, Mehta A, Alonso Solis-Lemus J, et al. Predicting atrial fibrillation recurrence by combining population data and virtual cohorts of patient-specific left atrial models. Circ Arrhythm Electrophysiol 2022;15:e010253.
[53] Tang S, Razeghi O, Kapoor R, Alhusseini MI, Fazal M, Rogers AJ, et al. Machine learning-enabled multimodal fusion of intra-atrial and body surface signals in prediction of atrial fibrillation ablation outcomes. Circ Arrhythm Electrophysiol 2022;15:e010850.
[54] Rodrigo M, Climent AM, Hernández-Romero I, Liberos A, Baykaner T, Rogers AJ, et al. Noninvasive assessment of complexity of atrial fibrillation: correlation with contact mapping and impact of ablation. Circ Arrhythm Electrophysiol 2020;13:e007700.
[55] Alhusseini MI, Abuzaid F, Rogers AJ, Zaman JAB, Baykaner T, Clopton P, et al. Machine learning to classify intracardiac electrical patterns during atrial fibrillation. Circ Arrhythm Electrophysiol 2020;. Available from: https://doi.org/10.1161/CIRCEP.119.008160.
[56] Seitz J, Bars C, Théodore G, Beurtheret S, Lellouche N, Bremondy M, et al. AF ablation guided by spatiotemporal electrogram dispersion without pulmonary vein isolation: a wholly patient-tailored approach. J Am Coll Cardiol 2017;69:303–21.
[57] Seitz J, Durdez TM, Albenque JP, Pisapia A, Gitenay E, Durand C, et al. Artificial intelligence software standardizes electrogram-based ablation outcome for persistent atrial fibrillation. J Cardiovasc Electrophysiol 2022;. Available from: https://doi.org/10.1111/jce.15657.
[58] Liu C-M, Chang S-L, Chen H-H, Chen W-S, Lin Y-J, Lo L-W, et al. The clinical application of the deep learning technique for predicting trigger origins in patients with paroxysmal atrial fibrillation with catheter ablation. Circ Arrhythm Electrophysiol 2020;13:e008518.
[59] Boyle PM, Zghaib T, Zahid S, Ali RL, Deng D, Franceschi WH, et al. Computationally guided personalized targeted ablation of persistent atrial fibrillation. Nat Biomed Eng 2019;3:870–9.
[60] Levy AE, Biswas M, Weber R, Tarakji K, Chung M, Noseworthy PA, et al. Applications of machine learning in decision analysis for dose management for dofetilide. PLoS One 2019;14:e0227324.
[61] Saltzman HE. Arrhythmias and heart failure. Cardiol Clin 2014;32:125–33 ix.

[62] Al-Khatib SM, Stevenson WG, Ackerman MJ, Bryant WJ, Callans DJ, Curtis AB, et al. AHA/ACC/HRS guideline for management of patients with ventricular arrhythmias and the prevention of sudden cardiac death: a report of the American College of Cardiology/American Heart Association Task Force on Clinical Practice Guidelines and the Heart Rhythm Society. Heart Rhythm 2017;2018(15):e73–e189.
[63] Sudden cardiac death. Report of a WHO Scientific Group. World Health Organ Tech Rep Ser 1985;726:5–25.
[64] Tseng ZH, Olgin JE, Vittinghoff E, Ursell PC, Kim AS, Sporer K, et al. Prospective countywide surveillance and autopsy characterization of sudden cardiac death: POST SCD study. Circulation. 2018;137:2689–700.
[65] Monkaresi H, Calvo RA, Yan H. A machine learning approach to improve contactless heart rate monitoring using a webcam. IEEE J Biomed Health Inf 2014;18:1153–60.
[66] Chan J, Rea T, Gollakota S, Sunshine JE. Contactless cardiac arrest detection using smart devices. npj Digit Med 2019;2:52.
[67] Schober P, van den Beuken WMF, Nideröst B, Kooy TA, Thijssen S, Bulte CSE, et al. Smartwatch based automatic detection of out-of-hospital cardiac arrest: study rationale and protocol of the HEART-SAFE project. Resusc Plus 2022;12:100324.
[68] Lee H, Shin S-Y, Seo M, Nam G-B, Joo S. Prediction of ventricular tachycardia one hour before occurrence using artificial neural networks. Sci Rep 2016;6:32390.
[69] Taye GT, Shim EB, Hwang H-J, Lim KM. Machine learning approach to predict ventricular fibrillation based on QRS complex shape. Front Physiol 2019;10:1193.
[70] Ming Y, Taihu W, Pengcheng Y, Meng L, Feixiang H, Guang Z, et al. Detection of shockable rhythm during chest compression based on machine learning. In: 2019 IEEE 8th joint international information technology and artificial intelligence conference (ITAIC); 2019. p. 365–370.
[71] Raghunath S, Ulloa Cerna AE, Jing L, vanMaanen DP, Stough J, Hartzel DN, et al. Prediction of mortality from 12-lead electrocardiogram voltage data using a deep neural network. Nat Med 2020;26:886–91.
[72] Meng F, Zhang Z, Hou X, Qian Z, Wang Y, Chen Y, et al. Machine learning for prediction of sudden cardiac death in heart failure patients with low left ventricular ejection fraction: study protocol for a retroprospective multicentre registry in China. BMJ Open 2019;9:e023724.
[73] Alis D, Guler A, Yergin M, Asmakutlu O. Assessment of ventricular tachyarrhythmia in patients with hypertrophic cardiomyopathy with machine learning-based texture analysis of late gadolinium enhancement cardiac MRI. Diagn Interv Imaging 2020;101:137–46.
[74] Mjahad A, Rosado-Muñoz A, Bataller-Mompeán M, Francés-Víllora JV, Guerrero-Martínez JF. Ventricular fibrillation and tachycardia detection from surface ECG using time-frequency representation images as input dataset for machine learning. Comput Methods Prog Biomed 2017;141:119–27.
[75] Mohanty M, Biswal P, Sabut S. Machine learning approach to recognize ventricular arrhythmias using VMD based features. Multidimens Syst Signal Process 2020;31:49–71.
[76] Yokokawa M, Liu T-Y, Yoshida K, Scott C, Hero A, Good E, et al. Automated analysis of the 12-lead electrocardiogram to identify the exit site of postinfarction ventricular tachycardia. Heart Rhythm 2012;9:330–4.
[77] Zheng J, Fu G, Abudayyeh I, Yacoub M, Chang A, Feaster WW, et al. A high-precision machine learning algorithm to classify left and right outflow tract ventricular tachycardia. Front Physiol 2021;12:641066.
[78] Zhao W, Zhu R, Zhang J, Mao Y, Chen H, Ju W, et al. Machine learning for distinguishing right from left premature ventricular contraction origin using surface electrocardiogram features. Heart Rhythm 2022;. Available from: https://doi.org/10.1016/j.hrthm.2022.07.010.
[79] Zheng J, Fu G, Struppa D, Abudayyeh I, Contractor T, Anderson K, et al. A high precision machine learning-enabled system for predicting idiopathic ventricular arrhythmia origins. Front Cardiovasc Med 2022;9:809027.
[80] Ghannam M, Cochet H, Jais P, Sermesant M, Patel S, Siontis KC, et al. Correlation between computer tomography-derived scar topography and critical ablation sites in postinfarction ventricular tachycardia. J Cardiovasc Electrophysiol 2018;29:438–45.
[81] Computed tomography targets for efficient guidance of catheter ablation in ventricular tachycardia (MAP-IN-HEART). <https://clinicaltrials.gov/ct2/show/NCT04747353>.
[82] Rogers AJ, Selvalingam A, Alhusseini MI, Krummen DE, Corrado C, Abuzaid F, et al. Machine learned cellular phenotypes in cardiomyopathy predict sudden death. Circ Res 2021;128:172–84.

[83] Attia ZI, Kapa S, Lopez-Jimenez F, McKie PM, Ladewig DJ, Satam G, et al. Screening for cardiac contractile dysfunction using an artificial intelligence-enabled electrocardiogram. Nat Med 2019;25:70–4.
[84] Lyon A, Ariga R, Mincholé A, Mahmod M, Ormondroyd E, Laguna P, et al. Distinct ECG phenotypes identified in hypertrophic cardiomyopathy using machine learning associate with arrhythmic risk markers. Front Physiol 2018;9:213.
[85] Tokodi M, Schwertner WR, Kovács A, Tősér Z, Staub L, Sárkány A, et al. Machine learning-based mortality prediction of patients undergoing cardiac resynchronization therapy: the SEMMELWEIS-CRT score. Eur Heart J 2020;41:1747–56.
[86] Kalscheur MM, Kipp RT, Tattersall MC, Mei C, Buhr KA, DeMets DL, et al. Machine learning algorithm predicts cardiac resynchronization therapy outcomes: lessons from the COMPANION trial. Circ Arrhythm Electrophysiol 2018;11:e005499.
[87] Feeny AK, Rickard J, Patel D, Toro S, Trulock KM, Park CJ, et al. Machine learning prediction of response to cardiac resynchronization therapy: improvement versus current guidelines. Circ Arrhythm Electrophysiol 2019;12:e007316.
[88] Sweatt AJ, Hedlin HK, Balasubramanian V, Hsi A, Blum LK, Robinson WH, et al. Discovery of distinct immune phenotypes using machine learning in pulmonary arterial hypertension. Circ Res 2019;124:904–19.
[89] Cantwell CD, Mohamied Y, Tzortzis KN, Garasto S, Houston C, Chowdhury RA, et al. Rethinking multiscale cardiac electrophysiology with machine learning and predictive modelling. Comput Biol Med 2019;104:339–51.
[90] Yang P-C, DeMarco KR, Aghasafari P, Jeng M-T, Dawson JRD, Bekker S, et al. A computational pipeline to predict cardiotoxicity: from the atom to the rhythm. Circ Res 2020;126:947–64.
[91] Wu JC, Garg P, Yoshida Y, Yamanaka S, Gepstein L, Hulot J-S, et al. Towards precision medicine with human iPSCs for cardiac channelopathies. Circ Res 2019;125:653–8.
[92] Jolaei M, Hooshiar A, Dargahi J, Packirisamy M. Toward task autonomy in robotic cardiac ablation: learning-based kinematic control of soft tendon-driven catheters. Soft Robot 2021;8:340–51.
[93] Razeghi O, Solís-Lemus JA, Lee AWC, Karim R, Corrado C, Roney CH, et al. CemrgApp: An interactive medical imaging application with image processing, computer vision, and machine learning toolkits for cardiovascular research. SoftwareX 2020;12:100570.
[94] Feeny AK, Chung MK, Madabhushi A, Attia ZI, Cikes M, Firouznia M, et al. Artificial intelligence and machine learning in arrhythmias and cardiac electrophysiology. Circ Arrhythm Electrophysiol 2020;13:e007952.
[95] Goto S, Solanki D, John JE, Yagi R, Homilius M, Ichihara G, et al. Multinational federated learning approach to train ECG and echocardiogram models for hypertrophic cardiomyopathy detection. Circulation. 2022;146:755–69.
[96] Bracic A, Callier SL, Price 2nd WN. Exclusion cycles: reinforcing disparities in medicine. Science. 2022;377:1158–60.

Index

Note: Page numbers followed by "f," "t," and "b" refer to figures, tables, and boxes, respectively.

B

C

D

E

H

M

N

O

Q

R

S

T

U